T0199820

Routledge Revivals

The History of the Study of Landforms
or
The Development of Geomorphology

Volume 1

This re-issue, first published in 1964, is the first of a seminal series analysing the development of the study of landforms, from both the geographical and geological point of view, with special emphasis upon fluvial geomorphology. Volume 1 treats the subject up to the first important statement of the cycle of erosion by W. M. Davis in 1889, and attempts to identify the most significant currents of geomorphic thought, integrating them into the broader contemporary intellectual frameworks with which they were associated.

The History of the Study of Landforms
or
The Development of Geomorphology

Volume 1: Geomorphology before Davis

Richard J. Chorley,
Antony J. Dunn
and
Robert P. Beckinsale

Routledge
Taylor & Francis Group

First published in 1964
by Methuen

This edition first published in 2009 by Routledge
2 Park Square, Milton Park, Abingdon, Oxon, OX14 4RN

Simultaneously published in the USA and Canada
by Routledge
270 Madison Avenue, New York, NY 10016

Routledge is an imprint of the Taylor & Francis Group, an informa business

Publisher's Note
The publisher has gone to great lengths to ensure the quality of this reprint but points
out that some imperfections in the original copies may be apparent.

Disclaimer
The publisher has made every effort to trace copyright holders and welcomes
correspondence from those they have been unable to contact.

ISBN 13: 978-0-415-56369-7 (set)
ISBN 13: 978-0-415-55278-3 (hbk)
ISBN 13: 978-0-415-55994-2 (pbk)
ISBN 13: 978-0-203-87137-9 (ebk)

ISBN 10: 0-415-56369-0 (set)
ISBN 10: 0-415-55278-8 (hbk)
ISBN 10: 0-415-55994-4 (pbk)
ISBN 10: 0-203-87137-5 (ebk)

THE HISTORY OF THE STUDY OF LANDFORMS
OR THE DEVELOPMENT OF GEOMORPHOLOGY

VOLUME ONE

Portrait of James Hutton by Raeburn
(*By permission of Lord Bruntisfield*)

THE HISTORY OF THE STUDY OF
LANDFORMS
OR
THE DEVELOPMENT OF GEOMORPHOLOGY

VOLUME ONE: GEOMORPHOLOGY
BEFORE DAVIS

BY
RICHARD J. CHORLEY
ANTONY J. DUNN
ROBERT P. BECKINSALE

METHUEN & CO LTD
JOHN WILEY & SONS INC

First published in 1964
© *R. J. Chorley & R. P. Beckinsale, 1964*
Printed in Great Britain by Butler & Tanner Ltd,
Frome and London

Contents

List of Illustrations

Portrait of James Hutton   *frontispiece*

PART FOUR

Preface

There does not exist, as far as we can discover, a history of landform study, by which we mean the history of the evolution of ideas relating to the development of the physical landscape. Before the early nineteenth century this study formed the bulk of geology, but thereafter, largely because of the rapid growth of stratigraphy and palaeontology, it occupied a decreasing share of that science, although its fundamental necessity in general geological knowledge was never in doubt. After about 1860 the study of landforms became part of both geology and physical geography and was later also known as physiography or geomorphology. Today it forms the link between geology and geography and probably its association with the latter discipline is one of the main reasons why it has failed to develop along more strictly scientific lines. Even in 1963 it is still a 'pseudo-science' or a 'quasi-scientific art' and its historian is faced with the portrayal of its early rise both as an artistic perception and as a gradual transition towards its goal of truly scientific explanation which it has so far not attained. In this task we have been faced with difficulties that might well have daunted wiser men! The obstacles, it is true, were those common enough in historical and geographical research but the time-scale was rather long and the literature extraordinarily varied.

To choose a beginning proved virtually impossible as we had to try to assess the influence of early writings on later authors. We finally decided, as had Von Zittel in 1899, that 'the geological writings of antiquity have little scientific value' (1899, Preface, p. VI). The classical authors, in spite of their appreciable knowledge of landscapes, did not greatly influence landform studies in later centuries. In contrast, the traditional accounts of the Creation, and particularly the Mosaic account as later incorporated into the Bible of the Christian Church, had a great control over scientific ideas especially in the early nineteenth century when the study of landforms occupied the bulk of geological—and not a little of theological—thought. On this and on the other great formative periods of geomorphology from the time of Abraham Werner and James Hutton up to about 1890 we have given detailed accounts in this volume.

In interpreting the ideas of authors we have encountered numerous difficulties in meaning and language. The texts in French, German and Italian have been translated into reasonably modern English, but the texts in English, where quoted, have been left as in the original. Unfortunately, landscape studies seem to lend themselves to vagueness and verbosity as well as to precision and elegance. We have tried not to read into works of early writers, or of the many vague authors, more than they themselves clearly intended or understood; but to leave the reader in no doubt as to what an author actually said we have quoted, wherever possible, the most important statements in the original text. We hope thereby to save the reader the necessity of searching through the vast amount of often indifferent or irrelevant matter in which these 'gems' are hidden, and also to supply him with a balanced and reliable 'source-book'.

We have necessarily paid much attention to the interaction of national ideas on the development of landscapes; and the degree of influence exerted by one nation on another in the geomorphological sphere has proved especially fascinating. Here, however, we must confess that partly owing to language difficulties we have confined ourselves mainly to European and American writers. That is, of course, far from implying that the text is concerned only with the study of European and American landforms, since the Europeans and the Americans were great travellers and wrote extensively on other lands. Nor could we be sure that in, for example, Persia or China some eminent early geomorphologist may not yet be resurrected, but the chapter on Geology in the sumptuous volume by Dr Joseph Needham and Wang Ling on *Science and Civilisation in China* (Vol. 3 (1959), pp. 591–623) seems to indicate that we are over-cautious. In any event, the impact of such peripheral writing on the main currents of geomorphic thought was undoubtedly insignificant. Even in Europe vital ideas on landscapes percolated very slowly to and from Britain and the Continent; it is never safe, before 1800, to assume that the existence of a topographic book or geomorphic idea in one country means that it was known elsewhere. What is more, occasionally the same idea on landforms developed quite independently at about the same time in widely separated countries. Hence, in the absence of concrete evidence, it is seldom wise to assume international influence or absolute priority.

In this Volume we have dealt mainly with the rise of ideas on the development of landforms created largely by running water, ice and the sea. The growing interest in the study of desert and tundra landscapes and the other enormous strides made in all aspects of geomorphology during the twentieth century will be dealt with in Volume Two. It is because of the fundamental and introductory nature of the first volume that we have already drawn attention to its quotations. However, in choosing them we wished to do more than exemplify the author's main findings; they are meant also to be examples of the author's literary style. They will enable the reader to compare the halting sentences of Hutton with the eloquent phrases of Lyell and Playfair and to enjoy in the descriptions of the American West some of the finest topographical literature ever written. The word-paintings of some of the nineteenth-century geologists are an enduring delight. To many of them artistic expression was as important as scientific accuracy. On the other hand, to our regret we have, after much sifting, had to omit purely artistic writers, such as Ruskin, who added little to the scientific study of landforms. Yet it was Ruskin who wrote, 'The eye of knowledge is as a telescope to the common eye, whereby more is apprehended than is immediately seen.' Unfortunately, we shall have no difficulty in showing in these volumes that the 'eye of knowledge' may sometimes distort and sometimes bring into focus.

Throughout the text we have tried to illustrate notable themes and important people by means of diagrams and portraits respectively. The pictorial illustrations have been taken, wherever possible, from the original works and serve to demonstrate, as with style, the importance of diagrammatic or artistic skill in appealing to the public. The pen-sketches of Henry de la Beche and of Louis Agassiz, the geometrical drawings by Tylor, and the etchings of the professional draughtsman who first sketched the Colorado, remind us of a time when artistic skill was more widely practised. Some early photographs are also included together with several modern photographs, including aerial views which give striking panoramas of some of the problem-landscapes under discussion. The personal portraits and photographs will, we hope, add some reality to the personalities involved. They represent many nations and many ages and, although we do not suggest even the slightest connexion between personal appearance and geological thought, they may help the reader to associate the right person with the right idea.

It seemed advisable to divide Volume One into four parts, each of which ends with a bibliography that should more than suffice the needs of the normal student. In it, we have noted the page-length of the major texts if only as a warning, or more rarely as an encouragement, to those whose time is limited. Each part also contains at least one brief chapter on the quantitative approach to landform studies. The amateur or beginner may not find this more scientific work easy reading, but we can assure him that if he skips it he is in good company as many important writings of this kind were virtually ignored by the contemporaries of the unfortunate quantitative authors. However, to the more dedicated reader we would recommend greater perseverance as some of the early quantitative assessments are extraordinarily modern and not a few even futuristic.

It is a pleasure to express our thankfulness and gratitude to numerous generous helpers for quotations, diagrams, and criticisms, and to our predecessors, the authors of the chief existing histories of scientific thought, and more particularly of geology.

We are especially grateful to the following authors for kind permission to quote extracts from their works: Kirtley F. Mather and Shirley F. Mason and Harvard University Press for several quotations from *A Source Book in Geology* (1939; copyrighted by the President and Fellows of Harvard College); Carroll Lane Fenton and Mildred Adams Fenton and Doubleday and Company Inc. of New York for several quotations from their *Giants of Geology* (copyright 1945 and 1952); George P. Merrill and Yale University Press for several quotations from *The First One Hundred Years of American Geology* (1924); William Culp Darrah and Princeton University Press for many quotations from his *Powell of the Colorado* (1951); Aurèle La Rocque and the University of Illinois Press for several quotations from his translation of *The Admirable Discourses of Bernard Palissy* (1957); Frederick J. North and the editors of the Geologists' Association of London for many quotations from his 'Centenary of the Glacial Theory' (1943); Frank D. Adams and Dover Publications for a quotation from *The Birth and Development of the Geological Sciences* (1938; Dover edn., 1954); William D. Thornbury and John Wiley & Sons for a quotation from *Principles of Geomorphology* (1954); Julian M. Drachman for two quotations from his *Studies in The Literature of Natural Science* (Macmillan, 1930); Sergei I. Tomkeieff of Durham University for two quotations (1946 and 1950); Sir Edward B. Bailey for one passage (1927); Louis MacNeice for a passage from his *Translation of Goethe's Faust* (Faber, 1951); Wilma B. Fairchild, editor, and *The Geographical Review* of New York for a quotation from an article by Isaiah Bowman (1934); Robert S. Platt, editor, and the Association of American Geographers for quotations from their *Annals* from W. M. Davis (1925) and L. Martin (1950); Hesketh Pearson for a quotation from *The Smith of Smiths* (Penguin, 1948); and last, but by no means least, Karl Alfred von Zittel for quotations from his *History of Geology and Palaeontology* (translated by Maria M. Ogilvie-Gordon: London, 1901). Brief acknowledgements for the use of extracts from the above authors and from other authors, ancient and modern, are given below the quotations, the full reference being added in the bibliographies.

We are equally pleased to express our thanks to the following publishers, organizations and individuals for their kind permission to reproduce the illustrations numbered as follows in our text:

The Royal Society of Edinburgh: Frontispiece, 10, 11; Deutscher Kunstverlag, Munich and Berlin: 5; George Allen and Unwin Ltd, London: 3; Dover Publications Inc, New York: 4; Edward Arnold (Publishers) Ltd, London: 6, 8, 9; The National Portrait Gallery, London: 12, 20, 34, 35; The University of Illinois Press: 19; Dr F. J. North and the Editors of the *Annals of Science*: 22; Dr F. J. North and the Editors of the Geologists' Association of London: 37; The Department of Mineralogy and Petrology, Cambridge University, England: 33; The Controller of H.M. Stationery Office, London (from *British Regional Geology, The Grampian Highlands*; H.M.S.O., 1948): 41; The University of Chicago Press; 43, 44; The Director, United States Geological Survey, Washington, D.C.: 47, 94, 98, 99, 100, 101, 115, 116, 117, 119, 120, 124, 127; Yale University Press: 46, 49, 65, 67, 95, 96, 97, 104, 105; Princeton University Press (from *The Principles of Historical Geology from the Regional Point of View* by R. M. Field, Del. E. T. Raisz): 51, 90; Dr Erwin Raisz of Cambridge, Massachusetts (from *The Physiographic Provinces of North America*, by W. W. Atwood; Ginn and Co., Boston, 1940): 106; Aerofilms and Aero Pictorial Ltd, London: 54, 75, 76; The Air Ministry, London, and the Curator in Aerial Photography, Cambridge University (photographs by

J. K. St. Joseph; Crown Copyright reserved): 57, 60, 63; Richard Cave Esq., of Richmond, Surrey: 70; Edward Stanford Ltd, London: 79; The Scottish National Portrait Gallery, Edinburgh, and Dr R. E. Hutchison: 87; Ginn and Co., Boston (from *The Physiographic Provinces of North America*, by W. W. Atwood): 122; The late R. Mott Davis Esq., of Silver Spring, Maryland: 130; W. M. Davis II of Bass River, Massachusetts: 131; Dodd, Mead and Co, New York: 94, 101; Dr J. C. Allen, The John Hopkins University, Baltimore: 19; G. P. Putnam's Sons, New York: 89, 102, 103; The British Association for the Advancement of Science: 25, 62; The Royal Society of London: 74; The Geological Society of London: 30; Mary Potter and the Department of Geography, University of Oxford, for the drawing of 27, 77, 78.

For helpful suggestions and information we are especially indebted to Professor George W. White of the University of Illinois, Urbana; Professor Maurice Pardé of the University of Grenoble; Madame G. Steinmann of the SOGRÉAH, Grenoble; Dr V. A. and Mrs J. M. Eyles of Great Rissington, Gloucestershire; and Frederick E. Leese, Librarian at Rhodes House, Oxford.

For assistance with the selection and translation of German texts we express our thanks to Dr Gerhard Aymans of the University of Bonn, and similarly for German and Italian texts to Drs K. F. Mather and S. F. Mason, and through them to Professor A. O. Woodford who translated the passages from Hermann Sternberg's article of 1875. The allusions to Russian authors were kindly improved and augmented by Professor Chauncy D. Harris of the University of Chicago. The translations from French and Latin are, unless otherwise stated, by R. P. Beckinsale and D. R. Stoddart.

The *Informative Index* is intended to be a complement and supplement to the main text. It will, we hope, answer most of the queries that arise in the text, such as the identity of persons mentioned incidentally in quotations. It also includes some mention of most of the authors incorporated in the bibliographies as we considered that the historians or writers on a science have a worthy place among its practitioners. This Index is largely the research of Monica Beckinsale to whose literary sense and skill we are deeply grateful in many other ways. We will not discourage would-be emulators by disclosing the great amount of time and energy spent in compiling it.

One of us would like to acknowledge gratefully the generous aid of the British Association for American Studies and a year's hospitality and use of the facilities of the Department of Geography at the University of Chicago. We all unite in gratitude to the person who having purloined the only copies of the notes and text of Part 3 from Victoria Station (the Margate line) subsequently returned them. We hope that most of our readers will not consider this second act more unpardonable than the first.

Finally, we make an inadequate gesture of thanks to the numerous students at the Universities of Cambridge, Oxford, Columbia, and Brown (Providence, R.I.), who over the years have acted as critics and guinea-pigs for parts of the original manuscript of this volume. We trust that they will not miss too grievously the quarter of a million words jettisoned in the published version.

R. J. CHORLEY   A. J. DUNN   R. P. BECKINSALE
University of Cambridge   *University of Oxford*

September 1963

PART ONE

'Worlds Without End'

CHAPTER ONE

Introduction

To unravel the secrets of nature requires the intuitive wisdom, deductive capacity and experimental ability of generations of scientists. The landmarks of discovery arise from the refined product of their labours. Sometimes research advances by great bounds as a theory is propounded and the many pieces appear momentarily to fall into place. More often it proceeds step by step with one man concentrating on a small fragment and making it his life's work. The result of his research, and that of others like him, has often to await the entry of the theorist who takes all the parts and builds them into one grand scheme, the very nature of which may point to new lines of research. Study of fluvial and other geomorphic processes has proved no exception to this rule and as the work went forward the task of comprehension assumed increasing proportions.

Through all the ages Man, as he moved within the boundaries of his immediate environment, could not have failed to notice the differences in the surface features or to experience the wilder antics of the elements. On every side of him stood the mighty monuments of nature confounding and prompting reasonings which ranged to the outermost limits of his mental capacity.

Yet throughout the periods of history these varied phenomena have given rise to a series of widely-different explanations. Each man in his turn in exploring nature's secrets has used his especial skill, together with the acquired ideas and prejudices of his age. From the example of the great scientists it appears that the monuments of nature are best deciphered by a vigorous combination of patient exploration, cautious generalization and constant re-appraisal.

This, we hope, will become apparent in the following chapters where we trace the growth of ideas on the shape and nature of landforms (or geomorphology) with special emphasis on the origin of water-worn landscapes. As in the history of some other branches of scientific knowledge, this tale of evolution, to quote words we shall use so often later, reveals 'no vestige of a beginning – no prospect of an end'. Yet the account has fascinating peculiarities.

First, from earliest times man's interest in his physical environment has assured a great and enduring popularity for the study of landscape and of natural forces. Therefore, personal investigations of landscape features have proceeded continuously within historic times.

Second, as this environmental study is mentioned in many parts of the Bible, it became irrevocably linked with theories of the Creation and with theology.

3

Third, being connected with universal matters, landscape study became peculiarly liable to be dominated by generalizations of universal extent. Such generalizations often preceded and took control of detailed deductions. It would be hard to imagine a more stultifying situation for a youthful 'science'.

Fourth, it was inevitable that the steady progress of practical work or field-observations would lead to the overthrow of many ill-founded universal generalizations, which would then be replaced by universal laws based on more scientific principles.

Fifth, even to this day, in spite of countless spasmodic and short-term observations, there has been a marked absence of continuous measurements or long-term recordings of landscape processes, except perhaps on floods, rainfall, earthquakes and violent vulcanism. Thus geomorphology remains a pseudo-science with a largely unscientific tradition. No doubt a century hence it could, if given today a proper basis of physics and physical chemistry, be called an *earth science*.

The story we have to tell begins with the interest of man in the work of nature, more particularly through the agency of water, and in his inheritance of the tradition of a great flood, a tradition derived no doubt from human experience in the last Ice Age and the pluvial epoch following it. The present volume ends with the global generalizations of William Morris Davis, who throwing caution to the winds and ignoring many contemporary scientific observations, published a most attractive theory of landscape evolution. In venturing to suggest that Davis took only a partial view, we maintain the ancient geomorphological tradition of opposing inadequate generalizations.

It should be noticed that all generalizations on landscape evolution fail largely because of relative lack of knowledge of erosional processes and because of ignorance of the interaction between the earth's crust and the earth's interior. Ignorance of vulcanism and earth movements proved the weak links in the ideas of the Wernerians, Huttonians, and Davisians alike. Modern geophysicists are only just beginning to delve into the problem. These twentieth-century aspects of the study of landscape evolution we intend to treat in further volumes. Here we end with Davis not only because he, by his magical generalizations, lulled the geomorphological world into a temporary slumber, but also because he was a great figure, who occupied a worthy place in a long line of fascinating personalities stretching back to Hutton and beyond.

CHAPTER TWO

Early Ideas on Erosion and Rock Strata

INTRODUCTION

Faced by the daily testimony and record of the massive operations of nature's agents and the colossal proportions of their structures, it is not surprising that man's first explanations of the World were visionary and in the form of fables. To the Christian peoples of the west the story of creation in the Book of Genesis is the best-known example. The eastern peoples had their own versions (Suess, 1904, I). All the stories taught that the world was created by supernatural persons or forces in a relatively short space of time and that it had since retained, to a substantial extent, its original form. This is not to say that the ancients did not recognize the existence of recurrent events like floods and tempests—the Bible is full of such topics—but that these occurrences were regarded as being only of local or temporary significance. During the Middle Ages, when perhaps the Church had its strongest hold over the laity and the advancement of learning, the notion of a single act of creation became rigidly enforced on the principle that it was a religious truth.

Consequently, in the early history of landscape studies we are faced with a strange dichotomy: the abundance of observers who recognized that erosive forces were actively fashioning the earth's surface; and the superabundance of people who considered that the earth's features were created in the six days of the Creation. These ideas might well have remained separate but for the great interest taken in fossils. Indeed the sea-shell on the mountain-top has been of supreme importance in the history of landscape study and may even have been of more significance than the countless early observations on the obvious erosive force of rivers and waves. It happens, however, as we have already hinted, that ideas on erosion need not be applied universally and need not offend cosmogonists whereas ideas on sea-shells in mountain rocks involve the Creation and world floods and discussion upon them soon opens the flood-gates of theology. In this chapter we shall attempt to show briefly how some writers restricted their observations to erosion and others to fossils and how any who tried to combine speculations on both were eventually forced to reconcile landscape evolution with theological cosmogony.

ANCIENT EROSIONISTS

Among the Greeks one of the chief geological figures is Aristotle (384–322 B.C.), who theorized that the cold temperature of mountains condensed the atmospheric moisture and gave rise to the origin of streams. From the presence of rock strata he surmised a regular series of mutations between dry land and sea, and between tropical and polar temperatures; and here he anticipated Hutton. As a Greek, Aristotle was not hampered by religious inhibitions and his field of speculation was not blurred by *a priori* reasonings. However, unfortunately, he did not find the uncertainties of the physical processes of nature a fruitful source for logical hypotheses, so dear to the classical Greeks, and he pursued his observations no further. This neglect by the Greeks of the natural sciences contributed notably to the retardation of the growth of those studies in Britain where well into the nineteenth century many establishments for education were mainly concerned with the teaching of classical knowledge.

In fact, none of the ancient classical authors progressed in the natural sciences as far as the Arab philosopher and physician, Avicenna (A.D. 979–1037), who seemed to have glimpsed the vision of a continuous succession of landscape processes and pointed out that the natural topography could result from the action of running water.

THE FIRST MODERN EROSIONISTS AND PALAEONTOLOGISTS

The later medieval period was marked by a great interest in earth features as contemporary paintings reveal in their striking landscape backgrounds. Several authors now showed a remarkable advance in the study of landscape formation and their writings began to combine ideas on erosion and on fossils into universal generalizations of a cosmogonical kind.

It is in a sense unfortunate that we have to begin our account with Leonardo da Vinci (1452–1519) whose work was of such a quality that it might well have represented the culminating thoughts of a long line of illustrious thinkers. As it was, in landscape studies Leonardo seems to have outstripped in intelligence his predecessors and immediate successors alike. It would be hard not to say that he grasped firmly the ideas both of local erosion and of world-wide landscape evolution.

The following quotation demonstrates the quality, width and stupendous foresight of his observations on the evolution of topography.

> I perceive that the surface of the Earth was from of old entirely filled up and covered over in its level plains by the salt waters, and that the mountains, the bones of the earth with their wide bases, penetrated and towered up amid the air, covered over and clad with much high-lying soil. Subsequently, the incessant rains have caused the rivers to increase, and by repeated washing, have stripped bare part of the lofty summits of these mountains, leaving the site of the earth so that the rock finds itself exposed to the air, and the earth has departed from these places. And the earth from off the slopes and the lofty summits of the mountains has already descended to their bases, and has raised the floors of the seas which encircle these bases, and caused the plain to be uncovered, and in some parts has driven away the seas from there over a great distance. (*Translated by McCurdy, 1906, pp. 94–95*)

In the following vivid passages he refutes the possibility that the existing fossil remains found on mountain and plain could have been the product of the biblical forty days and forty nights of rain and instead he asserts that the presence of fossil beds at high altitudes is proof of changes in level between the land and the sea:

If you should say that the shells which are visible at the present time within the borders of Italy, far away from the sea and at great heights, are due to the flood having deposited them there, I reply that granting this flood to have risen seven cubits above the highest mountain, as he has written who measured it, these shells which always inhabit near the shores of the sea ought to be found lying on the mountain sides, and not at so short a distance above their bases, and all at the same level, layer upon layer.

Should you say that the nature of these shells is to keep near the edge of the sea, and that as the sea rose in height the shells left their former place and followed the rising waters up to their highest level:—to this I reply that the cockle is a creature incapable of more rapid movement than the snail out of water, or is even somewhat slower, since it does not swim, but makes a furrow in the sand, and, supporting itself by means of the sides of this furrow it will travel between three or four braccia in a day; and therefore with such a motion as this it could not have travelled from the Adriatic sea as far as Monferrato in Lombardy, a distance of two hundred and fifty miles in forty days—as he has said who kept a record of that time.

And if you say that the waves carried them there—they could not move by reason of their weight except upon their base. And if you do not grant me this, at any rate allow that they must have remained on the tops of the highest mountains, and in the lakes which are shut in among the mountains . . .

If you should say that the shells were empty and dead when carried by the waves, I reply that where the dead ones went the living were not far distant, and in these mountains are found all living ones, for they are known by the shells being in pairs and by their being in a row without any dead, and a little higher up is the place where all the dead with their shells separated have been cast up by the waves, near where the rivers plunged in mighty chasm into the sea. (*Translated by McCurdy, 1906, pp. 106–7*)

Where the valleys have never been covered by the salt waters of the sea, there the shells are never found. (*Translated by McCurdy, 1906, p. 109*)

The water wears away the mountains and fills up the valleys, and if it had the power, it would reduce the earth to a perfect sphere. (*Translated by McCurdy, 1906, p. 101*)

The shells of oysters and other similar creatures which are born in the mud of the sea, testify to us of the change in the earth round the centre of our elements. This is proved as follows:—the mighty rivers always flow turbid because of the earth stirred up in them through the friction of their waters upon their bed and against the banks; and this process of destruction uncovers the tops of the ridges formed by the layers of these shells, which are embedded in the mud of the sea where they were born when the salt waters covered them. And these same ridges were from time to time covered over by varying thicknesses of mud which had been brought down to the sea by the rivers in floods of varying magnitude; and in this way these shells remained walled up and dead beneath this mud, which became raised to such a height that the bed of the sea emerged into the air. And now these beds are of so great a height that they have become hills or lofty mountains, and the rivers, which wear away the sides of these mountains, lay bare the strata of the shells and so the light surface of the earth is continually raised, and the antipodes draw nearer to the centre of the earth, and the ancient beds of the sea become chains of mountains. (*Translated by McCurdy, 1906, pp. 104–5*)

Leonardo was not alone in his disbelief that the biblical Flood could account for the high-level inland distribution of fossils. Fracastoro (1517) ridiculed the suppositions that fossils could have been produced by the antics of stars or by the temporary and local inundation of the flood and was firmly convinced that they were derived from the remains of former animals (Geikie, 1905).

Georg Bauer, who was born in Saxony in 1494 and began medical practice in the mining region of Bohemia in 1527, wrote (under the *nom-de-plume* of Agricola) in 1546:

> We can plainly see that a great abundance of water produces mountains, for the torrents first of all wash out the soft earth, next tear away the harder earth, and then roll down the rocks, and thus in a few years excavate the plains or slopes to a considerable depth; . . . By such excavation to a great depth through many ages there rises an immense eminence on each side. When an eminence has thus originated, the earth slips down loosened by constant rain and split away by frost, and the rocks, unless they are extremely firm, since their seams are similarly softened by the damp, roll down into the excavations below. This continues until the steep eminence is changed into a slope. . . . Moreover, streams, and, to a far greater extent, rivers, effect the same results by their rushing and washing; for this reason they are frequently seen flowing either between very high mountains, which they have produced, or close to their foot. . . . Nor did the hollow places which now contain the seas all formerly exist, nor yet the mountains which curb and break their advance, but in many parts there was a level plain until the force of winds let loose upon it a tumultuous sea and a seething tide. By a similar process the impetus of water completely demolishes and flattens out hills and mountains. (*Agricola, De Ortu, 1546, lib. 3, pp. 1–2*)

In France, the famous Huguenot potter, Bernard Palissy (1510?–1590) made it quite clear that he defied established opinion and that he based his scientific work on his own observations of the sky and ground. He described in some detail the use of a soil auger and the nature of soils ('earths'), ground water, and fossils (La Rocque, 1957; Thompson, 1954). Among Palissy's interesting geological statements are the following:

> . . . trees planted along the mountain . . . will serve much to reduce the violence of the waters . . . I would advise you to plant some if there were none: for they would serve to prevent the waters from excavating the ground, and by such means grass will be preserved, and along this grass the waters will flow quietly straight down to your reservoir. (*1580, p. 61, trans. by La Rocque*)

> Those who say that the rocks were created at the very beginning of the world are mistaken and do not understand the matter. (*1580, p. 195*)

> Those who think that rocks acquire their full hardness as soon as they are first formed, do not understand the matter. (*1580, pp. 245 and 246*)

All clays are the beginnings of rocks. (*1580, p. 301*)
Those who have written that the shells found in rocks date from the Flood
have erred clumsily. (*1580, p. 212*)

Geologically Palissy was in a crude way an early Neptunist and it is perhaps
not surprising that the first of several re-printings of his complete works
appeared in 1777.

FIG. 1. *Nicolaus Steno*

Yet Fracastoro, Agricola and Palissy approached less closely to Leonardo's
ideas than did Nicolaus Steno (1638–86), who was born in Copenhagen but
settled in 1666 in Florence as Court Physician to the Grand Duke of Tuscany.
Steno not only held that fossils were of animal origin but he believed that they
and rock strata formed a catalogue of the earth's history. In other words, he

envisaged the basic principles of the science of stratigraphy (Garboe, 1958). He went slightly further than Leonardo by propounding in his most famous work of 1667 a scheme of six phrases of earth history (*Geikie, 1905, p. 59*):

1. Submergence of the earth's surface and deposition of strata without fossils.
2. Emergence of the surface as a dry plain.
3. The horizontal strata disrupted into hills, mountains and valleys by the sudden up-thrusting of gases and the collapse of air-filled cavities.
4. Re-submergence of valleys and plains.
5. Filling of these low areas by river sedimentation.
6. Re-elevation of the plains accompanied by some disruption of the strata and followed by erosion by running water.

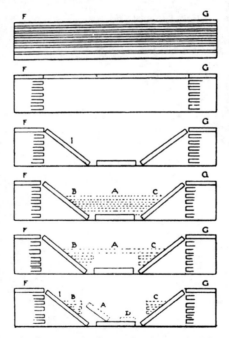

FIG. 2. *Steno's six stages in the production of the Plain of Tuscany*

EARLY CATASTROPHIC RECONCILERS

Nicolaus Steno, with his six earth phases, marks the transition to the geological thinkers of the late seventeenth and early eighteenth centuries who tried to reconcile their theories with biblical history. Typical of these reconcilers are Thomas Burnet, John Woodward, and the Comte de Buffon.

Burnet (1635?–1715), a Cambridge M.A., had an eloquent style and rather crude scientific notions but his publications (1681–9) on the origin of the earth attracted much attention in his day. He considered that man's own wickedness brought about the end of a spring-time on earth, a misfortune which occurred when the sun's rays cracked the shell of our egg-like globe and released a great mass of waters from a central abyss. The broken fragments of the outer shell then formed the mountains. Although this catastrophe was stated to be the same as the biblical Flood and although Burnet professed to reconcile all his theories closely with Genesis, he was ridiculed in a popular ballad as saying

> That all the books of Moses
> Were nothing but supposes.

Woodward (1665–1728) had more definite geological learnings. Before being made professor of physic, at Gresham College in 1692, he had collected fossils in various parts of England and had made many inquiries of acquaintances abroad as to whether similar fossils were to be found there. In his *Essay towards a Natural History of the Earth* he tells us that he examined the earth for evidence of the means of its construction in those places:

> where the Entrails of the Earth were laid open, either by Nature (if I may so say), or by Art, and humane Industry. And wheresoever I had notice of any considerable natural Spelunca or Grotto, any digging for Wells of Water, or for Earths, Clays, Marle, Sand, Gravel, Chalk, Cole, Stone, Marble, Ores of Metals, or the like, I forthwith had recourse thereunto; and taking a just account of every observable Circumstance of the Earth, Stone, Metal, or other Matter, from the Surface quite down to the bottom of the Pit, I entered it carefully into a Journal, which I carry'd along with me for that purpose. (*Woodward, 1695*)

Many of Woodward's observations bear the stamp of quality.

> the stone and other terrestrial Matter was distinguished into Strata, or Layers; that those Strata were divided by parallel Fissures; that there were enclosed in the Stone, and all the other denser kinds of terrestrial Matter, great numbers of Shells, and other Productions of the Sea.
>   Shells, and other marine Bodies, found at Land, were originally generated and formed at Sea; . . . they are the real spoils of once living Animals. (*Woodward, 1695*)

Yet Woodward failed to arrive at the true nature of stratigraphical deposition, probably largely because he was circumscribed in his theoretical outlook by the force of biblical teachings. Inevitably he was driven towards a catastrophic explanation as being the most likely instrument, and one acceptable to the Church. He had his own variation on the biblical story but substantially his theory remained the same as that of the Flood in Genesis. His version made the centre of the earth a spherical cavity filled with water, which by God's will was released and the whole globe 'was taken to pieces and dissolved'. All the rocks, minerals, animal and vegetable bodies were swallowed up, severed and carried about in the flood waters. When God relented, the

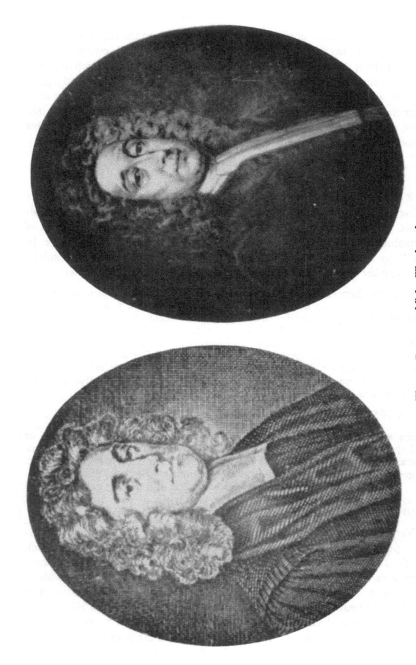

FIG. 3. *Thomas Burnet and John Woodward*

waters subsided and the various bodies carried in suspension successively sank down in proportion to their specific gravity, the heaviest elements falling first formed the bottom strata and the lighter elements afterwards to form the upper strata. Woodward fitted his various fossils into their parent strata by supposing that they had a specific gravity similar to that of the containing strata. Though the argument contains obvious errors, for the period in which it was made it represents a genuinely intelligent explanation and had the prime practical merit of avoiding conflict with the Church.

The arguments of Georges Louis Leclerc Buffon (1707–88) are altogether more grandiose and more ingenious than those of Woodward. They demonstrate clearly the growing credibility and complexity of the explanations put forward by catastrophic reconcilers in the eighteenth century. The two main problems facing the would-be exponent of earth evolution remained the same: the presence of fossils high up on mountains; and the impossibility of fitting in the growth of the earth and the moulding of its surface to a dogmatic timescale of six days' creation and forty days of flood. The latter difficulty was not lessened by the fact that the date now, and henceforth, assigned by biblical scholars to the birth of the earth was 4,004 B.C. Archbishop James Ussher's chronology, calculated about 1650, was assumed to be definitive and was widely accepted. Dr John Lightfoot, vice-chancellor of Cambridge University in 1654, followed him in stating that 'Heaven and Earth, centre and circumstance were made in the same instance of time, and clouds full of water and man were created by the Trinity on the 26th October 4004 B.C. at 9 o'clock in the morning.'

Buffon, however, was more concerned with reconciling his theories of earth evolution with the Creation.

He supposed the earth to have originated as a molten mass broken off from the sun after its collision with another planet. The detached fragment then continued in the sun's orbit. His explanations of the later phases of Earth history were accommodated to Church teachings by means of the argument that the space of six days allowed for the Creation was merely figurative. The six days were periods each requiring a fairly long interval of time for its completion.

The first, immediately after the earth's detachment from the sun, was a period of extreme incandescence during which the earth remained a fiery ball.

The second period was one of cooling, during which the molten mass solidified and crinkled into the primitive mountain chains.

The third period was marked by further cooling when the water vapour of the atmosphere condensed to form an ocean nine to twelve thousand feet higher in level than it is today (thereby partly accounting for the presence of marine fossils at such heights). During this inundation, ocean currents scooped out the present valleys.

From what had been advanced, we may conclude, that the flux and reflux of the ocean have produced all the mountains, valleys, and other inequalities on the surface of the earth; that currents of the sea have scooped out the valleys, elevated the hills, and bestowed on them their corresponding directions; that the same waters of the ocean, by transporting and depositing earth etc. have given rise to the parallel strata; that the waters from the heavens gradually destroy the effects of the sea, by continually diminishing the height of the mountains, filling up the valleys, and choking the mouths of rivers; and by reducing everything to its former level, they will, in time, restore the earth to the sea, which, by its natural operations will again create new continents, interspersed with mountains and valleys, every way similar to those which we now inhabit. (*Buffon, 1785, I, pp. 57–58*)

During the fourth period the cooling progressed to the deeper interior parts of the earth; the inner mass contracted in places causing cavities to open in the earth's surface and the water was drawn down through them until it reached its present level.

In the fifth period the continents, which had been formed by this time, became divided and the Old World (Eurasia and Africa) was separated from the New (The Americas), an event presumed to have happened because the similarity of certain fossils found in the various continents indicated the former continuity of the land masses.

During the sixth and last period there was a final cooling of the surface and the gradual erosion of the higher areas.

It was no sooner suspected that our continent might formerly have been the bottom of the sea, than the fact became incontestible. The spoils of the ocean found in every place, the horizontal position of the strata, and the corresponding angles of the hills and mountains, appeared to be convincing proofs; for, when we examine the plains, the valleys, and the hills, it is apparent that the surface of the earth has been figured by the waters. When we descend into the bowels of the earth, it is equally evident, that those stones which include sea shells, have been formed by sediments deposited by the waters, since the sea shells themselves are impregnated with the same matter that surrounds them. And, in fine, if we consider the corresponding angles of the hills and mountains, we cannot hesitate in pronouncing that they received their configuration and direction from currents of the ocean. It is true, that, since the earth was first left uncovered with water, the original figure of its surfaces has been gradually changing; the mountains have diminished in height; the plains have been elevated; the angles of the hills have become more obtuse; those bodies which have been rolled along by the rivers have received a roundish figure; new beds of tufa, of soft stone, of gravel, etc. have been formed. But every thing has remained essentially the same. The ancient form is still recognizable; and I am persuaded that every man may be convinced by his own eyes of the truth of all that has been advanced on this subject; and, that whoever has attended to the proofs I have given, must be fully satisfied that the earth was formerly under the waters of the ocean, and that the surface which we now behold received its configuration from the currents and movements of the sea. (*Buffon, 1785, I, pp. 484–5*)

The rivers generally occupy the middle of the valleys, or the lowest ground between two opposite hills: If the two hills have nearly an equal declivity, the river runs nearly in the middle between them, whether the intermediate valley be broad or narrow. If, on the contrary, the declivity of one of the hills be greater than that of the other, the river will not occupy the middle of the valley, but will approach to the steepest hill, in proportion to the superiority of its declivity. . . . In process of time, however, the declivity of the steepest hill is diminished by the rains, the melting of snow, etc. The steeper any hill is, it loses greater quantities of earth, sand, and gravel, by the operation of rains, and these substances are carried down into the plain with a proportionably greater rapidity. . . . (*Buffon, 1785, I, p. 255*)

EIGHTEENTH-CENTURY EROSIONISTS

Although many of the early catastrophic reconcilers were well acquainted with the powers of river erosion they always returned to the Flood as the main agency in the formation of landscape features. Hence it was inevitable that the erosionists, being directly concerned with earth history, should be critical of reconciliatory explanations.

It is difficult, if not impossible, to distinguish modern from medieval erosionists but we may well begin the modern era with the work of John Ray (North, 1943), a noted naturalist, who just before the opening of the eighteenth century, described how rain and river action

by little and little wash away . . . the mountain, and fill up the lower places of the valleys, whereby it appears that what the mountain loseth the valley gains, and consequently, that on the whole of the earth nothing is lost, but only removed from one place to another, so that in process of time the highest mountains may be humbled into valleys. (*Ray, 4th edn, 1721, p. 245*)

Lyell describes him as

. . . one of the first of our writers who enlarged upon the effects of running water upon the land, and of the encroachment of the sea upon the shores. So important did he consider the agency of these causes, that he saw in them an indication of the tendency of our system to its final dissolution; and he wondered why the earth did not proceed more rapidly towards a general submersion beneath the sea, when so much matter was carried down by the rivers, or undermined in the sea-cliffs. (*Lyell*, Principles, *11th edn, 1872, Vol. 1, p. 45*)

As regards fossils, Ray noted their position and distribution in definite groups, which, he argued, showed

the effects of the animals breeding there for a considerable time. (*Ray, 4th edn, 1721, p. 171*)

But he thought that this time must have been vastly more prolonged than the period of the Flood, and was driven to two conclusions neither of which satisfied him. Either the fossils must have been in the land when the land and the waters were separated, and to this he objected:

If we stick to the letter of the Scripture history of the Creation, that the Creation of Fishes succeeded the Separation of the land and the sea, and that the six days wherein the world was created were six Natural Days, and no more, it is very difficult to return a satisfactory answer to this Objection. (*Ray, 4th edn, 1721, p. 172*)

or else it may have been that

possibly in the first Creation the whole earth was not all at once covered, but only those parts whereabout Adam and the other animals were created, and the rest gradually afterwards . . . during which time these shellfish might breed abundantly all the sea over, the bottom whereof being elevated and made dry land, the beds of shellfish must necessarily be raised together with it. (*Ray, 4th edn, 1721, p. 172*)

On the continent of Europe, erosionists were probably more advanced than in England. In Italy, Giovanni Targioni-Tozzetti (1712–84) ridiculed Buffon's theory of Flood erosion. From his own observations Targioni-Tozzetti decided that the origin of valleys must be ascribed to the work of the present rivers. He marked as conclusive evidence of this the orderly and systematic pattern of the river network in contrast to the chance passage of the waters' course which one would expect if a sea or flood had been the cause:

From the Ponte a Moriano to the mouth of the Valley of Anchiano, the Serchio runs through a narrow and winding channel and by the impact of its waters is shaping, eroding and dividing the vast mountain masses. It is Signor Buffon's opinion that the winding channels which are found on the surface of the earth and through which flowing waters today make their way have been formed by the rushing currents of the sea when our terraqueous globe was covered by its waters. Surely the deep channel of the Serchio from the Ponte a Moriano clear to Anchiano, that of the same river to Ripafratta, those of the Arno through the Valle dell' Inferno and from the Incisa to the Ponte a Rignana, and all the other channels of rivers that I have not yet seen were not thus excavated when these regions were covered with the waters of the sea. On the contrary, they were not formed until after the lowering of the level of the sea, when the fresh water, flowing toward the lowered sea, began to descend and acquired velocity. Then it was that the mountain streams, arranged in a network, began to erode and have continued to deepen their channels until, on approaching the level of the modern sea, they lose their velocity and the force of their impact. It might seem to some, that, according to the laws of hydrostatics, erosion and channelling in the primitive mountains ought to have been accomplished in straight lines, rather than twisted and at angles as we find it. However, that is dependent on the diverse resistances of the materials composing the primitive mountains which it must cut and channel. . . . Other important causes of variation in direction, of course, have been the torrents which, with different directions and different amounts of waters, were discharging into the main river, particularly when, on account of flood, they have more impetus than it. Finally, another cause of this winding is due to the fact that the looseness of the material of one mountain is greater than that of another adjoining. (*Translated in Mather and Mason, 1939, pp. 74–75*)

In France at about the same time (1752), Jean Étienne Guettard (1715–86) was extolling the rapidity of surface erosion. Guettard, when physician to the

Duke of Orleans, spent much time in constructing geological maps of northern France and was impressed with the variability of strata represented. In his 200-page article 'On the degradation of mountains effected in our time by heavy rains, rivers and the sea' he emphasized the ceaseless changes of the earth's surface but concluded that of all the erosive forces now in operation the sea was the most important (*Guettard, 1768–83, III, pp. 209–403*).

Half a century later Nicolas Desmarest (1725–1815) wrote a memoir for the Academy of Sciences in Paris on the rocks of Auvergne, which Guettard (1752) had previously recognized to be volcanic in origin. In this memoir Desmarest noted that the amount of fluvial erosion of these lavas seemed to be proportional to the relative age of the lava-flows. Thus he anticipated the work of Lyell and Murchison by over fifty years. (*See* Desmarest, 1806.)

THE EARLY STRATIGRAPHERS

The erosionists discussed above were interested mainly in the forces that tend to erode a landscape. But it happens that a full knowledge of landscape processes also involves the influence of geological structure or of the resistance of the rocks themselves. This structural aspect was greatly enriched in the eighteenth century by scholars who were primarily or solely interested in the deposition and disposition of strata, for which reason we have dubbed them stratigraphers.

Among those worthy of notice is John Strachey (1671–1743) who in his *Observations on the strata in the Somersetshire coalfields* (1719) drew attention to the existence of a relationship between rock structure and surface features, an idea which was not finally accepted and certainly not fully recognized until about a century later.

> This is all I can say in relation to the different veins of coal and earth in the coal-works in these parts; wherein all agree in the oblique situation of the veins; and every vein hath its cliff or clives lying over it, in the same oblique manner. All of them pitch or rise about twenty-two inches in a fathom, and almost all have the same strata of earth, malm, and rock over them, but differ in respect to their course or drift, as also in thickness, goodness, and use.
>
> Now as coal is here generally dug in valleys, so the hills, which interfere between the several works before mentioned, seem also to observe a regular course in the strata of stone and earth found in their bowels: for in these hills . . . we find on the summits a stony aragle mixt with a spungy yellowish earth and clay; under which are quarries of lyas, in several beds, to about eight or ten feet deep, and six feet under that thro' yellowish loom, you have a blue clay, enclinable to marle, which is about a yard thick: under this is another yard of whitish loom, and then a deep blue marle soft, fat, and soapy, six feet thick; only at about two feet thick, it is parted by a marchasite about six inches thick. . . . (*Strachey, 1719, p. 972*)

The main progress in stratigraphy and the influence of structure on landscapes was, however, to take place in the mining academies of continental Europe. Here four men were outstanding: Giovanni Arduino of Italy, and

Johann Gottlob Lehmann, Georg Christian Füchsel, and Peter Pallas, all of Germany.

Arduino (1713–95) in the capacity first as Inspector of Mines in Tuscany and later as Professor of Mineralogy and Metallurgy at Venice was able to carry out researches in the central mountain region of his country. He classified rocks into four grades (Arduino, 1760):

Primitive, which formed the centre of mountains and were distinguished by the absence of fossils;

Secondary, which consisted of stratified sedimentary formations containing all kinds of fossils;

Tertiary, which was also sedimentary and contained many land and marine fossils but was less consolidated (for example, gravels, sands and marls) than the secondary group;

Volcanic.

The derivation of the tertiary rocks from the older secondary rock series could be readily inferred. In addition, alluvial material washed down from the mountains by the present streams was recognized as resting on all of the four grades or formations. The importance of this classification lies in the initial appreciation that there was a key by means of which rock strata could be distinguished and possibly their relative age determined.

In Germany, Lehmann (1719–67), a teacher of mineralogy and mining at Berlin, advanced on similar lines. In 1756 he published a work based on his observations of the rocks of the Harz and Erzgebirge mountains which he subdivided into three stratigraphical groups or orders. In the first order (Primitive : Urgebirge) he considered the rocks to be coeval with the world's creation and characterized by their greater height, durability and limitless extension into the bowels of the earth. The second order (Flötzgebirge) he differentiated by their sedimentary formation in horizontal strata and by the abundant existence of fossils. The third order, of unconsolidated gravels and sands (Angeschwemmigebirge), he explained as resulting from local accidents. Lehmann did not restrict these ideas to the origin of mountains but extended them to speculations on the beginnings of the Earth.

> The Earth began as a vast blob of muddy matter uniformly dispersed through water. At the first stroke of creation the mud is supposed to have settled forming a shell whose high parts became continents and islands while the surplus water drained away into an abyss within the earth. At a later time there occurred Noah's flood which washed the sediments from his primitive mountains and deposited it on their slopes. Later other formations appeared, some as volcanoes or lava flows and some as stratified deposits laid down by still later floods. (*Fenton and Fenton, 1952, p. 42*)

Füchsel (1722–73) was possibly the most remarkable of these four German stratigraphers. Although the son of a humble baker, he managed to study at the Universities of Jena and Leipzig. At Erfurt he became a Doctor of Medicine and later obtained the post of physician to the Prince of Rudolstadt. With the security of this position he was able to follow his particular hobby of the study of rocks and minerals, and in 1762 published his conclusions in a treatise written in Latin (Füchsel, 1762). He recognized nine different types of strata in the Thuringen and Harz mountains, with sequences of derived sedimentary strata resting against the central mass of hard primitive rocks. He visual-

FIG. 4. *Lehmann's types of earth structure*
Above: *Parts of two Primitive Mountains between which lie the bedded deposits of the Mosaic deluge from which Secondary Mountains may be developed.*
Below: *A Primitive Mountain (31) overlain by a succession of thirty beds composing a group of Secondary Mountains south of Hannover.*
(*Originals in 'Essai d'une Histoire naturelle de la terre'; Paris, 1759*)

ized all the rock formations including the central mass as having been laid down below the sea in horizontal beds, and to him must go the credit for defining 'formation' as a sequence of strata formed under approximasely similar conditions. Where the strata were inclined or tilted he believed this to be due to the effects of some subsequent violent disturbance.

Peter Simon Pallas (1741–1811) had all the material advantages that the youthful Füchsel lacked. His father was a surgeon professor and his mother French (Geikie, 1905). He studied medicine and natural science at Berlin,

Halle, Göttingen, and Leyden and at twenty-seven was already widely travelled and had, as was customary for educated men of that age, a mastery of French, English and Latin, as well as of his own native tongue. His life reached a notable turning point by his appointment to lead a scientific exploration into Siberia, and there he spent six years (1768–74) in company with a small band of fellow-scientists. Though the results of the expedition (Pallas, 1777) covered the whole field of the then known sciences, geology benefited greatly from his generalizations on the forms and structure of mountain chains. He remarked that in all cases granite formed the centres of such chains; that in turn these were bounded by various unfossiliferous rocks in highly tilted positions; that these were laterally succeeded by schists, shales and fossil-bearing masses of limestone, the successive outer strata becoming less highly tilted and containing an increasing number of fossils; and that towards the outer limits of the mountains the low hills and plains were made up of sandstones, clays and other loose disorganized materials. From this evidence he concluded that the granite cores of the mountain chains had never lain below the surface of a primeval ocean but had risen to their present position as a result of upheaval by some violent volcanic forces which forced upwards the younger secondary and tertiary formations which were lying on their margins into their present tilted position.

It is worth while to analyse collectively the contribution of these four men. Though they said nothing to contradict the popular theory of the Flood as being a contributory cause of the earth's surface features, they were the first to draw attention to the principal distinctions between the various rock groups. Instead of treating rocks as being of uniform origin, they discerned both a succession of distinctly separate deposits and a fundamental difference between the hard, crystallized strata, seemingly formed either chemically or by heat, and the rough, granular, sedimentary beds which owed their origin to deposition. This basic classification was eventually to lead to such conflicting notions as Werner's cosmogonical concept of a universal deposition (precipitation) and Hutton's theory of an endless succession of dying worlds. Thus these four men very accurately portrayed much of what the later scientists were to teach and for which their successors were to be heralded by their contemporaries as noble and original thinkers. In fact, it is certain that Werner borrowed extensively from Lehmann's theory and today no one doubts that the tremendous outburst of geological science in the late eighteenth century can be traced back clearly to somewhat neglected stratigraphers such as Arduino, Füchsel, Lehmann and Pallas.

In this respect, we could hardly do better than quote the appreciative summary written in 1946 by Professor Tomkeieff.

> This sudden rise of geology as a systematic science was predetermined by the work accomplished since the middle of the eighteenth century. The remarkable progress achieved during this short interval was due to many

and various causes, among which one may be singled out for mention – the rapid development of mining and agriculture in connection with those industrial changes normally labelled 'the industrial revolution'. The development of systematic mineralogy – the prelude to systematic geology – was definitely conditioned by the intensive development of mining in such countries as Sweden and Saxony. It was in Freiberg in Saxony that Werner began in 1775 to teach geognosy in the Mining Academy, thus creating an impulse which pushed geology forward on its road of progress. The magnificent descriptive works of Guettard, Desmarest, Dolomieu, Arduino, Pallas, Saussure and Whitehurst, testify to the energy and acumen of the eighteenth century observers, but it was already obvious, even to them, that the facts were outrunning the theories. These theories, hoary with age, were survivors of a mythological, pre-scientific age and were absolutely incapable of explaining newly acquired facts. The theory of aqueous origin of all rocks, nearly as old as Thales, the doctrine of the universal flood, even older, volcanoes as living mountains, earthquakes as piling up mountains and rending valleys, all these extraordinary and often supernatural phenomena undermined as they were by the new empirical currents, yet cluttered up the main stream of the science of geology and made the navigation difficult, if not impossible. (*Tomkeieff, 1946, pp. 322–3*)

CHAPTER THREE

Abraham Werner and the Universal Ocean

And Noah he often said to his wife when he sat down to dine,
'I don't care where the water goes if it doesn't get into the wine.'

(G. K. Chesterton, *The Flying Inn*)

Werner and his more famous contemporary James Hutton brought forth their ideas at a moment of history when the intellectual atmosphere was most favourable to science, for this was the eve of the French Revolution and a period when the cult of liberty of the mind was freeing itself from the religious dogmatism characteristic of the Middle Ages. Also at this time the needs of the Industrial Revolution were providing great encouragement to the pursuit of scientific knowledge. As the Industrial Revolution regenerated existing sciences, so, also, industry gave birth to new sciences and spread the influence of the older ones. William 'Strata' Smith was a canal engineer, and Werner a mining expert. By the late eighteenth century the English Church was beginning to feel the effects of Methodism and, as a consequence, was torn between deciding whether to close its ranks or to revitalize its doctrines. The country suffered from lack of public education, as we know it today, and even at the universities the natural sciences were almost totally neglected. Oxford and Cambridge were largely training grounds for young gentlemen and politicians most of whom were given a knowledge of the classics and little more; London University did not then exist, and does not appear until 1827 in the form of what is now University College. Consequently, most advances in science stemmed from men engaged in industry and were made as a direct result of the practical necessities of their work. But industry, though rapidly gathering strength, was still socially unacceptable and young men dissatisfied with the sedentary way of life of the country gentleman chose to enter either the army or politics. Science as the hand-maiden of the machine largely suffered the same fate as industry. If a gentleman were fortunate enough to possess a scientific bent he generally directed his energies to archaeological or botanical researches, and if he were not a gentleman he joined the medical school!

It was into such an atmosphere that Werner and Hutton entered and it is partly because of the confident vigour of their approach that their theories stand out so prominently in comparison with those of their predecessors. Both were to assume eminent positions in the history of geology, and they were contemporaries yet they must be treated separately because, in the early stages of their studies, they followed different paths without any cognizance of the results of the other.

Straightaway it must be admitted that Werner is more memorable for the force of his personality than for the quality of his written work. Geikie (*1905, p. 236*) says of him:

> It was not his writings, nor even his opinions and theories in themselves, that gave him his unquestioned authority among the geologists of his time. His influence and fame sprang mainly from the personality of the man. His unwearied enthusiasm and eager zeal in the furtherance of his favourite studies, his kindness and helpfulness, his wide range of knowledge, and the vivacity, perspicuity, and eloquence with which he communicated it, his absolute confidence in the solidity of his theoretical doctrines – these were the sources of his power rather than the originality and importance of his own contribution to geology.

The contents of Werner's quarto booklet *Kurze Klassifikation und Beschreibung der Verschiedenen Gebirgsarten* (1787) are relegated summarily by the Fentons in the following words:

> We may doubt whether 28 pages ever covered a greater field, displayed more obvious errors, or contained more self contradictions. (*Fenton and Fenton, 1952, p. 42*)

No doubt such statements are partly intended to distract attention from Werner in order to give proper stature to Hutton but to dismiss Werner's ideas as being of no consequence is to overlook the potential purpose and intrinsic value of any idea, correct or false. Although Werner's hypothesis of the origin of the earth's crust as a series of layers derived from mechanical and chemical precipitations or deposits of sediment from a vast ocean laid down one upon the other in a regular sequence (or, as Von Zittel (1901) describes them, a succession of 'onion skins') is now discredited, it had an original value in gathering together a series of hitherto unrelated phenomena in a grandiose scheme which for the first time could be easily understood by others. Moreover it was an idea built upon readily observable data which could easily be examined and tested while the type sequence of rocks was held to apply in other areas. If these data carried with them the very seeds of destruction of Werner's theory, the scheme gave to the geological world a method of study which delimited the boundaries of the science and pointed out further paths to be followed. Moreover, it had the contemporary advantage of not clashing directly with theological doctrines relating to the Flood.

Abraham Gottlob Werner (1749–1817) was born into a mining tradition; his father was inspector of the iron foundry at Wehrau and his ancestors had been associated with the iron industry for nearly three centuries. Even as an infant Abraham showed interest in rocks and this early enthusiasm was assiduously fostered by his father who taught him the names of the various minerals and allowed him to examine and handle his own collection of rocks.

When 15 years old Abraham joined his father at the foundry. After serving five years, his devotion to the study of mineral characteristics could no longer be contained and he entered upon a course of study at the Freiberg Mining Academy. During his time there his obvious enthusiasm and genius were so outstanding that the mining authorities offered him a place in the Saxon Corps

FIG. 5. *Abraham Gottlob Werner*

of Mines. He declined this post and entered Leipzig University where for two years he studied law and in his last year divided his attention between history, philosophy and foreign languages before finally returning to his first interest, mineralogy. Shortly after leaving the university he was offered the post of lecturer at the Freiberg Academy, where he remained virtually till his death. During his forty years of office at the Academy he raised its status from that of a provincial institution for technical instruction in mining to the rank of a university of the highest standing and repute throughout Europe. It is reported that when the news of his death became known, the enrolment of the Academy fell away to such a staggering extent that the authorities became alarmed and set up a committee to investigate means of remedying the situation (Fenton and Fenton, 1952).

What particularly characterized Werner's work was his orderly pattern of ideas leading to one final conclusion. From the evidence of a few pieces of rock brought up from some mine he could guide his listeners through a catalogued sequence of strata demonstrating how each deposit had its own particular fossils, and how the character of the rock exerted an influence on the surface features; all culminating and leading on to a terrestrial conception of the birth of the world in the bowl of a mighty ocean, each different layer of rock marking a temporary advance of the waters and the laying down of a fresh stratum by the deposition on to the surface below of the heaviest sediments from the massive aqueous solution.

This theory, in itself enough to fire the imagination, was made more colourful when used as a background for an interpretation of the art, history and philosophy of nations and cited as one of the controlling factors in their otherwise seemingly accidental rise and fall in power. The main defects grew out of the fact that Werner built his whole concept on the rather narrow foundation of his own mineralogical experience which was mainly confined to Germany. This knowledge he expanded into a grand theory by fitting it into Lehmann's original idea, in which the earth was supposed to have begun as a fluid compound of earth and water revolving in space. First the heavier masses of the watery solution settled down by a process of chemical crystallization as the hard central core of granite. Upon this initial base all the other formations were deposited in their turn. Simultaneously with this ordered procession of strata separating and precipitating from the parent solution, the size of the containing ocean was undergoing gradual but intermittent reduction until, by the time the last deposit had been laid down, it had assumed its present magnitude.

Werner defined the scope of the subject which he termed *geognosy* as embracing the identification of mineral deposits, their stratigraphical location, the historical relationship of a stratum with others above and below and, finally, the mode of origin of all such deposits. Under his scheme of a universal precipitation, universal not in time but in extent, Werner divided the rocks in the following chronological order of appearance (*Von Zittel, 1901, p. 58*): (1) *Primitive* (granite, gneiss, schist, certain greenstones, etc., with no fossils); (2) *Transitional* (crystalline schist, slate and greywacke, greenstone, gypsum, etc., with first organic remains); (3) *Flötz* (limestone, ore-bearing rocks, coal, chalk, basalt); (4) *Derivative* (sand, clay, pebbles, soapstone); (5) *Volcanic rocks*. Each type represented different periods and separate circumstances, and the main basis of the division was of course their special mineral characteristics.

In the initial period the primitive rocks were supposed to have been chemically crystallized into their present form during the chaotic turmoil of the
earth's birth.

The Transitional period marked a reduction in the violent processes of
creation and the drawing off of the waters. The deposits were laid down
either by chemical precipitation, as in the case of slates and shales, or mechanically, as in the case of greywacke.

The Flötz period saw alternations between peaceful intervals of deposition
when the waters again spread over the land surface and occasional resurgences
of the earlier causes of upheaval. Mechanical precipitation was at this time
the more dominant process.

These conditions continued throughout the Derivative period with a gradual
diminution in the amount of the ocean waters. There followed, after a fairly
long interval of time, a final epoch of violent volcanic outbreaks which were
motivated by the ignition of subterranean coal beds and ejected only obviously
superficial lavas, ash and tuffs.

The occurrence of valleys Werner, like Buffon, attributed to marine action
and to other powerful influences which carved the earth into its existing form.

Thus Werner's system was based on the subsiding of, and the precipitation
of rocks chemically in a once universal ocean. This subsidence was intermittent, often rapid and accompanied by winds, storms and currents.

> Powerful and shifting currents set up by the winds and by the draining
> away of the subsiding waters, cut deep channels through the sediments in all
> directions and by their erosive power gave rise to deep valleys separated by
> high mountains. (*Adams, 1938, p. 223*)

The principal features of Werner's theory are the emphasis on mineralogy
and the reliance on processes outside existing experience. Historically he
belonged to the catastrophists who believed that present natural processes
could not have wrought the changes or brought about the structures which
exist as part of the earth's surface. These catastrophic ideas held undisputed
sway up to the time of James Hutton.

> One of the favourite theories was that a planet had passed close to the earth
> causing the tides to inundate the lands—perhaps in connexion with Noah's
> flood. Others supposed that a radical shift of the earth's axis of rotation had
> caused a catastrophic shifting of the oceans. (*Williams, 1904, p. 3*)

The weakness of Werner's theory as we see it today lies in his picture of a
gigantic ocean containing in solution the totality of the material that was later
to constitute the earth's crust. This idea has rightly been criticized because it
refused to face up to the question of what happened to the ocean after the
strata had been deposited. Did the ocean evaporate into space or was it
absorbed by the new deposits? Did it subside into the bowels of the earth, or
was it drawn off by some passing planet? Playfair in his *Illustrations* (1802)

criticizes Werner's idea on the grounds that water alone is an agent of solution but not of consolidation and that some other agent would be necessary (he follows Hutton in suggesting heat) before the earth could have assumed its present form. Playfair doubts whether an aqueous solution can ever become anything other than an aqueous solution without the intervention of some fresh force. Viewed in retrospect it is easy to point out that Werner's chosen agents of creation are as mythical and catastrophic as those of Buffon or Burnet and other medievalists, and in this respect he cannot claim to have advanced any further than his less illustrious predecessors. Two quotations from the *Edinburgh Review* on Werner's *The New Theory of the Formation of Veins* (1791) are symptomatic of this type of criticism:

> We believe that the vast power of generalization which the Wernerian system appears to possess is the basis upon which its popularity is founded. (*Fitton, 1811, p. 93*)
> Perhaps, however, the greatest fault in Werner's system remains yet to be mentioned. It is the theoretical language which it employs, by which hypothesis is interwoven with the description of every phenomenon. (*Fitton, 1811, p. 97*)

In reply, it can be contended that just because Werner could not explain satisfactorily the disappearance of the ocean, that he should for that reason alone have abandoned his own carefully built scheme is a demand more than unreasonable. For example, the theory of the Ice Ages has and will continue to play an important part in the science of geology despite the fact that no one has yet provided a completely satisfactory explanation for the formation of ice in such colossal quantities. Werner may have generalized in order to substantiate his theory but when speculating on such a vast scale and on cosmogonical lines generalization seems unadvoidable.

As a mineralogist Werner's repute is not questioned. While still a student at Leipzig he wrote a description of mineral characteristics which introduced descriptive methods then unknown to practitioners of the science. He was certainly no armchair theorist for he sought his information both from the hills and down the mines, even at times indulging in the ordinary work of a miner in order to glean first-hand material. At the same time he was a prolific reader and could absorb a wealth of learned treatises, extracting the few valuable portions for his own future usage. Added to this tremendous adsorptive capacity, he was meticulous and fastidious in all that he did whether it was the treatment of ideas of facts or the preparations involved in choosing the places and the menu for a dinner party.

He was not essentially a strong man and it is reported that often his lectures used to leave him weak and his clothes drenched in sweat. Yet, despite this, his devotion to the subject was such that if there were too many students he would divide the class and repeat the same lecture twice. It was a peculiarity of his that, despite his unflagging powers of oratory, he wrote almost nothing.

This became an obsession towards the later stage of his life, when he would neglect the opening of his correspondence – even that from his close relatives. During his whole life only twenty-six publications are known and the majority of these were short magazine articles. When past students threatened to publish his lecture notes he was greatly annoyed and forbade such publication, promising that his books would:

> appear forthwith, one after another, enriched by his latest observations and discoveries.

The sole result was a 'table of contents' which appeared in 1811 and this had already been printed in 1783! On this point the *Edinburgh Review* (Fitton, 1811, p. 80) says tartly:

> With all his high reputation, Werner has written but few books; and it is not clear that his fame would have been greater if he had written more.

Viewed in finality, might not an impartial judge ask in what respect is Werner worthy of mention? Many of his ideas were second-hand and his theories mostly incorrect; the man seems an orator and nothing more. All this must be admitted. Yet on the credit side he undoubtedly drew attention to a subject that before his day was largely neglected. He made it one worthy of academic recognition as a separate science and discipline. He brought to it an ordered pattern of ideas and made the whole into a recognizable picture of universal extent and application which could be comprehended and tested by lesser minds. But by the very interest he aroused and the disciplined approach to the study which he taught, he brought about the destruction of his own idea. The implausibility of the disappearing ocean was soon pointed out and other anomalies rapidly became apparent. Marked variations in thicknesses of strata from place to place revealed deficiencies in his scheme, as also did the presence of igneous intrusions through the Transitional and Flötz rocks, and the existence of some seemingly anomalous rock successions in areas outside Germany. In the case of volcanic deposits occurring on or between Flötz deposits, his followers, the Neptunists, even attempted to hold to the theory by temporarily calling back to their aid the ocean that was supposed to have permanently disappeared.

Werner has been castigated as unreasonable for not accepting these inconsistencies as being irrefutable evidence of the wrongness of his notions. Whether he was obstinate or whether he still genuinely believed in his scheme will never be known, for he died without making any admission. As late as 1791 in his book on the origin of metalliferous veins it is evident he had in no way altered his ideas. He still asserted that as the rocks of the earth's central core settled down cracks would appear on the surface which the ocean would fill from above with mineral deposits. When the ocean was drawn off these deposits would solidify into mineral veins.

In recapitulating the state of our present knowledge, it is obvious that we know with certainty, that the flötz and primitive mountains have been produced by a series of precipitations and depositions formed in succession; that they took place from water which covered the globe. . . . We are also convinced, that the solid mass of our globe has been produced by a series of precipitations formed in succession, (in the humid way); . . . that the pressure of the materials, thus accumulated, was not the same throughout the whole; and that this difference of pressure and several other concurring causes have produced rents in the substance of the earth, chiefly in the most elevated parts of its surface. (*Werner, 1791, translated by Charles Anderson in New Theory of the Formation of Veins, Edinburgh, 1795*)

Even after his death his ideas existed by their own strength and that of the magic of his name, though they became increasingly discredited as one by one his disciples, like Leopold Von Buch and D'Aubisson de Voisins, were won over to the school of Vulcanists – those who believed that not all rocks were of marine origin. Jameson, a product of the Wernerian school, was one of the last upholders of the Neptunist tradition. Darwin, who attended Jameson's lectures on geology, found them 'incredibly dull' and afterwards wrote:

The sole effect produced on me was the determination never as long as I lived to read a book on Geology, or in any way to study the science. (*Darwin, 1887, I, p. 41*)

CHAPTER FOUR

James Hutton and 'Worlds Without End'

As it was in the beginning, is now,
and ever shall be, world without end.

The Wernerian doctrine contributed little to the study of the processes of fluvial erosion and the real stepping-stone between Werner and Hutton was a Swiss, Horace-Bénédict De Saussure (1740–99; Freshfield, 1920). Indeed Hutton may have been directed in finally formulating his theory by De Saussure's researches, many of which he speaks of with warm approval.

It was De Saussure who popularized mountain climbing by his detailed accounts of his own intrepid explorations. The western Alps were his particular laboratory of inquiry and experiment. He was one of the first to climb Mont Blanc (8 August 1786) and spent a hazardous sixteen days taking scientific observations on the Col du Géant.

His geological and geomorphological ideas were an interesting mixture of medieval and modern. Like the followers of Werner he conceived the rock strata, including granite, to be the result of successive, sub-aqueous depositions. On glaciers, he was quite modern, and gives an excellent account of their formation and movement and of the solid matter they carried. However, he considered the sub-glacial streams and much of the water issuing from the snout was due to melting caused by (subterranean) earth-heat. He believed strongly in the erosive action of running water and paid much attention to rounded pebbles.

> The naturalist who travels on high mountains, where rivers have their sources, sees stones that are by nature angular lose their angularity and change into rounded pebbles almost as he watches.
> But it is especially at the end of great glaciers, whence issue impetuous torrents that are violent from their birth, that I have had the great pleasure of observing this; for example at the sources of the Aar, Rhône . . . etc. When the rivers break forth from glaciers at heights where other currents (of water) have not passed, all the stones not in the beds of the rivers, have a natural angular shape. Thus, on the glacier from which the torrent issues and on the mountain sides bordering it, there is not a single stone which has not sharp angles and cutting edges. But in the river bed the same kind of stones have all their angles blunted, with rounded shapes. These are truly rounded pebbles. (*De Saussure, 1779, I, p. 147*)

It happened, however, that the distribution of water-worn channels and of fluvial deposits high above the present rivers in the Geneva neighbourhood

FIG. 6. *Map of the Mont Blanc region, showing De Saussure's route to the summit*

caused De Saussure to depart widely from Huttonian principles. He was, in fact, almost a diluvialist although he does not refer to the biblical Flood. For example, he imputes the high-level deposits and the erratics brought from very distant mountains and the size of the main valleys to a 'violent debacle of water'.

FIG. 7. *View of the Aiguille des Charmoz from Le Montanvert, near Chamonix*
(From De Saussure, 1779)

FIG. 8. *De Saussure's descent from the Col du Géant, 1788*
(From a contemporary print)

FIG. 9. *Portrait of H. B. De Saussure about 1777*
(By Jens Juel)

The waters of the ocean in which our mountains were formed still covered a part of these mountains when a violent earthquake suddenly opened great cavities that had previously been empty and caused the rupture of a great number of rocks.

The water rushed towards these abysses with extreme violence, proportionate to the elevation that they then had, cutting deep valleys and sweeping along immense quantities of earth, sand, and fragments of all sorts of rock. This semi-liquid mass, pushed by the weight of the waters, accumulated up to the height where we still see many of these scattered fragments.

Then the waters, continuing to flow but with a speed that diminished gradually in proportion to the decrease of their height, swept away the lightest parts, little by little, and cleared the valleys of this mass of mud and debris, leaving behind only the heavier lumps and those whose position or more solid seat protected them from this action. (*De Saussure, 1779, I, pp. 151–2*)

This vacillation between the idea of river erosion and the catastrophic notions of the Neptunists and later diluvialists is noticeable throughout much of De Saussure's writings. Hutton quotes extensively from him and relied heavily upon his descriptions of mountain scenery and of river action. Thus we notice, and translate, the following:

> We see a great many valleys created on the flanks of a mountain. We can see them widen and deepen in proportion to the water that flows in them. A stream issuing from a glacier or an alpine meadow cuts a furrow, small at first but which widens successively in proportion as the flow increases because of the junction with other springs or other torrents. (*De Saussure, II, p. 920: in Hutton, II, pp. 394–5*)
>
> The flanks and base of the mountain (the Jura) have also been degraded by torrents that rain and melting snow produce and which have formed wide and deep excavations. (*De Saussure, I, p. 274; in Hutton, II, p. 107*)

Yet it is noticeable that in the last quotation Hutton does not continue with De Saussure's words

> 'If to all these destructive agents we add the great currents, which have anciently undermined and eroded the flanks of the Jura; the earthquakes . . .'

In fact, Hutton gently reproves the Swiss naturalist for imagining a current of water (*débâcle*) which 'however in the possibility of things, is not in nature', and which 'I do not understand'. Hutton explains the high-level channels and pot-holes as due to river-action and he shows that the granite blocks on the Jura were not due to a debacle of water but to 'a gradual declivity' from the Alps to the Jura in former times, the transportation being done naturally 'by means of water and ice adhering to those masses of stone' (Vol. II, p. 174). De Saussure seemed in later life to be quite certain that the narrower valleys and small valley-systems were the work of rivers but he apparently attributed the great longitudinal valleys and major transverse valleys mainly to ocean currents and in some cases to faulting. It remained for the simple logic of Hutton to attribute even the larger valleys to river-action.

THE FIRST GREAT FLUVIALIST

For students of fluvial erosion James Hutton is the first great prophet and his clear exposition of the pattern of geological evolution has proved the permanent foundation of the subject. He was born in 1726, the son of a comparatively well-to-do Edinburgh merchant at one time the City Treasurer (Playfair, 1805). When his father died Hutton inherited a small and poorly run estate in Berwickshire, which with the early financial success of his chemical experiments provided him with a sufficiency of means. At seventeen he commenced his working life as a law clerk in a city office. The musty routine of copying and the insistence on legal hair-splitting did not satisfy his inquiring mind and he left the learned profession to take up the study of medicine at the University, primarily as a means of studying chemistry. Here he passed three years before going on to Paris and Leyden. Even the study of medicine did not seem to satiate his curiosity, for, although he qualified, he never practised as a doctor. Chemistry was his most absorbing occupation but he indulged in it only as a part-time hobby.

His next venture was farming (Bailey, 1950), and for the purpose of informing himself as to the best practices he passed two years in Norfolk and made a tour of Flanders. In 1754, armed with this experience, he returned to

FIG. 10. *Sligh Houses Farmhouse, north-east of Duns, Berwickshire, where James Hutton lived between 1754 and about 1762*

his Berwickshire estate and began to transform it into a modern experimental farm. By this time his interest had fastened firmly on to the observation of rocks and minerals and their mode of origin. As early as 1753 he noted the relation of soils to the composition of the underlying rock and while touring northern Scotland he first saw crystals in basalt and observed where rocks had been dissolved and gullied. Quietly he pursued his speculations. The early experiments with sal ammoniac which he had undertaken in conjunction with James Davie proved commercially successful and in 1765 Hutton became a co-partner in the resulting business. By 1768 his scientific promptings began to gain mastery and in that year he gave up his farm in order to move into Edinburgh where his obvious talent, education and financial independence permitted him to enter the select social circle of the Edinburgh group, consisting of 'giants' such as Doctor Black, who is known for his work on carbonic acid (Ferguson, 1805), and Sir James Hall, the first scientist to conduct laboratory experiments in geology in order to assist the interpretation of physical processes. The interest they showed in his ideas and their own example may well have given him the necessary confidence to bring his theory to the notice of the public.

Edinburgh was then the cultural centre and seedbed of the country's intellect (MacGregor, 1950). As a capital and university city it attracted the social life and intelligentsia of Scotland, as well as not a few Englishmen. At that particular time it was fortunate in possessing some of the most liberal and vigorous minds in Britain. Yet it should not be regarded as a powerful centre of propagation in the same way as Freiberg had been for the Wernerian School. This was because a great deal of its intellectual vigour was diffused. The so-called Edinburgh 'group' was in fact a collection of small groups which coalesced and separated at will. At their head was the august Royal Society of Edinburgh which existed for the presentation and discussion of scientific questions; as offshoots there were numerous small literary or social clubs which served as convivial meeting places for the exchange of both light-hearted and serious ideas. The Oyster Club to which Hutton and Black belonged was one of the latter. It seemed to be the pressure of the various influential groups of persons with whom Hutton mixed that persuaded him to put his theories into print. Eventually his article, *Concerning the System of the Earth, its Duration and Stability*, was read at two successive meetings of the Royal Society of Edinburgh in March and April 1785. A few copies of a brief *Abstract* were printed in this or the next year. The full-length paper first appeared in 1788 in the first volume of the *Transactions* of the Society as *Theory of the Earth; or an Investigation of the Laws observable in the Composition, Dissolution, and Restoration of Land upon the Globe.* It was not heralded with the skirl of bagpipes and aroused little immediate enthusiasm except of hostility.

We can appreciate, now time has sifted the essential ingredients, that the value of Hutton's work rests on his discovery of the simple, but radically new and unifying principle that the cause of the earth's formations and underlying strata should be sought not in pre-supposed cataclysmic agents of destruction but in agents which had always existed and could actually be seen in daily operation. Of the needlessness of postulating unnatural causes Hutton says:

> Not only are no powers to be employed that are not natural to the globe, no action to be admitted of except those of which we know the principle, and no extraordinary events to be alleged in order to explain a common appearance, the powers of nature are not to be employed in order to destroy the very object of those powers; . . . Chaos and confusion are not to be introduced into the order of nature, because certain things appear to our partial views as being in some disorder. Nor are we to proceed in feigning causes when those seem insufficient which occur in our experience. (*1795, II, p. 547*)

In examining things which actually exist, and which have proceeded in a certain order, it is natural to look for that which had been first; man desires to know what has been the beginning of those things which now appear. . . . A theory of the earth, which has for object truth, can have no retrospect to that which had preceded the present order of this world. . . . A theory, therefore, which is limited to the actual constitution of this earth, cannot be allowed to proceed one step beyond the present order of things. (*1795, I, pp. 280–1*)

It is in the philosophy of nature that the natural history of the earth is to be studied; and we must not allow ourselves ever to reason without proper data, or to fabricate a system of apparent wisdom in the folly of a hypothetical delusion. (*1795, II, p. 564*)*

It was his denial of catastrophic forces that made his theory so distinct from all the others that had preceded it. In contrast to the Catastrophists his theory was to give birth to the school of 'Uniformitarianism', where a continuing uniformity of existing processes was regarded as being the key to the history of the earth. What caused Hutton to seize upon this theory it is impossible to say for certain. In his wanderings through Scotland and Norfolk, he observed the same phenomena of river, valley, terrace, gravel, alluvium, mountain and plain as his predecessors had but for him alone they pointed to a scheme of continuous creation. He alone saw in their form and arrangement an infinity of inorganic bodies which, under the action of observable physical agencies, were slowly changing their relationships. He alone recognized that the present form of the earth was not a feature created long ago in the past which had maintained the same shape ever since, but was an ephemeral form, one of a procession of similar forms going back in time beyond the limit of man's vision.

One pointer in favour of Hutton's scheme was the occurrence of beds resting unconformably upon others which in some cases had been truncated. This argued against Werner's idea of a succession of layer upon layer and suggested interruptions in the process during which the lower beds were tilted and then eroded before the upper beds were laid down. Werner did not believe this feature to be contradictory of his scheme and passed it off as being the result of inclined deposition or slipping. Hutton, on the other hand, felt that such features represented clear evidence of a succession of revolutions, unconformities marking the positions of each.

The second pointer was to be discerned in the composition of many of the sedimentary rocks. For Hutton the sands and gravels in the river beds or on their banks and the proliferation of fossils in almost all strata did not represent the relics of the Flood or the work of Werner's ocean but were clear evidence of the constant processes of change which were taking place and had occurred all over that part of the globe which lies above sea level. His chemical training caused him to realise that as sand resulted from the breaking down of parent rocks, so if a rock stratum was composed of a compound of sand grains it

---

* Quotations from Hutton's later elaboration of his *Theory* (1795) are interspersed at this early stage to allow a more complete explanation of his ideas in his own words.

followed that each individual stratum was not in its original state but must have been derived from the erosion and breaking up of an older series of rocks. In his *Theory of the Earth* he says:

> The solid parts of the globe are, in general, composed of sand, of gravel, or argillaceous and calcareous strata, or of the various compositions of these with some other substances, which it is not necessary now to mention. Sand is separated and sized by streams and currents; gravel is formed by the mutual attrition of stones agitated in water; and marly, or argillaceous strata, have been collected, by subsiding in water with which the earthy substances had

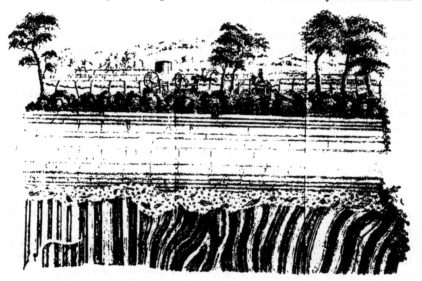

FIG. 11. *Hutton's plate showing the unconformity in the valley of the Jed, near Jedburgh. Devonian sandstones unconformably overlie steeply dipping Silurian rocks*

> been floated. Thus, so far as the earth is formed of these materials, that solid body would appear to have been the production of water, winds and tides. (*1788, p. 218*)
>
> ... in finding masses of gravel in the composition of our land, we must conclude that there had existed a former land, on which there had been transacted certain operations of wind and water, similar to those which are natural to the globe at present, and by which new gravel is continually prepared, as well as old gravel consumed or diminished by attrition upon our shores. (*1788, p. 289*)

In the case of limestones and chalks, he considered that their main characteristics clearly indicated that they had been laid down below the sea and were in a great proportion built of the remains of dead sea life:

> In all regions of the globe, immense masses are found, which, though at present in the most solid state, appear to have been formed by the collection of the calcareous exuviae of marine animals. (*1788, p. 219*)

He goes on to say that there are many ways in which rocks are eroded and the detached material borne away to the bottom of the ocean, there to be transformed into new strata. These processes, he explains, have been overlooked previously because of their extreme slowness.

> In like manner, the co-relative, or corresponding operation, the destruction of the land, is an idea that does not easily enter into the mind of man in its totality, although he is daily witness to part of the operation. We never see a river in a flood, but we must acknowledge the carrying away of part of our land, to be sunk at the bottom of the sea, we never see a storm upon the coast, but we are informed of the hostile attack of the sea upon our country; attacks which must, in time, wear away the bulwarks of our soil, and sap the foundation of our dwellings. (*1788, p. 295*)

> The natural operations of this globe, by which the size and shape of our land are changed, are so slow as to be altogether imperceptible to men who are employed in pursuing the various occupations of life and literature. (*1795, II, p. 563*)

Attention is drawn to the orderly system of the wearing down of the earth's surface and its transport to the ocean bottom, there to be stratified as the raw material from which a new land mass might be fabricated:

> Our land has two extremities; the tops of the mountains, on the one hand, and the sea shores, on the other:—While there is a sea shore and a higher ground, there is that which is required for the system of the world: Take these away, and the world would remain an aqueous globe, in which the world would perish. But, in the natural operations of the world, the land is perishing continually. (*1788, p. 296*)

In fairness to Hutton's opponents we should not overlook the fact that Werner had also observed these operations and had included such deposits in his 'derivative' group. Where Hutton differed here was in his emphasis. He would place the origin of most surface rocks in one category. Even the schists which Werner thought represented the crust of the primitive earth, Hutton did not consider to be in their original state but to have been made up of clay, sand, and other materials formed from the break down of earlier mineral bodies:

> The general amount of our reasoning is this, that nine-tenths, perhaps, or ninety-nine hundredths of this earth, so far as we see, have been formed by natural operations of the globe, in collecting loose materials, and depositing them at the bottom of the sea; consolidating those collections in various degrees, and either elevating those consolidated masses above the level on which they were formed, or lowering the level of that sea. (*1788, p. 221*)

From this hypothesis he was quick to elaborate the principle of the cycle of erosion and sedimentation. He visualized the process of water as continually attacking the land and tending to cut it down to shore level, the debris being borne away into the sea to be set down as future sedimentary strata. The waters of the land, sea and atmosphere were the servants of God carrying out their daily task as part of an ordered plan of destruction and re-creation.

Once again he draws attention to what he deems to be God's place in the evolution of the world, explaining in the first extract how the breaking down of the rock into soil is necessary to promote the growth of plants and thus providing food for man and the animals. In the second passage he shows how all this material is by various processes carried away from the mountains and gradually borne down towards the sea.

A solid body of land could not have answered the purpose of a habitable world; for a soil is nothing but the materials collected from the destruction of the solid land. Therefore, the surface of this land, inhabited by man, and covered with plants and animals, is made by nature to decay, in dissolving from that hard and compact state in which it is found below the soil; and this soil is necessarily washed away, by the continual circulation of the water, running from the summits of the mountains towards the general receptacle of that fluid. (*1788, p. 214*)

The heights of our land are thus levelled with the shores; our fertile plains are formed from the ruins of mountains; and those travelling materials are still pursued by the moving water, and propelled along the inclined surface of the earth. These moveable materials, delivered into the sea, cannot, for long continuance rest upon the shore; for by the agitation of the winds, the tides and currents, every moveable thing is carried further and further along the shelving bottom of the sea, towards the unfathomable regions of the ocean. (*1788, p. 215*)

Our solid earth is everywhere wasted, where exposed to the day. The summits of the mountains are necessarily degraded. The solid weighty materials of those mountains are everywhere urged through the valleys, by the force of running water. The soil, which is produced in the destruction of the solid earth, is gradually travelled by the moving water. . . . This travelled soil is at last deposited on the coast, where it forms most fertile countries.

But the billows of the ocean agitate the loose materials upon the shore and wear away the coast with the endless repetitions of this act of power, or this imparted force. Thus the continent of our earth, sapped in its foundation, is carried away into the deep and sunk again at the bottom of the sea from whence it had originated.

We are thus led to see a circulation in the matter of the globe, and a system of beautiful economy in the works of nature. (*Hutton, 1795, II, pp. 561-2*)

Unlike Werner, Hutton ascribed the formation of mountain and valley almost exclusively to the action of the rivers as they erode the surface and carry the material to the ocean.

We have but to consider the mountains as formed by the hollowing out of the valleys, and the valleys as hollowed out by the attrition of hard materials coming from the mountains. (*1795, II, p. 401*)

In his scheme the soil, coasts, rivers, valleys and mountain tops were transient forms particular only to that one point in time and which would, after the passage of a millennium or so, in turn be replaced by other forms and they in their time by yet others. This evidence of constant demolition did not fill him with a sense of impending doom for he regarded the carrying away of the soil not as a loss but as a necessary adjustment in a finely ordered sequence much as one half of an egg timer replaces the other:

> Such as has a constitution in which the necessary decay of the machine is naturally repaired, in the exertion of those productive powers by which it had been formed. (*1788, p. 216*)

Having followed the passage of the sediments to the river mouths he then applied his reasoning to the bottom of the ocean. Here for the first time he must indulge in speculation, though he does not forget to adhere to his cherished principle of arguing deductively from known causes.

> In what follows, therefore, we are to examine the construction of the present earth, in order to understand the natural operations of time past; to acquire principles, by which we may conclude with regard to the future course of things, or judge of those operations, by which a world, so wisely ordered, goes into decay; and to learn, by what means such a decayed world may be renovated, or the waste of habitable land upon the globe repaired. (*1788, p. 218*)

In the earlier passages Hutton was able to illustrate how the rocks were worn down. However, if this process were to go on indefinitely the earth would at some time disappear below the sea and the mechanism would run down. It was necessary therefore to show how the earth so destroyed could be rebuilt. As mentioned earlier, he had noted the presence of marine organisms in many sedimentary strata. This led him to argue that after the debris had been re-assembled as strata below the sea it was consolidated by heat and then by some violent force thrust up above sea level as a new land mass. Wherever crystalline calcareous rock formations do not appear to contain fossils he postulates that when the rock underwent the process of consolidation evidence of the marine organisms was destroyed. He therefore cites the mineral characteristics of limestone as evidence of its marine origin:

> There are, in all the regions of the earth, huge masses of calcareous matter, in that crystalline form of sparry state, in which no vestige can be found of any organized body, nor any indication that such calcareous matter had belonged to animals; but as, in other masses, this sparry structure, or crystal-line state, is evidently assumed by the marine calcareous substances, in operations which are natural to the globe, and which are necessary to the consolidation of the strata, it does not appear that the sparry masses, in which no figured body is formed, have been originally different from other masses, which being only crystallized in part, and in part still retaining their original form, leave ample evidence of their marine origin. (*1788, p. 221*)

When searching for a means of consolidation, he states that man-made experiments are not necessarily conclusive in deciding the capabilities of natural processes:

> But if, even in our operations, water, by means of compression, may be made to endure the heat of red hot iron without being converted into vapour, what may not the power of nature be able to perform? The place of mineral

operations is not on the surface of the earth; and we are not to limit nature with our imbecility or estimate the powers of nature by the measure of our own. (*1788, p. 250*)

He points to known examples of the consolidation and subsequent decay of derivative materials:

Field-spar is a compound of siliceous, argillaceous and calcareous earth, intimately united together. This compound siliceous body being, for ages, exposed to the weather, the calcareous part of it is dissolved, and the siliceous part is left in form of a soft white earth. (*1788, p. 231*)

He sought therefore a force with capacity to expand and thrust up the layers deposited and, at the same time, to fuse them by heat into the hardened masses of rock such as are often found in nature. His answer is characteristic of his reasoning. Look for it in the same place as the material is deposited!

If such a power is to be reasonably concluded as accompanying those operations which we have found natural to the globe, and situated in the very place where this expansive power appears to be required, we should thus be led to perceive in the natural operations of the globe, a power as efficacious for the elevation of what had been at the bottom of the sea in to the place of the land, as it is perfect for the preparation of those materials to serve the purpose of their elevation. (*1788, p. 262*)

There is nothing so proper for the erection of land above the level of the ocean, as an expansive power of sufficient force, applied directly under materials in the bottom of the sea . . . (*1788, p. 262*)

In support of the need for an expansive force, he reasoned that without it sediment deposited must rest conformably on the surface of the ocean bottom and unless strongly acted upon would for ever retain that horizontal position. This he knew was not the case, for strata are found in all positions and angles and subject to complex faulting and fracture. Moreover he stresses the fact that uninterrupted deposition by itself could in no way account for the occurrence of the cracks and fissures normally found in sedimentary rocks, as deposition is a process which does not permit of irregularities but tends to fill in and gloss over any such cavities or faults. His deduction drawn from the facts was expressed in these words:

The strata formed at the bottom of the ocean are necessarily horizontal in their position, or nearly so, and continuous in their horizontal direction or extent. They may change, and gradually assume the nature of each other, so far as concerns the materials of which they are formed; but there cannot be any sudden change, fracture or displacement naturally in the body of a stratum. But, if these strata are cemented by the heat of fusion, and erected with an expansive power acting below, we may expect to find every species of fracture, dislocation and contortion, in those bodies, and every degree of departure from a horizontal towards a vertical position.

The strata of the globe are actually found in every possible position; For from horizontal, they are frequently found vertical; from continuous, they are broken and separated in every possible direction; and, from a plane, they are bent and doubled. It is impossible that they could have originally been formed, by the known laws of nature, in their present state and position, and the power that has been necessarily required for their change, has not been inferior to that which might have been required for their elevation from the place in which they had been formed. (*1788, p. 265*)

He considers the type of force required:

Let us now consider what power would be required to force up, from the most unfathomable depth of the ocean, to the Andes or the Alps, a column of fluid metal and of stone. This power cannot be much less than that required to elevate the highest land upon the globe. (*1788, p. 271*)

And he found an agent of sufficient proportions in the explosive force of volcanic action (1788, p. 271):

A stream of melted lava flows from the sides of Mount Etna. Here is a column of weighty matter raised an immense height above the level of the sea, and rocks of an enormous size are projected from its orifice some miles into the air. (*1788, p. 272*)

For an example of the credibility of his supposition he cites Edinburgh and the occurrence of intrusive veins of whinstone penetrating sedimentary strata:

On the south side of Edinburgh, I have seen, in little more than the space of a mile from east to west, nine or ten masses of whinstone interjected among the strata . . . which have flowed by means of heat among the strata of the globe, strata which had been formed by subsidence at the bottom of the sea. . . . (*1788, p. 279*)

It was on this point of the volcanic origin of granite that Hutton seemed to his contemporaries to be diametrically opposed to the theory of Werner and the Neptunists who, as we have seen, ascribed the formation of virtually all rocks to precipitation from the material carried in the ocean waters. In contrast, Hutton and the Vulcanists, or Plutonists as his followers were called, distinguished between sedimentary rocks formed from mechanical and chemical deposits on the ocean bottom and intrusive rocks which they felt to be of volcanic origin, or upwellings from the innermost recesses of the earth. Both the Neptunists and the Vulcanists in their respective theories of the earth had to account for the fact that igneous rocks are often found below sedimentary rocks. Under the Neptunist scheme the igneous rocks were either made to exist at the appearance of the universal ocean and hence the sediments could be conveniently deposited directly on top of them, or were themselves chemical marine precipitates. Hutton in his volcanic theory also accepted igneous rocks as being below the sediments but he differed in treating them as an agent still actively engaged in shaping the earth's surface. His igneous rocks were considered to have been in a molten state and when, therefore, the sediment laid down below the sea was associated with the igneous matter the heat generated from the contact was supposed to consolidate the sediments as fire does potter's clay. In addition he felt that the fluid state of the igneous rocks would, after the nature of volcanoes, make them liable periodically to violent expansion and intrusion at the expense of the overlying sedimentary rocks. By this means he visualized the newly consolidated sediments being thrust up from beneath the oceanic deep as new continents. That is why he considered the examples of intrusive veins of igneous rocks penetrating through sedimentary formations to be a supreme justification of his idea. For, by the logic of the Neptunist theory, such an occurrence could not take place.

Though geologically Hutton's new theory of Vulcanicity ranks as of equal importance with that of natural causes it is not treated here in detail because the fierce arguments it provoked do not directly concern the student of fluvial erosion. Yet for the sake of accuracy we must note that on this question of the consolidation of sediments by heat alone Hutton was wrong. It is now known that sediments assume their durable and solid structure by the many processes of lithification. As the sediments increase in depth of burial, the pressure downwards increases in proportion until a stage is reached when most of the water is squeezed out of the granular material and, circulating between the granular spaces, precipitates the minerals which it previously carried in solution. The precipitation of these chemicals aids in a perfect cementation of a whole mass, without the aid of igneous action.

Having suggested his scheme of nature Hutton rests for a moment and compares his own ideas with those that have gone before and suggests the reason for the failure of the others to observe what he has seen.

> Philosophers observing an apparent disorder and confusion in the solid parts of this globe, have been led to conclude, that there formerly existed a more regular and uniform state, in the constitution of this earth; that there had happened some destructive change; and that the original structure of the earth has been broken and disturbed by some violent operation, whether natural, or from a supernatural cause . . . there is no occasion for having recourse to any unnatural supposition of evil, to any destructive accident in nature, or to the agency of any preternatural cause, in explaining that which actually appears. (*1788, p. 285*)

He emphasized the weakness of deducing facts from the division of rocks into only two series, primary and secondary, and considered it to be a fundamental fallacy of the works of his opponents that they conceived it necessary to postulate what happened in the beginning and from that hypothetical premise to trace the world's evolution down to the present day. He was convinced it was essential to obtain a thorough understanding of the present natural processes and having mastered that knowledge to trace back where possible from these examples to the elucidation and discovery of past processes and changes.

> We are not at present to enter into any discussion with regard to what are the primary and secondary mountains of the earth; we are not to consider what is the first and what is the last, in those things which are now seen; whatever is most ancient in the strata which we now examine, is supposed to

be collecting at the bottom of the sea, during the period concerning which we are not to enquire. (*1788, p. 288*)

The key to the past he felt must be sought from an examination of present processes and not from any presupposed succession of events following upon a hypothetical initial situation;

> Therefore, in knowing how these (sea) animals had lived, or with what they had been fed, we shall have learned a most interesting part of the natural history of this earth; a part which it is necessary to have ascertained, in order to see the formed operations of the globe, while preparing the materials of the present land. (*1788, p. 289*)

He concluded by reiterating his own ideas of a family of continents, each new-born surface arising from the sea and being built upon and from the body of its predecessor. The existence of unconsolidated clays and sands having close affinities with their consolidated counterparts of shales and sandstones he saw as the main clue to a succession of processes; the unconsolidated materials merely being at a different stage of continuous evolution in one general scheme of creation and destruction.

> Therefore, from the consideration of those materials which compose the present land, we have reason to conclude, that, during the time this land was forming, by the collection of its materials at the bottom of the sea, there had been a former land containing materials similar to those which we find at present in examining the earth. We may also conclude, that there had been operations similar to those which we now find natural to the globe, and necessarily exerted in the actual formation of gravel, sand and clay. (*1788, p. 290*)

He goes on to describe where the action of contemporary processes is most clearly seen and indicates the various means by which the rock surface is broken off and borne away.

> Upon the one extremity of our land, there is no increase or there is no accession of any mineral substance. That place is the mountain top, on which nothing is observed but continual decay. The fragments of the mountain are removed in a gradual succession from the highest station to the lowest. Being arrived at the shore, and having entered the dominion of the waves, in which they find perpetual agitation, these hard fragments, which had eluded the resolving powers natural to the surface of the earth, are incapable of resisting the powers here employed for the destruction of the land. By the attrition of one hard body upon another, the moving stones and rocky shore, are mutually impaired. And that solid mass, which of itself had potential stability against the violence of the waves, affords the instruments of its own destruction, and thus gives occasion to its actual instability. (*1788, p. 296*)

By this slow wasting action whole mountain ranges may be worn down to solitary stumps, whilst the loose materials thus removed assist the erosive forces in attacking the parent body. In order to learn the measure of geological time it is necessary to study these erosive processes, for by an understanding of how they work, and at what rate, the geologist may be able to arrive at the knowledge that a portion of the earth is at a certain stage in its life history, has taken so many years to reach its present state, and is the 3rd, 4th or Nth mass in a succession of such masses.

> The highest mountain may be levelled with the plain from which it springs, without the loss of real territory in the land; but when the ocean makes encroachment on the basis of our earth, the mountain unsupported, tumbles with its weight; and with that accession of hard bodies, moveable with the agitation of the waves, gives the sea the power of undermining further and further into the solid basis of our land. This is the operation which is to be measured; this is the mean proportional by which we are to estimate the age of worlds that have terminated, and the duration of those that are but beginning. (*1788, p. 298*)
>
> The world which we inhabit is composed of the materials, not of the earth which was the immediate predecessor of the present, but of the earth, which is ascending from the present, we consider as the third, and which had preceded the land that was above the surface of the sea, while our present land was yet beneath the water of the ocean. Here are the distinct successive periods of existence, and each of these is, in our measurement of time, a thing of indefinite duration. (*1788, p. 304*)

He admits that the main difficulty in such a study is the slowness with which nature proceeds, immense periods of time being involved before an existing land mass will be worn flat and brought down to the level of the sea:

> To sum up the argument, we are certain, that all the coasts of the present continents are wasted by the sea, and constantly wearing away upon the whole; but this operation is so extremely slow, that we cannot find a measure of the quantity in order to form an estimate. Therefore, the present continents of the earth, which we consider as in a state of perfection, would, in the natural operations of the globe, require a time indefinite for their destruction. (*1788, p. 301*)

He clearly recognizes that land movements take place which form no part of his scheme but are due to local or extraneous causes. These local events, he considers, do not alter the substance of his scheme.

> Parts of the land may often sink in a body below the level of the sea, and parts again may be restored without waiting for the general circulation of land and water, which proceeds with all the certainty of nature, but which advances with imperceptible progression. (*1788, p. 302*)

Finally Hutton is driven to the conclusion that this succession of world out of world is such a perfectly ordered system that it is only reasonable to suppose this is the way our world is constantly growing and dying. Consequently there is no need to seek a different state of things as had the early geologists.

> But if the succession of worlds is established in the system of nature, it is in vain to look for anything higher in the origin of the earth. The result, therefore, of our present enquiry, is that we find no vestige of a beginning, – no prospect of an end. (*1788, p. 304*)

CHAPTER FIVE

The Initial Reaction and the Second
Appearance of Hutton's *Theory*

Once more search undismay'd, the dark profound
Where Nature works in secret: view the beds
Of mineral treasure, and the eternal vault
That bounds the hoary Ocean; trace the forms
Of atoms moving with incessant change
Their elemental round;

(*The Pleasures of Imagination* (1744) by Mark Akenside, five
years Hutton's senior and medical student of Edinburgh
and doctor of Leyden)

Despite the brilliance of its ideas, Hutton's work seemed doomed to be buried
quietly and unobtrusively in the *Transactions* of the Royal Society of Edinburgh. There were many reasons for its cold reception. Hutton was a relatively modest man and would have been content merely with addressing the
Society and with having his views in print. He was simple and stubborn by
nature with strong likes and dislikes and dressed in the garb of a Quaker with
the addition of an old-fashioned cocked hat. As a member of the City's distinguished social circle he was renowned for his joviality and powers of
ordinary conversation but, unlike Werner, he did not impress his beliefs by
the inner vitality of his personality. It is characteristic of him that he felt and
left his work to speak for itself by the power of its own simple logic. When it
came to self-advertisement and the conversion of others he was clearly less
interested and for all his scientific acuteness it is highly probable that he was
rather unworldly in these respects and in many of his more ordinary doings,
as the following reminiscence tends to suggest.

Several highly respectable literary gentlemen proposed to hold a convivial
meeting once a week, and deputed two of their number, Doctors Black and
Hutton, to look out for a suitable house of entertainment to meet in. The two
accordingly sallied out for this purpose, and seeing on the South Bridge a sign
with the words, 'Stewart, vintner, down stairs', they immediately went into
the house and demanded a sight of their best room, which was accordingly
shown to them, and which pleased them much. Without further inquiry, the
meetings were fixed by them to be held in this house; and the club assembled
there during the greater part of the winter, till one evening Dr Hutton, being
rather late, was surprised, when going in, to see a whole bevy of well-dressed
but somewhat brazen-faced young ladies brush past him, and take refuge in
an adjoining apartment. He then, for the first time, began to think that all

was not right, and communicated his suspicions to the rest of the company. Next morning the notable discovery was made, that our amiable philosophers had introduced their friends to one of the most noted houses of bad fame in the city! (*Kay, 1838, I, p. 57*)

But Hutton's lack of personal magnetism was one of many reasons for the inauspicuous launching of his masterpiece. The title was unfortunate as it resembled that of many fantastic and fabulous speculations on the earth's origin then in print. The appearance of seemingly just such another rigmarole was not calculated to excite general attention.

In addition, some of its early neglect must be attributed to Hutton's clumsy style which would discourage and weary the interest of the chance reader. To describe metamorphism he needed a sentence one hundred and thirty-six words long and even Charles Lyell, a greedy reader, admitted to having read only half of Hutton's works and of skimming over the remainder.

The scientific stage seemed set fair for the reception of the new *Theory* but in fact the general cultural environment was not friendly. Events on the Continent and social conditions in a Britain that was being rapidly urbanized were causing much disquiet. Hence it is recognized that

> Hutton . . . was unfortunate in his choice of a time for springing new conceptions on the world. The heresy hunting and heretic baiting which were characteristic sports of the latter half of the eighteenth century were intensified by the reaction to the French Revolution in this country. (*Tomkeieff, 1946, p. 324*)
>
> The French Revolution lasted from 1789–95, and removed Lavoisier's head in 1794. As Lyell points out, the bitterness with which Hutton's novel ideas were received was partly due to the political situation. (*Bailey, 1927, p. 185*)

There may well be other reasons, not least the small circulation of the *Transactions*, but the result was undeniable: Hutton's discourse was given a poor reception by the audience on the opening night and upon printing it passed almost unnoticed. Two things rescued the work from oblivion: the directness of the opposition; and the quality of Hutton's friends. The two were inseparable as the opposition threw its objections at all Vulcanists. Whereas Hutton himself might have ignored in silence any form of indifference or obscurity he, as with all Scotsmen, would not tolerate a direct affront. It was therefore highly fortunate that his *Theory* was bluntly contradicted by Jean André De Luc. De Luc (1727–1817) was a Swiss geologist and meteorologist who had settled in England in 1773 and soon afterwards became Reader to Queen Charlotte. He published a popular geological text in 1778 which explained the six days of the Mosaic creation as epochs preceding the present state of the globe. In his zeal to defend these traditional ideas De Luc in 1790 totally denied the Huttonian theory. Seizing on the focal point of Hutton's scheme that gravel pebbles represented one stage in a cycle of change from strata to pebble and pebble to strata, he put forward the challenging assertion:

> Then, inversely, if I demonstrate, by actual observation, that the operation of wind and water upon our continents, do not prepare any gravel for future ones; I shall have proved, that the whole of your Theory is without foundation. (*De Luc, 1790, p. 584*)

He set about his task by questioning the notion of consolidation and uplift by 'fire', preferring the traditional diluvial explanation. In answer to the central theme of the continual wasting away of the land and its regrowth below the ocean, he attacked along two lines. First he denied the soil moved at all and supported this statement by the following remarks:

> When solid strata are found under that almost general soil, their nature bears seldom any nature to it; which, however, ought to be the case, if it proceeded from their decomposition. (*De Luc, 1790, p. 586*)
> ... stratum is as distinct from the original soil, as oil paint from cloth, wood or metal, over which it is laid: ... This is evident proof, that the soil is not washed away; for if it had been the case, in respect of those grounds, the vegetable earth had not gathered over them. (*De Luc, 1790, p. 588*)

This idea that the soil is an unchanging feature clearly demonstrates the fundamental difference between Hutton and his opponents. They could not accept the startling thought that daily changes were taking place, while Hutton could see no order to the world without this universal succession of surface by surface. Their reluctance to accept his novel concept stemmed not only from the fact that it would mean the total abandonment of traditional teaching but because they sincerely believed the evidence of their senses led unfailingly to contrary explanations.

Secondly, De Luc would not credit rivers with the power suggested by Hutton, putting forward as proof the phenomenon of rivers cutting down through gravel deposits:

> The consequences of these facts are so obvious in themselves, as great in the theory of the earth. 1st, it is evident, that the gravel deposited in the bed of rivers flowing in the plains, was before in that part of the ground which they have furrowed in forming their channels. 2nd, since those rivers have no other gravel but what proceeds from the plains themselves, it is evident also, that the plains cannot have been formed from the ruins of the mountains; for these ruins ought to have been spread over them by these very rivers which, on the contrary, rob them in cutting their channels. 3rd, consequently all the gravel of the plains is come out of the sea, as well as every other of their strata, when our continents were left dry. 4th, but no gravel goes into the sea from our continents; since rivers, which have gravel in some parts of their beds, do not even move it along their whole course. (*De Luc, 1790, p. 595*)

On this point he understandably failed to appreciate what we now know of the law of attrition and the previously high melt-water discharge of glacial times. This assault was quickly followed by criticism from Richard Kirwan, a staunch disciple of the Wernerian school. Born in County Galway in 1733 (*D.N.B.* XI, pp. 228–30), Kirwan entered the Jesuit novitiate at St Omer in 1754, but left it in the following year and two years later was married, only to be arrested on his wedding day for his wife's debts. In 1766 he was called to the Irish Bar, but abandoned the law in 1768 in favour of scientific pursuits. Between 1777 and 1787 Kirwan lived in London, becoming a Fellow of the Royal Society in 1780. At this time he was mainly interested in chemistry and also in meteorology, making estimates of the temperatures of different latitudes designed to pave the way for the theory of global wind circulation. In 1787 he moved to Dublin, becoming President of the Royal Irish Academy in 1791, where 'he was consulted as a weather prophet by half the farmers of Ireland'. In a paper written in 1793, Kirwan concentrated his arguments on endeavouring to refute Hutton's heat mechanism for the consolidation of strata and their elevation from the sea floor. As a student of Werner and a mineralogist, his initial arguments were naturally directed against any heat theory in direct antithesis to the idea of the universal ocean. But he also briefly denied Hutton's claim that soil moves and that the action of the rivers is responsible for most surface features. In this aspect of his attack he was content to rest his disbelief on the opinion of others:

> He next tells us that this soil is necessarily washed away by the continual circulation of water running from mountains to the sea. Here are two suppositions, neither of which is grounded on facts. Soil is not constantly carried away by the water, even from mountains as Mr De Luc has clearly shown in his nineteenth and twentieth letters to the Queen, and if it were, it would be deposited on the plains, for there are plains as well as mountains on the dry parts of the globe. All water does not flow into the sea, much of it is carried off by evaporation. Most of the earth swept off by rivers is deposited at their mouths; of that which is carried into the sea, much, if not all, is rejected on the shore. (*Kirwan, 1793, p. 55*)

THE SECOND APPEARANCE OF HUTTON'S 'THEORY'

The above criticisms drew some attention to a theory that otherwise would have been largely ignored and, above all, goaded Hutton into publishing a spirited reply. Since presenting his scheme in 1785 he had occupied himself in travel and observation and in finishing a voluminous work on metaphysics and principles of knowledge. The following extract from this long *Investigation* typifies his approach to Natural Science.

> To a philosopher, who sees the evidence of this principle, that all things have been ordained in perfect wisdom, a preternatural event is a contradictory proposition.
> If we shall thus scientifically believe in the existence of a superintending Being, who has conceived everything in wisdom and justice, and who has executed everything in power and goodness, then, that which is commonly termed the laws of nature must appear to be the decrees of God; and therefore, whenever we are led, in reasoning from the appearances of things, to the

conclusion of general principles by which these appearances are explained, we may rest satisfied, so far as this intellectual process is without error, that there is a species of truth which is above all suspicion, being no other than the visible word of God enlightening the rational mind of man. (*Hutton, 1794, II, p. 310*)

He had also planned to expand his original idea but postponed doing so largely because of increasing ill-health. He was already suffering from the illness of which he was to die four years later, when he heard of Kirwan's attack. The result was spontaneous.

> The very day, however, after Mr Kirwan's paper was put into his hands, he began the revisal of his manuscript, and resolved immediately to send it to the press. (*Playfair, 1805, p. 86*)

Hutton defended his views in their entirety and his second publication includes virtually a reprint of his original *Theory* in a broader setting. The work appeared in 1795 in two volumes as *Theory of the Earth, with Proofs and Illustrations*. The first volume included almost a word for word repetition of his 1785 paper; the second is more diversified in its themes and deals with such topics as: (1) the formation of rocks beneath the sea and their subsequent elevation; (2) the formation of soil from rock; (3) the planing down of all slopes to sea level; (4) the action of glaciers; (5) the proof of erosive processes by the calculation of the mass of material removed; and (6) the natural and harmonious construction of mountain and valley. These new ideas did not detract from the main theme of his original theory but provided further proof of its correctness by application to specific examples.

There was a third volume, but this was not published during his lifetime. It, or the greater part of it, was discovered by F. D. Adams (Adams, 1934; Eyles, 1950) almost a century later and published by Geikie in 1899. It was concerned with Hutton's travels in Scotland and De Saussure's work in the Alps and emphasized, with numerous field-studies, the granitic, or vulcanist, portion of his theory and the formation of stratified rocks. When the first two volumes were published in 1795 it was advertised as a four-part work. The fourth volume, if it ever existed, has never come to light.

Hutton dealt with his critics in general before dealing with separate points of criticism. The following passage illustrates how strongly he disapproved of the traditional theories and explains partly why he moved so quickly to defend his own scheme.

> But when, in forming a theory of the Earth, a geologist shall indulge his fancy in framing, without evidence, that which had preceded the present order of things, he then either misleads himself, or writes a fable for the amusement of his reader. (*Hutton, 1795, I, pp. 280–1*)

Each of the main criticisms were dealt with in turn. In answer to the charge that soil was a permanent attachment he wrote:

> I have not said that it (soil) is constantly washed away, while it is soil in

which plants grow, it is not travelling to the sea, although it be on the road, and must arrive in time. (*1795, I, p. 206*)

. . . the apparent permanency of this earth is not real or absolute; and that the fertility of its surface, like the healthy state of animal bodies, must have its period, and be succeeded by another. (*1795, II, p. 90*)

It will be seen that Hutton's main difficulty in meeting these criticisms lay in the radical implications of his theory. His ideas were so novel and so far in advance of those of his contemporaries that it seems almost as if they were arguing on different premises. What to his contemporaries seem inconsistencies and misconceptions are known today to be the main bulwarks of his scheme. Thus in the passage below where, in summing up De Saussure's work, he mentions the wearing down of mountains and the continual shifting of the position of the plains, he is dealing with advanced notions of erosion, the full implications of which have only recently been understood.

In this picture of the Alps, there is presented to our view the devastation of the solid rocks by agents natural to the surface of the earth; here is the degradation of the mountains in the course of time, of these ruins plains are formed below; and these plains are continually shifting in their place, in affording materials to be washed away and rolled in the rivers, and in receiving from the higher grounds the spoils of ruined rocks and mountains. (*1795, II, p. 141*)

Replying to De Luc's contention that examples of streams cutting down through gravel appear to contradict his scheme, Hutton explains that this is not an inconsistency but a natural stage in the life cycle of the surface. He sees the rivers in the early period of erosion when the mountains are at their highest, sweeping down to the plains masses of fragmented materials which will in time form a deep flood plain. As the mountains are denuded and the rivers carry down less material they will begin to erode the softer lower reaches and thus cut into and sweep away their own deposits.

When the river has enlarged its bed by preying upon one side whether of the mountain or the haugh, the water only covers . . . (the haugh) in a flood; at other times, it leaves it dry. Here, . . . the seeds of plants, . . . spring up and grow; and, in little floods, some sand and mud is left among these plants; . . . In this state, the haugh is always deepening or increasing its soil . . . At last, when this soil becomes so high as only to be flooded now and then, it becomes most fertile, . . but this operation, . . . has a period, at which time the river must return again upon its steps, and sweep away the haugh which it had formed. (*1795, II, pp. 206–7*)

Of De Luc generally he concludes:

M. de Luc in his *Histoire de la Terre*, would make the mountains last forever, after they have come to a certain slope. (*1795, II, p. 202*)

Hutton's answers to his critics should not be allowed to distract attention from the importance of his enlarged work as a whole. Contemporary criticism had neither disconcerted him nor persuaded him to alter any of his scheme.

His careful repetition and expansion of his earlier ideas was in the manner of a patient teacher, puzzled by the obstinate blindness of his pupils.

We need not here reiterate that he had provided a new basic premise and had revealed vast fresh fields of study. Before leaving his direct contributions, it is tempting and interesting to notice how many of his observations seem to foreshadow modern theories on landscape evolution. No doubt, as with the writings of Leonardo da Vinci, more can be read into some of these brief statements than was ever intended or fully understood, but the germ of the suggestion seems there. The following extracts, preceded by the modern terminology for the idea, must suffice.

(Superimposition of streams)

. . . (beds of rivers) take winding courses around the hills which they cannot surmount; sometimes again they break through the barrier of rocks opposed to this current; thus making gaps in places by wearing away the solid rock over which they formerly had run upon a higher level. (*1795, II, p. 100*)

(Glaciation)

There would have been immense valleys of ice sliding down in all directions towards the lower country, and carrying large blocks of granite to a great distance; (*1795, II, p. 218*)

(W. M. Davis's concept of structure, process and stage)

I have always found the appearances . . . (of hills &c.) most intelligible, and strictly corresponding with the general principle of atmospheric influence acting upon the particular structure of the earth below. (*1795, II, p. 402*)

(Cuesta topography)

. . . a horizontal bed of rock forms a table mountain, . . . An inclined rock of this kind forms a mountain sloping on the one side, and having a precipice on the upper part of the other side, with a slope of fallen earth at the bottom; (*1795, II, p. 413*)

(Horton's law of stream numbers)

. . . it is these channels, increasing in their size as they are diminished in number by the uniting of their waters, that give so clear a prospect of the operations of time past and prove the theory of the land being in a continual state of decay. (*1795, II, p. 446*)

Needless to say, the above suggestions were outside the comprehension of most of Hutton's contemporaries. Yet his teachings were now the subject of controversy and between 400 and 500 copies of his two-volume work had been sold. When he died on 26 March 1797, 'having written a good deal in the course of the same day', his theory was still unaccepted by the main authorities and he had no real conception of the change his ideas were to cause in geological thought.

POSTHUMOUS CRITICISM OF THE 'THEORY'

The reappearance of Hutton's *Theory* called forth renewed criticisms, particularly from Wernerians and theologians. In 1799, Richard Kirwan took up arms again with the publication of his *Geological Essays*, which was largely a setting down in print of Werner's ideas; supporting the Mosaic Deluge (Essay II), advocating the greater efficiency of past processes – citing misfit streams as evidence of previous humidity (Essay III), and reiterating his attack on Hutton (Essay X). Playfair criticizes it in the following manner:

> His (Mr Kirwan's) *Geological Essays* have also for their object to explain the first origin of things; and to say that he has not succeeded, in an attempt where no man ever can succeed, implies no reproach on the execution of his work, whatever it may do on the design. (*1802, p. 480*)

Kirwan's publication was motivated by a desire to provide a suitable answer to Hutton's theory and this could obviously not be done while Werner continued to neglect to formulate in writing his own ideas. In 1802 Kirwan returned to the charge with two papers read to the Royal Irish Academy. In the first (1802A) he again contested Hutton's theory about igneous rocks and disagreed with Sir James Hall's experiments relating to the flowage of granite in a fused state. His second paper (1802B) dealt with the differences sometimes found between the slopes on each side of a mountain range. He believed such evidence supported the Wernerian idea of the previous existence of a universal ocean. He visualized the sediment-laden ocean sweeping across the intervening submerged mountain range and on meeting the obstacle, depositing its sediments on one side of the range only.

> Thus the mountains that stretch from N to S must have opposed the motion of the waters from E to W this opposition diminishing the motion of that fluid disposed it to suffer the earthy particles with which in those early periods it must have been impregnated to crystallize or be deposited on these eastern flanks, and particularly on those of the highest mountains, for over the lower it could easily pass, these deposits being incessantly repeated at heights gradually diminishing as the level of the waters gradually lowered, must have reduced the eastern declivities or descent, gentle, gradual and moderate, while the western sides receiving no such accessions from depositions, must have remained steep and craggy. (*Kirwan, 1802B, p. 46*)

Today we find it hard to understand how Kirwan's firm belief in the diluvial theory of creation could make him overlook the difference in resistance between various rock-outcrops.

Opposition to Hutton's views was not likely to be restricted to Neptunists. Theologians also showed real concern and eventually Dr William Richardson entered the conflict in defence of the sanctity of Christian truth. Richardson (1740–1820) was rector of an Antrim parish and was particularly interested in the basalts of north-eastern Ireland. He was, had he known it, a diluvialist dwelling in a Vulcanist paradise. Between 1802 and 1808 he issued several papers attacking the views of Desmarest, Hutton and others who believed basalt to be of volcanic origin. His vigorous writing represents faithfully the sharp reaction of a Church already aware of the onslaughts on its security by

Methodism, the French Revolution and the political demands for reform. The doctor belaboured Hutton for his mortal presumption and desire for self-advertisement:

> An impeachment of the credibility of Moses has of late, it appears, been a favourite topic; and the mode of attack most frequently adopted is, by contradicting his chronology in the date of the creation. (*Richardson, 1803, p. 438*)

> Philosophers seem to have been seized with a sort of rage for inventing and supporting theories, and for explaining the operations of nature, and the phenomena she exhibits, upon the principles discovered by themselves; they seem to have considered it as humiliating to admit they were not privy to her secrets, and that they were unable to explain the manner in which her various works were executed.

> The mere adoption of the opinion of others could not procure celebrity; hence it became necessary, that those who thirsted for fame should strike out something new, which, while it showed their own sagacity, gave them an opportunity also of displaying their ability in support of the systems they invented.

> Thus the attention of mankind was diverted from the study of nature to the discussion of opinions; for even those who did not invent, thought it necessary to adopt some theory, for which they soon acquired a strong partiality, and became zealous to procure proselytes to it.

> Hence, the progress of natural history was small. (*Richardson, 1803, pp. 429–30*)

The passage of Hutton which most particularly called forth the foregoing academic vituperation is the one known to every student:

> But if the succession of worlds is established in the system of nature, it is in vain to look for anything higher in the origin of the earth. The result, therefore, of our present enquiry is, that we find no vestige of a beginning, – no prospect of an end. (*Hutton, 1788, p. 304*)

Richardson referred to it thus:

> Every reader will probably form a conjecture as to Dr Hutton's intentions, when he reads his assertion, that he could find no traces of a beginning of the world. (*1803, p. 438*)

Waning in his religious ardour he continued, like De Luc, to deny the constant movement of the soil and the force of river erosion, considering both to be minor features having relatively little effect on the main pattern of the landscape.

> I consider this stratum of soil, with its vegetable coat, as a suit of armour, with which nature, in her wisdom, clothes the world, to protect its loose, moveable materials, and to prevent their being carried off by the rain and winds. (*1803, p. 443*)

> All depredations committed upon our surface are at the expense of this soil, its abrasions discolour our rivers in a flood, and are the source of all our muddy depositions; the original earth is rarely encroached upon, except in a few gullies and ravines of little consequence. (*1803, p. 444*)

> I would remind our geologist, that the long chains of mountains bounding

the valley of the Nile on both sides, seem to have valleys and defiles exactly like our own; yet, in that country they have neither rain nor rivers. (*1803, p. 451*)

Some measure of the scientific standard of these criticisms can be gauged from the fact that Richardson drew support from such varied sources as Bacon, Voltaire, Lucian, Livy, Xerxes and Oliver Goldsmith's *Vicar of Wakefield*.

Fortunately, or perhaps inevitably, contemporary criticism as exemplified by Kirwan and Richardson again defeated its own ends. It drew attention to a book that was already hard to come by and compelled Hutton's friends to take up the cudgels on his behalf. In passing, we must again notice that few writers can have had more need of a Boswell. Indeed Hutton's scholarly biographers fail to do full justice to his literary short-ccmings and appear to have quite overlooked a major characteristic of the *Theory*. Few books of such originality contain so high a proportion of extracts in a foreign language. Of Vol. 1 (620 pp.) well over one-eighth of the text consists of quotations in French. Vol. 2 is even less digested and more illustrative: of its 567 pages over 250, or 44 per cent, are filled with quotations in French, mainly from De Saussure and De Luc. Of its first 200 pages no less than 113 are in that language and they form in fact a French text with a very brief English commentary. It seems strange that the first great work on fluvialism is, in a sense, also almost a masterpiece of literary imperspicuity.

CHAPTER SIX

Playfair's *Illustrations of the Huttonian Theory of the Earth*

> the geological system of Dr Hutton, resembles in many respects, that which appears to preside over the heavenly motions. In both, we perceive continual vicissitude and change, but confined within certain limits, and never departing far from a certain mean condition, which is such, that, in the lapse of time, the deviations from it on the one side, must become just equal to the deviations from it on the other.
>
> (J. Playfair, 1802, p. 440)

Hutton's death had a marked and unexpected influence on the fortunes of his work. By his passing John Playfair, his daily companion and confidant, was prompted to write both an account of his life, as well as a justification and clarification of his ideas.

Hutton was most fortunate in Playfair, who held the post of Professor of Natural Philosophy at Edinburgh, and had achieved considerable attainment in the field of mathematics. What was more he thoroughly understood and believed in his friend's theory and had the intellect to expand the original scheme and to translate it into passages of unsurpassed literary felicitude, without in any way detracting from its content. Even now it is easier and certainly more pleasant to cull the essential points from Playfair's *Illustrations* rather than to venture through the tortuous circumlocutions of Hutton's prose. Playfair's genius was such that he could follow and express erudite conceptions in the clearest and most harmonious manner. He combined with a natural literary skill his stern university training in the spartan disciplines of logic and mathematics, and his writing is characterized by the deductive method – a polished construction of premise upon premise crowned by the final conclusion. Yet it must not be assumed that Playfair was merely another Boswell, for not only did he relate Hutton's ideas but added greatly to them by his own conclusions – as for instance with his 'principle of accordant junctions'. Undoubtedly the spread of his ideas was helped by his connexion with the *Edinburgh Review*, which first appeared as a magazine on the 10th October 1802 and was soon finding many readers throughout Scotland and England. Playfair, indeed, was an intimate member of the intellectual society of Edinburgh as is demonstrated in Pearson's biography of Sydney Smith:

> It was largely due to Sydney's presence in Edinburgh that Scott and a few others started the Friday Club. It was founded on the model of Johnson's

FIG. 12. *John Playfair*
(*National Portrait Gallery*)

Club and the members met for dinner (at £2 a head) every Friday in Fortune's
Tavern. Among the original members were Professors Dugald Stewart and
John Playfair, Sydney Smith, Walter Scott, John Archibald Murray, Henry
Cockburn, Thomas Campbell, Francis Jeffrey, Francis Horner and Henry
Brougham. The Club severely discouraged strangers, with the single excep-
tion of old James Watt, who was always welcome when visiting Edinburgh.
(Pearson, 1948, p. 43)

In the *Illustrations*, with the arguments of Hutton's opponents as a guide
and a challenge, Playfair re-drew the outlines of his friend's ideas on a larger
scale, made them more readily believable by means of illustrative examples,
and combined the whole in a forthright argument that, in the clearest and most
pleasing language, did summary violence upon the contentions of the critics.

In his support of Hutton he seems to have considered that two points required special emphasis. The first in importance was the action of heat in consolidating strata and raising them to their present elevation. He is at pains to labour this point because it contradicts to a great extent the theory of Werner; despite the fact that, as we have already mentioned, Hutton had in this instance drawn the wrong conclusion.

In stressing the action of subterranean heat Playfair draws attention to the complex and broken pattern of strata as evidence of subsequent disturbance, and to the unconformable relationships indicating significant breaks in earth history.

> Now, it is certain that many of the strata have been moved angularly, because that, in their original position they must have been all nearly horizontal. (*1802, p. 58. All pages refer to the edition of White, 1956*)

> Disturbance and removal of strata from their original place of formation is by a force directed from below upwards. We are therefore to assume that the power of the same subterraneous heat, which consolidated and mineralized the strata at the bottom of the sea, has since raised them up to a height at which they are now placed, and has given them the various inclinations to the horizon which they are found to actually possess. (*1802, pp. 69–70*)

> The evidence of the different formation of the primary and secondary strata, and of the changes which the former have undergone is best seen at the points where these strata come into contact with one another. The leading facts to be remarked are:— (1) The vertical or very upright position of the primary or lower strata (2) The superstratification of the secondary, in a position nearly horizontal, so as to be at right angles to those on which they rest (3) The interposition of the breccia between them; or, as happens in many cases, the transition of the lowest of the secondary beds into a breccia, containing fragments sometimes worn, sometimes angular of the primary rock. (*1802, pp. 209–10*)

Playfair repeatedly appeals to the occurrence of intrusive veins, such as the whin sill in north England, as proof of Hutton's theory and as definite denial of the Wernerian idea that all igneous rocks must be older than the sedimentary strata.

> That these (metallic or mineral) veins are of a formation subsequent to the hardening and consolidation of the strata which they traverse, is too obvious to require any proof; and it is no less clear, from the crystallized and sparry structure of the substances contained in them, that these substances must have concreted from a fluid state. Now, that this fluidity was simple like that of fusion by heat, and not compound, like that of solution in a menstruum, is inferred from many phenomena. It is inferred from the acknowledged insolubility of the substances that fill the veins, in any one menstruum whatsoever; from the total disappearance of the solvent, if there was any; from the complete filling up of the vein by the substances which that solvent had deposited; from the entire absence of all the appearances of horizontal or gradual deposition; and, lastly, from the existence of close cavities, lined with crystals, and admitting no egress to any thing but heat. (*1802, pp. 57–8*)

> Again, if it be true that the masses of whin thus interposed among the

strata, were introduced there, after the formation of the latter, we might expect to find, at least in many instances, that the beds on which the whinstone rests, and those by which it is covered, are exactly alike . . . now, this is precisely what is observed . . . this similarity of the strata that cover the masses of whinstone, to those that serve as a base on which they rest, and again the dissimilitude of both to the interposed mass, are facts which can hardly receive any explanation, on the principles of the Neptunian theory. (*1802, pp. 75–6*)

Elevation of the land is chosen in preference to the subsidence of the sea level because:

In order to depress or elevate the absolute level of the sea, by a given quantity, in any one place, we must depress or elevate it by the same quantity over the whole surface of the earth: whereas no such necessity exists with respect to the elevation or depression of the land. It is evident, therefore, that the simplest hypothesis for explaining those changes of level, is, that they proceed from the motion, upwards or downwards, of the land itself and not from that of the sea. (*1802, pp. 446–7*)

The second bastion requiring reinforcement was the proposition that the rivers were the chief agents of terrestrial erosion and that they had carved out the general pattern of mountain, valley and plain. This again was diametrically opposed to the propositions of Werner and the Neptunists, who considered that rivers were quite unable to perform the mammoth tasks demanded by the Huttonians. Playfair therefore felt that if he could convince a majority of opinion by the logical nature of his arguments of the verity of Hutton's scheme of erosion, the acceptance of the remainder of the theory would follow as a matter of course. By his treatment of this portion of the theory, and particularly by his attention to fluvial action, he probably made his greatest contribution to the defence of his friend's ideas. In supporting Hutton's scheme of continual decay he mentions the host of agents daily employed in this work of destruction, gives details of their various forms and shows how the eroded material can play its part in the erosion cycle.

Here again water appears as the most active enemy of hard and solid bodies; and, in every state from transparent vapour to solid ice, from the smallest rill to the greatest river, it attacks whatever has emerged from above the level of the sea and labours incessantly to restore it to the deep. The parts loosened and disengaged by the chemical agents, are carried down by the rains, and, in their descent, rub and grind the superficies of other bodies. (*1802, p. 99*)

It is sufficient to remark, that the consequence of so many minute, but indefatigable agents, all working together, and having gravity in their favour, is a system of universal decay and degradation, which may be traced over the whole surface of the land, from the mountain top to the sea shore. (*1802, p. 100*)

On such shores, the fragments of rock once detached, become instruments of further destruction, and make a part of the powerful artillery with which the ocean assails the bulwarks of the land. (*1802, p. 101*)

The idea of a succession of continents is presented without embellishment, presumably being regarded as unquestionable:

> The series of changes which fossil bodies are destined to undergo, does not cease with their elevation above the level of the sea; it assumes, however, a new direction and from the moment that they are raised up to the surface, is constantly exerted in reducing them again under the dominion of the ocean. The solidity is now destroyed which was acquired in the bowels of the earth; and as the bottom of the sea is the great laboratory, where loose materials are mineralized and formed into stone, the atmosphere is the region where stones are decomposed, and again resolved into earth. (*1802, p. 109*)

Playfair's answer to the earlier critics of the carrying power of rivers befits a trained mathematician: the amount of sediment which can be borne by some streams has, he says, been measured:

> The quantity of earth thus carried down, varies according to circumstances; it has been computed, in some instances, that the water of a river in a flood, contains earthy matter suspended in it, amounting to more than the 250th part of its own bulk. (*1802, p. 106*)

Taking his idea of the quantitative measurement of forces a step further, he explains how quite large rocks can be propelled along in a fast-moving stream.

> Next in force to the glaciers, the torrents are the most powerful instruments employed in the transportation of stones. The fragments of rock which oppose the torrent, are rendered specifically lighter by the fluid in which they are immersed, and lose by that means at least a third part of their weight; they are at the same time impelled by a force proportional to the square of the velocity with which the water rushes against them, and proportional also to the quantity of gravel and stones which it has already put in motion. (*1802, pp. 389–90*)

Whereas Hutton concerned himself more with the cardinal principle of successively dying continents, Playfair's attention was concentrated upon the work of rivers and particularly the finely balanced adjustment of each basin which he regarded as unanswerable proof to the rightness of Hutton's contentions regarding the work of rivers, and thus logically leading to a demonstration of the truth of the whole theory. Playfair was the first to observe that a river was not a chance feature on the surface of the landscape but a finely ordered organism, each tributary being perfectly adjusted in height at its junction with the main trunk of the river and each branch joining in the common task of eliminating all irregularities. From this observation he was drawn to the inevitable conclusion that the only possible cause, of such a remarkable uniformity of level and purpose throughout a river basin, was the river itself. If the valleys had been carved by any other agent he felt that the assumption of such a coincidence between the original form and that taken up by the river and its tributaries placed too great a strain on the limits of possibility. The radical nature of this concept was such that even the later

proselytes of the Huttonian creed of natural causes like Charles Lyell did not
accredit rivers as being the most important process in the formation of valleys
and were inclined to call upon the power of an auxiliary agent such as the sea
to do the main work of erosion.

If we proceed in our survey from the shores, inland, we meet at every step
with the fullest evidence of the same truths, and particularly in the nature
and economy of rivers. Every river appears to consist of a main trunk, fed
from a variety of branches, each running in a valley proportioned to its size,
and all of them together forming a system of valleys, communicating with
one another, and having such a nice adjustment of their declivities, that none
of them joins the principal valley, either on too high or too low a level; a
circumstance which would be infinitely improbable, if each of these valleys
were not the work of the stream that flows in it. (*1802, p. 102*)

If indeed a river consisted of a single stream without branches, running in a
straight valley, it might be supposed that some great concussion or some
powerful torrent, had opened at once the channel by which its waters are
conducted to the ocean; but, when the usual form of a river is considered, the
trunk divided into many branches, which rise at a great distance from one
another, and these again subdivided into an infinity of smaller ramifications,
it becomes strongly impressed upon the mind, that all these channels have
been cut by the waters themselves; that they have been slowly dug out by the
washing and erosion of the land; and that it is by the repeated touches of the
same instrument, that this curious assemblage of lines has been engraved so
deeply on the surface of the globe. (*1802, p. 103*)

The structure of the valleys among mountains shows clearly to what cause
their existence is to be ascribed. Here we have first a large valley, communicating directly with the plain, and winding between high ridges of mountains,
while the river in the bottom of it descends over a surface, remarkable, in
such a scene, for its uniform declivity. Into this, open a multitude of transverse valleys, intersecting the ridges on either side of the former, each bringing a contribution to the main stream, proportioned to its magnitude; and,
except where a cataract now and then intervenes, all having that nice adjustment in their levels, which is the more wonderful, the greater the irregularity
of the surface. These secondary valleys have others of a smaller size opening
into them; and, among mountains of the first order, where all is laid out on
the greatest scale, these ramifications are continued to a fourth, and even a
fifth, each diminishing in size as it increases in elevation, and as its supply of
water is less. Through them all, this law is in general observed, that where a
higher valley joins one, of the two angles which it makes with the latter, that
which is obtuse is always on the descending side; a law that is the same with
that which regulates the confluence of streams running on a surface nearly of
uniform inclination. This alone is a proof that the valleys are the work of the
streams; and indeed what else but the water itself, working its way through
obstacles of unequal resistance, could have opened or kept up a communication between the inequalities of an irregular and alpine surface? (*1802, p. 114*)

These three passages, and particularly the first, contain the essence of 'Playfair's law'. In them can clearly be seen three main ideas; (1) that rivers cut
their own valleys; (2) that the angle of slope of each river shows an adjustment of equilibrium with the velocity and discharge of water and the amount

of material carried; and (3) that a whole river system is integrated by the mutual adjustment of the constituent parts.

The first idea is important because it shows the definite stand that Hutton and Playfair took on the effectiveness of river erosion, and contrasts strongly with most of the arguments put forward by others during the period 1820–45. The second idea has become known in modern technology by the term *grade*. The third idea represents the famous principle of accordant junctions. It had in a sense been appreciated by Hutton as the following lines show:

> Here, therefore, is infinitely more than a single river, and a valley corresponding to the river; here is a system of rivers and valleys, things calculated in perfect wisdom, or properly adapted to each other. (*1795, II, p. 293*)

It is not surprising that some of Playfair's most penetrating observations occur in his description of rivers. Thus he recognized the process of attrition in which stones rolling against each other gradually diminish in size by having pieces chipped off and gradually becoming rounded.

> It is a fact very generally observed, that where the valleys among primitive mountains open into large plains, the gravel of those plains consists of stones, evidently derived from the mountains. The nearer that any spot is to the mountains, the larger are the gravel stones, and the less rounded is their figure, and, as the distance increases, this gravel, which often forms a stratum nearly level, is covered with a thicker bed of earth or vegetable soil. (*1802, p. 381*)

He also recognized the modern geological principle of *facies*, or, briefly, that currents do not carry all types and sizes of rock to the same place of deposit but act as sorting agents. In early stages of geology its practitioners used to believe that each stratum was laid down under distinct conditions and during a distinct age. Therefore the occurrence of interdigitated conglomerates, sandstones and stratified muds in close proximity and with all the appearance of having been formed simultaneously was very puzzling. With the acceptance of the principle of facies and the realization that several different types of rock could be formed at the same time in close juxtaposition, another obstacle to the understanding of the nature and correlation of rock deposits was removed.

> When the detritus of the land is delivered by the rivers into the sea, the heaviest parts are deposited first, and the lighter are carried to a greater distance from the shore. (*1802, p. 413*)

Because Playfair understood the appeal and plausibility of the traditional catastrophic notions he judged the best means of counter-attack was to persuade his readers to exercise the logic of their senses. Therefore, with favourable examples, he continually returns to the feature of the river system as being the corner stone of his proof. He asks how any reasonable person could imagine a universal torrent or flood carving out the delicate intricacies of the ordinary river system:

> The general structure of valleys among mountains, is highly unfavourable

to the notion that they were produced by any single great torrent, which swept over the surface of the earth. In some instances, valleys diverge, as it were from a centre, in all directions. In others, they originate from a ridge, and proceed with equal depth and extent on both sides of it, plainly indicating, that the force which produced them was nothing, or evanescent at the summit of that ridge, and increased on both sides, as the distance from the ridge increased. The working of water collected from the rains and the snows, and seeking its way from a higher to a lower level, is the only cause we know of, which is subject to this law. (*1802, p. 401*)

The whole point is hammered home by citing the example of an incised meander. Even if a flood could carve out river valleys he was convinced that it was beyond all reason to imagine that it could bring about such an incision.

A river, of which the course is both serpentine and deeply excavated in the rock, is among the phenomena, by which the slow waste of the land, and also the cause of that waste, are most directly pointed out. (*1802, p. 104*)

Playfair is able to answer Kirwan's puzzle of differences in the gradient and smoothness of mountain slopes by pointing out that such variations arise directly from the differing resistances of the rocks acted upon by the erosive forces. Dr Richardson's argument that valleys occur above the source of the stream is disposed of by two arguments: first, that many dry valleys are found on the higher parts of mountain sides because they have been deserted temporarily by the rivers which originally cut them; second, that the peaks of mountains provide the most obvious examples of denudation as there erosive forces of many kinds, including rain and weathering, are strongest and nothing which has been removed can be replaced.

This decomposition of all mineral substances, exposed to the air is continual, and is brought about by a multitude of agents, both chemical and mechanical, of which some are known to us, and many, no doubt, remain to be discovered. (*1802, p. 97*)

Again, wherefore is it, that among all mountains remarkable for their ruggedness and asperity, the rock, on examination, is always found of very unequal destructibility, some parts yielding to the weather, and to other causes of disintegration, much more slowly than the rest, and having strength sufficient to support themselves, when left alone, in slender pyramids, bold projections and overhanging cliffs? Where, on the other hand, the rock wastes uniformly the mountains are similar to one another; their swells and slopes are gentle and they are bounded by a waving continuous surface. The intermediate degrees of resistance which the rocks oppose to the causes of destruction, produce intermediate forms. It is this which gives to the mountains, of every different species of rock, a different habit and expression, and which, in particular, has imparted to those of granite that venerable and majestic character, by which they rarely fail to be distinguished. (*1802, p. 112*)

It is not, in the greatest rivers, that the power to change and wear the surface of the land is most clearly seen. It is at the heads of rivers, and in the feeders of the larger streams, where they descend over the most rapid slope, and are most subject to irregular or temporary increase and diminution, that the causes which tend to preserve, and those that tend to change the form of the earth's surface, are farthest from balancing one another, and where, after every season, almost after every flood, we perceive some change produced, for which no compensation can be made, and something removed which is never to be replaced. When we trace up rivers and their branches towards their source, we come at last to rivulets, that run only in the time of rain and they are dry at other seasons. No other person can examine the valley of the rivulet without seeing that the rivulet carries away matter which cannot be repaired, except by wearing away some other part of the surface of the place upon which the rain that forms the stream is gathered. It requires but little study to replace the parts removed, and to see nature at work, resolving the most hard and solid masses, by the continued influences of the sun and atmosphere. We see the beginning of a long journey by which heavy bodies travel from the summit of the land to the bottom of the ocean, and we remain convinced, that, on our continents, there is no spot on which a river may not formerly have run. (*1802, pp. 351–2*)

Thus by the clarity of his word pictures Playfair disposed forcibly with the argument that rivers are not the cause of valleys but valleys the cause of rivers. In fact the accordance of river junctions continued to prove an unanswerable point and an enduring bulwark in support of Hutton's theory.

The claim of the critics that soil is permanent was also answered in the clearest fashion:

The soil, therefore, is augmented from other causes, just as much, at an average, as it is diminished by that now mentioned; and this augmentation evidently can proceed from nothing but the constant and slow disintegration of the rocks. In the permanence therefore of a coat of vegetable mould on the surface of the earth, we have a demonstrative proof of the continual destruction of the rocks. (*1802, pp. 106–7*)

Having considered the more scientific criticisms, Playfair proceeds to defend the Huttonians against the charge of being atheists. He argues, as did Hutton, that no denial of the Bible was intended but that the findings of science supported the infinite wisdom of God's work. He suggests that the Bible should not be believed in literally as a scientific document and that its bold statements of the process of creation should be considered against the background of proved and accepted facts and, if necessary, treated as fanciful illustrations only. Why, he asks, in this respect should geology be treated differently from mathematical sciences?

The author of nature has not given laws to the universe, which, like the institutions of men, carry in themselves the elements of their own destruction. He has not permitted, in his works, any symptom of infancy or of age, or any sign by which we may estimate either their future or their past duration. He may put an end, as he no doubt gave a beginning, to the present system, at some determinate period; but we may safely conclude that this great catastrophe will not be brought about by any of the laws now existing, and that it is not indicated by anything which we perceive. (*1802, pp. 119–20*)

To assert, therefore, that, in the economy of the world, we see no mark, either of a beginning or of an end, is very different from affirming, that the world had no beginning, and will have no end. (*1802, p. 120*)

We see everywhere the utmost attention to discover, and the utmost disposition to acquire, the instances of wise and beneficent design manifested in the structure, or economy of the world. The enlarged views of these, which his geological system afforded, appeared to Doctor Hutton himself as its most valuable result. They were the parts of it which he contemplated with greatest delight; and he would have been less flattered, by being told of the ingenuity and originality of his theory, than of the addition which it had made to our knowledge of *final causes*. (*1802, pp. 121–2*)

On the other hand, the authority of the Sacred Books seems to be but little interested in what regards the mere antiquity of the earth itself; nor does it appear that their language is to be understood literally concerning the *age* of that body, any more than concerning its *figure* or its *motion*. The theory of Doctor Hutton stands here precisely on the same footing with the system of Copernicus; for there is no reason to suppose, that it was the purpose of revelation to furnish a standard of geological, any more than of astronomical science. It is admitted, on all hands, that the Scriptures are not intended to resolve physical questions, or to explain matters in no way related to the morality of human actions; and if, in consequence of this principle, a considerable latitude of interpretation were not allowed, we should continue at this moment to believe, that the earth is flat . . . It is but reasonable, therefore, that we should extend to the geologist the same liberty of speculation, which the astronomer and mathematician are already in possession of. (*1802, p. 126*)

We have already noticed (p. 53) how Hutton, while outlining his main theme, foreshadows a few relatively modern geological or geomorphological theories. So it was with Playfair. He explained better than Hutton the feature of stream superimposition.

On observing the Patowmack, where it penetrates the ridge of the Alleghany mountains, or the Irtish, as it issues from the defiles of Altai, there is no man, however little addicted to geological speculations, who does not immediately acknowledge, that the mountain was once continued quite across the space in which the river now flows. (*1802, p. 105*)

He was the first to stress the importance of river terraces in the history of a river, and seemed to possess an advanced understanding of them. He even attempted to explain their origin.

The changes which have taken place in the courses of rivers, are also to be traced, in many instances, by successive platforms of flat alluvial land, rising one above the other, and marking the different levels at which the river has run at different periods of time. . . . each change which the river makes in its bed obliterates at least a part of the monuments of former changes. . . . only a small part of the progression can leave any distinct memorial behind it. . . . there is no reason to think, that in the part which we see, the beginning is included. . . . (*1802, pp. 103–4*)

This happens, especially, when successive terraces of gravelly and flat land are found on the banks of a river. Such platforms, or *haughs* as they are called in this country, are always proofs of the waste and *detritus* produced by the river, and of the different levels on which it has run; but they sometimes lead us farther, and make it certain, that the great mass of gravel which forms the successive terraces on each side of the river, was deposited in the bason of a lake. (*1802, p. 355*)

Playfair also has an important statement on the removal of breaks-of-slope or nick-points and the establishment of 'grade' in a stream.

> In order to give uniform declivities to the river, the lakes must not only be filled up or drained, but the cataract, wherever there is one, must be worn away. The latter is an operation in all cases visible. The stream, as it precipitates itself over the rocks, hurries along with it, not only sand and gravel, but occasionally large stones, which grind and wear down the rock with a force proportioned to their magnitude and acceleration. (*1802, p. 361*)

We have already quoted numerous other extracts which show Playfair's acute understanding of river-work but the fact still remains that his prime importance lay in his recognition and championing of the whole of Hutton's theory. He had no doubt whatsoever of its intrinsic qualities.

> It is impossible to look back on the system which we have thus endeavoured to illustrate, without being struck with the novelty and beauty of the views which it sets before us. The very plan and scope of it distinguish it from all other theories of the earth, and point it out as a work of great and original invention. The sole object of such theories has hitherto been, to explain the manner in which the present laws of the mineral kingdom were first established, or began to exist, without treating of the manner in which they now proceed, and by which their continuance is provided for. The authors of these theories have accordingly gone back to a state of things altogether unlike the present, and have confined their reasoning, or their fictions, to a crisis which never has existed but once, and which never can return . . . (Dr Hutton's) theory, accordingly, presents us with a system of wise and provident economy, where the same instruments are continually employed, and where the decay and renovation of fossils being carried on at the same time in the different regions allotted to them, preserve in the earth the conditions essential for the support of animal and vegetable life. We have been long accustomed to admire that beautiful contrivance in nature, by which the water of the ocean, drawn up in vapour by the atmosphere, imparts, in its descent, fertility to the earth, and becomes the great cause of vegetation and of life; but now we find, that this vapour not only fertilizes, but creates the soil; prepares it from the solid rock, and after employing it in the great operations of the surface; carries it back into the regions where all its mineral characters are renewed. Thus, the circulation of moisture through the air, is a prime mover, not only in the annual succession of the seasons, but in the great geological cycle, by which the waste and reproduction of entire continents is circumscribed. Perhaps a more striking view than this, of the wisdom that presides over nature, was never presented by any philosophical system, nor a greater addition ever made to our knowledge of final causes. (*1802, pp. 127–9*)

Nor had Playfair the slightest doubt of the tremendous possibilities inherent in Huttonian principles.

> If indeed this theory of the earth is as well founded as we suppose it to be, the lapse of time must necessarily remove all objections to it, and the progress

of science will only develope its evidence more fully. As it stands at present, though true, it must be still imperfect; and it cannot be doubted, that the great principles of it, though established on an immoveable basis, must yet undergo many modifications, requiring to be limited, in one place, or to be extended in another. A work of such variety and extent cannot be carried to perfection by the efforts of an individual. Ages may be required to fill up the bold outline which Dr Hutton has traced with so masterly a hand; to detach the parts more completely from the general mass; to adjust the size and position of the subordinate members; and to give to the whole piece the exact proportion and true colouring of nature. (*Playfair, 1802, pp. 138–9*)

The emphasis on the lapse of time thought necessary 'to remove all objections' to Hutton's theory is typical of Playfair's wisdom. When he died in 1819 the Huttonian cause was at best making desultory progress. His enthusiasm and academic standing had saved the day but had by no means won it. Even his friends regretted that so learned a mathematician and physicist should have devoted 'so much of his time and so large a proportion of his publications . . . to the subjects of the Indian Astronomy; and the Huttonian Theory of the Earth'.

CHAPTER SEVEN

The Wernerian Counter-Offensive, 1802-20

> By supposing the order fixed and determined, when it really is not, further inquiry is prevented; and propositions are taken for granted, on the strength of a theoretical principle, that require to be ascertained by actual observation. It has happened to the Wernerian system, as it has to many other improvements: that they were at first inventions of great utility; but by being carried beyond the point to which truth and matter of fact could bear them out, they have become obstructions to all further advancement, and have ended with retarding the progress which they began with accelerating.
>
> (Fitton, 1881, p. 96)

The period from 1780 to 1820 has been called the 'Golden Age' of geology yet the last eighteen years of it were virtually barren of geomorphology. Throughout, Hutton's ideas are generally ignored or intermittently discredited and the majority of geologists still accepted Werner's doctrines. Some of the reasons for this stagnation have already been mentioned; for instance, the extreme novelty of Hutton's ideas made it unreasonable to expect their rapid acceptance. In addition the Napoleonic wars, which did not end till 1815, prevented easy travel to and on the Continent and largely inhibited Hutton's ideas from becoming known there. Similarly the second British–United States conflict between 1812 and 1814 hindered the passage of his ideas to that country. Also at home there were many unfavourable influences militating against the spread of Huttonian principles: the universities were backward; the newly founded Geological Society was biased; the Wernerians were active; and the better textbooks were antagonistic or cautious.

BACKWARDNESS OF THE NATURAL SCIENCES AT ENGLISH UNIVERSITIES

About this time, Cambridge and Oxford, the two English seats of learning, were wallowing in a medieval atmosphere of monastic contemplation and lethargy. They were woefully inadequate in the natural sciences, some of which were not even recognized and in none of which did a degree-course exist. The main science courses consisted of mathematics backed up by a smattering of philosophy. The classical tripos at Cambridge was not established until 1822 and triposes in theology, history, law and natural sciences did not come into being until much later. In addition, the courses of studies were restricted because the various college statutes imposed religious tests before candidates could claim their academic honours. The following gloomy picture is drawn of the contemporary social life at Cambridge by Sedgwick's biographers:

> . . . it must be remembered in the first place that foreign travel was impossible, that communication with other parts of England was slow and costly, and that therefore journeys were seldom undertaken. Many Fellows made Cambridge their home, which they rarely left, and died, as they had lived, in their college rooms. Newspapers – such as they were – travelled as slowly as individuals, and the arrival of a letter was a rare event. . . . (*Clark and Hughes, 1890, I, p. 105*)
>
> Again, the refining influence of ladies' society was almost wholly absent. . . . Nor was it the custom for Fellows of Colleges to see anything of the undergraduates. It was rather the fashion to ignore their existence. . . .
>
> Few Fellows of Trinity College, except the officials, had any definite occupation. With two exceptions, they were bound to take Holy Orders. Some held small livings in Cambridgeshire, tenable with their fellowships; others, who had been appointed to the office of College Preacher, held more lucrative pieces of preferment. But in neither case was residence compulsory. . . . Men of ambition went out into the world and boldly courted fortune, as soon as they had obtained their fellowships – some without even waiting for that assistance. Those who despaired of success, or had no energy to strive after it, remained behind. (*Clark and Hughes, 1890, I, pp. 106-7*)

Sedgwick himself, when discussing his reasons for applying for a Professorship, confirms many of these obvious failings:

> If I succeed I shall have a motive for active exertion in a way which will promote my intellectual improvement, and I hope make me a happy useful member of society. . . . I am not such a fool as to suppose that my present employment is useless; and my pecuniary prospects are certainly better than they would be if I were Woodwardian Professor. Still, as far as the improvement of the mind is considered, I am at this moment doing nothing. Nay I often very seriously think that I am doing worse than nothing; that I am gradually losing that little information I once had, and very sensibly approximating to that state of fatuity to which we must all come if we remain here long enough. (*Clark and Hughes, 1890, I, pp. 153-4*)

Before Sedgwick's appointment as Woodwardian Professor in 1818 geology was not taught systematically at Cambridge. Moreover, it must be remembered that he himself was professionally innocent of geology at the time of his appointment, as his Fellowship had been gained in mathematics. It is assumed he must have learnt his own geology from his endeavours in the field after his appointment, aided by guidance received by attending the lectures delivered by Professor E. D. Clarke on mineralogy. Though this professorship had existed for some decades and was created with the avowed object of propagating geological knowledge, it does not seem that this was followed. When Woodward in his will had created the lectureship, he had inserted sound provisions to ensure geological teaching but unfortunately they were not enforced by the trustees.

> The Lecturer is to be a bachelor; . . . a layman is to be preferred to a divine; . . . he is not to hold any preferment, office, or post, whatsoever, that shall any ways so employ and take up his time as to interfere with his duty herein set forth, and in particular that shall require his attendance out of the University; . . . he is not to be absent from Cambridge for more than two months in the year; . . . he is there to read at least four Lectures every year . . . on some one or other of the subjects treated of in my Natural History of the Earth, my Defence of it against Dr Camerarius, my Discourse of Vegetation, or my State of Physick, at his discretion; . . . the said Lectures, or at least one of them, at the Lecturer's own free choice and election, to be published in print every year. (*Clark and Hughes, 1890, I, pp. 182–3*)

Even John Michell and John Hailstone two of the more illustrious previous holders of the office do not appear to have delivered any lectures, although the former had some geologic talent. At other times the post became a mere title for whoever won the favour of the trustees. The Reverend Ogden, whose 'uncivilized appearance and bluntness of demeanour were the great obstacles to his elevation in the Church', was an outstanding example.

> The Lectureship being vacant for a fourth time, Colonel King, now a very old man, appointed the Rev. Samuel Ogden, D.D., Fellow and President of St John's College. Dr Ogden had been master of the grammar-school at Halifax from 1743 to 1753, when he returned to Cambridge, where he resided until his death, 22nd March, 1778. He held the livings of Stansfield in Suffolk and of Lawford in Essex, and was vicar of St Sepulchre's, Cambridge, from March, 1759, to May, 1777. (*Clark and Hughes, 1890, I, p. 193*)
> Ogden had a turn for writing verse, and his name appears in three of those volumes which, in the seventeenth and eighteenth centuries, the University used to address to the sovereign on important occasions. In 1760 he mourned the death of George the Second in Latin elegiacs; in 1761 he hailed the marriage of George the Third in English stanzas; and, in the following year, the birth of George Prince of Wales in Arabic. . . . It is almost needless to add that the Woodwardian Lectureship was a sinecure during the fourteen years that it was held by Mr Ogden. For two or three years before his death he was 'much broken with Gout and other Complaints'. (*Clark and Hughes, 1890, I, pp. 193–4*)

At Oxford the situation was not much better. There, Sir Christopher Pegge lectured on anatomy and Dr John Kidd lectured first on chemistry and then on mineralogy. Round the latter formed the Oxford group of Buckland, Conybeare, Fitton and Greenough, famous later as prominent members of the Geological Society. Even so, geology was not taught as a main subject in the same way that it is today. It was only in 1819 as a result of Buckland's example that a readership was created in geology and a Treasury stipend of £100 granted. With this picture in mind it is easy to understand why the Huttonian theory made little impact on the Universities. What geological science was taught concentrated on mineralogy and naturally tended to favour an adherence to Werner's theories. Although, on the other hand, in 1825 only two scholars at Oxford were said to know German (Liddon, 1895, p. 72). It was fortunate that the backwardness of the Universities did not dampen the ardour of the non-academic geologists.

ENTHUSIASTIC AMATEURS AND THE LONDON GEOLOGICAL SOCIETY

The collecting of minerals and fossils remained exceedingly popular in the period 1780–1820. It was fashionable for gentlemen to display their specimens to their guests and to use such objects in much the same way as we use photographs and travel films. Some collections reached extraordinary proportions. The natural history collections of Peter Pallas were bought by Empress Catherine II of Russia for 20,000 roubles (5,000 more than he asked for them) who allowed him to keep them for life and gave him a large estate in the Crimea as well as a considerable sum for its upkeep. We do not hear of British patronage on this scale but it is certain that a common interest in rocks led to the foundation of at least one surviving 'society'. The London Geological Society arose when a few more enthusiastic members of a Club that had been gathering informally banded together at a meeting at the Freemason's Tavern on the 13th November 1807 (Woodward, 1907).

The thirteen original and other early members varied widely in character and calling, although financial independence or sufficiency and a superior social position were probably common characteristics. Not one of them had started life as a geologist. Conybeare, Buckland, and Sedgwick were ministers; Fitton, Wollaston, and Mantell were medicals; De la Beche and Murchison were retired soldiers; William Phillips was a printer and bookseller; and Greenough, who became the Society's first president, was a wealthy M.P. with a large private collection of geological specimens (Woodward, 1907 and 1911).

From the start the Geological Society, as its first *Transactions* (1811) avers, aimed to set itself above narrow theoretical conflict and planned as a guiding principle to study geological phenomena with scientific exactitude and impartiality and as a subsidiary task to lay 'the foundation of a general geological map of the British territory'.

It is ironic that although the Society included many talented persons, it was dominated at least in its early years by a majority who by their training were likely to favour Werner's mineralogical propositions. Greenough had actually been a student of Werner and was, according to the pen-portrait below, not likely to favour innovations.

> Amiable, yet shy, and somewhat hesitating in manner, full of all kinds of miscellaneous knowledge, obstinately sceptical of new opinions, a kind of staunch geological Tory, and playing the part of objector general at the evening discussions. (*Geikie, 1875, p. 114*)

As it transpired this pro-Wernerian attitude of the Geological Society persisted until about 1830 when the first real changes of allegiance began to appear. The dominance of Neptunist opinion may have prompted Dr Mitchell's strictures on the monopoly of publication within the Society by the older members. However, we must point out that the members were much more interested in the compilation of geological maps than in landscape evolution.

THE WERNERIAN NATURAL HISTORY SOCIETY OF EDINBURGH

While the English remained indifferent to Hutton and Playfair, the Scots afforded them sporadic recognition, mostly antagonistic. In 1804 the mineralogist Robert Jameson (1774–1854) became Professor of Natural History at Edinburgh. He had just spent two years at Freiberg and his head was still 'positively humming' with Wernerian assertions. The city at once became the centre of a violent geological controversy as Benjamin Silliman discovered

FIG. 13. *Robert Jameson*
(*From a miniature*)

during his visit in 1805–6 (Fulton and Thomson, 1947). With Playfair the Professor of Mathematics and after 1805 the Professor of Natural Philosophy at the same university, the battle became almost hand-to-hand and reached an extraordinary ferocity as may be judged from an ill-fated theatrical performance:

> A minor and more whimsical cause of disruption in the Edinburgh society was the controversy going on between the Huttonians and Wernerians, as they were then called, the respective advocates of fire and water, as agents concerned in moulding the crust of the earth. The basaltic and other rocks clustering round Edinburgh furnished ample local materials for discussion. No compromise of combined or successive agency, such as reason might suggest, was admitted into this scientific dispute, which grew angry enough to show itself even within the walls of a theatre. A play written by an ardent Huttonian, though graced with a prologue of Walter Scott, and an epilogue by Mackenzie . . . , was condemned the first night, as many persons alleged, by a packed house of the Neptunian school. (*Holland, 1872, p. 81*)

In this theatrical connexion it is interesting that Goethe, who was engaged at this time on writing the second part of his *Faust*, saw fit to include a dialogue between a Vulcanist and a Neptunist. He did, however, make the error of associating the Huttonian vulcanist with catastrophism, whereas the Wernerian neptunist holds to the slow operation of natural forces!

ANAXAGORAS	Your mind is stiff and will not bow.
	What further argument is needed now?
THALES	The wave will bow to all the winds that play
	But from the rugged cliff it holds away.
ANAXAGORAS	It was volcanic gas produced this cliff
THALES	In moisture is the genesis of life.

. . . .

ANAXAGORAS	Have you, O Thales, ever in one night
	Produced from slime a mountain of such height?
THALES	Never were nature and her living floods
	Confined to day and night and periods.
	Each form she fashions with due providence,
	Even in great things there's no violence.
ANAXAGORAS	But here there was! A fierce Plutonic fire,
	Explosive gases of Aeolia, dire,
	Burst through that ancient crust of level ground
	That a new mountain might at once be found.

(*MacNeice, 1951, pp. 219–20*)

Jameson, however, was less confused than Goethe. He was excessively biased against 'fire philosophers' as he called in ridicule all Huttonians. We have already noticed that Hutton's one real mistake was his heat mechanism for the consolidation and uplift of strata. Jameson, with his great mineral knowledge, was not slow to attack this weakness.

He had one great advantage over his opponents. Accurate mineralogical knowledge enabled him to discriminate rocks with a precision to which they could make no pretension, and although this was an accomplishment of little real moment in the theoretical questions chiefly in dispute, he did not fail to make the most of it, nor they to betray their consciousness of their inferiority in that respect. (*Geikie, 1875, p. 108*)

His desire to crush Huttonian heresies goaded him into constant activity and two months after the formation of the London Geological Society he brought into being the Wernerian Natural History Society of Edinburgh. For three decades this Wernerian Society maintained a strong counter-offensive against Hutton and Playfair. Its influence should not be belittled because after 1813 not a single Fellow of the Royal Society and only two university professors were included among its new members. Nor because the Society's total output of publications from 1811 to 1839 amounted only to eight *Memoirs*, later described as 'dreary geognostical communications' (*Geikie, 1905, p. 327*). Jameson alone wrote enough for a whole society. In 1808 his *Elements of Geognosy*, the third volume of his *System of Mineralogy*, formed a violent attack on Huttonians and an unqualified re-assertion of the principal Wernerian notions.

> We should form a very false conception of the Wernerian Geognosy, were we to believe it to have any resemblance to those monstrosities known under the name of Theories of the Earth. . . . Armed with all the facts and inferences contained in these visionary fabrics, what account would we be able to give of the mineralogy of a country, if required of us, or of the general relations of the great masses of which the globe is composed? (*Jameson, 1808, p. 42*)

His complete faith in Werner's ideas is exemplified by the following passage where he endeavours to counter the criticism that the universal ocean had not been satisfactorily disposed of:

> Although we cannot give a very satisfactory answer to this question, it is evident that the theory of diminution of water remains equally probable. We may be fully convinced of its truth, and are so, although we may not be able to explain it. To know from observation that a great phenomenon took place, is a very different thing from ascertaining how it happened. (*Jameson, 1808, pp. 82–3*)

We ought to repeat here that in this controversy the attacks largely originated from the Neptunist entrenchments and that Hutton's original theory was not put forward with an antagonistic motive.

> The doctrines of the Wernerian School were unknown to Dr Hutton, nor were they introduced effectively into England till the publication of Professor Jameson's *Elements of Geognosy*, in 1808. (*Fitton, 1839, p. 455*)

Only when attacked by Kirwan and De Luc did Hutton and Playfair reply in kind. Yet Jameson continued to make bloody sorties and seemed to gain vigour from the opposition's passiveness. His final onslaught was the translation in 1815 of Cuvier's *Essay on the Theory of the Earth*. Georges Cuvier (1769–1832) was a student of Werner and primarily a zoologist who from a study of fossils recognized among them a progression, though not by a continual 'evolutionary' process. He found that each rock formation appeared to possess an assemblage of fossil species different in type from those in the formations above or below. However, unlike Darwin, he did not visualize any intermediate links or evolutionary progression, for the evidence of his

FIG. 14. *Georges Cuvier*

eyes disclosed no such links and each species seemed to come to an end with the particular rock formation to which it belonged, to be replaced later in time by an entirely different set of species peculiar to the next formation. This observation of the apparent and sudden extinguishings of life made him enlarge the Flood theory and postulate a series of catastrophes in the form of successive uplifts and depressions of the crust which caused fluctuations in sea level:

> We are therefore forcibly led to believe not only that the sea has at one period or another covered all our plains, but that it must have remained

there for a long time, in a state of tranquility; which circumstance was necessary for the formation of deposits so extensive, so thick, in part so solid, and containing exuviae so perfectly preserved. (*Cuvier, 1815, p. 8*)

Finally, if we examine with greater care these remains of organized bodies, we shall discover, in the midst even of the most ancient secondary strata, other strata that are crowded with animal or vegetable productions, which belong to the land and to fresh water; and amongst the more recent strata, that is, the strata which are nearest the surface, there are some of them in which land animals are buried under heaps of marine productions. Thus the various catastrophes of our planet have not only caused the different parts of our continent to rise by degrees from the basin of the sea, but it has also frequently happened, that lands which have been laid dry have been again covered by the water. . . . (*1815, p. 14*)

These repeated irruptions and retreats of the sea have neither been slow nor gradual; most of the catastrophes which have occasioned them have been sudden. . . . (*1815, p. 15*)

The breaking to pieces and overturnings of the strata, which happened in former catastrophes, show plainly enough that they were sudden and violent like the last; and the heaps of debris and rounded pebbles which are found in various places among the solid strata, demonstrate the vast force of the motions excited in the mass of waters by these overturnings. (*1815, p. 16*)

It is easy to see why Jameson thought Cuvier's work important for its main principle was completely Wernerian; furthermore, Cuvier was one of the grand figures of continental learning and his support of any theory inevitably enhanced its prestige.

Robert Jameson was never a man of half-measures. Long afterwards at an evening meeting of the Royal Society, he acknowledged that Wernerism was doomed to a deserved death. But this frank admission of his conversion to the views of Hutton came far too late to have any influence on the historic progress of geologic thought.

SIGNS OF HUTTONIAN SURVIVAL

A more objective analysis of the influence of Hutton's theory is obtained from a brief examination of the geologic textbooks which appeared during the period. Their contents reveal little progress and the usual omission of Hutton's name is damningly significant.

At least two of them are worthy of discussion. First, Robert Bakewell's *An Introduction to Geology* (1813), which is important as being one of the first textbooks to attack Wernerism, 'the dark lantern of German geognosy'.

It is scarcely possible for the human mind to invent a system more repugnant to existing facts. (*1813, p. 217*)

Nothing but the obscure language in which this doctrine of Werner was advanced, could have prevented its absurdity from being instantly perceived and acknowledged. (*1813, p. 218*)

Yet, though Bakewell derided Werner, this did not mean that he fully grasped the principles of Hutton's doctrines:

> The disintegration of rocks and mountains is constantly taking place by the incessant operation of the elements; all bare and lofty cliffs and eminences are gradually wearing down; and the process will go on, until they are covered with soil and vegetation, which protect them from further decay. Beside the causes which at present operate to reduce the most exposed and prominent parts of the earth's surface, and transport their materials into plains, or to the sea shore, there are evident indications of the destructive effects of ancient inundations, which have swept over the surface of the present continents, have excavated new valleys, torn off the summits of the loftiest mountains, and spread their ruins in immense fragments over distant regions. (*1813, pp. 421–2.*)

He vaguely refers the formation of valleys to rivers which flow through them and to marine currents; thus, in spite of his protestations, his geomorphic views generally are Wernerian rather than Huttonian.

Second, we will deal with *Outlines of Geology of England and Wales* (1822), an influential textbook written by William Daniel Conybeare and William Phillips. In it the views of Hutton are disclaimed and a catastrophic standpoint taken on the formation of valleys – although the authors also specifically deride Wernerian theories. These two writers well represent the growing school of non-theoretical, down-to-earth geologists:

> We must begin, however, by remarking that many of the theories advanced on this subject (the origin of valleys) appear defective in two points. First because, ascribing every thing in the formation of vallies to the agency of running waters, they entirely overlook the effect which must have been produced by the violent convulsions which appear in many instances to have broken and elevated the strata, and must in so doing have necessarily formed a surface diversified by many and great inequalities: . . . although on these grounds we may refer the original formation of the vallies of such districts, in part at least, to the convulsions alluded to, yet there are the strongest proofs that even here also the vallies have subsequently been greatly modified by the rush of mighty currents of water through them; and in lower countries, where the horizontal and undisturbed position of the strata shew that other convulsions cannot very sensibly have affected the figure of the surface, we must refer its present inequalities almost exclusively to the excavating action of such currents.
>
> The second defect which calls for animadversion in some of the theories which in other respects have given the clearest views of the phaenomena under discussion, is that while they correctly ascribe the excavation of vallies to the agency of aqueous currents, they look to no other supply of that agency than the streams (often inconsiderable rills) which now flow through them, borrowing liberally from time what they confessedly want in force. (*1822, pp. xxii–xxiii*)
>
> . . . The continents, now dry land, were once covered with the ocean . . . (which) could never have drained off the surface of the lands it has deserted, without experiencing violent currents in its retreat. (*1822, p. xxiv*)

(Werner's) attempts at theorizing must now appear to all but his devoted adherents, among the most unsuccessful and unphilosophical ever made. (*1822, p. xiv*)

Generally speaking the Vulcanists or Plutonists gained most of the ground that the Neptunists lost, although a few of the popular textbooks were of indirect help to the Huttonian cause mainly because they derided Werner. Yet there was a small number of contemporary writers who either tolerated or, in rarer instances, actually praised the whole or part of Hutton's *Theory*.

Among these was the Rev. Joseph Townsend of Pewsey, who was partly responsible for discovering the geological genius of William Smith. In describing the country around Guadix, near Granada, Townsend admitted great stream erosion of an extensive upland surface;

> It is evident, that the formation of these hills is of recent date, and subsequent to the general revolution, which took place when the horizontal strata, for ages covered by the waters of the ocean, were lifted up to view, and became the habitable portion of our globe. At that period, the whole of this country appears to have been one extensive plain, but being composed of soft materials, and subject to violent and heavy rain, it was soon torn in every possible direction by gullies, which, in process of time, became deep ravines, till, the mouldering angles of high cliffs being washed away, the wide expanse was left covered with hills, whose pointed tops, as we may here observe, are all on the same level. (*Townsend, 1792, II, p. 219*)

Twenty years later Townsend still admits the power of river-work, although in the same volumes he strongly criticizes Hutton's uniformitarian views and continued to believe implicitly in the Mosaic time-scale.

> . . . in every part of the globe, the gravel near to a river shows, even to a considerable depth, the strata of the country through which all the tributary streams now pass. (*Townsend, 1813, I, p. 229*)

On the other hand, Sir James Hall supported Hutton's ideas on petrology but disagreed with him on erosional matters. He believed in cataclysmic torrents resulting from the violent uplift of continents from beneath the sea, and supported his views by drawing analogies with the huge tidal waves associated with the earthquakes at Lima and Lisbon in the eighteenth century.

In 1809, a printed argument between John Farey, consulting surveyor and geologist, and J. Carr on the relative merits of the theory of the Flood and of river erosion emphasizes the growing balance between anti- and pro-Huttons. Farey, the traditionalist, explains all landscape features by reference to supernatural forces. 'Outliers' had been isolated by 'denudation' in the form of a mighty catastrophe.

> . . . the intervening parts of the same stratum, once continuous, have been torn off from our globe. (*1809, p. 238*)

Likewise he attributes faults to the heaving up of the surface by the gravitational attraction of a dense satellite moving in close proximity to the earth.

Rev.ᵈ Joseph Townsend A.M.

from a Sketch by William Smith.

FIG. 15.

Carr on the other hand, though not mentioning Hutton or Playfair, was obviously acquainted with certain of their principles. He speaks of mountains being the more durable remnants of rocks, the remainder of which has been swept away thereby providing

masses of diversified material for the formation of other stratified countries in other situations. (*1809, p. 386*)

In speaking of water he comes out clearly as a disciple of Huttonian doctrines of river erosion.

> . . . to this powerful and incessant operator allow but a sufficiency of duration, and a suitable diversity of fluctuating circumstance, and he will have a bold and arduous task to perform who shall undertake to advocate its limitations in geological efficacy. (*1809, p. 388*)
>
> To me, there are few things more evident than that 'the irresistible forces' which have affected the excavation of valleys 'are no other than the identical streams which now flow through them'. (*1809, p. 452*)

He adds, in phrases worthy of Playfair, that the characteristic elements of 'uniform direction', 'general connection' and 'admirable subserviency of the whole'

> are so palpable, that we are irresistibly led to one of two conclusions, – either that the several ranges of valleys had been purposely and specifically formed for the streams which now flow through them, or that the streams themselves have scooped out their own peculiar valleys. The former opinion is too absurd to merit a moment's attention. (*1809, p. 453*)

We may also add in favour of Hutton the comments of John Kidd (1775–1851), a physician and professor of chemistry at Oxford, who was extremely tolerant although a cleric. Kidd had already written a book on mineralogy and given lectures in that subject and geology when he published an essay on the *Imperfect Evidence in support of a Theory of the Earth*. He advises against mixing science and geology and admits that his own field-observations show that valleys, or at least 'transversal vallies', can be cut by streams.

> . . . it is preposterous to suppose that that high degree of moral evidence on which the credibility of Scripture rests can with any justice be weakened by our interpretation of phenomena, the connection of which among themselves even, we certainly are at present, and probably ever shall be, incapable of explaining. (*1815, pp. 13–14*)
>
> . . . he (the reader) will at once be compelled to allow that rivers are among the most extensive and powerful of those causes by which the surface of the globe is altered; and will be inclined to admit the probability of Dr Hutton's proposition, that the materials of even the natural strata of this earth have been originally prepared by similar agents. (*1815, pp. 182–3*)
>
> . . . though not disposed to agree with Dr Hutton in the opinion, that the main or longitudinal vallies of an alpine district have been excavated by the rivers which flow through them, I think it cannot be denied that most of the transversal vallies are in a great measure the work of their several streams. (*1815, p. 192*)

Although in Britain these signs of survival seem sporadic and flickering they were lasting and brilliant compared with Huttonian progress on the Continent where no islands of hope rose above Werner's universal ocean. Playfair's *Illustrations* were translated into French in 1815 but Wernerian notions remained supreme till at least 1820. Amidst all the Wernerian thundering the French Hutton had literally gone unheard. Jean Baptiste Lamarck had in

1802 produced his *Hydrogéologie*, in which, apparently without acquaintance with the works of Desmarest, De Saussure and Hutton, he had reached some remarkably Huttonian conclusions.

> . . . every mountain which has not been erupted by volcanic action or some other local catastrophe, has been cut out of a plain, . . . so that the mountain-summits represent the relics of the ancient plain, save in so far as its level has been lowered in the general degradation. (*Lamarck, 1802, p. 14; Packard, 1901, pp. 101–2*)
>
> Because of this movement of freshwater, the plains by slow degrees lose their unity and levelness, gullies form and deepen and are converted into large valleys; the slopes accompanying and enclosing river basins are changed into elevated crests which are then divided into separate blocks which, in their turn, are cut and sharpened into mountains. (*Lamarck, 1802, p. 11*)
>
> But though he was mistaken in supposing that the doctrine was novel, Lamarck was not surpassed by any of his predecessors in the firmness with which he expressed it, and in the clearness with which he presented it to the world. Realizing so fully as he did the magnitude of the scale on which the surface of the land undergoes disintegration, he could see no difficulty in conceiving that the mere passage of running water over the surface must inevitably lead to the erosion of a system of drainage-lines from the mountain-crests down to the margin of the sea. (*Geikie, 1906, pp. 193–4*)

There can be no doubt that Lamarck was strongly opposed to any theory of catastrophes on a general scale although he was convinced that the oceans experienced a gradual displacement and successive transgression over the various points of the surface of the globe. The following translations barely do credit to his stimulating arguments and style.

> If in numerous mountains, we notice beds inclined at various angles instead of being horizontal, these inclinations generally prove that the localities once formed part of a shore that was always sloping toward the sea. Nonetheless, often they are due to local subsidence. (*1802, p. 22*)
>
> On our globe everything inevitably suffers continual changes . . . which work at varying rates, according to the nature, state or situation of objects. . . . For Nature, time is no problem and is an ever-available limitless resource with which she can achieve the greatest as well as the smallest of tasks. (*1802, p. 67*)
>
> Oh! How great is the age of the terrestrial globe! And how small are the ideas of those who attribute to it an existence of six thousand and several hundred years from its origin to our times! (*1802, p. 88*)

Such non-catastrophic notions courted disaster on continental Europe before the days of Waterloo.

THE CLOSE OF AN EPOCH

This indecisive period of argument and counter-argument virtually ends with the death of the main controversial figures. Werner died in 1817 and Playfair in 1819 and with their departure geological thought entered on a new course not devoid of conflict but rather more concerned with newly defined issues. The period to come was generally characterized by a marked concentration on

FIG. 16. *William Smith at the age of 69*

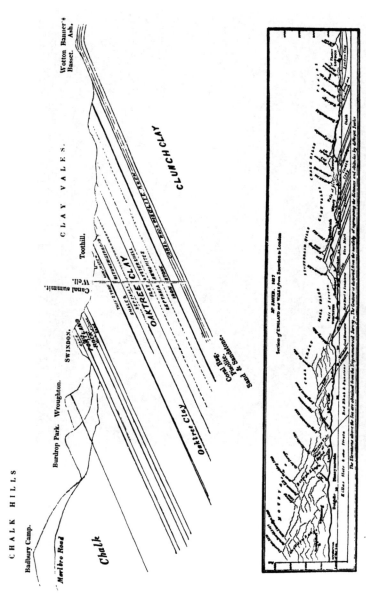

FIG. 17. *Geological cross-sections by William Smith, showing the relation of lithology and relief*
Above: *Section of Jurassic and Cretaceous rocks (Oxford (Clunch) Clay to Upper Chalk) near Swindon, Wiltshire.*
Below: *Section from Snowdon to London, published in 1817.*

stratigraphy and palaeontology. Every geologist of note interested himself in the findings of both of those subjects and the teachings of Hutton and Playfair slipped into the background. The publication of William Smith's *Map of the Strata of England and Wales* in 1815 invigorated the Geological Society's original intention to produce a stratigraphical map. Everywhere the rapid development of stratigraphy and palaeontology led to a neglect of what the Victorians called 'physiographical geology, or the study of the origin of the present external features of the land' (*Geikie, 1875, I, p. 110*).

The characteristics of a young geologist in the 1820's were said to be:

1. He would almost certainly, with some reservations, be a Huttonian.
2. He would depart from Hutton by avoiding theoretical questions and would concern himself with the observation and accumulation of data.
3. He would show little taste for experimental research (despite the recent work of Sir James Hall) and not much regard for the contributions of physicists to geology.
4. He would be ignorant of mineralogy.
5. He would neglect the external features of the earth and entertain erroneous views regarding their formation.
6. His main interest would be in the stratigraphical succession in England and the associated fossils (*Geikie, 1875, I, pp. 112–13*).

Thus, from the point of view of the study of landscape evolution, there was no call for undue pessimism or optimism. The Huttonians had managed to survive even if in a subsidiary role and for those who think the progress slow we can but suggest that it should be measured against the fate on the Continent of Lamarck and of the other brilliant hydrologists discussed in the following chapter.

CHAPTER EIGHT

The First Quantitative Work on Landscape Processes

> . . . Our reasons, which suffice
> Ourselves, be ours alone; our piece of gold
> Be, to the rustic, reason and to spare!
> We must translate our motives like our speech
> Into the lower phrase that suits the sense
> O' the limitedly apprehensive. Let
> Each level have its language! . . .
> (Robert Browning, *The Ring and the Book*, lines
> 1,495–1,501)

Landscape evolution results from a complicated series of physical processes which are, or at least should be, capable of scientific evaluation. Yet whereas most sciences originally adopted calculation and measurement as a natural form of discipline, their rigid application does not appear to have been generally introduced into geomorphology until comparatively modern times. The reasons for the belated growth of scientific facts on physical processes was partly due to distractions; first, the vertical superimposition of strata proved of more interest to geologists; and second, the horizontal distribution of rocks and of landscapes also attracted much attention. This is almost as true today as of the early nineteenth century.

Consequently, it seemed inevitable that details of landscape-forming processes should await the completion of the more general stratigraphical and distributional studies. But as geologists began more commonly to confine their researches to restricted areas, the possibility of defining and of studying small landscape features greatly increased. The modern knowledge of processes grew up with the knowledge of minor landforms. Thus in the nineteenth century Lyell's mention of oxbow lakes is a typical advance in landform recognition while Buckland's discussion of the weathering of limestone by acidulated water is a characteristic advance in the recognition of processes.

This method of working back from the result to the cause, or of putting the study of landscape first and of physical processes second, seemed inevitable as long as geologists controlled geomorphological thought. But there was a time, previous to about 1800, when advances in the knowledge of landscape processes exceeded those in geology proper. In fact, the mensurative and quantitative approach to landscape-formation pre-dates Hutton and Playfair.

During the seventeenth and eighteenth centuries the remarkable understanding of river-flow and river-work could well have led to the full acceptance of Huttonian fluvialism but this store of hydrological knowledge was either utterly ignored or quite overlooked by contemporary geologists. Hydraulic engineers might then have founded a tradition of truly scientific investigation into landscape development had not geologists been so determined that rock-structure and rock-composition were to assume the dominant role in their studies. The flow and work of rivers especially attracted the attention of Italian and French engineers who were concerned with the construction of fountains and of canals and with the retention or improvement of navigable channels in streams.

The father of the 'science of river mechanics' is generally held to be the Italian Domenico Guglielmini (1655–1710) who scientifically analysed river-flow and in 1697 stated the following fundamental propositions:

1. Streams erode or build up their beds until equilibrium is attained between force and resistance;
2. Slope therefore will vary inversely with velocity;
3. The less the resistance of the bed materials the less the slope of the channel;
4. Slope varies inversely as the normal discharge; and
5. A mobile bed is modified to give a longitudinal profile which is concave upwards (*Baulig, 1926*).

Guglielmini at first accepted Torricelli's incorrect theorem ($V = \sqrt{2gX}$; where X is the depth below the stream surface) which assumed that the greatest velocity was near the stream-bed but soon rectified this by experiment and clearly saw that the resistance of the bed strongly retarded the speed.

In France H. Pitot in 1728 also considered that bed-resistance greatly lowered the stream's velocity and in his notable *Principes d'Hydraulique* he dealt in detail with the intensity of water-friction and described a tube for measuring fluid velocity which allowed discharge to be estimated.

A well-known equation describing continuous fluid-flow conditions was published in 1738 by Daniel Bernoulli, a professor at Basle. In 1753, A. Brahms noted that the almost constant velocity of rivers in a downstream direction suggested that channel friction was equal to and opposed to the acceleration of gravity (Woodford, 1951). He further supposed that friction is proportional to the area of the stream cross-section divided by the wetted perimeter (later termed the HYDRAULIC RADIUS).

In 1771 there appeared the most valuable work as yet, the *Traité Élémentaire d'Hydrodynamique* of the Abbé Charles Bossut who conducted experiments on behalf of the French government. It dealt mainly with flow in pipes and open conduits and showed indisputably that rivers of the same depth, width, slope and bed-material had the same speed, but if one of the elements changed the speed varied in a uniform way.

Four years later Antoine de Chézy presented his famous formula for the velocity of pipe-flow:

$$Vm = C(RS)^{\frac{1}{2}}$$

(where Vm = mean velocity; C = Chézy's coefficient; R = hydraulic radius; S = slope of flow)

Chézy developed his ideas as the result of his work on the Regole de Cour-palet, the main feeder-channel of the Briare Canal linking the Loire and Seine river systems (Singer et al., 1957, p. 464). It was not until a century later that W. R. Kutter of Bern developed a refined method for obtaining Chézy's coefficient, by then recognized not to be a constant, in terms of the experimental coefficient of fluid friction (Ganguillet and Kutter, 1869).

To this wealth of eighteenth-century hydraulic research was added, in 1779, a quite remarkable work by Comte Louis Gabriel Du Buat, the *Principes d'Hydraulique, vérifiés par un grand nombre d'Expériencies faites par ordre du Gouvernement*. The author, a Lieutenant-Colonel in the Royal Corps of Engineers, set out to discover experimentally the factors controlling and in-fluencing the flow and windings of a river. His quantitative findings are so often quoted in mid-nineteenth-century works on hydrology that it seems strange that his general observations were apparently unknown to con-temporary geologists, especially as the book, although costly (the second edition of 1786 was priced at £12 unbound for the two volumes), went into a third edition, of three volumes, in 1816.

The introductory remarks are remarkably advanced and, if widely known, would surely have been a clarion call for Huttonian principles and a stepping-stone to Davisian cycles.

'A river, from its source to the sea, depicts the different ages of man. Its BEGINNING is a mere nothing; it rises from the earth but originates in heaven. Its INFANCY is frolicsome and capricious; it turns mills and eddies playfully beneath the flowers. Its YOUTH is impetuous and hasty; it buffets, uproots and overturns. Its MIDDLE COURSE is serious and wise; it makes detours and yields to circumstances. In OLD AGE its step is measured, peaceful, majestic and silent; its tranquil waters roll softly and soon lose themselves in the immense ocean.' (*Du Buat, 1786, I, pp. xxxii–xxxiii*)

Du Buat's idea of the original landscape is that of a flat continent with numerous interior depressions and lakes, like the Great Lakes, which fill and overflow into each other and so eventually into the sea. The overflows form deep valleys which by their incision give rise to springs, hitherto absent. The valleys become deepest near the sea, where the volume and erosive force of the rivers is greatest. Consequently,

The bed of rivers does not form a uniformly sloping plane but is the combination of several continuous inclined planes the gradients of which constantly decrease towards the sea. (*1786, I, p. 109*)

There is no doubt in Du Buat's mind that the valleys have been cut by the rivers, although he mentions the Deluge!

When we consider the width and depth of valleys, in the bottom of which all the world's rivers flow today, we see clearly that they have been excavated gradually by the running water, and that although their beds now have some degree of stability, it is due solely to the enormous work done by water flowing ceaselessly over a very long period, that the sacred historians teach us to be 4,000 years, that it is say since the general overthrow of the globe's surface by the deluge. Without adopting any particular plan, it is certain that the surface of highlands or areas raised above sea-level, changes continually and that the soil underfoot today is not that on which our ancestors trod. Rainwash and rivers . . . lower the hills, fill the valleys and expose the interior rocks of the mountains; and the low lands . . . will in their turn, but later, be destroyed and deposited in the ocean. The land then, reduced to a frightful level, will in future present merely an immense and uninhabitable marsh. . . . (*1786, I, pp. 105–6*)

Du Buat expresses various aspects of river-flow and -work by means of statistics and equations obtained from numerous experiments. He was especially interested in the *vitesse de régime* (stability; equilibrium; or grade) which is the state acquired by a current when it neither deepens its bed nor fills it up with deposits, but his practical mind took into account the action of floods.

A river has equilibrium (stability), or its speed is that of grade, when in times of greatest flood its rapidity is such that the tenacity of its bed is equal to the force and opposes the erosion not only of the bottom but also of the sides of the bed and the foot of its banks. If the velocity is too great then erosion and transport occur, and the speed is greater than that of equilibrium (*régime*). Thus by equilibrium we mean the relation between velocity of flow and re-sistance of bed-rock. The speed of equilibrium will vary with the nature of the bed, and the Rhône, with a bed of sizeable stones, ought to have a greater speed than that of the Meuse or Seine which flow over gravels. . . .(*1786, I, p. 110*)

On stability or equilibrium Du Buat came to the following conclusions.

The bed of a river has true stability only when it can discharge all the water of extraordinary floods with a velocity compatible with the tenacity of its bottom and sides; that as this tenacity is usually greater in the bed than in the banks the channel-section must eventually become much wider than it is deep; that the point of exact equilibrium (grade) begins at the mouth and migrates steadily upstream; that in this respect some rivers are more mature than others but grade will be established on all only after a very long period of time; that the more grade has progressed, the longer the floods take to arrive at the mouth, and *vice versa*; and, apart from high floods, the bed could survive with a speed less than that of grade without being filled up (aggraded) or narrowed appreciably, because the water is clear and does not carry coarse sediment; in any event sediment deposited there is carried away at each flood. (*1786, I, pp. 115–16*)

Du Buat made a special study of water-flow at bends.

> Everything in nature is subjected to the law of equilibrium; this law is the universal moderator of movement. It occurs in the determination of the speed of a river, in the establishment of its bed and the cross-section of its channel, and of the angle in which it departs from a straight line in forming a bend. If equilibrium is broken, the river strives to re-assert it. (*1786, I, p. 117*)

> PROPOSITION 86. For the bed of a river to achieve stability (grade) at the curvature of a bend it is necessary that,
> 1. the depth of water there is greater than elsewhere.
> 2. that the current of water in the middle of the bed, after striking the concave bank, must be reflected at the same angle into the middle of the bed below the bend.
> 3. that the angle of incidence must be proportionate to the tenacity of the terrain.
> 4. that there is at the bend an increase of slope or of flow capable of overcoming the resistance of the bend. (*1786, I, p. 121*)

One example given by Du Buat showed clearly that

> An actual slope, equal to 1 in 16,200 in reality produces only a speed due to a slope of 1 in 21,600 because of the resistance of the bends; which confirms my contention that the windings of rivers are a means used by nature to hasten the establishment of grade, in spite of the apparent excess of the gradient. (*1786, I, p. 150*)

Du Buat's table of the relation between sizes of solid materials and the velocity required to carry them was widely used later and is given on p. 438. His principle that the motive force of each particle of water in a river is entirely due to the surface slope and that the resistances are due to viscosity and friction with the bed was not improved upon until about 1800 when Coulomb added the idea that channel (or wetted perimeter) resistance to flow is proportional to the velocity at low speeds and to the square of the velocity at moderate and high speeds.

More than a century after the first appearance of Du Buat's treatise, many of his ideas were echoed by De La Noë in his *Les Formes du Terrain* (see pp. 627–34) but De La Noë, also a Lieutenant-Colonel of Engineers, now *au Service Géographique de L'Armée,* does not mention his illustrious predecessor. In the meanwhile an upheaval, as great as the French Revolution, had overthrown the progress of geomorphological thought; the DELUGE had come!

References: Part One

ADAMS, F. D. (1934) 'The Scottish School of Geology' *Science*, Vol. 80, pp. 365–8.

— (1938) 'The Birth and Development of the Geological Sciences' (page numbers refer to the Dover Edn of 1954).

AGRICOLA, G. (1546) 'De Ortu et Causis Subterraneorum' (Basle).

ANDREWS, J. (1797) 'Historical Atlas of England' (London).

ARDUINO, G. (1760) 'Saggio Fisico-Mineralogico di Lythogomia e Orognosia' Atti dell' Accademia delle Scienze di Siena, T.V.

BAILEY, E. B. (1927) 'James Hutton, father of modern geology' *Edin. Geol. Soc. Trans.*, Vol. 12, pp. 183–6.

— (1950) 'James Hutton, founder of modern geology' *Roy. Soc. Edin. Proc.*, Section B, Vol. 63, Part 4, pp. 357–68.

BAKEWELL, R. (1813) 'Introduction to Geology' (Harding, London).

BAULIG, H. (1926) 'La notion de profil d'équilibre; histoire et critique' *Congrès Internat. Géog.* (1925), *C.R.*, Vol. 3, pp. 51–63.

BECK, C. R. (1918) 'Abraham Gottlob Werner, . . .' (Berlin).

BERNOULLI, D. (1738) 'Hydrodynamica' (Strasbug).

BESSON, J. (1596) 'L'Art et Science de trouver les eaux et fontaines' (Orleans).

BOSSUT, C. (1775) 'Traité Élémentaire d'Hydrodynamique' 2 vols. (Paris).

BRAHMS, A. (1753) 'Anfangsgründe der Deich- und Wasserbaukunst' (Aurich).

BUFFON, Comte de (1749) 'The Natural History of Animals, Vegetables, and Minerals, with the Theory of the Earth in General', translated in 1776 (London).

— (1785) 'Natural History, General and Particular', Vol. 1 (London), 2nd Edn. Trans. by W. Smellie.

BURNET, T. (1681–89) 'Telluris Theoria Sacra' or 'The Sacred Theory of the Earth' (London).

CATCOTT, A. (1768) 'Treatise on the Deluge', 2nd Edn.

CHALLINOR, J. (1954) 'The early progress of British Geology, III; From Hutton to Playfair' *Annals of Science*, Vol. 10, pp. 107–48.

CHENEVIX, R. (1810) 'Reflections on some mineralogical systems' *Phil. Mag.*, Vol. 36, pp. 286–303 and 378–391 *et seq.*

CHÉZY, ANTOINE DE (1775) Manuscript report on the Canal de l'Yvette, 'Mémoire sur la vitesse de l'eau conduite dans une regole' (reprinted in *Annals des Ponts et Chaussées*, Vol. 60, 1921).

CLARK, J. W. and HUGHES, T. M. (1890) 'The Life and Letters of the Rev. Adam Sedgwick', 2 vols. (Cambridge).

CONYBEARE, W. D. and PHILLIPS, W. (1822) 'Outlines of the Geology of England and Wales' (London), 470 pp.

CUVIER, G. (1815) 'Essay on the Theory of the Earth' (Trans. by R. Jameson), (Edinburgh), 2nd Edn, 322 pp.

DARWIN, F. (1887) 'The Life and Letters of Charles Darwin', 3 vols. (London).

DE LUC, J. (1790) 'Letters to Dr Hutton on the Theory of the Earth' *Monthly Rev.*, enlarged, Vol. 2, pp. 206–27 and 582–601.

DE SAUSSURE, H. B. (1779–96) 'Voyages dans les Alpes', 4 vols. (Neuchâtel).

DESMAREST, N. (1806) 'Mémoire sur clef de 3 epoques de la nature par les produits des volcans' *Mém. de l'Institut*, Tome VI (first presented 1775).

DU BUAT, L. G. (1779 and 1786) 'Principes d'Hydraulique', 2 vols. (Paris); 3rd edn, 3 vols. (Paris), 1816.

EYLES, V. A. (1950) 'Note on the original publication of Hutton's Theory of the Earth' *Roy. Soc. Edin. Proc.*, Section B, Vol. 63, Pt. 4, pp. 377–86.

FAREY, J. and CARR, J. (1809) 'Correspondence relating to the origin of valleys' *Phil. Mag.*, Vols. 33 and 34.

FAREY, J. (1813) 'Notes and observations on the introduction and first three chapters of Mr. Robert Bakewell's "Introduction to Geology" ' *Phil. Mag.*, Vol. 42, pp. 246–61.

FENTON, C. L. and FENTON, M. A. (1952) 'Giants of Geology' (Doubleday, New York).

FERGUSON, A. (1805) 'Minutes of the Life and character of Dr Black' *Roy. Soc. Edin. Trans.*, Vol. 5., pp. 101–17.

FITTON, W. H. (1811) Review of Werner's 'New Theory of the formation of veins' *Edin. Rev.*, Vol. 18, pp. 80–97.

— (1839) Review of Lyell's 'Elements of Geology' *Edin. Rev.*, Vol. 69, pp. 406–66.

FRESHFIELD, D. W. (1920) 'The Life of Horace Bénédict De Saussure' (Arnold, London), 479 pp.

FREYBERG, B. (1955) 'Johann Gottlob Lehmann . . .' *Erlanger Forschungen*, Bd. I (Erlangen), 159 pp.

FÜCHSEL, G. C. (1762) 'Historia terrae et maris ex historia Thuringiae per montium descriptionen erecta' *Acta, Acad. Elect. moguntinae zu Erfuhrt*, Vol. 2, pp. 44–209.

FULTON, J. F. and THOMSON, E. H. (1947) 'Benjamin Silliman' (New York).

GANGUILLET, E. and KUTTER, W. R. (1869) 'A General Formula for the Uniform Flow of Water in Rivers' (Engl. trans. by R. Hering and J. C. Trautwine, London, 1889, 240 pp.).

GARBOE, A. (1958) 'The Earliest Geological Treatise (1667) by Nicolaus Steno' (London), 51 pp.

GEIKIE, A. (1875) 'Memoir of Sir R. Murchison', 2 vols. (London).

— (1895) 'Memoir of Sir A. C. Ramsay' (London), 397 pp.

— (1905) 'The Founders of Geology', 2nd Edn (London). Dover Edn 1963.

— (1906) 'Lamarck and Playfair; a geological retrospect of the year 1802' *Geol. Mag.*, N.S., 5 Dec., Vol. 3, pp. 145–53 and 193–202.

GILLISPIE, C. G. (1951) 'Genesis and Geology' (Harvard), 315 pp.

GUETTARD, J. E. (1752) 'Mémoire sur quelques montagnes de la France qui ont été des volcans' *Mém. de l'Academie Royale des Sciences*, pp. 27–59 (see also pp. 1–8).

— (1768–83) 'Mémoires sur différentes parties des Sciences et Arts' (Paris), 5 vols.

GUGLIELMINI, G. D. (1697) 'Della Natura de Fiumi, Trattato Fisico-Mathematica' (Bologna).

HALL, J. (1805) 'Account of a series of experiments, shewing the affects of compression in modifying the action of heat' *Trans. Roy. Soc. Edin.*, Vol. 6, pp. 71–185.

— (1812) 'On the revolutions of the earth's surface' *Trans. Roy. Soc. Edin.*, Vol. 7, pp. 139–67 and 169–212.

HOLLAND, H. (1872) 'Recollections of Past Life' (New York).

HUTTON, J. (1788) 'Theory of the Earth' *Roy. Soc. Edin. Trans.*, Vol. 1, pp. 209–304.

— (1794) 'An Investigation of the Principles of Knowledge, and of the Progress of Reason, from Sense to Science and Philosophy' (Edinburgh), 3 vols.

— (1795) 'Theory of the Earth', 2 vols. (Edinburgh).

— (1899) 'Theory of the Earth', Vol. 3 (London). Edited by A. Geikie.

JAMESON, R. (1808) 'Elements of Geognosy', Vol. 3 of 'System of Mineralogy', 368 pp.

KAY, J. (1838) 'A Series of Original Portraits', Vol. 1 (Black, Edinburgh).

KEFERSTEIN, C. (1840) 'Geschichte und Litteratur der Geognosie' (Halle).

KIDD, J. (1815) 'A Geological Essay on the Imperfect Evidence in support of a Theory of the Earth' (Oxford), 269 pp.

KIRWAN, R. (1793) 'Examination of the supposed igneous origin of stony substances' *Roy. Irish Acad. Trans.*, Vol. 5, pp. 51–81.

— (1799) 'Geological Essays' (London), 502 pp.

— (1802A) 'On the proofs of the Huttonian Theory of the Earth' *Roy. Irish Acad. Trans.*, Vol. 8, pp. 3–27.

— (1802B) 'An essay on the declivities of mountains' *Roy. Irish Acad. Trans.*, Vol. 8, pp. 35–52.

LA ROCQUE, A. (1957) 'The Admirable Discourses of Bernard Palissy' (Univ. of Illinois, Urbana), 264 pp.

LAMARCK, J. B. P. (1802) 'Hydrogéologie' (Paris), 268 pp.

LEHMANN, J. G. (1756) 'Versuch einer Geschichte von Flötzebürgen' (Berlin). (See Adams, 1938, pp. 374–8.)

LEONARDO DA VINCI 'Leonardo da Vinci's Note-books', edited by Edward McCurdy (London, 1906), 289 pp.

LIDDON, H. P. (1893) 'Life of Edward Bouverie Pusey', Vol. 1 (Longman's, London), 479 pp.

MACGREGOR, M. (1950) 'Life and Times of James Hutton' *Roy Soc. Edin. Proc.*, Section B, Vol. 63, Pt. 4, pp. 351–6.

MACNEICE, L. (1951) 'Translation of Goethe's Faust' (Faber, London), 306 pp.

MATHER, K. F. AND MASON, S. L. (1939) 'A Source Book in Geology' (McGraw-Hill, New York), 702 pp.

PACKARD, A. S. (1901) 'Lamarck, The Founder of Evolution: His Life and Works' (London), 451 pp.

PACKE, C. (1743) 'Convallium Descriptio' (Canterbury).

PALISSY, B. (1580) 'Discours Admirables, de la nature des eaux et fonteines, . . . des pierres, des terres . . .' (Paris), 361 pp.

PALLAS, P. S. (1777) 'Observations sur la formation des montagnes et les changements arrivés au globe, particulièrement de l'Empire Russe' (St Petersburg).

PEARSON, H. (1948) 'The Smith of Smiths' (Penguin), 379 pp.

PHILLIPS, W. (1816) 'Outlines of Mineralogy and Geology' (London).

— (1818) 'An Outline of the Geology of England and Wales' (London), 240 pp.

PITOT, H. (1728) 'Principes d'Hydraulique'.

PLAYFAIR, J. (1802) 'Illustrations of the Huttonian Theory of the Earth' (Edinburgh).

— (1805) 'Biographical account of the late Dr James Hutton' *Roy. Soc. Edin. Trans.*, Vol. 5, pp. 41–99.

RAY, J. (1692) 'Miscellaneous Discourses concerning the Dissolution and Changes of the World' (London); later edns as 'Three Physico-Theological Discourses,' 1693, 1713, 1721, 1732.

RICHARDSON, W. (1803) 'Inquiry into the consistency of Dr Hutton's Theory of the Earth' *Roy. Irish Acad. Trans.*, Vol. 9, pp. 429–87; Dublin, 1803, 4to.

SINGER, C. *et al.* (1957) 'A History of Technology', Vol. 3 (Oxford), 766 pp.

STENO, N. (1669) 'De solido intra solidum naturaliter contento dissertationis prodromus' (Florence).

STRACHEY, J. (1719) 'Observations on the strata in the Somersetshire Coal Fields' *Phil. Trans. Roy. Soc.*, No. 360, pp. 972–3.

SUESS, E. (1904–9) 'The Face of the Earth', trans. Sollas (Oxford), 4 vols.

TARGIONI-TOZZETTI, G. (1752) 'Realizioni d'alcuni viaggi fatti in diverse parti della Toscana', Vol. 3 (Florence). Translated in Mather and Mason, 1939, pp. 74–75.

THOMPSON, H. R. (1954) 'The Geographical and Geological Observations of Bernard Palissy the Potter' *Annals of Science*, Vol. 10, pp. 149–65.

TOMKEIEFF, S. I. (1946) 'The Theory of the Earth' *Procs. Geol. Assn.*, London, Vol. 57, pp. 322–8.

— (1950) 'James Hutton and the Philosophy of Geology' *Roy. Soc. Edin. Proc.*, Section B, Vol. 63, Pt. 4, pp. 387–400.

— (1962) 'Unconformity – An Historical Study' *Procs. Geol. Assn.*, London, Vol. 73, pp. 383–416.

TOWNSEND, J. (1792) 'A Journey Through Spain in the Years 1786 and 1787' (Dublin), 2 vols.

— (1813–15) 'The Character of Moses established for veracity as an Historian; Recording events from the Creation to the Deluge', 2 vols. (Bath).

USSHER, J. (1650) 'Annalium Pars Prior'; 1654, 'Annalium Pars Posterior'; trans. as 'The Annals of the World . . .', 1658.

VON ZITTEL, K. (1901) 'History of Geology and Palaeontology' (London), 562 pp.

WEGMANN, E. (1958) 'Das Erde Werner's und Hutton's *Geologie*', I, pp. 531–59.

WERNER, A. G. (1787) 'Kurze Klassifikation und Beschreibung der verschiedenen Gebirgsarten' (Dresden).

— (1791) 'Neue Theorie von der Entstehung der Gänge' (Freiberg).

WHITE, G. W. (1956) Facsimile of Playfair's 'Illustrations'; (Univ. Illinois Press), with introductory note.

WILLIAMS, H. S. (1904) 'Everyday Science', Vol. 3 (New York).

WOLF, A. (1950) 'A History of Science, Technology, and Philosophy in the 16th and 17th Centuries' (Allen and Unwin, London), 2nd Edn, 692 pp.

WOODFORD, A. O. (1951) 'Stream gradients and Monterey Sea Valley' *Bull. Geol. Soc. Amer.*, Vol. 62, pp. 799–852.

WOODWARD, H. B. (1907) 'History of the Geological Society of London' (London), 336 pp.

— (1911) 'History of Geology' (London).

PART TWO

The Age of Lyell
1820-45

FIG. 18. *Noah's Ark, as described in Genesis Chapter 7*
(Illustration published by Thomas Kelly in 1814)

FIG. 19. *The Deluge, as painted by Hoet*

ACCEPTED VERSION AND
CHRONOLOGY OF THE OLD TESTAMENT

GENESIS: *The Creation.* 4004 B.C.

In the beginning God created the heaven and the earth (*Verse One*). And the earth was without form and void; and darkness was upon the face of the deep· And the spirit of God moved upon the face of the waters (*Verse Two*).

1. And God said Let there be light and there was light . . . and God divided the light from the darkness, And God called the light day and the darkness he called night, And the evening and the morning were the first day.

2. And God said Let there be a firmament in the midst of the waters, and let it divide the waters from the waters, And God made the firmament, and divided the waters which were under the firmament from the waters which were above the firmament . . ., and God called the firmament Heaven, and the evening and the morning were the second day.

3. And God said Let the waters under the heaven be gathered together unto one place, and let the dry land appear . . . And God called the dry land earth, and the gathering together of the waters called he seas . . . And God said Let the earth bring forth grass, the herb yielding seed, and the fruit tree . . . And the evening and the morning were the third day.

4. And God said Let there be lights in the firmament of the heaven, to divide the day from the night; and let them be for signs, and for seasons, and for days, and years. And God made two great lights, the greater light to rule the day, and the lesser light to rule the night; he made the stars also . . . and the evening and the morning were the fourth day.

5. And God said Let the waters bring forth abundantly the moving creature that hath life and the fowl that may fly above the earth in the open firmament of heaven . . . And the evening and the morning were the fifth day.

6. And God said Let the earth bring forth the living creature after his kind, cattle, and creeping thing, and beast of the earth after his kind; . . . And God said Let us make man in our image . . . And the evening and the morning were the sixth day.

7. And on the seventh day God ended his work which he had made, and he rested. . . .

GENESIS: *The Deluge or Noah's Flood.* 2348 B.C.

In the six hundredth year of Noah's life, in the second month, the seventeenth day of the month, the same day were all the fountains of the great deep broken up, and the windows of heaven were opened. And the rain was upon the earth forty days and forty nights. . . . And the flood was forty days upon the earth . . . and the waters prevailed exceedingly upon the earth, and all the high hills, that were under the whole heaven, were covered, fifteen cubits upward did the waters prevail and the mountains were covered, . . . And the waters prevailed upon the earth an hundred and fifty days.

And God remembered Noah . . . and the rain from heaven was restrained . . . and the ark rested in the seventh month, on the seventeenth day of the month, upon the mountains of Ararat, and the waters decreased continually until the tenth month; in the tenth month, on the first day of the month, were the tops of the mountains seen. . . . And it came to pass, in the six hundredth and first year . . . in the second month, on the seven and twentieth day of the month, was the earth dried.

CHAPTER NINE

Dean Buckland and the Diluvialists

The beginning of the discrediting of Neptunist ideas on the deposition of all rocks beneath a universal ocean did not necessarily lead to a noticeable weakening in the popularity of catastrophism. Catastrophes took many major forms such as volcanic eruptions, subterranean upheavals, the gravitational effects of a passing comet, and the Deluge. But as only the Deluge was recorded at length in the Bible, its explanation of geological events was sanctioned and championed by the Church.

It happened that this traditional belief was eventually undermined by supporters of the catastrophic idea. Considering that the Church dated creation at 4004 B.C., for the world to assume its present shape in the brief interval allowed, the application of catastrophes was inevitable and yet fossil evidence pointed to slow processes and immense intervals of time. We have already noticed (pp. 15–16) how John Ray as early as 1692 stated that the evidence of fossils was strongly opposed to the idea of a Creation of 'six Natural Days'; and how Buffon, in order to escape the impossibility of having to conceive the world created in six days, imagined instead that the term 'six days' was a figurative expression which could be interpreted as vast periods of time (see pp. 13–15). The introduction of a long interval resolved many of the geological difficulties involved and soon the futile explanations of all fossil formation by the action of the Flood were being omitted altogether.

Many of the successors of Ray and Buffon also began to realise that the Bible is not a complete geologic record but is silent or ambiguous on many points. These gaps seemed to provide a means for the inclusion of many catastrophes which would bring the scientific explanations in closer accord with known facts.

In addition, however, there were superficially good scientific grounds for believing in the existence of the Flood. Over parts of both Europe and North America there are innumerable glacial features. These include the erratics or far-travelled boulders foreign to their present resting place; the large patches of clay and of ill-sorted gravel and sand, known collectively as 'drift'; and the parallel scratches or striations on hard rock surfaces. The presence of river terraces and of marine and lake beaches also seemed to indicate a recent radical drop in water-level. At this time, of course, the idea of glacial action had not even been considered and as the (glacial) deposits were obviously new and made up of a heterogeneous collection of materials, what more natural explanation could there be than that they were tokens of the passing of the Flood?

In addition, the smallness of many rivers wandering in valleys apparently far larger than their needs, valleys in some cases strewed with gravel deposits well above the level of the existing rivers, seemed to demonstrate the hopeless inadequacy of the rivers to accomplish the necessary erosion and pointed to the work of some faster and more powerful agent. The erratic boulders could not have been moved by the rivers, for in some cases it was determined, by tracing the path of the boulders from their place of origin, that this would have meant pushing them hundreds of feet uphill. A gigantic deluge seemed the only possibility that could explain all these phenomena. The Flood, with giant waves and tremendous currents, covered the whole earth, leaving at one stage only the peaks and the mountain ridges exposed; the mighty rush of waters moved boulders and these in their quick passage over the sea-floor left parallel scratches; the swirling waters tore off rocks and ground them up and strewed them about as gravel deposits; the waters subsided suddenly, leaving exposed their beaches and terraces; the rush of the retreating torrents down the mountain slopes carved out large valleys, many of them now dry. Such an explanation suited scientist and theologian alike. Its mutual advantages were marshalled by Dr William Buckland, who revived in modern guise the earlier school of diluvialists.

THE REVEREND DR WILLIAM BUCKLAND

Buckland was born at Trusham in Devon on the twelfth of March 1784, the son of a clergyman (Gordon, 1894). The area immediately around his birth-place is rich in organic remains, and his energies received an early direction towards fossil collecting as a result of his father taking him hunting for shells. His talents and powers of observation were evident even in those early days and the elder Buckland was persuaded by friends to send his son to Win-chester. In 1801 he passed on to Corpus Christi, Oxford, where he took his B.A. and began to show a definite interest in geology and went on geologic expeditions with the Rev. Townsend. He also attended Dr Kidd's lectures on mineralogy and those of Sir Christopher Pegge on anatomy. His first major geologic tour, along the chalk hills of Berkshire, Wiltshire, and Dorset to the Isle of Purbeck, was made on horseback, a mode of transport which he in-variably adopted when accompanying his students and which was later to call forth the derision of Lyell who believed in the need for a slower approach to geology! By 1809, when Buckland was elected a Fellow of his college and ordained, he had clearly determined on studying geology. He explored the granite of Dartmoor in that year, toured central and northern England in the next, and made a similar expedition to Ireland in 1813 in the company of W. Conybeare. In the same year, on Doctor Kidd's retirement, he succeeded his old teacher as Reader in Mineralogy and entered upon his new duties with

FIG. 20. *William Buckland*
(*National Portrait Gallery*)

vigour and enthusiasm, even if his performance of them was tinged with some charming eccentricities. Unlike his predecessor, Buckland did not limit his lectures to mineralogy but was fond of launching out into broad geological themes. The following poem and Murchison's description provide private glimpses of his apartments, his character and devotion to his studies:

> Here see the wrecks of beasts and fishes,
> With broken saucers, cups and dishes;
> The prae-Adamic system jumbled,
> With sub-lapsarian breccia tumbled,
> And post-Noachian bears and flounders
> With heads of crocodiles and founders;
> Skins wanting bones, bones wanting skins,
> And various blocks to break your shins:

No place is this for cutting capers
'Midst jumbled stones and books and papers,
Stuffed birds, portfolios, packing-cases,
And founders fallen upon their faces.
He'll see upon the only chair
The great Professor's funeral fare,
And over all behold illatum
Of dust and superficial stratum.
The sage amidst the chaos stands,
Contemplative, with laden hands, –
This grasping tight his bread-and-butter,
And that a flint, whilst he doth utter
Strange sentences that seem to say,
'I see it all as clear as day'.

(Gordon, 1894, p. 9.)

On repairing, from the Star Inn to Buckland's domicile, I never can forget the scene that awaited me. Having, by direction of the janitor, climbed up a narrow staircase, I entered a long, corridor-like room, which was filled with rocks, shells, and bones in dire confusion, and in a sort of sanctum at the end was my friend in his black gown, looking like a necromancer, sitting on one rickety chair covered with some fossils, and clearing out a fossil bone from the matrix. (*Related by Murchison in Gordon, 1894, p. 10*)

By 1814 we are able to detect Buckland's future plans when he spoke of setting Moses against Hutton. Like many other geologists he took the opportunity at the end of the continental war to travel through Germany, travelling with Greenough and J. Conybeare, and paying a visit of homage to Werner on the way. In the same year he also managed to see parts of Austria, Poland and Italy. By 1819 his lectures had aroused such general interest both within and outside the university that a Readership was created in geology and by order of the Prince Regent an annual course of lectures was commanded, to be assisted by the grant of a Treasury stipend. This commendation immediately marked Buckland in the public eye as a man of exceptional merit.

Like Werner before him, Buckland delighted his listeners by his glowing enthusiasm and graphic powers of description. J. H. (afterwards Cardinal) Newman attended his lectures in 1821 and wrote of them:

To tell the truth, the science is so in its infancy that no regular system is formed. Hence the lectures are rather an enumeration of facts from which probabilities are deduced, than a consistent and luminous theory of certainties, illustrated by occasional examples. It is, however, most entertaining, and opens an amazing field to imagination and to poetry. (*Quoted in Faber, 1954, p. 71*)

At times Buckland's humour and drollery took him to extremes as happened at a British Association meeting at Bristol where in that part of a discourse dealing with fossil footprints, 'the Doctor exhibited himself as a cock or a hen on the edge of a muddy pond, making impressions by lifting one leg after another'.

Buckland's particular interest, and the one which probably best captured

FIG. 21. *Buckland lecturing in the Ashmolean at Oxford in 1822*
(From Gordon, 1894)

Awful Changes.

Man found only in a fossil state —— Reappearance of Icnthyosauri!
"A change came o'er the spirit of my dream" Byron

... A Lecture, — "You will at once perceive, continued Professor Ichthyosaurus, that the skull before us belonged to some of the lower order of animals the teeth are very insignificant the power of the jaws trifling, and altogether it seems wonderful how the creature could have procured food".

FIG. 22. *Cartoon drawn by H. T. De La Beche in 1830 depicting Buckland, in the guise of an Ichthyosaurus, lecturing upon a human skull to an audience of Jurassic reptiles*
(*From a lithograph given to Dr F. J. North by Dr Mervyn H. Gordon*)

the imagination of his pupils, was in the fossil skeletons of awesome monsters like the ichthyosaurus or pterodactyl, as well as in more recent remains which were then being unearthed in cave deposits. With a few pictures on a board in the Ashmolean and the skeletal remains before him he indulged his audience in a world of swamps, convulsions and battling titanic creatures. A contemporary impression of the quality of his lectures appears in a poem, written by Dr Shuttleworth, later Bishop of Chichester:

> In Ashmole's ample dome, with look sedate,
> 'Midst heads of mammoths, heads of houses sate;
> And tutors, close with undergraduates jammed,
> Released from cramming, waiting to be crammed.
> Above, around, in order due displayed,
> The garniture of former worlds was laid;
> Sponges and shells in lias moulds immersed,
> From Deluge fiftieth, back to Deluge first;

And wedged by boys in artificial stones,
Huge bones of horses, now called mammoths' bones;
Lichens and ferns which schistose beds enwrap,
And – understood by most professors – trap.
Before the rest, in contemplative mood,
With sidelong glance, the inventive Master stood,
And numbering o'er his class with still delight,
Longed to possess them cased in stalactite.
Then thus with smile suppressed: 'In days of yore
One dreary face Earth's infant planet bore;
Nor land was there, nor oceans' lucid flood,
But, mixed of both, one dark abyss of mud;
Till each repelled, repelling by degrees,
This shrunk to rock, that filtered into seas;
Then slow upheaved by subterranean fires,
Earth's ponderous crystals shot their prismy spires;
Then granite rose from out the trackless sea,
And slate, for boys to scrawl – when boys should be –
But earth, as yet, lay desolate and bare;
Man was not then, – but Paramoudras were.
'Twas silence all, and solitude; the sun,
If sun there were, yet rose and set to none,
Till fiercer grown the elemental strife,
Astonished tadpoles wriggled into life;
Young encrini their quivering tendrils spread,
And tails of lizards felt the sprouting head.
(The specimen I hand about is rare,
And very brittle; bless me, sire, take care!)
And high upraised from ocean's inmost caves,
Protruded corals broke the indignant waves.
These tribes extinct, a nobler race succeeds:
Now sea-fowl scream amid the plashing reeds;
Now mammoths range, where yet in silence deep
Unborn Ohio's hoarded waters sleep.
. . . .
The earth, what is it? Mark its scanty bound, –
'Tis but a larger football's narrow round;
Its mightiest tracts of ocean – what are these?
At best but breakfast tea-cups, full of seas,
O'er these a thousand deluges have burst,
And quasi-deluges have done their worst.

Of this enough. Of Secondary Rock,
To-morrow, gentlemen, at two o'clock.
(*Gordon, 1894, pp. 32–33*)

Buckland's ever-present and robust sense of humour peeps through in an anecdote which relates how he gave

a new class of equestrian listeners a practical lesson in geology, by sticking them all in the mud to make them remember the Kimmeridge clay. (*Gordon, 1894, p. 30*)

It goes without saying that he was a tireless collector of rocks and fossils. The famous large blue bag in which he carried these specimens became almost an inseparable part of himself and once accompanied him in a brief imprisonment. When in Italy he made:

> a rich collection of the shells of the Sub-Apennine Hills, many of which resemble those of Hampshire and Sheppey Island, and it would have been more perfect had he not been arrested in the act of making it and sent back fifteen miles to prison at Parma! (*Gordon, 1894, p. 19*)

If the following account is to be believed, the reasons for this imprisonment lay in his visit to the shrine of Rosalia, the patron saint of Palermo, who died in the twelfth century and revealed herself again in the early seventeenth century in order to save the city from plague.

> When Dr Buckland was at Palermo on his wedding tour in 1826 he, as all strangers did, visited the shrine, and with his keen eyes saw in a moment that the bones never belonged to Rosalia. 'Those are the bones of a goat,' he said, 'not of a woman!' Of course the priests were greatly scandalized, and declared that the saint would not permit him to see what only the faithful could discern. From that time, however, the bones were enclosed in a casket, and neither faithful nor heretics were any longer permitted to scan the sacred relics too closely. (*Gordon, 1894, pp. 95–96*)

In spite of his quaintness and not unconscious humour, Buckland took his duties seriously and was quite as capable of excelling in the philosophic sphere as in the normal give and take of pleasantry. He must have been the perfect lecturer. Col. Portlock sums it up neatly as the 'union of the most playful fancy with the most profound reflections.' (Gordon, 1894). In between his other numerous activities Buckland found time to marry. Fortunately his wife was interested in geology and fossils and before her marriage had done some drawings for Cuvier. She was able to be of great assistance to her husband in his work and during the completion of his 'Bridgewater Treatise' worked late into the night writing to his dictation (Tuckwell, 1900, pp. 35–41). For some reason neither she nor their nine children seemed to have been disturbed by her husband's peculiarities. One of his particular fancies was to try out gastronomic dishes on his visitors and we hear of a meal of crocodile proving an utter failure, but the persistent story of his having eaten the heart of Louis XIV one hopes to be apocryphal!

Although much of his writings and teachings have long since been forgotten, Buckland was a considerable force in his time and fairly represented a majority of contemporary geologic thought. He was a devout Churchman in thought and deed and made a conscious effort to reconcile the best of religion and of science. He genuinely believed that science did not contradict religion but, on the contrary, tended to magnify the glory of God by bringing to the public eye the wondrous contrivances of an ordered nature. Similarly, to him

FIG. 23. *Impressions of Buckland's palaeontological expeditions to
Shotover Hill, near Oxford
(From Gordon, 1894)*

any ordered method, mechanical or otherwise, surely revealed a small part of God's eternal design.

On his appointment to the Oxford Readership in Geology Buckland, in his inaugural address, lectured on this theme and put forward a suggested reconciliation of geology and theology, pointing out how the evidence of the former served to prove the writings of the latter. His main point was that existing geological evidence indicated that the world had recently been subject to some form of watery catastrophe, which fact very well coincided with the scriptural record. The contents of the lecture were published in 1820 and as the title *Vindiciae Geologicae* suggests, assume the nature of an apology for geology. Buckland appreciated that a general feeling in favour of the subject could not be expected until the apparent anti-religious tendencies had been eliminated or softened by persuasive explanation. He takes pains to emphasize the worthiness of geology and stresses its immense scope and its direct connexion with astronomy (already a well-established science) and other allied subjects:

> The real question then, more especially in this place, ought surely to be, how far the objects of Geology are of sufficient interest and importance to be worthy of this large and rational species of curiosity, and how far its investigations are calculated to call into action the higher powers of the mind.
>
> Now when it is recollected that the field of the Geologist's inquiry is the Globe itself, that it is his study to decipher the monuments of the mighty revolutions and convulsions it has suffered, convulsions of which the most terrible catastrophes presented by the actual state of things (Earthquakes, Tempests, and Volcanos) afford only a faint image, (the last expiring efforts of those mighty disturbing forces which once operated;) these surely will be admitted to be objects of sufficient magnitude and grandeur, to create an adequate interest to engage us in their investigation. (*Buckland, 1820, p. 5*)
>
> And by its connexion with such subjects as the origin of Aerolites, calculations on the depth of the sea and mean density of the earth, and the investigation of the second causes that were employed in the gradual arrangement of the matter of which our planet is composed, and in producing the over-whelming convulsions that appear at distant intervals to have affected it, Geology becomes associated with Astronomical speculations. So that while she herself receives assistance from many sciences, she on the other hand imparts her light to others; and by means of this constant and extensive reciprocation becomes intimately connected with them all. (*Buckland, 1820, p. 10*)

He then embarked on his main task of giving a reasoned explanation of Creation on catastrophic lines in deliberate opposition to the ideas of Hutton. The following quotation regarding the formation of strata is a mixture of the original teachings of Werner on deposition in concentric layers, and of Woodward on sorting out by gravitation. Though Buckland appreciated that the adoption of this mechanism would result in horizontal beds, whereas in reality strata are not often found in a purely horizontal position, he does not here offer any detailed explanation of this:

> A great majority of the strata having been formed under water, and from materials evidently in such a state as to subject their arrangement to the operation of the laws of gravitation; had no disturbing forces interposed, they must have formed layers almost regularly horizontal, and therefore investing in concentric coats the nucleus of the earth. But the actual position of these beds is generally more or less inclined to the horizontal plane, though often under an angle almost imperceptible. (*Buckland, 1820, p. 11*)

The next extract illustrates his main theme, that out of nature's apparent confusion and contradiction there can be discerned a wonderful unity of purpose which can only be ascribed to God. When explaining the workings of the machine, he marvels at the perfect mechanism of what we call today the hydrological cycle.

> In the whole machinery also of springs and rivers, and the apparatus that is kept in action for their duration, through the instrumentality of a system of curiously constructed hills and valleys, receiving their supply occasionally from the rains of heaven, and treasuring it up in their everlasting storehouses to be dispensed perpetually by thousands of never-failing fountains; we see a provision not less striking or less important. So also in the adjustment of the relative quantities of sea and land in such due proportions as to supply the earth by constant evaporation, without diminishing the waters of the ocean; and in the appointment of the atmosphere to be the vehicle of this wonderful and unceasing circulation; in thus separating these waters from their native salt, (which, though of the highest utility to preserve the purity of the sea, renders them unfit for the support of terrestrial animals or vegetables,) and transmitting them in genial showers to scatter fertility over the earth, and maintain the never-failing reservoirs of those springs and rivers, by which it is again returned to mix with its parent ocean: in all these we find such undeniable proofs of a nicely balanced adaptation of means to ends, of wise foresight and benevolent intention and infinite power, that he must be blind indeed, who refuses to recognize in them proofs of the most exalted attributes of the Creator. (*Buckland, 1820, pp. 12–13*)

There follows a more direct attack on the ideas of Hutton. Buckland is convinced of the need for catastrophic causes and cannot concede to rivers, or even torrents, the power to have eroded the valleys in which they flow:

> Many of these valleys and basins are drained by chasms and precipitous gorges of enormous depth, which could not have been produced by the most violent torrents that now flow through them, but must be referred to the disruption of mountain masses at the epoch of ancient revolutions that have overturned the globe, not to establish thereon the kingdom of disorder and confusion, but to produce that variety of surface which should be most pleasant to the eye, and best adapted to the support of animal and vegetable life, and that disposition which is best calculated to supply the various wants of those multitudes of being that were destined to become its future inhabitants. (*Buckland, 1820, pp. 16–17*)

The next passage attempts to answer Hutton's contention that the constitution and mechanism of the earth was such that he failed to find any traces of a beginning. It misfires not only because Buckland had no real argument to put forward but also because his approach showed a lack of understanding of what Hutton actually meant. Hutton never denied that the world had a beginning; he said merely that he failed to see any sign of a beginning and made the point primarily because he believed it foolish to presuppose catastrophes to be the cause of the present form of the earth. Buckland, a catastrophist, was forced to criticize this argument on both scientific and theological grounds:

> With what acuteness of argument, and what obstinacy of perseverance, the extraordinary notion of an eternal succession was maintained in ancient times, even by some of the greatest philosophers, it is quite unnecessary here to state: and if some writers on Geology in later times have professed to see in the earth nothing but the marks of an infinite series of revolutions, without the traces of a beginning; it will be quite sufficient to answer, that such views are confined to those writers who have presumed to compose theories of the earth, in the infancy of the science, before a sufficient number of facts had been collected; and that, if possible, they are still more at variance with the conclusions of Geology, (as a science founded on observations,) than they are with those of Theology. (*Buckland, 1820, pp. 21–22*).

Buckland is far happier when he passes on to the Deluge for here, as already mentioned, geologic evidence appeared to support the occurrence of this scriptural phenomenon:

> Again, the grand fact of a universal deluge at no very remote period is proved on grounds so decisive and incontrovertible, that, had we never heard of such an event from Scripture, or any other authority, Geology of itself must have called in the assistance of some such catastrophe, to explain the phenomena of diluvian action which are universally presented to us, and which are unintelligible without recourse to a deluge exerting its ravages at a period not more ancient than that announced in the Book of Genesis. (*Buckland, 1820, pp. 23–24*)

However in the next quotation he is again countering uniformitarianism and draws attention to the concern of religious persons at the implications involved by the acceptance of such an idea:

> When it was attempted to explain every thing by the sole agency of second causes, without any reference whatever to the first; when nature was set up as an original source of being, distinct and independent of the Almighty; when it was taught that matter possessed an existence which he never gave it, and that the elements had differences and qualities independent of him; these surely were grounds sufficient to excite alarm in all persons who were zealous for the cause of religion, and the preservation of the best interests of mankind. (*Buckland, 1820, pp. 27–28*)

Yet, almost in the same breath, he shows complete inconsistency in admitting that the formation of strata must have been a gradual process and that their constituent parts point to an origin from an older succession of rocks. In fact this uniformitarian idea seems to have received general approval from an early date and was now incorporated as a traditional part of geological learning, without any recognition whatsoever being given to Hutton. The passage as it stands might well have been written by Hutton himself!

> These strata do not appear to have been deposited hastily and suddenly; on the contrary, the phenomena attendant on them are such as prove that their formation was slow and gradual, going on during successive periods of tranquillity and great disturbance; and being in some cases entirely produced from the destruction of more ancient rocks, which had been consolidated, and again broken up by violent convulsions antecedent to the deposition of those more modern or secondary strata which are sometimes in great measure derivative from their exuviae. (*Buckland, 1820, p. 29*)

The obvious fact that an immense length of time has been necessary to deposit the strata drives Buckland to abandon the Flood as the cause of all surface features and he makes the important step of distinguishing between what had been deposited over long periods of time and what had been eroded within a comparatively recent period. To the work of the Deluge he ascribes the features of the more recent erosion and deposition. Unlike many of the writers before him he sees the Deluge as an agent of destruction rather than of construction. To it he apportions the wide expanses of the valleys, the isolated outliers and the drift deposits. To explain the formation of the mass of sedimentary rocks he resorts to the scriptural gap provided by the period which occurred between the Creation and the Deluge. He speculates on the current theory that the rocks and fossils were accumulated below the sea during this period and were raised above the surface at the time of the Flood, but, for reasons he gives at length, the idea does not satisfy him and he inclines towards the opinion that the evidence is more conclusive of the action of a retreating flood. To rivers he denies any substantial power to erode.

> It seems therefore impossible to ascribe the formation of these strata to a period so short as the single year occupied by the Mosaic deluge; . . . the strata we have been considering, although they bear on their surface unequivocal marks of the agency of that convulsion, were evidently not produced, but partially destroyed by it, and must be referred for their origin to periods of much higher antiquity.
>
> It has been supposed therefore by others, with greater plausibility, that these strata have been formed at the bottom of the antediluvian ocean during the interval between the Mosaic Creation and the Deluge; and that, at the time of that Deluge, portions of the globe, which had been previously elevated above the level of the sea, and formed the antediluvian continents, were suddenly submerged with their inhabitants, while the ancient bed of the ocean rose to supply their place. This hypothesis, it has been said, has the advantage of explaining the cause why the remains imbedded in the strata are principally those of marine animals: but it labours under considerable objections. (*Buckland, 1820, pp. 30–31*)

NUMBERED POINTS WHICH TEND TO PROVE THE RECENT ACTION
OF THE DELUGE

1. The general shape and position of hills and valleys; the former having their sides and surfaces universally modified by the action of violent waters, and presenting often the same alternation of salient and retiring angles that mark the course of a common river. And the latter, in those cases, which are called valleys of denudation, being attended with such phenomena as show them to owe their existence entirely to excavation under the action of a retiring flood of waters.

2. The almost universal confluence and successive inosculations of minor valleys with each other, and final termination of them all in some main trunk which conducts them to the sea; and the rare interruption of their courses by transverse barriers producing lakes.

3. The occurrence of detached isolated masses of horizontal strata called outliers, at considerable distances from the beds of which they once evidently formed a continuous part, and from which they have been at a recent period separated by deep and precipitous valleys of denudation.

4. The immense deposits of gravel that occur occasionally on the summits of hills, and almost universally in valleys over the whole world; in situations to which no torrents or rivers such as are now in action could ever have drifted them.

5. The nature of this gravel, being in part composed of the wreck of the neighbouring hills, and partly of fragments and blocks that have been transported from very distant regions. (*Buckland, 1820, p. 37*)

7. The total impossibility of referring any one of these appearances to the action of ancient or modern rivers, or any other causes, that are now, or appear ever to have been in action since the last retreat of the diluvial waters. (*Buckland, 1820, p. 38*)

The *Vindiciae Geologicae* was a sensational success and its author stood out in the popular mind as a brilliant catastrophist. He had rejected almost all of Hutton's theory and had presented a revitalized notion, that the Flood was responsible for all recent landscape features and particularly the formation of valleys and 'glacial' deposits. The diluvialists had become scientific.

Yet Buckland did not escape criticism as the following anonymous contribution will show.

I have . . . just perused Professor Buckland's 'Inaugural Lecture', delivered in May 1819, at Oxford, and observed, that he therein labours to prolong those errors and delusions, respecting the evidences, which Geological phenomena were so confidently said to present or afford, of the occurrence and circumstances attending the Mosaic Deluge.

I remember having seen Mr Bakewell commended in your Work, for having in the year 1813 abstained, from introducing the Deluge of Moses into his *Introduction to Geology*, as the previous Writers had almost invariably done, to the manifest injury of Geology on the one hand, and of Religion on the other; since which, the practice has almost entirely grown into disuse, while the number of writers on Geological subjects, have been greatly on the increase; and I regret therefore to see, the new Geological Professor at Oxford, attempting now to revive the exploded notion, that any of the phenomena at this time visible, on or within the Earth, are, with any proper regard to probability, referable to the Deluge of which Moses writes.

It is undoubted that the surface of the Earth, almost universally presents the evidences, of a most violent and over-whelming Torrent, or rather, perhaps, a succession of such, the Waters of which (assisted perhaps by some Tidal reversion of the action of Gravity, as has been maintained by Mr Farey in your work) were able to move vast masses of earthy Matters, mixed with gravel Stones and even with large Bowlders, and very heavy Blocks of Stone, and to lodge them on tops of Hills, and on the surfaces of Plains of considerable elevation; such Hills and Plains, and the Valleys which intersect them, having most evidently existed in their present form and shape, at the time of these early or gravel Floods, which most evidently did not excavate the Valleys, or in any material degree abrade or alter the contour of the Hills. (*Anon, 1820, pp. 10–11*)

The anonymous critic proceeded to point out that the Bible revealed that the Flood 'retired from the surface of the land by very slow degrees and in the most quiet manner' and that it left 'enough vegetable earth or mould for the growth of useful plants' (*Anon, 1820, p. 11*).

A more pungent criticism came from the Bishop of Chichester.

> Some doubts were once expressed about the Flood:
> Buckland arose, and all was clear as mud.
> (*Quoted in White, 1920, p. 232*)

There were no reasons why the above and other criticisms should have impressed Buckland as they presented no vital scientific points against his thesis. Moreover the discovery of fossil remains in the caves of Kirkdale and Paviland strengthened his theory for he thought that in some cases the bodies of animals had been carried into caves and buried by the Flood and in other cases the caves had served as the last haunts of beasts escaping the rising waters. He published a volume on this subject in 1823 under the title *Reliquiae Diluvianae*. The main content of the book hardly concerns us but certain portions of the appendix are interesting because they discuss the exact mechanisms of the Flood.

The first passages quoted here describe the erosive and carrying power of water, but the reader will be quick to notice that the examples are introduced for the sole purpose of leading up to the grand conclusion that if rivers were capable of producing such minor changes what more could be done by the colossal power of a Flood:

> Our present rivers excavate but little, as they flow through valleys already formed by an overwhelming ocean; and the destructive action of the present sea is limited to the partial cutting away of cliffs by the slow undermining of the waves in storms and at high tides. Yet we know from the effect of a mountain torrent in cutting ravines and drifting gravel; from the blocks of granite which were lifted to an elevated point on the side of a mountain by the bursting of a small lake in the Val du Bagnes, in Switzerland; and from the excavation of the Zuyderzee, by the bursting of a dyke in Holland; that

the force of water in rapid motion is competent both to transport masses of gravel and granite blocks as we have been tracing over the world, and to excavate valleys which though many miles in breadth, and many hundred, and in some cases perhaps, some thousand feet in depth, still bear a due proportion to the bulk and power of the agent that produced them. (*Buckland, 1823, p. 236*)

An agent thus gigantic appears to have operated universally on the surface of our planet, at the period of the deluge; the spaces then laid bare by the sweeping away of the solid materials that had before filled them, are called valleys of denudation; and the effects we see produced by water in the minor cases I have just mentioned, by presenting us an example within tangible limits, prepare us to comprehend the mighty and stupendous magnitude of those forces, by which whole strata were swept away, and valleys laid open, and gorges excavated in the more solid portions of the substance of the earth, bearing the same proportion to the overwhelming ocean by which they were produced, that modern ravines on the sides of mountains bear to the torrents which since the retreat of the deluge have created and continue to enlarge them. (*Buckland, 1823, p. 237*)

In the next passage he makes a closer examination of his evidence and draws attention to the coincidence between the angle of the strata and the area of material removed by the act of excavation. He correctly understood, like Catcott, Hutton and Playfair before him, that if we were hypothetically to join the strata exposed on the opposite sides of a valley, the restored beds would roughly correspond with the volume of rock removed. This of course clearly proved that the valleys had not always existed, as had been suggested by some geologists, but must have been excavated at some time. But again Buckland discounts the present rivers, and relies entirely on the agency of the Flood:

When a gorge or valley takes its beginning, and continues its whole extent within the area of strata that are horizontal, or nearly so, and which bear no mark of having been moved from their original place by elevation, depression or disturbance of any kind; and when it is also inclosed by hills that afford an exact correspondence of opposite parts, its origin must be referred to the removal of the substances that once filled it; and as it is quite impossible that this removal could have been produced in any conceivable duration of years by rivers that now flow through them, (since all the component streams, and consequently the rivers themselves, which are made up of their aggregate, owe their existence to the prior existence of the valleys through which they flow,) we must attribute it to some cause more powerful than any at present in action, and the only admissible explanation that suggests itself is, that they were excavated by the denuding force of a transient deluge. (*Buckland, 1823, pp. 237–8*)

Some of the best examples I am acquainted with of valleys thus produced exclusively by diluvial denudation occur in those parts of the coasts of Dorset and Devon which lie on the east of Lyme, and on the east of Sidmouth; and the annexed views and map will illustrate, better than any description, the point I am endeavouring to establish. In passing along this coast . . . we cross, nearly at right angles, a continual succession of hills and valleys, the

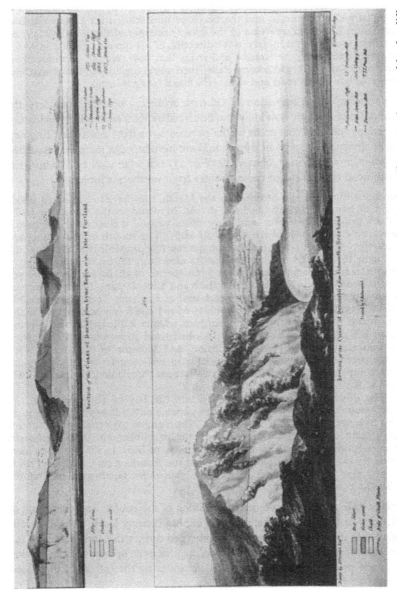

FIG. 24. *Buckland's (1823) illustrations of the Dorset and Devon coasts, showing the valley systems intersected by the cliffs*

southern extremities of which are abruptly terminated by the sea; the valleys gradually sloping into it, and the beach or undercliff, with a perpendicular precipice. The main direction of the greater number of these valleys is from north to south; that is, nearly in the direction of the dip of the strata in which they are excavated: the streams and rivers that flow through them are short and inconsiderable, and incompetent, even when flooded, to move anything more weighty than mud and sand. (*Buckland, 1823, p. 239*)

Continuing his examination of diluvial evidence, he next considers the significance of what are now known to be glacial pebbles or erratics. Buckland's mineralogy told him that many of the pebbles came from north-east England and, as obviously the present rivers could not have brought them, he visualized the Warwick–Gloucester countryside (*Fig. 27*) as a wide sea bay to which the vast diluvial current swept these pebbles from northern sources:

Between Shipston and Moreton in the Marsh, they have been drifted into a kind of bay, formed by the horn-shaped headland of the Campden Hills, which projects like a pier-head some miles beyond the ordinary line of the great limestone chain of the Cotswold Hills. The mouth of this bay opens directly to the north-east, from which quarter it is probable the current which brought the pebbles in question had its direction; for on the south-east of Shipston there are pebbles of a hard red species of chalk, which occurs not unfrequently in the Wolds of Yorkshire and Lincolnshire, but is never met with in the chalk of the south or south-east of England. The nearest possible point, therefore, to which these pebbles of red chalk can be referred, is the neighbourhood of Spilsby, in Lincolnshire, whence a diluvial current flowing from the north-east would find an unobstructed passage across the plains of Leicestershire to the Bay of Shipston, and Moreton in the Marsh. With these pebbles of red chalk are others of hard and compact white chalk, such as accompanies the red chalk in the two last mentioned counties, and occurs also at Ridlington, in Rutlandshire.

The diluvian current thus impelled into the Bay of Shipston, from the north-east, appears to have continued its course onwards beyond the head of this bay, near Moreton in the Marsh, . . . bursting in over the lowest point of depression of the great escarpment of the limestone; and being deflected thence south east-wards by the elevated ridge of Stow in the Wold, to have gone forward along the line of the vale of the Evenlode by Charlbury, till it joined that of the Thames at Ensham, five miles north-west of Oxford. (*Buckland, 1823, pp. 250–1*)

From differences between the composition of the pebble deposits, some of which are of foreign origin and others almost entirely local, Buckland concluded that the Thames valley was submerged during the early stages of the Flood and that the destruction of the oolitic limestone upland and the carving out of the present Thames valley took place during the retreat of the Flood waters:

There is another strong fact tending to prove the excavation of the valleys of the Evenlode and Cherwell, and of the Thames (in part) near Oxford, to have been subsequent to the transport of the Warwickshire pebbles, namely, the absence of pebbles of oolite in the beds of gravel just mentioned as crowning the summits of Wytham Hill and Bagley Wood (near Oxford). Hence we may infer that the destruction of the oolite strata was not so much the effect of the advancing deluge as of its retiring waters, cutting out valleys in the table-lands, and sides of the higher ridges, and covering them with gravel, composed partly of the wreck of the strata immediately enclosing them, and partly of pebbles, which their first rush had transported from more distant regions; and thus it will appear that the lower trunks of the valleys of the Thames, Cherwell, and Evenlode, (i.e. those portions of them which may be fairly attributed to the exclusive action of denudation, and which lie below the average level of the tablelands which flank their course), did not exist at the time of the first advance of the waters, which brought in the pebbles from Warwickshire, but were excavated by the denuding agency which they exerted during the period of their retreat. (*Buckland, 1823, pp. 253–4*)

Having argued throughout in favour of the Flood, Buckland now suddenly introduces a new factor, structure. He suggests two ways in which the nature of the underlying rocks might affect the form of the surface; first, where soft beds alternate with hard, thereby encouraging erosion to carve out the less resistant beds and leave the harder upstanding; and second, as quoted below where convulsions of underground forces have contorted the strata into ridges and valleys which erosive forces could then attack.

Though traces of diluvial action are most unequivocally visible over the surface of the whole earth, we must not attribute the origin of all valleys exclusively to that action; . . . in mountain districts, (where the greatest disturbances appear generally to have taken place), the original form in which the strata were deposited, the subsequent convulsions to which they have been exposed, and the fractures, elevations, and subsidences which have affected them, have contributed to produce valleys of various kinds on the surface of the earth, before it was submitted to that last catastrophe of an universal deluge which has finally modified them all. (*Buckland, 1823, p. 258*)

In this acknowledgement of the influence of structure we begin to realize that the controversy is no longer simply a battle between catastrophists and uniformitarians. The conflict has led to a closer examination of the evidence which is now revealing the complexity and interdependency of landscape-forming processes. There is no absolute cause but several causes each of which may dominate under a particular group of circumstances. This diversification of ideas is borne out in Buckland's subsequent publications (e.g. Buckland, 1824) and particularly in his paper on 'The Valley of Kingsclere' (1829). He suggests that this anticlinal valley could not have been formed by 'denudation' alone but must have resulted from a combination of factors: the chalk was elevated locally, causing along the line of greatest uplift a fracture or 'gaping apart of the strata with reverse dips away from it'; this 'valley of elevation' was then modified by diluvial action.

Buckland's ideas on anticlinal valleys were contested by George Scrope (1825) who was then studying the Weald, a much larger 'valley of elevation'. Scrope admitted that some part of the apparent denudation might be explained by the slipping of the beds in the direction of the dip during uplift but he thought that most of the valley had been fluvially eroded:

> These (anticlinal) flexures present a remarkable diversity of character on different points. On some, the upper remaining beds still cover over the whole breadth of the convexity, . . . While on other points of the same flexure or anticlinal ridge, the upper beds have in part disappeared, leaving a longitudinal valley, at the bottom of which the inferior beds shew themselves. . . . In the former case the upper chalk beds possessed sufficient tenacity to sustain their elevation and curvature without splitting; in the latter a longitudinal crack opened across them, parallel to the axis of elevation. The chalk resting on beds of clayey marl (the upper strata of the greensand formation) slipped away on either side from the axis, leaving bare the lower strata of greensand. Again, the partial subsidence of this formation upon the slippery beds of the Weald Clay, disclosed in turn the ironsand, which forms the visible axis of the ridge. (*Scrope, 1825, pp. 212–13*)
> . . . transverse fractures appear to have been formed in the subsiding strata, but these will have remained more or less narrow crevice-like gorges; . . . These transverse chinks or gaps are frequently the channels, through which the drainage of the interior valley has been since effected, while they have been enlarged, and have lost the angular roughness of their fracture edges, by denudation, or meteoric abrasion. . . . The waters of the ocean in their rapid and violent retreat from the surfaces elevated above its level, would naturally occupy these fissures, and still further enlarge and deepen them, . . . Many other transverse vallies, however, were no doubt originally scooped out by these retiring waters alone, without the previous existence of any directing fissure; . . . The vallies of either kind have been subsequently enlarged and otherwise modified; and many others, perhaps indeed a far greater number were wholly and entirely excavated by the slow but constant and powerful action of the same causes which are still continually in force; amongst which the fall of water from the sky, and its abrasive power as it flows over the surface of the land from a higher to a lower level, is the principal. (*Scrope, 1825, pp. 213–14*)

Unhappily Scrope was in the minority and most prominent geologists vied in echoing praises of Buckland's thesis.

CONTEMPORARY DILUVIALISTS

Among Buckland's contemporaries was the Rev. Adam Sedgwick, Woodwardian Professor of Geology at Cambridge (pp. 70–71) who also had a strong sense of humour, as may be judged from his own account of an incident during the experiments he and Airy, the astronomer, undertook down a Cornish mine. These activities had already excited the suspicions of the local inhabitants when –

> We gave them some cause for their suspicions. Our lampbox, marked outside 'Deville, Strand,' stood well for a formal address to his infernal majesty. We were clambering down one day, when, to keep up the joke, I asked a sturdy miner who was guiding us, 'How far is it to the infernal regions?' He was a match for me – for he replies – 'Let go the ladder, Sir, and you'll be there directly.' (*Clark and Hughes, 1890, I, p. 332*)

On his appointment in 1818, Sedgwick set out vigorously to make good his lack of geological knowledge and his learning acquired a Wernerian tinge as little other material was readily available to him. But he set himself the praiseworthy task of observing in the field and of forming his own conclusions. Yet so closely did Sedgwick's ideas coincide with those of Buckland in these early years, that he remains comparatively unimportant from a geomorphological

FIG. 25. *Adam Sedgwick*

viewpoint (Sedgwick, 1825). The only way he differs appreciably from Buckland is in his cautious attitude to the supposed coincidence between the findings of science and the truths of revelation. Sedgwick suggests that although science and revelation do not contradict each other, the methods and modes of dealing with each are so different that confusion is likely to arise if they are too closely mixed.

Another contemporary of Buckland was the Rev. William Daniel Conybeare (1787–1857) who expresses much the same opinions as Sedgwick although his arguments are often better thought out and based more securely on field-evidence. Conybeare graduated in 1808 from Christ Church, Oxford, where he was a friend of Buckland as well as an early member of the Geological Society (Conybeare, 1905). He was undoubtedly acclimatized to a Wernerian way of thinking and sceptical of Huttonian ideas. His paper 'On the Hydrographical Basin of the Thames' (1829) departs little from the lead given by Buckland. Yet the following passage shows an improvement on Buckland's thesis, for not content with one flood Conybeare feels forced to call upon several such cataclysmic agents to account for the varied phenomena which had come to his notice. Like other diluvialists he was misled into thinking that all gravel deposits were indicative of Flood movement; and that conglomerates by their rough and disunited appearance were laid down at the time of turmoil, whereas we now know that conglomerates are as 'natural' a sedimentary formation as are sandstones and shales:

> He proceeds to distinguish several different geological epochs, at which it is probable that currents must have taken place calculated to excavate and modify the existing surface. I. In the ocean, beneath which the strata were originally deposited. II. During the retreat of that ocean. III. At the periods of more violent disturbance, which are evidenced by the occurrence of fragmentary rocks, the result of violent agitations in the waters of the then existing ocean propagated from the shocks attendant on the elevation and dislocation of the strata. – Four such periods are enumerated as having left distinct traces in the English strata. 1. That which has formed the pudding-stone of the old-red-sandstone, ascribed to the elevation of the transition rocks. 2. That which has formed the conglomerates of the new-red-sandstone, ascribed to the elevation of the carboniferous rocks. 3. That which has formed the gravel beds of the plastic clay. 4. That which has produced the superficial gravel, spread alike over the most recent and oldest rocks as a general covering, and which is found to contain bones of extinct mammalia: this (it is agreed) may be identified as the product of one aera, by the same evidence which is employed to demonstrate the unity of any other geological formation. (*Conybeare, 1829, pp. 145–6*)

Conybeare then begins to examine the Thames valley in some detail. The depth of the valleys and the steepness of their declivities is such that he concludes some violent force must have brought this about. He is reinforced in this view by the evidence presented by British and Roman camps, upon the earthen ramparts of which the present erosive forces have made no apparent impression.

> Each of these valleys is separately described, and the general features of denudation presented by the Cotteswold chain are pointed out; these, it is asserted, bear traces of the most violent action, and they are contrasted with the state of repose which has evidently prevailed in the same districts from the period to which our earliest historical monuments ascend. In the most exposed situations, and those which appear to have suffered most from the action of the denuding causes, earth works of British and Roman antiquity are frequently found, attesting by their perfect preservation that the form of the surface has remained unaltered since the time of their construction. The

FIG. 26. *William D. Conybeare*

drainage of the atmospherical waters has here produced no sensible effect for more than fifteen centuries: it is inferred, therefore, that to assign to this cause the excavation of the adjoining valleys, 600 or 700 feet deep, is to ascribe to it an agency for which we have no evidence; the evidence, indeed, as far as it can be examined, being adverse. (*Conybeare, 1829, p. 146*)

In later pages Conybeare comments upon the profusion of deposits above or beyond the limits of the present-day flood waters and states that their presence can only be explained by existing processes if the river had continually changed its course. Such movement he asserts is directly contradicted by the survival of the historical remains.

The river collected from these head-waters flows through the plain of Oxford, which is covered to a great extent by water-worn debris; these are diffused over situations inaccessible to the present floods, and if produced by the actual streams, we must suppose that they have repeatedly changed their channel so as to have flowed successively over every portion of the plain where these debris are now found: the oldest historical monuments attest, however, the permanence of the actual channels, and the floods at present bring down no pebbles whatsoever. (*Conybeare, 1829, p. 147*)

The next portion of Conybeare's paper is by far the most interesting. In looking at the upland of the Cotswolds, the Oxford Heights and the dividing vale he ponders on the question why a river like the Thames should have cut right through the Oxford hills in its passage to the sea when it would apparently have been far easier for it to have followed the lowland line of the vale. In fact, in this context, the idea of river erosion is so improbable that he is driven to fill up the vale and make the land between the Cotswolds, the Oxford chain and the Thames estuary into a gently inclined surface. Even then, he argues, the fall is so gradual that the river would lack the necessary erosive power, and for that reason he does not pursue his supposition of an inclined plane. Instead, he adheres to Buckland's idea of a retreating Flood:

> The author inquires how this configuration of the valleys could have been produced on the fluvialist's theory. He argues, that if the Oxford chain originally (as at present) formed a barrier of superior elevation to the tract intervening between itself and the Cotteswolds, that barrier must have turned all the drainage of the Cotteswolds into the vale of Ouse: under those circumstances the crest of the Oxford chain could never have been eroded by waters which would have flowed off in another direction. There is, however, another alternative; and the interval between these chains may be supposed to have formed originally a uniformly inclined plane, from the summits of the one to those of the other, along which the waters once flowed, and which they have since furrowed (by perpetually deepening their channels) into the present valleys. The author calculates the mass of materials which must on this supposition have been excavated and washed away, and contends that the drainage of atmospherical waters along such an inclined plane (which would have a fall of 10 feet per mile) does not afford an agent adequate to such vast operations. (*Conybeare, 1829, p. 147*)

He backs up his argument with the oft-cited example of dry valleys which he considered to be definitely of diluvial origin because they are in every respect like other valleys and yet no longer contain any water.

> The Chilterns, like most other chalky districts, abound with dry valleys, the rifted and absorbent structure of that rock not permitting the rain waters to collect into streams: these valleys agree in every other feature with those containing water courses, and have been obviously excavated by the same denuding causes, which, in this case, it is self-evident could not have been river waters. The surface of the chalk has been deeply and violently eroded, and is deeply covered with its own debris; ... this action appears, in part, to have taken place during the epoch of the plastic clay formation. (*Conybeare, 1829, p. 148*)

Conybeare ends by combining two of the points he has already made. In order to explain why the plains about London are covered by foreign gravels

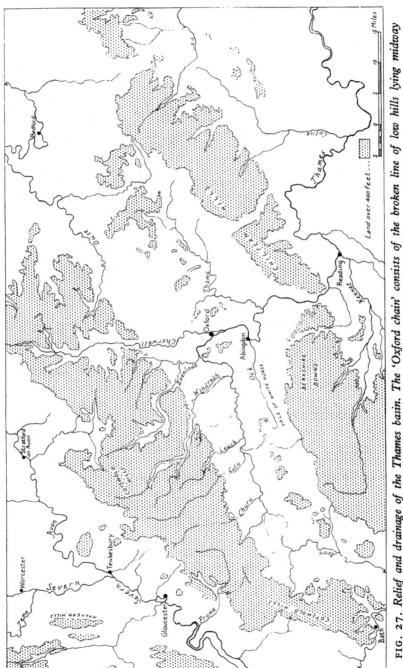

FIG. 27. *Relief and drainage of the Thames basin. The 'Oxford chain' consists of the broken line of low hills lying midway between the Cotswold and Chiltern Hills and crossed by the Thames at Oxford*

at all levels he resorts to his original hypothesis of an inclined plane and of a
river continually changing its bed:

> The plains of London are covered with enormous accumulations of water-
> worn debris, chiefly of chalk-flints, and often abounding in fossil remains of
> elephants, hippopotami, etc.: the gravel is not confined to the low grounds,
> but caps the highest summits of the district; e.g. Highgate on the north, and
> Shooter's Hill on the south of the river. To explain this distribution of this
> gravel by the operation of the actual rivers, the author observes that it is
> necessary, first, to suppose that a uniform plane originally existed from the
> summit of Highgate to the Hertfordshire chalk downs, and from the top of
> Shooter's Hill to those of Kent; on the surface of which the rivers once
> flowed. Secondly, that having worn down that surface into nearly its present
> form, the rivers perpetually shifted their channels so as to distribute the
> gravel equally over the whole plain of London, yet remained long enough in
> each channel to lodge there deposits of this gravel 20 or 30 feet thick. (*Cony-
> beare, 1829, pp. 148–9*)

But the great length of time and the enormous amount of river erosion
required caused Conybeare to discard what happens to be the correct explanation of these gravel terraces. Once again he chose the Flood.

THE HIGH TIDE OF DILUVIALISM

There is perhaps no need to add that during the 1820's Buckland and his
associates largely displaced the Neptunists and raised diluvialism to its high-
water mark. Their catastrophic theory proved acceptable to most traditional
thinkers and their strong religious tone and clerical vocations assured the
favour of the less bigoted members of the Church.

Yet as a reconcilist creed diluvialism was unlikely to escape for ever the
criticisms of theologians (White, 1920, pp. 233–4) and as a field-study it was
sure to encounter serious difficulties with advances in the scientific knowledge
of landscapes. The diluvialists themselves diverged in their views on points
of detail but they were united in their adamant denial of the claims of the
Huttonians, whom they dubbed the 'fluvialists'. Yet strangely enough while
denying rivers great powers of erosion and transportation, they accepted
freely the Huttonian principle that all sedimentary strata were derived from
the breaking down of older rocks.

Buckland, Dean of Westminster, survived until 1856; Conybeare, Dean of
Llandaff, until 1857; and Sedgwick, Canon of Norwich, until 1873. Hence in
the following chapters we shall describe not only how they lived to witness the
inevitable turn of the diluvial tide but also how they contrived to sail out
upon its retreating waters.

CHAPTER TEN

The Early Anti-Diluvialists

The early nineteenth century was a period of active field-work, and the ruling principle of the day was to gather from actual observations sufficient data to form the basis of a reasonable hypothesis. Much of the inspiration for these detailed labours stemmed from the work of William Smith on stratigraphy and palaeontology. Consequently most contemporary geological studies tended to concentrate on the interpretation and correlation of strata and fossils. For this reason geomorphic processes were neglected and if it had not been for the energies of Buckland, and later of Lyell, they might have languished indefinitely while the major problems of stratigraphy were worked out. Fortunately their intensive labours and those of their opponents brought to light some important aspects of landscape evolution. The early objections and corrections to Buckland's attractive diluvialism especially concern us here. This opposition may be grouped into two main categories: the fluvialists; and the structuralists.

THE EARLY FLUVIALISTS

After the end of the Napoleonic wars, many British geologists began to indulge in the examination of continental deposits, and the volcanic region of Central France proved a favourite locality. George Julius Poulett Scrope, a former student of Sedgwick and a remarkably intelligent man, was among these visitors. Between 1826 and 1828 Scrope wrote much on volcanoes in general and on the geology of Central France and demonstrated that the Wernerian ideas on the formation of volcanic rocks were utterly inaccurate. Even in these years he seems to have been a confirmed uniformitarian and the following quotations, on river erosion in an area of volcanic eruption and deposition, are strongly in favour of the fluvialists.

> It is indeed impossible to observe the many strips of originally continuous freshwater formation which rise from the plain of Limagne in long tabular hills, without being convinced that they owe their preservation from the destruction which has swept away the remainder of the formation, to the capping of basalt which all alike possess, and which by reason of its superior hardness would naturally protect the underlying strata from the rains, frosts, and other meteoric agents, to which the uncovered materials of the marly plain left by the emptying of the lake were permanently exposed. Such a capping on the other hand, would afford very inefficient protection against the denuding force of any violent deluge or general current of waters. . . . (*Scrope, 1827, p. 160*)
> Again, had the whole excavation effected in the freshwater formation of the Limagne been produced at once, by the debacle mentioned above, or by any diluvial or other violent catastrophe, it is clear that the remnants of the lava

currents which had flowed into the freshwater basin before this epoch, would be necessarily all found at one level, or nearly so, corresponding to the average level of the bottom of the lake-basin at that time; while on the other hand, all the lava streams which have flowed since the debacle or supposed deluge, would be found at another nearly uniform, but much lower level, viz. that of the lowest places of the excavated valley. But, as we have seen, no marked distinction of this sort exists; . . . They are found at all heights from 1500 to 15 feet above the water channels of the proximate valleys. (*Scrope, 1827, p. 161*)

. . . the immense abstraction of matter which has occurred in the freshwater formation of the Limagne was for the most part effected gradually and progressively, and went hand in hand with the occasional flooding of parts of this valley and its tributary ravines by lavas emitted in the eruptive paroxysms of the volcanoes on the neighbouring heights. Even were it allowable to have recourse to vague and hypothetical conjectures, we can conceive no gradual and progressive excavating forces, other than those which are still in operation wherever rains, frosts, floods and atmospheric decomposition act upon the surface of the earth. To these agents then we must refer the effects in question, of which, with an unlimited allowance of time, no one will pronounce them to be incapable. (*Scrope, 1827, p. 162*)

Charles Lyell reviewed Scrope's book on Central France and justly called it 'the most able work which has appeared since Playfair's *Illustrations of the Huttonian Theory*'(*Quart. Rev.*, Vol. 36, p. 437). Soon afterwards Lyell in company with Murchison visited the Auvergne and together they published a paper '*On the excavation of valleys*' in which they revealed positive evidence of the quantity of erosive work done by rivers in an area that disclosed no sign of the action of a Flood. It was, as the following quotations show, a clarion blast in favour of the fluvialists.

In the new Valley, about 250 feet in depth, opened at the Etang de Fung by the waters of the Sioule, after the stream had been diverted from its course by the lava of Come, the matter removed, and still continually carried away by the river, consists of alluvial clay and sand, and in some cases of the subjacent gneiss, which has thus been excavated to the depth of forty feet. That no general inundation contributed to this effect, is inferred from the total absence of sand, mud, or pebbles, on the surface of the lava of Come; although that current has occupied a low and exposed situation, ever since the period when the Sioule began to open for itself its present channel. (*Lyell and Murchison, 1828-9, p. 90*)

The authors conceive . . . that a satisfactory explanation of these phaenomena may be derived from the effects of the latest volcanic eruptions of Central France. For the more recent lavas appear to have dammed up the channels of several rivers, and converted ancient valleys into lakes; wherein as at Aidat and Chambon, alluvial matter is continually accumulating at present. The modern lava of Montpezat, in the Vivarais, has thus obstructed the course of the Fontaulier, and given origin to a lake, since filled with river alluvion and volcanic ashes; and these deposits themselves, together with a part of the volcanic barrier, have been subsequently cut through, by the action of the river and the waters of the lake. The early and more copious lava-currents of Auvergne must have occasioned larger lakes than those of

FIG. 28. *George Poulett Scrope*

recent formation; and these, as has been stated by other authors, seem to have been gradually filled up, with materials introduced by rivers, and occasionally by floods from the sides or craters of volcanoes, probably during their moments of eruption; through which accumulations new valleys were excavated by the continued action of the rivers: – as at Mont-Perrier, to the depth of about 100 feet; and at Maurice on the Allier, to 400 feet, below their original bottoms. (*Lyell and Murchison, 1828–9, p. 91*)

The above paper was followed by a more theoretical article which provided an answer to many of the contentions of the diluvialists. In it Lyell and Murchison observe that, by the fortunate occurrence of successive lava outflows, the landscape had been preserved in its original state thus demonstrating not only the stage it had reached when it was overwhelmed by the lava but what is more important the fact that, at the time of burial, valleys had already been eroded without the intervention of a Flood. From this the two writers assumed that river erosion is a real and continuous process. Of evidence at any period of diluvial action they could find none. They realized that fluvial erosion is successive and cumulative, beginning as soon as the area has been raised above sea level, and that its constant action confuses for the observer the origin of the different features because the work of many different periods and processes may be compressed within one landscape. This complexity tends quite naturally to suggest that the whole result has been brought about by the work of a catastrophe. However, under the peculiar conditions in Auvergne each surface is crystallized, so to speak, at the stage when a particular phase of erosion temporarily ceased to have effect while the lava outflowed. A curious, but not particularly relevant, feature of the publication of this paper is that Jameson edited the journal:

> In concluding this article, we may observe, that Auvergne, Velay, and the Vivarais, throw peculiar light on the theory of valleys, because the volcanic rocks having been introduced upon the surface successively, and sometimes at intervals immensely distant from each other, have preserved portions of the surface in the state in which it existed at those several periods. Hence it becomes impossible to confound the effects of erosive action of one epoch with those of another. But for this circumstance, events the most remote in point of time, – the waste of floods or violent torrents of most distant eras, would have been regarded as simultaneous. Thus the conglomerates, several hundred feet thick, on which rests nearly the whole series of alternating beds of trachyte, basalt, and scoriae of Mont Dor, would have remained without any distinct line of separation from the latest alluvions; and the same would, in many cases, have happened in Cantal. . . . If the debris of all these various periods were now strewed over the country at the various elevations where they are at present observed, and if we possessed not the means, which we now have, of pointing out their different ages, they would present the appearance of having been the result of one sudden and dreadful catastrophe, whereby rocks of different ages were violently broken up, and carried, without the least reference to existing levels, to vast distances, often across deep ravines, and spacious intervening valleys.
>
> Are there not, however, some other signs, we shall be asked, upon those older basaltic plateaus of Dor and Cantal, which are unquestionably more ancient than the great valleys of the same district (by whatever theory the origin of those valleys be explained), – are there not some decided proofs of a violent flood, which transported thither fragments of foreign rocks from distant regions? We answer there are none. (*Lyell and Murchison, 1829, pp. 45–46*)

These two papers prompted a sharp reaction from the opposing diluvial camp, which is reported by Lyell in letters to friends:

> My letters from geological friends are very satisfactory as to the unusual interest excited in the Geological Society by our paper on the excavation of valleys in Auvergne. Seventy persons present the second evening, and a warm debate. Buckland and Greenough furious, contra Scrope, Sedgwick, and Warburton supporting us. (*Quoted in Bonney, 1895, p. 38*)

In April, 1830 Lyell writes thus to Dr Mantell:

> A splendid meeting (at the Geological Society) last night, Sedgwick in the chair. Conybeare's paper on Valley of the Thames, directed against Messrs Lyell and Murchison's former paper, was read in part. Buckland present to defend the 'Diluvialists', as Conybeare styles his sect; and us he terms 'Fluvialists'. Greenough assisted us by making an ultra speech on the importance of modern causes. . . . Murchison and I fought stoutly, and Buckland was very piano. Conybeare's memoir is not strong by any means. He admits three deluges before the Noachian! and Buckland adds God knows how many catastrophes besides; so we have driven them out of the Mosaic record fairly. (*Bonney, 1895, p. 42*)

Again, in the month of June, he writes to the same correspondent about the second portion of the same paper:

> The last discharge of Conybeare's artillery, served by the great Oxford engineer against the Fluvialists, as they are pleased to term us, drew upon them on Friday a sharp volley of musketry from all Sides, and such a broadside, at the finale, from Sedgwick as was enough to sink the *Reliquiae Diluvianae* for ever, and make the second volume shy of venturing out to sea. (*Bonney, 1895, pp. 42–43*)

The opinions of Murchison and Lyell obtained excellent corroboration from a further paper by Scrope (1829-30), who had been studying the valleys of the Meuse and Moselle. Scrope was already a confirmed fluvialist and the title of his new paper, 'the gradual excavation of valleys . . .', shows no weakening. He takes as his main argument the feature of incised meanders, and argues that, though he can allow a flood to carve out relatively straight valleys, he cannot admit that a tempestuous current could for no accountable reason suddenly adopt a deep winding course. He is sure on the other hand that a slow-moving river could have effected such a phenomenon. This belief is strengthened by the fact that deep meanders only occur in the harder rocks, for where the strata are less resistant the river tends slowly to eliminate its own curves.

> The author finds another equally valuable test in the extreme sinuosities of some valleys. Any sudden, violent, and transient rush of water of a diluvial character, could only produce straight trough-shaped channels in the direction of the current, but could never wear out a series of tortuous flexures, through which some rivers now twist about, and often flow for a time in an exactly opposite direction to the general straight line of descent, which a deluge or debacle would naturally have taken. Curvatures of this extreme kind are frequent in the channels of rivers flowing lazily through flat alluvial plains; and the author shows the mode in which the curves are gradually deepened and extended, till the extreme of aberration is corrected at once, and the direct line of descent restored, by the river cutting through the isthmus, which separates two neighbouring curves.
>
> But examples must be infinitely rarer of whole valleys characterized by extreme sinuosity; because, in the author's opinion, the frequent shiftings of the channels of streams tend to obliterate their windings, and reduce the sum

of the several successive excavations or valley to a more or less straight form. Still there are occasional instances where the bias of the river, or direction of its lateral force of excavation, has remained so constant as to give to the valley itself the utmost degree of sinuosity. (*Scrope, 1829–30, p. 170*)

Valleys which like these twist about in the same regular curves as the channel of a brook meandering through a meadow, can, according to the author, only be accounted for by the slow and long-continued erosion of the streams that still flow in them, increased at intervals by wintry floods. To attribute them to a transient and tremendous rush of water in the main direction of the valley, is in his opinion impossible. He contends that whilst these valleys were slowly excavated, other rivers could not have been idle during the same protracted period; but will have produced likewise an amount of excavation proportioned to their volume and velocity, and the nature of the rocks they flowed over. In the examples quoted, the rocks are mostly hard transition strata, yet the valleys are wide and deep. Where softer strata, as sands, clays, and marls, were the materials worked upon, the valleys excavated may be expected, as they are found to be, far wider in proportion to the volume of water flowing through them. The comparative softness of the materials also, by accelerating the lateral erosion of the stream, will have multiplied the shiftings of its channel, and reduced their sum with greater certainty to one average direction. Hence the deeply sinuous valleys are only found penetrating the more solid rock formations. The author thinks that a certain subdued velocity in the stream is also necessary to produce this result; and, therefore, in mountainous districts, where the torrents and rivers are most rapid, their course is nearly straight; thus confirming the author's opinion, that extreme curvature of channel can only be produced by a slow and comparatively tranquil process of excavation. (*Scrope, 1829–30, p. 171*)

The more one reads of these writings of Scrope, a Member of Parliament from 1833 to 1868, it becomes increasingly hard not to feel that had he devoted as much time to geology as did his more eminent contemporaries he might well have outshone them all. As it was, Thomas Bonney thought that none of them excelled him 'in closeness and accuracy as an observer or in soundness of induction, and firm grasp of principles as a reasoner'.

THE STRUCTURALISTS

Whereas the rival schools of Fluvialists and Diluvialists stood firmly opposed to each other, there was a small body of thinkers who accepted a mixture of both theories, and whose thought was often tinged or in parts dominated by considerations of structural control over topography. Henry Thomas de la Beche and John Phillips may be taken as typical of those who tried to utilize and reconcile the precepts of both schools and strengthened the advocacy of structural control.

In 1829 in a well-planned article De la Beche neatly summarized the existing ideas on the formation of valleys without fully joining forces with either diluvialists or fluvialists. Valleys, he says, may be formed by the action of violent catastrophes; violent catastrophes modified by river action; river action alone; river action after the bursting of a lake; and of underlying struc-

FIG. 29. *De la Beche inspecting an unconformity*

ture. He justifies the fluvialists when he states that rivers do a sizeable proportion of the work of erosion. But concludes that valleys excavated by floods or catastrophes may be distinguished from those produced by rivers alone, the former being wide with rounded outlines, and the latter narrow and more especially confined to gorges and chasms:

It would, I think, be almost impossible to deny that rivers, more particularly those discharged from the many lakes that probably once existed, have cut deeply into the land, and have formed gulleys, ravines, and gorges: but again, it seems utterly at variance with the relations of cause and effect, to suppose that valleys, properly so called, could have been formed either by the discharge of lacustrine waters, or by the rivers that now run, or could ever have run, in them.

In the discussion of this subject, we should consider only such valleys as, by the correspondence of horizontal or nearly horizontal strata on their opposite sides, show that the strata were once continuous, and that their continuity has been destroyed by the removal of the intermediate portions; – of course, the very numerous valleys formed by rents and contortions, and such as have been termed valleys of elevation and depression, as well as those of original formation, do not enter into our present consideration.

It seems to me that aqueous excavations are of two kinds: 1. Those produced by vast and violent causes not now in action; and, 2. Those resulting from the continuous and gradual operation of lakes, rivers, and other agents that have been termed meteoric: the latter series of causes operating upon valleys that most frequently owe their prior existence to the former series, and both offering very distinct appearances. Excavations of the second kind, or those produced by actual streams, present cliffs, gorges, and ravines: while the first are marked by grand and extensively rounded outlines, and by valleys of a breadth and magnitude which would seem only referable to a voluminous mass of moving waters. (*De la Beche, 1829, pp. 241–2*)

He proceeds to examine the valleys of excavation (those produced by a flood) in more detail. Evidence in favour of excavation by a flood is to be deduced from the horizontal and undisturbed position of the strata which disallows any possibility that the valleys may have arisen from the dislocation of the rocks. The streams running in the valleys because of their small size are dismissed as quite improbable agents:

Valleys of the first class, which have been usually termed valleys of denudation, are very common in districts where rocks are not far removed from an horizontal position; these, to take examples from our own country, are very abundant in Dorsetshire and the east of Devon. In these valleys, the former continuity of the strata on either side is most apparent, and neither elevations nor depressions could have caused them: they are exclusively due to the excavation of the materials by which their sides were connected. The question then arises, what has excavated them? At the bottoms of each of these valleys we find a small stream, the natural drain of the land. Could these streams have cut out such valleys as they now flow through? If there be any true relation between cause and effect, they could not. (*De la Beche, 1829, p. 242*)

As an actual example he mentions the amount of cutting achieved by the small River Char in Dorset:

The actual force of this stream, even with every assistance from floods and rains, has not accomplished more than a cut varying from four to fifteen feet deep, bounded by perpendicular walls. (*De la Beche, 1829, p. 243*)

De la Beche mentions, as had the earlier critics of the Huttonian theory, the existence of dry valleys in limestone areas, which he believes to be true 'valleys of denudation' or entirely Flood-made. He next considers what might be accomplished by 'existing causes', particularly how far a deep valley of denudation might be modified by the present drainage.

> These changes are often very considerable, and sometimes so modify the valleys that their features derived from denudation are nearly obliterated. When the original valley has been scooped out of soft substances, such as soft sandstone or conglomerate, a river resulting from the drainage of the land will have cut a gorge or ravine with cliffs of greater or less height on either side according to circumstances. (*De la Beche, 1829, p. 244*)

On the other hand, where the rivers run through wide, shallow valleys he finds that little downward erosion takes place, though some lateral cutting occurs where the stream meets an obstruction:

> Rivers when flowing through extensive and nearly level valleys seem to effect little beyond an occasional change of bed; but when a bank, a small hill, or the foot of a mountain, opposes their progress, they assail it, and form cliffs, the materials of which, if soft, fall into the stream, or make undercliffs, which are in time removed, and the work of destruction slowly continued. (*De la Beche, 1829, p. 244*)

However, the reader ends with the feeling that De la Beche does not permit himself to indulge too far in fluvial theories, and despite his statement of the considerable modifications wrought by rivers, he explains the occurrence of steep-sided gorges by the special circumstance of a river draining a lake:

> It is by no means uncommon to find plains of greater or less extent bounded on all sides by high land, and through which a principal river meanders, entering at one end by a valley, and passing out through a gorge at the other, augmented by tributary streams from the surrounding hills; . . . It can I think be scarcely doubted that this gorge has been formed by the river that now rushes along it, and still continues its excavations. It has cut below the ancient bed of the lake, as may be seen where the gravel level has been destroyed and torn away at the higher extremity of the gorge. (*De la Beche, 1829, pp. 245–7*)
>
> The celebrated falls of Niagara afford an example of a river now in the act of cutting a gorge, which, if time be allowed, may let out the waters of the lake above it. If this should ever be accomplished, the gorge will resemble those we have been describing, and show equally with them, that existing rivers may excavate gorges and precipitous channels, but that these excavations are entirely distinct from valleys of denudation. (*De la Beche, 1829, pp. 247–8*)

In his analysis of divergent opinion, De la Beche employs the guile of a diplomat by touching upon everything and pleasing everybody without eventually saying anything very definite on his own account. He was already marked as the first Director-General of the Geological Survey!

Another noted structuralist in this period was John Phillips, an uncle of William Smith, with whom he was living when Smith published his famous geological map of England. Phillips first comes to our notice in an article of 1828 which treats in more detail the point mentioned by Buckland that

FIG. 30. *Henry De la Beche*
(*From a portrait in the possession of the Geological Society*)

structure could influence surface features. In the oolitic area which he is
studying Phillips has observed the upward dip of the strata and the formation
of an escarpment along the face of the beds so exposed. He rightly appreciates
that the escarpment has been produced partly by the initial dip bringing up
the strata and partly by the forces of denudation working along the face of the
scarp.

Here then we first find indications of a denudation of a very interesting
character: a denudation on the dipping edge of the strata which has produced
here and there escarpments looking towards the formations which cover their
continuous slopes. This principle will, I am satisfied be admitted by all who

care to fully consider the facts; and it avoids the impropriety of supposing a dislocation without evidence. According as the denudation has extended more or less towards the 'rise' of the strata, more or fewer beds of the oolitic series are exposed: where the denudation had produced the greatest effect, the lowest stratum exposed is Oxford clay. (*Phillips, 1828, pp. 245–6*)

It thus appears that from Helmsley to Pickering the effects of denudation on the dipping edge of the oolitic series are scarcely at all observable; the beds declining regularly, and almost without interruption, into the Vale of Pickering. The denudation of Helmsley appears to be connected with the formation of the valleys about that town; but from Pickering eastward, a more extensive operation of similar causes has produced a long range of escarpments crossing the direction of the little existing valleys, but parallel to the general line of the great Vale of Pickering. (*Phillips, 1828, p. 247*)

I have no doubt that an impartial examination of this line of country will lead to the adoption of my opinion, that the pseudo-escarpments of the oolitic rocks on the north side of the Vale of Pickering are caused by denudation nearly parallel to their ranges, and that no general dislocation is at all concerned in the appearances. (*Phillips, 1828, p. 248*)

In the opening passages of a book published on the geology of Yorkshire in the following year, Phillips is clearly an advocate of the Deluge, though Hutton does get a passing mention in the midst of some classical allusions. Phillips follows Buckland in claiming that there are many features pointing towards the occurrence of the Flood, and follows Hutton in his description of the various sub-aerial processes. However, though the author acknowledges the daily action of the rivers and tides in breaking up and distributing rocks, this admission is really prefatory to the main statement that the achievement of rivers demonstrates the potential of a great Flood.

. . . the deluge must be admitted to have happened, because it has left full evidence in plain and characteristic effects upon the surface of the earth . . . the whole earth is covered by pebbles, the wreck of a general flood. Filling the valleys, overspreading the plains, and covering the hills, rounded stones, of. all sizes and all kinds, mixed together in as much confusion as pebbles on the sea-shore, (fragments of all the known rocks which compose the interior of the earth,) are profusely scattered on its surface. (*Phillips, 1829, pp. 16–17*)

As he traces in detail the nature of the diluvial evidence, he falls into the traditional error of attributing to the Deluge the distribution of glacial pebbles and erratics.

As to the height of this flood in our own country, the sides of Ingleborough, on which rest fragments of rocks transported from Keswick; the brow of Stainmoor, which supports large masses of granite; and the top of Carrock fell, from which so large a quantity of sienite has been removed, demonstrate that our proudest hills were overflowed; and as to the extent, all countries acknowledge the wide-spreading visitation: – the deluge covered the whole earth. (*Phillips, 1829, pp. 17–18*)

Phillips, after dating the Deluge to a time after the stratification of the earth and the advent of animals, attempts to define the topographical changes brought about by it. He credits it with considerable erosive powers but is inclined to account for the shape of many elevations and depressions by the influence of the underlying strata. Diluvial valleys of denudation he restricts to valleys cut through horizontal strata which appear as similar beds on either side and thus prove that the rock-structure can have played no part in the valley-formation.

> No one can doubt that great alterations were occasioned in the features of the earth's surface, at the period of the deluge, who considers the extensive tracts formed of the diluvial detritus. All the solid land of Holderness is an accumulation of this kind, from the ruins of others parts of England and Scotland and perhaps Norway. If hills were known before the flood, their present peculiar shapes must be dated from that event; and if vallies were then in existence, they must have been deepened and widened, or possibly filled up and obliterated. (*Phillips, 1829, pp. 19–20*)
>
> Many great natural depressions or wide vales are produced, evidently by the convergence of opposite declinations of strata; as the great vale of the Thames is occasioned by meeting dips from Hertfordshire and Surrey; and such are, doubtless, antediluvian. Many geologists believe that, from some unexplained causes operating during their deposition, some strata were originally deposited at higher elevations than others; that, for example, the lower part of the coal series was made to attain elevations not reached by the upper part of the same series; and that the new red sandstone was never in England placed at so great an altitude as some of the strata which lie above it and below it. In these instances, therefore, it has been concluded that the antediluvian features of the earth were not very different from what we now witness; and these instances admitted to their full extent, actually include the most striking variations in the surface of the earth; for it is certainly true, that the great mountain ranges which seem to compose the skeleton of the earth; the wide oceans, plains, and level tracts, and even the remarkable lines of secondary hills and most extensive vallies are placed in accordance to the interior structure of the earth. Hence it follows that we must limit our inquiry, as to the changes produced on the surface of the earth by the deluge, to the vallies and hills which seem evidently to have derived their peculiar features from currents of water, since the consolidation of the strata. Even thus limited, the subject is ample, fertile, and instructive. Many vallies in a secondary country are excavated through several strata, as limestone, clay, and sandstone, which appear on the opposite sides in most exact agreement as to thickness, composition, and mode of arrangement. That such rocks were originally deposited in continued planes, and therefore, once connected across the chasm or valley which now divides them, can hardly be doubted. The vallies themselves bear marks of their origin; their bottom is a continued plane: their sides correspond with answering sinuosities; and their every peculiarity suggests the action of recurrent water. (*Phillips, 1829, pp. 20–21*)

There is, Phillips continues, also evidence that diluvial action has cut valleys through vertical strata:

> Some valleys cross and cut through vertical strata, which must necessarily have been at first deposited nearly horizontal. Therefore, such valleys were not produced till after the displacement of the rocks. (*Phillips, 1829, p. 21*)

Later Phillips speaks approvingly of the denudation and deposition accomplished by rivers and the sea although most of his other statements have been concerned with proving that such operations had been carried out by the Flood alone.

> The natural agents now employed in altering the face of the globe, are fire and water. The former forces fluid matter from the interior, and spreads it around the volcanic mountains; the latter is incessantly occupied in lowering heights, wasting and smoothing precipices, filling up valleys, and equalizing the surface.
>
> . . . The records of history declare what large tracts of inhabited country have been lost in the sea, and what extensive surfaces of new land have arisen to contract the dominion of water. Observation shews on our own shores much of the reciprocal process of demolition and augmentation; and thus we are enabled to form a correct estimate of the effects of this 'war of sea and land'. (Phillips, 1829, p. 24)

He also begrudgingly admits that the 'imperceptible' erosion of the landscape by rain and rivers involves a great amount of material. Yet he misapplies his observations by emphasizing deposition and by stating that stream action is employed mainly in filling up the valleys denuded by the Flood.

> Imperceptible as is the reduction of mountains and hills by rains and rivulets, yet the matter thus collected, by constant attrition, assumes an important character, when concentrated along the margins of rivers, and changes the appearance of the valleys. In proportion to the magnitude of the stream, the altitude of its sources, and the nature of the country through which it flows the effects are more or less considerable. But they every where tend to the same result; the raising of the level of the valley by horizontal layers of sediment. (Phillips, 1829, pp. 25–26)

There were other, less prolific, contemporary authors who were impressed with the possibility that the pattern and much of the form of certain valley systems could be attributed to structural lines, which might have guided later diluvial, marine or fluvial eroding agencies. Peter John Martin, in his description of Western Sussex (1828), first noted the alignment of the transverse valleys on either side of the Weald and postulated their relationship to fissures formed during upheaval of the dome and later modified by diluvial action. Gideon Algernon Mantell, a prolific palaeontologist, also invoked a marine inundation to modify structural lines in order to explain the present relief of the Weald.

> The hills and vales around the (Tunbridge) Wells have been produced by the displacement which the strata have suffered during their emersion from the depths of the waters. This elevation was in all probability not sudden, but gradual; and in the changes here contemplated, some of the highest peaks would first appear above the waves, and constitute a group of islands. . . . During this process valleys would be scooped out by currents, and the sharp edges of the rocks rounded by diluvial action, and accumulations of debris be formed in the undisturbed depths of the ocean. This action would go on

during the gradual elevation of the land, the difference of level creating different systems of valleys, till the whole of the south-eastern part of England was elevated above the ocean, and presented the picturesque surface by which it is now characterized. (*Mantell, 1832, p. 105*)

The next era is that during which the Crag, and the tertiary strata, and the chalk on which they repose, were lifted up to their present situations; the channel which separates England from France was broken through, and the transverse valleys of the north and south downs were produced or enlarged; for, although these valleys are now river courses, yet it is obvious that they originated in disruption, for the strata, in every instance which I have observed, diverge from the line of fracture. . . . These elevatory movements and convulsions were manifestly of great intensity, and naturally changed the physical geography of the south-east of England . . . the waters resulting from the drainage of the land, and which, before the existence of the transverse fractures, probably flowed through the longitudinal valleys towards the east, would be thrown into different channels, and find their way to the ocean by the existing river courses . . .

Subsequently to the last-mentioned changes the surface of the country appears to have undergone no material alteration. . . . (*Mantell, 1833, pp. 350–2*)

GENERAL PROGRESS, 1820–30

In retrospect this decade brought for landscape studies in Britain slow but steady progress although Conybeare, a diluvialist, writing in 1832, thought that since 1821 'geology has received scarcely any valuable additions, and not a single fundamental one' (*Conybeare, 1832, p. 375*). Yet it is obvious that much useful progress had been made in field-observations, particularly in structural influences, and that diluvialism was weakening before the advance of fluvialism. Moreover, intellectual conditions in this subject were much less stagnant than on the Continent where the Wernerians still controlled geological thinking. In Germany, the erstwhile Wernerians, Von Buch (1774–1853) and Humboldt, who had been at Freiberg together, were directing the vision of the scientific intelligentsia. Von Buch, after visiting the Auvergne (1802) and Campagna, where he saw Vesuvius in eruption (1805), refuted the Neptunist ideas on volcanic rocks but the general faith in Wernerism was slow to weaken. A more distinct veer away from catastrophism became apparent in 1818 when

the Royal Society of Sciences in Göttingen offered a prize for the best 'investigation of the changes that have taken place in the earth's surface conformation since historic times, and the application which can be made of such knowledge, in investigating earth revolutions beyond the domain of history.' (*Von Zittel, 1901, p. 187*)

The prize was won by K. E. A. Von Hoff with a three-volume work that made a considerable impression on Charles Lyell (p. 144). The first volume (1822) dealt with changes of land and sea; the second (1824), with volcanoes, earthquakes and geysers; and the third (1834), with the degradation of the land, including the gradual destruction of a continent by water, snow, ice and the sea. Unfortunately, this third volume, in which Von Hoff 'discredits Buckland's hypothesis of a universal flood in a learned and convincing chapter' (Von Zittel, 1901, p. 188), was not published until after the appearance of Lyell's *Principles* which influenced it profoundly.

In France, the Huttonian position was, if possible, worse than in Germany, with Cuvier, the catastrophist, holding court to all prominent continental scientists. As already mentioned, Lamarck's *Hydrogéologie* (1802), with its strong uniformitarianism, was largely overlooked. Here, the only other person to make a notable attack upon traditional thought was Constant Prévost who in 1825 and 1828 dared to criticize Cuvier's ideas. Prévost favoured a uniformitarian approach to geology but he also admitted that certain forces may have been more powerful in the past than at present. In this he has, of course, our approval but such half-hearted attacks were not likely to make much progress in either Neptunist or diluvialist strongholds.

Yet, as we shall now see, the time produced the man, and upon this field of desultory skirmishes there now strode a veritable giant, the most imposing figure since Hutton and Playfair.

CHAPTER ELEVEN

Lyell's *Principles of Geology*: The First Reactions

Why, man, he doth bestride the narrow world
Like a Colossus.

(W. Shakespeare: *Julius Caesar*, I, ii, 135)

THE PREPARATORY YEARS

The early life of Charles Lyell gave little hint that he was to become the Napoleon of the geological world. He was born on the 14th of November 1797, the year of Hutton's death, at Kinnordy in Forfarshire of fairly well-to-do parents (Eyles, 1947). The family did not use their Scottish home as a regular habitation and most of Lyell's childhood was spent in the south of England. From his own account he appears to have shown a marked precocity in pursuit of his hobby of insect collecting and dissection, particularly of Lepidoptera and aquatic species.

After going through his school days as a rather sensitive and better-than-average pupil, he entered Exeter College, Oxford, at the age of 17. Though his purpose was to study classics, it is known from his continuous flow of letters that he attended the geological lectures of Dr Buckland and that he read some geology texts, including Bakewell's. At the end of his stay at the university he succeeded in obtaining a second class honours degree in Classics. With a view to adopting the Bar as a profession, he was entered at Lincoln's Inn but his eyes having suffered from over-studying, he was sent on a tour of France, Switzerland and Italy in order to rest them.

The weakness of his eyes troubled Lyell all his life and it is amazing how, by resting them periodically, he was able to go on and produce such a voluminous quantity of published matter. The same defect must also have made his field-work more difficult and makes the clear interpretations of his observations all the more remarkable.

> He is anxious for instruction, and so far from affecting the big-wig, is not afraid to learn anything from anyone. The notes he takes are amazing; many a one he has had from me today. He is very helpless in the field without people to point things out to him; quite inexperienced and unable to see his way either physically or geologically. He could not map a mile, but understands all when explained, and speculates thereon well. . . . He wore spectacles half the day, and looked ten years older in consequence. (*Diary of A. C. Ramsay quoted in Geikie, 1895, pp. 180–1*)

It was only in the last ten years of his life that his eyes really failed him and he was forced to employ a full-time secretary.

Despite his interest in the subject, Lyell did not devote time to active geology on his first continental tour which was really a holiday. Yet in the following year (1819) he was elected Fellow of the Geological Society and of the Linnean Society. His election to the former was probably assisted by his acquaintance with Buckland, for it was a feature of Lyell that while young he attracted the attention of many prominent scientists and then kept in touch with them by a copious stream of correspondence.

On returning from the Continent, Lyell studied law for a short time but as his eyes weakened again, he joined his father on a second continental tour. Once more little geology was carried out, though Lyell was fortunate enough to meet Cuvier and Humboldt on an equal social footing.

In 1824, he turned seriously to geological observation and made another tour in France, principally in the Marne area, where he met Brongniart and Prévost. The latter may well have encouraged Lyell to persist with the Huttonian tendencies already in his mind, and Murchison later speaks as if this were so:

> I induced my wife . . . to accompany me as well as my associate, Charles Lyell. We were off in April, and on the 26th of that month were at work in the field with Constant Prévost, following his subdivisions of the Paris basin. The theoretical views of Prévost made a deep impression on Lyell. (*Quoted in Geikie, 1875, I, p. 148*)

The two struck up a firm friendship, and Lyell invited Prévost to England where they made a joint tour of the country between Bristol and Land's End. Later in the same year Lyell travelled, this time with Buckland, through some of the more rewarding geological areas of Scotland.

After this spasm of geologic activity he returned to law and made sporadic appearances on the Western circuit for two years. Yet gradually his scientific pursuits began to take pride of place and by 1827 he had been elected a Fellow of the Royal Society and had contributed his first articles to the *Quarterly Revue* and *Brewster's Edinburgh Journal of Sciences*.

Between 1828 and 1830 he made three tours abroad. The first, with Murchison, through the volcanic districts of France and Italy, resulted in the publication of the two joint papers already mentioned (pp. 126–8) and greatly influenced Lyell's future thought.

> The great flows of basalt . . . modern in a geological sense, but carved and furrowed by the streams which still were flowing in their gorges . . . showed that rain and rivers were most potent, if not exclusive, agents in the excavation of valleys. 'The whole tour', thus he wrote to his father, 'has been rich, as I had anticipated (and in a manner which Murchison had not), in those analogies between existing nature and the effects of causes in remote eras which it will be the great object of my work to point out'. (*Bonney, 1895, p. 37*)

The second tour, during 1828 and 1829, was to France and Switzerland; and the third, early in 1830, to the Pyrenees and a small part of northern Spain. Such extensive expeditions showed Lyell's intensity of purpose and enthusiasm as geologizing in those days was no occupation for the dilettante. Train travel was only beginning to be established in England and was almost absent on the Continent. Moreover, geology demanded more often than not visits to almost inaccessible spots, reached only by travelling long distances on foot, or by horse or mule. Naturally accommodation was not easily had and Lyell often bemoans its deficiencies:

> and as I shall be busy every moment till I sail for Naples, I will employ a few minutes, if but to tell you that in the honest peasant's bed above mentioned I slept not, though tired, ten minutes in the whole night. They had added, at my asking for covering, an old blanket, in which were more fleas than all the insects in our collection, and I suppose they had starved since last winter, for next morning I was marked all over from head to foot with red spots like one of the Ancient Britons with his body painted. (*Letter, 31 Dec. 1828, in Lyell, Mrs, 1881, I, pp. 226–7*)
>
> Talking of rough work, there were two things of which I deemed it advisable to break to my valet Rosario. First, not to bring in the salt in his hands, often none of the cleanest; and secondly, not to stuff the meat remaining after dinner, and intended cold for next morning, into my slippers, without at least some paper round it. (*Letter, 10 Jan. 1829, in Lyell, Mrs, 1881, I, p. 231*)

In 1827 Lyell gave up the Bar and we can trace in his writings the first definite hints of his plan to publish a book. In a letter to his friend Dr Mantell, (5 February 1828), he discusses the type of work he has in mind:

> I at first intended to write 'Conversations on Geology': it is what no doubt the booksellers, and therefore the greatest number of readers, are desirous of. My reason for abandoning this form was simply this; that I found I should not do it all, without taking more pains than such a form would do justice to. . . . But finally, I thought, that when I had made up my own mind and opinions in producing another kind of book, I might then construct conversations from it. In the meantime there is a cry among the publishers for an elementary work, and I much wish you would supply it. Anything from you would be useful, for what they have now is positively bad, for such is Jameson's Cuvier. (*Lyell, Mrs, 1881, I, p. 177*)

At the same time he seems to be drawing upon very varied sources of knowledge, and this width of reading remained one of Lyell's principal qualities throughout his long career. He was able, as few other men at that time were, both to keep abreast of the several sciences and to give a lead in them. He could absorb and understand new trends and discoveries in the leading subjects and then skilfully condense them into an intelligible analysis which could be applied unerringly to some other particular feature or process. As early as 1827 he was reading Lamarck's work on biological transmutations with an avid appreciation heightened by his own experiences.

FIG. 31. *Charles Lyell*

I devoured Lamarck en voyage. . . . His theories delighted me more than any novel I ever read, and much in the same way, for they address themselves to the imagination, at least of geologists who know the mighty inferences which would be deducible were they established by observations. But though I admire even his flights, and feel none of the odium theologicum which some modern writers in this country have visited him with, I confess I read him rather as I hear an advocate on the wrong side, to know what can be made of the case in good hands. I am glad he had been courageous enough and logical enough to admit that his argument, if pushed as far as it must go, if worth anything, would prove that men may have come from the Ourang-Outang. But after all, what changes species may really undergo! How impossible will it be to distinguish and lay down a line, beyond which some of the so-called extinct species have never passed into recent ones. That the earth is quite as old as he supposes, has long been my creed, and I will try before six months are over to convert the readers of the *Quarterly* to the heterodox opinion. (*Letter, 2 March 1827, in Lyell, Mrs, 1881, I, p. 168*)

Geikie, at a later date, summed up this rare faculty:

> Lyell's function was mainly that of a critic and exponent of the researches
> of his contemporaries, and of a philosophical writer thereon, with a rare
> faculty of perceiving the connection of scattered facts with each other, and
> with the general principles of science. As Ramsay once remarked to me, 'We
> collect the data, and Lyell teaches us to comprehend the meaning of them'.
> (*Geikie, 1905, p. 404*)

Lyell showed the same powers of analysis and exhaustive application in the
preparation of his major book and in all that he did, as may be judged from
the following extract from the long letter (on 14 June 1830) he sent to Scrope
who was about to review *Principles of Geology* (Scrope, 1930):

> . . . I am sure you may get into *Q.R.* what will free the science from Moses.
> . . . They see at last the mischief and scandal brought on them by Mosaic
> systems. Férussac has done nothing but believe in the universal Ocean
> up to the chalk period till lately. Prévost had done a little, but is a diluvialist,
> a rare thing in France. If anyone has done much in that way, I have not been
> able to procure their books. Von Hoff has assisted me most, and you should
> compliment him for the German plodding perseverance with which he filled
> two volumes with facts like tables of statistics; but he helped me not to my
> scientific views of causes, nor to my arrangement. The division into aqueous
> and igneous causes is mine, no great matter, and obvious enough. Von Hoff
> goes on always geographically. For example, he will take as a chapter 'changes
> in the boundaries of sea and land', . . . I have done mine from actual observa-
> tion, principally in coast surveys. My division into destroying and reproduc-
> tive effects of rivers, tides, currents, etc., is, as far as I know, new – my theory
> of estuaries being formed is contrary to Bakewell and many others, who think
> England is growing bigger. In regard to Deltas, many facts are from Von Hoff,
> but the greater part, not. All the theory of the arrangement of strata in Deltas
> and stratification, etc., is new, as far as I know, and the importance of spring
> deposits. Von Hoff thinks all that is now going on, a mere trifle comparatively,
> though he has done more than any other to disprove it. My views regarding
> gneiss, mica schist, etc., could not come in Vol. 1. Sedgwick found in the
> centre of Eastern Alps an encrinital limestone alternating with genuine mica
> schist and the white stone of Werner. This made a sensation here at G.S.
> this session. It was before known to E. de Beaumont. Think of this fact.
> Whether so made originally or not, it is clear that mica slate owes its stratifica-
> tion to deposition, because the limestone did. Ergo after a cool sea existed,
> with zoophytes enjoying the light of heaven, and feeding on some animalcules
> which lived as now – these rocks were formed no matter how. Graywacke in
> its most ancient mineralogical form is proved to be posterior to vertebrated
> animals at Glaris, and to fuci and some coal plants lately in Ireland. In con-
> troverting, just allude to 'having heard something of the Alpine Discovery',
> because Sedgwick was most unwilling last year to admit such a thing; also
> hint that my reasons are yet to come, as I say in several passages.
> Probably there was a beginning – it is a metaphysical question, worthy a
> theologian – probably there will be an end. Species, as you say, have begun
> and ended – but the analogy is faint and distant. Perhaps it is an analogy, but
> all I say is, there are, as Hutton said, 'no signs of a beginning, no prospect of
> an end'. Herschel thought the nebulae became worlds. Davy said in his last

book, 'It is always more probable that the new stars become visible and then invisible, and pre-existed, than that they are created and extinguished.' So I think. All I ask is, that at any given period of the past, don't stop inquiry when puzzled by refuge to a 'beginning', which is all one with 'another state of nature', as it appears to me. But there is no harm in your attacking me, provided you point out that it is the proof I deny, not the probability of a beginning. Mark, too, my argument, that we are called upon to say in each case, 'Which is now most probable, my ignorance of all possible effects of existing causes', or that 'the beginning' is the cause of this puzzling phenomenon?

It is not the beginning I look for, but proofs of a progressive state of existence in the globe, the probability of which is proved by the analogy of changes in organic life. 'Tis an easy come-off to refer gneiss to 'the beginning, chaos', &c., and put back finding an encrinite for half a century. That all my theory of temperature will hold, I am not so sanguine as to dream. It is new, brannew. Give Humboldt due credit for his beautiful essay on isothermal lines: the geological application of it is mine, and the coincidence of time 'twixt geographical and zoological changes is mine, right or wrong. Sedgwick and Murchison have found an intermedial formation in Eastern Alps 'twixt chalk and oldest tertiary, helping to break down that barrier, to fill that lacune. Until Rennel's posthumous work on currents is out, I could not have a good copper-plate of their course. Thanks for the hint, which shall not be lost, if your review helps me in spite of the saints to a second edition, and in spite of 1,500 copies, a number I regret but could not avoid. My labour has been greater than you would suppose, as I have really had so little guidance. Your little valley paper was one of my best helps. I mean as guide in classification of facts. I was afraid to point the moral, as much as you can do in *Q.R.*, about Moses. Perhaps I should have been tenderer about the Koran. Don't meddle much with that, if at all. (*Lyell, Mrs, 1881, I, pp. 268–70*)

We must admit that few authors have the good fortune of explaining their works to the critics before being damned or praised. Yet to fewer still must such an opportunity prove unnecessary. It happened that *The Principles* was an outstanding and lucid book, and that Scrope was a very intelligent reviewer.

'THE PRINCIPLES OF GEOLOGY'

The first volume of Lyell's masterpiece appeared in January 1830 and revealed itself as a damning indictment of the worst inadequacies of the catastrophic theories. Whereas previously Hutton's theory had only implicitly suggested the principle of uniform causes, Lyell set himself up, in the words of Geikie, as 'the great high priest of uniformitarianism'. For him everything could, and should, be explained by the operation of existing natural processes and to prove his contention he took his examples from zoological, botanical, geological, and human settings. Like Hutton before him, he saw the world evolving, growing and changing by a series of continuous processes and in the total absence of any, but the most local, cataclysms or convulsions.

This first volume constituted Lyell's main attack upon traditional ideas and in it, besides assailing catastrophists, he made a vigorous assault on theological prejudices concerning the span of geologic time and favoured the rejection of all attempts to interpret geology by reference to the scriptural record.

The second volume of *The Principles* appeared in January 1832. It echoed the same theme but paid more attention to details of erosion and denudation and dealt with all the physical processes ranging from the action of rivers and tides to the fiery contortions of volcanic cones. It included new theories of slow climatic changes and discarded the catastrophic idea of rapid variations in the distribution of land and sea, putting in its place the theory that sea-level changes are due partly to the growth and contraction of polar and mountain ice-sheets. Lyell also introduced many new terms, as in the case of 'metamorphic' for sedimentary rocks changed by contact with igneous activity. This wealth of ideas, supported by an abundance of concrete examples and always related in a most pleasing and coherent manner, could hardly have been more satisfying for the general reader. The beauty of Lyell's geological philosophy, the majesty of its presentation, and the high quality of his stratigraphical observations, caused his geomorphic ideas to exercise a control for the succeeding half century far in excess of their intrinsic value.

When the third volume, which was mainly stratigraphical and palaeontological, appeared in May 1833, second editions of the first and second volumes had already been published. In 1834, the whole work was reprinted in four smaller volumes and styled the third edition.

THE RECEPTION

As his viewpoint was so contrary to the principles then being taught by most geologists, Lyell fully expected that his book would meet with united condemnation. Instead it won an almost immediate and overwhelming popularity with the scientific-reading public and the restraint of the opposition must have seemed an anti-climax, even if highly gratifying. In one of his numerous letters, Lyell records that

> Scrope's review of the first volume of *The Principles* has been much admired; . . . the book is selling steadily, and is likely to prove 'as good as an annuity'; . . . it has not been seriously attacked by the 'Diluvialists', while it has been highly praised by the bulk of geologists. (*Quoted by Bonney, 1895, pp. 57–58*)

There is no doubt about the influence of Scrope's review, which was most favourable, although Scrope with rare insight remarks unfavourably upon the importance given to marine action in land sculpture and also appears slightly worried about the author's rigidly uniformitarian attitude.

> The amount of excavation and accumulation carried on by marine currents is considered by Mr Lyell to exceed very greatly that of running water on the land. (*Scrope, 1830, p. 440*)
> We should no doubt be going too far, were we to assume, that the succession of events which we perceive on the surface of the earth has not been going on as now from all eternity. But what we do say is, that all analogy is in

favour of such a supposition, and that the contrary assumption, that of the eternal permanency of the actual state of things on the globe, is decidedly more unphilosophical. (*Scrope, 1830, p. 466*)

Yet favourable reviewing forms only one of many reasons for the almost magical appeal of *The Principles*. Much of its success must be attributed to its vast comprehensiveness, to its superabundance of details (relished at the time) and to the beguiling method whereby all occurrences are explained simply as the outcome of the long continuance of present processes. But do these intrinsic qualities fully explain the easy success of Lyell compared with the lack of recognition of Hutton and Playfair? It has been suggested that the book's 'delicately heretical flavour' added to its charm but numerous contemporary heretical works failed to gain a fraction of its popularity. Surely some credit must be given to the shift in the climate of opinion since Hutton's day, to the growing popularity of geology as a whole, to the fortunate choice of a metropolitan birthright and, not least, to Lyell himself – to his high position in society, his particular circle of friends and his charming manner? This last factor is emphasized by Drachman.

Lyell was most fortunate in his temperament, which charmingly suited the best elements in the spirit of his age. He was neither a reactionary nor a radical, neither an unworldly laboratory-beagle like Darwin, a defiant revolutionary like Huxley, nor an embittered idealist like Butler. With grace and tact worthy of Lord Chesterfield, he combined rare learning, originality and force of intellect, literary gifts of no mean order, and a balanced sanity that was most British and most genuinely Victorian, in being at once courageously honest and quietly cautious. (*Drachman, 1930, pp. 47–48*)

Whatever the reasons, the result in the world of geologic thought must be ranked as a bloodless revolution, for the influence of catastrophism waned markedly from the moment *The Principles* appeared. Wallace, writing at a later date, stresses this change in public opinion.

. . . in 1830, while Cuvier was at the height of his fame, and his book was still being translated into foreign languages, a hitherto unknown writer published the first volume of a work which struck at the very roots of the catastrophe theory, and demonstrated, by a vast army of facts and the most cogent reasoning, that almost every portion of it was more or less imaginary and in opposition to the plainest teachings of nature. The victory was complete. From the date of the publication of *The Principles of Geology* there were no more English editions of *The Theory of the Earth*. (*Wallace, 1898, p. 112*)

It is not surprising that on the Continent the reversal of thought was far less complete, as the following judgement indicates.

He (Lyell) became the great high priest of Uniformitarianism – a creed which grew to be almost universal in England during his life, but which never made much way in the rest of Europe, and which in its extreme form is probably now held by few geologists in any country. (*Geikie, 1905, pp. 403–4*)

A direct outcome of *The Principles* was the promotion of a vastly increased interest in geology and the publication of several works on the subject. Lyell, in writing to his future wife (1 September 1831), says:

> I have been profitably employed this morning since breakfast in reading Conybeare's last little work, and several chapters of Omalius d'Halloy's new *Elements of Geology*. What a proof of the interest now excited by the science, that since my first volume appeared we have three works, Macculloch, two volumes, De la Beche, and this, on Geology, besides innumerable memoirs! (*Lyell, Mrs, 1881, I, p. 330*)

It says much for Lyell that he emerged unconceited from the public recognition and the rewards that followed rapidly upon the success of his book. In 1831 he was made Deputy-Lieutenant of Forfarshire and, after declining the appointment in 1830, he accepted in 1831 the professorship of geology at the newly launched King's College, London. Here is Lyell's own account (March 1831) of the clerical hesitation over granting him the post:

> I have been within this last week talked of and invited to be professor of geology at King's College, an appointment in the hands entirely of the Bishop of London, Archbishop of Canterbury, Bishop of Llandaff, and two strictly orthodox doctors, D'Oyley and Lonsdale. Llandaff alone demurred, but as Conybeare sent him (volunteered) a declaration most warm and cordial in favour of me, as safe and orthodox, he must give in, or be in a minority of one. The prelates declared 'that they considered some of my doctrines startling enough, but could not find that they were come by otherwise, than in a straightforward manner, and (as I appeared to think) logically deducible from the facts, so that whether the facts were true or not, or my conclusions logical or otherwise, there was no reason to infer that I had made my theory from any hostile feeling towards revelation'. Such were nearly their words, yet Featherstonhaugh tells Murchison in a letter that in the United States he should hardly dare in a review to approve of my doctrines, such a storm would the orthodox raise against him! (*Lyell, Mrs, 1881, I, pp. 316–17*)

Everything, however, seems to have gone smoothly and in his second lecture (May 1832) Lyell made a special point of placating the worst fears of the clerics.

> I worked hard upon the subject of the connection of geology and natural theology, and pointed out that the system which does not find traces of a beginning, like the physical astronomer, whose finest telescope only discovers myriads of other worlds, is the most sublime; and as there is no termination to the view as regards space of the acts of creation, and manifestations of divine power, so also in regard to time, &c., concluding with a truly noble and eloquent passage from the Bishop of London's inaugural discourse at King's College, in which he says that truth must always add to our admiration of the works of the Creator, that one need never fear the result of free inquiry, &c. (*Lyell, Mrs, 1881, I, p. 382*)

His lectures were popular and the only cloud that appears to have been created arose from the action of the college governors in banning ladies from attending. Lyell regretted this as it diminished his audience. He had, it should be noticed, recently married Miss Mary Horner, the daughter of an eminent geologist. Soon afterwards, in 1833, he resigned the professorship on the grounds that it interfered too much with the geological work he planned to carry out. He realised too that his eyes were weak and that constitutionally he was far from strong.

THE IMMEDIATE REACTION OF THE CATASTROPHISTS

Lyell's physical courage was never more needed than in the early 1830's when criticisms of his work began to appear in Britain and on the Continent. In Britain, considering that their reputations were at stake, the replies of the diluvialists were remarkably mild. By now, however, the mass of uniformitarian evidence, mostly based on extensive field-work, was so great that much effort and research would have been required even to attempt to discredit it. The best the catastrophists could hope to do was to ridicule or belittle the value of Lyell's main theme, and this they did.

Sedgwick (1830), one of the first attackers, began by striking at the one real weakness of Lyell's notions, his insistence on the *exclusive* operation of uniform processes throughout geologic time. This, Sedgwick says, cannot be so, because the complexity of most features results not from one cause but from the interaction of a multitude of causes. At the same time such profound differences occur between features, as in the case of horizontal and vertical strata, that it is foolish to deny the action of catastrophic forces. From the variations in strata he is prepared to assume an unknown number of catastrophes occurring at different places at different times. Therefore, the formation of features such as valleys must be sought in the inter-relation of many factors. He then went on to suggest a number of ways by which valleys may have been formed, and emphasizes the influence of structure or fractures.

> If I might give my own opinion on this debated question, I should say, that the existing river drainage of every physical region, is a complex result, depending upon many conditions – the time when the region first became dry land – its external form at the time of its first elevation above the sea – and all the successive disturbing forces which have since acted upon its surface. But none of these elements are constant; no wonder, then, that results derived from distant parts of the earth should be so greatly in conflict with each other. In the formation of valleys there is therefore little wisdom in attributing every thing to the action of one modifying cause. We know by direct geological evidence, that nearly all the solid portions of the earth were once under the sea, and were lifted to their present elevation, not at one time, but during many distinct periods. We know that elevating forces have not only acted in such variations of intensity, that the same formation is in one country horizontal, in another vertical; in one country occupies the plains, in another is found only at the tops of the highest mountains. Now every great irregular elevation of the land (independently of all other results) must have produced, not merely a rush of the retiring waters of the sea, but a destruction of

equilibrium among the waters of inland drainage. Effects like these must have been followed by changes in the channels of rivers, by the bursting of lakes, by great debacles, and in short by all the great phaenomena of denudation. In comparing distant parts of the earth, we may therefore affirm that the periods of denudation do not belong to one, but to many successive epochs; and by parity of reasoning we may conclude that the great masses of incoherent matter which lie scattered over so many parts of the surface of the earth, belong also to successive epochs, and partake of the same complexity of formation.

The excavation of valleys seems therefore to be a complex result, depending upon all the forces, which, acting on the surface of the earth since it rose above the waters, have fashioned it into its present form. We have old oceanic valleys which were formed at the bottom of the sea in times anterior to the elevation of our continents. Such is the great valley of the Caledonian canal, which existed nearly in its present form at a period anterior to the conglomerates of the old red sandstone. We have longitudinal valleys formed along the line of junction of two contiguous formations, simply by the elevation of their beds. To this class belong some of the great longitudinal valleys of the Alps. We have other valleys of more complex origin; where the beds through which the waters now pass have been bent and fractured with an inverted dip at the period of their elevation. Such is the valley of Kingsclere, described in a former volume by Dr Buckland. We have valleys of disruption, marking the direction of cracks and fissures produced by great upheaving forces. Such are some of the great transverse valleys of the Alps. Of valleys of denudation our island offers a countless number. Some are of simple origin; for example, the dry combes and valleys of the chalk, which appear to have been swept out by one flood of retiring waters during some period of elevation. Others are of complex origin, and are referable to many periods, and to several independent causes. Lastly, we have valleys of simple erosion: such are some of the deep gorges and river channels in the high regions of Auvergne, excavated solely by the long continued attrition of the rivers which still flow through them. (*Sedgwick, 1830, pp. 191–2*)

Sedgwick's argument is important because it shows how the diluvialists had shifted position until they were clearly accepting some of the despised ideas of Hutton. Indeed, Buckland's thesis of a flood now retained little of its original form and instead of the one universal flood a whole series of catastrophes was pressed into service. Though the valleys of denudation were still accepted as physical features, the origin of valleys was being increasingly ascribed to the influence of structure or the occurrence of fractures. Also for the first time since Buffon (pp. 13–15) there appeared the idea that valleys might have been fashioned while continents remained below the surface of sea, and this was one of the main geomorphic suggestions in Lyell's *Principles*. On the question of the formation of strata Sedgwick has become strictly Huttonian and we know that this partial conversion was hastened by his explorations on the Continent with Murchison:

Sedgwick and Murchison are just returned, the former full of magnificent views. Throws overboard all the diluvian hypotheses; is vexed he ever lost time about such a complete humbug; he says he lost two years by having also started a Wernerian. He says primary rocks are not primary, but, as Hutton supposed, some igneous, some altered secondary. (*Letter of Lyell in 1829, Bonney, 1895, p. 45*)

That Lyell felt the force of Sedgwick's criticism is apparent from changes he made in later editions; and when he issued his third volume (1833), he

. . . denied that he advocated that 'The existing causes of change have operated with absolute uniformity from all eternity'; nevertheless, his doctrine of Uniformity was somewhat extreme, and it was undoubtedly taken in too restricted a sense by some of his followers. (*Woodward, 1911, pp. 98–99*)

Conybeare (1830) was another diluvialist who put his objections into print. In his first reply he adopted a generalized and rather sarcastic tone against the whole principle of uniformitarianism. Reduced to its essence, he merely says that things just do not happen in the way Lyell imagined they did, and he bases his disbelief on the height and form of mountain chains, such as the Alps, which he cannot visualize as having been uplifted by a slow process.

I hope it will not be considered as an invidious remark, but merely as expressing the general impression which the book has left on my mind, that it is an expanded commentary on the celebrated Huttonian axiom, that 'in the economy of the world no traces of a beginning or prospect of an end can be discerned'. Now I would not doubt for a moment on the unfounded (as I am most willing to own) moral objections which have been urged against this axiom; but considered surely as it ought to be philosophically, I have ever regarded it, and I continue to regard it, as one of the most gratuitous and unsupported assertions ever hazarded. (*Conybeare, 1830, p. 218*)

Conybeare proceeds to put forward his alternative to Lyell's Huttonian method of working back from the present to the past. The investigation, he suggests, should commence at the beginning of things with an examination of each particular set of processes as it existed during the space of its own geologic epoch and should carry on the examination until the present processes were reached. Finally, each set of processes should be compared and the whole analysed in an attempt to discover whether ancient processes differed from those still going on and how earlier features may have influenced present developments. Conybeare was convinced that this latter method demonstrated undeniably the action of catastrophes, although inevitably he had to admit that the present landscape was the end-product of varying processes.

In the first place, a question of method presents itself; for we may, First, commence with the effects actually resulting from the causes still in operation and acting with their present power; and thus taking our departure from circumstances with which we are familiarly acquainted, we may proceed to the consideration of the geological changes produced at former periods, hoping to illustrate them by the light elicited in the course of this progress from the known to the unknown. This is the method which it is apparently the intention of Mr Lyell to adopt; as the portion of his work as yet published is confined to the present order of things, or, as Brongniart in his late Treatise somewhat affectedly calls it, the Jovian, as distinguished from the Saturnian period. I hope I have fairly stated the advantages which this method appears to offer. The other method is, Secondly, to survey the geological phenomena, in what may be called a chronological order; beginning with those which appear to have taken place at the earliest periods, classing as completely as we can, the effects which have been produced in the successive geological epochs in a regular order, of which the effects still in progress at the present time will of course form the last term; and finally, comparing the whole together, with the view of observing whether they all indicate a uniform and constant operation of the same causes, acting with the same intensity, and under the same circumstances; or rather evince that there has been a gradual change in these respects, and that the successive periods have often given rise to such new circumstances, as must have in a very great degree modified the original forces. (*Conybeare, 1830–1, pp. 359–60*)

The second part of Conybeare's article, published a year later, is more interesting. It shows that he had had time to consider the full implication of Lyell's views and, although the bases of his reasoning have altered little, some of his deductive propositions are rewarding for they reveal a foretaste of a new development in geologic thought, one which Lyell himself was to follow. Conybeare is still obsessed by the feeling that the immensity of time required by the fluvialists for their erosive mechanism was of an extent impossible and beyond all reasonable comprehension:

Mr Lyell says that the streamlets actually flowing through our valleys are adequate to account for all, if we will but throw all prejudice aside, and allow a sufficient number of millions (I should rather say infinit-illions) of ages since their continued action. This may be called the Fluvial theory: or more properly, the Atmospheric theory: for it evidently amounts to this, 'that the atmospherical waters falling on any given district and draining off from it are adequate to produce, by their continued action, all the phaenomena of water-worn gravel and excavation which we observe in that tract'. Now it will be my endeavour to show, from the arrangement and investigation of those phaenomena, that the atmospheric drainage, even if continued for ever and a day (that with the liberality of common parlance I may allow all the time I can), is altogether incapable of accounting for them; and that they indicate the effects not of drops and rills** but of violent currents and of vastly extended sheets of water. This I shall call the Diluvial theory, promising that I use the term diluvial only in a general and philosophical sense.

** The Atmospheric theory always reminds me of the celebrated line in Coleridge's tragedy:
'Drip, drip, drip, drip; there's nothing here but dripping'.

(*Conybeare, 1830–1, pp. 189–90*)

He illustrates his contention by reference to the scarplands of eastern Yorkshire and Lincolnshire.

Let the Fluvialist, however, so reconstruct the district in question; I next ask what it will require to reduce it from this 'its form ten million years ago' to its actual features? Why simply the excavation of the entire vales of

Lincoln and York (a district about 100 miles long and more than 15 broad) to a depth of 700 feet beneath its supposed original level. I will ask but one other question. How long would atmospherical drainage take to effect this? Seeing that since the Romans occupied Eboracum 1700 years ago, that agency has not effected a degradation of 7 inches on any one of the valla of their encampments, we may perhaps have sufficient data to calculate upon. I leave the Fluvialists to work out the question at leisure, offering in the meanwhile, as a mere approximation, an infinitillion of ages in the nth power. (*Conybeare, 1830–1, pp. 261–2*)

Conybeare repeats most of his old ideas on the puzzle of the Thames drainage – how, he asks, can the fluvial theory explain why a river such as the Thames flows straight across ranges of hills while some of its tributaries flow at right angles along the wide vales formed of softer rocks? He also repeats his contention that the elevation of strata must have caused local flood currents which accomplished extensive erosion and deposition. In addition he brings into play large lakes which by bursting their banks caused further minor floods. The lakes originated as remnants of the sea left in hollows after the raising of the ocean floor to form dry land. We have already dealt with most of Conybeare's other diluvialist notions and, as he and his contemporaries never seemed short of words and paper, must content ourselves with remarking upon his explanation of the major landscape features of the Weald. He explains the escarpments and the tilted strata as a direct result of the original deposition. The escarpments so formed stood out above the flood-waters as sea-cliffs and, assuming that sea level fell slowly, the soft rocks of the wide longitudinal vales would be eroded away over a considerable lateral extent. The narrow transverse valleys were carved out by diluvial currents furrowing the dip slopes; they are, in other words, normal 'valleys of denudation'.

We may easily conceive forces in action during the period of the original deposition of the strata, which may have caused the strata to terminate with truncated edges, facing towards the elevated ridges of the older rocks, against and upon which they were precipitated, instead of having allowed their planes uniformly to extend until they abutted against those older ridges: for we must suppose the oceans which deposited these strata to have possessed some lines of shore; these we may naturally conceive to be indicated by the most elevated crests of the older ridges; against such lines of coast, currents most probably have ranged. While therefore the depositions were proceeding quietly in the deeper and more tranquil waters, they would be interrupted in the range of these littoral currents, – may not the longitudinal valleys have originated in this cause? (*Conybeare, 1830–1, p. 264*)

Now the lines indicating the main direction towards which the waters in their subsidence must tend, being coincident with the dip of the strata over the backs of which the descent was taking place, must of course have been transverse to the bearing of those strata: the general currents of the so descending waters would therefore naturally tend to produce transverse furrows in the strata: hence would the transverse valleys originate; while at the same time the longitudinal valleys would be materially modified; the descending currents settling against the escarpments of the strata would naturally tend to undermine them, and from the direction of the inclination of their planes, would act to advantage, especially as we usually find the longitudinal valleys extending into the softer alternating strata, such as clay, and sand, and the harder rocks constituting the overhanging escarpment: hence the undermining agency of the waves operating with facility on these softer materials, would considerably increase the breadth of the longitudinal valleys and render the escarpments steeper and more abrupt. In proportion as the depression of the sea-level was gradual, there may have been a long continued reiteration of tidal waves sweeping over the same tracts. (*Conybeare, 1830–1, p. 265*)

The origin of the valleys of excavation, then, I am inclined to refer in part to the currents of the ocean in which they were first deposited, in part to those accompanying the gradual retreat of that ocean. (*Conybeare, 1830–1, p. 266*)

The importance given to this idea of marine action working upon structure shows the immediate impact of Lyell's ideas, and indicates how far the diluvialists were prepared to wander in their speculations, prompted largely by a desire to avoid accepting the Huttonian hypothesis.

While at home Lyell was meeting with relatively subdued opposition, on the Continent fresh catastrophic theories were being promulgated. Omalius D'Halloy in an article published in 1830 was reiterating that some valleys were formed by gaping due to elevation of the land from below. He classified five types of valleys, according to their supposed origin:

1. Vallées d'écartement – due to spreading or separating along a line of fracture during uplift, without necessarily any differential movement between the two sides.
2. Vallées de plissement – synclinal valleys due to folding.
3. Vallées d'éruption – caused by irregularities in or beside lava flows.
4. Vallées d'érosion – including all valleys due to the erosive action of running water.
5. Vallées de refoulement – due to the building up of deposits on the calmer flanks of marine currents, so forming submarine hills with bordering valleys, later elevated above sea level.

Also at this time Élie de Beaumont (1830 and 1831) published his theory of the parallel elevation of mountain chains, which had essentially a catastrophic basis as it assumed successive periods of intense earth-movements or 'feverish spasmodic energy'.

These anti-uniformitarian suggestions made Lyell redouble his literary output. At home and abroad he was determined to combat all geologic contentions in any way tainted with a catastrophic bias:

I daresay you may have heard of a certain case of diluvium of the Rotherberg, i.e. gravel on one side of the crater of one of these modern (tertiary) volcanos. The fact is, it proves nothing as to the post-diluvial, or rather post-alluvial origin of that volcano. The gravel was there before the eruption, and contains no volcanic matter. I never saw so clear a case, and it is one of ten thousand proofs of the incubus that the Mosaic deluge has been, and is I fear long destined to be on our science. Now I am fully determined to open my strongest fire against the new diluvial theory of swamping our continents by waves raised by paroxysmal earthquakes. I can prove by reference to cones (hundreds of uninjured cones), of loose volcanic scoriae and ashes, of various and some of great antiquity (as proved by associated organic remains), that no such general waves have swept over Europe during the tertiary era – cones at almost every height from near the sea, to thousands of feet above it. (*Letter 29 August 1831, in Lyell, Mrs, 1881, I, p. 328*)

So indefatigable was he in his desire to prove his case that he travelled widely in western Europe and pushed himself to extreme efforts. In 1833 he again visited the Rhine and Bavaria and in 1834 toured Sweden and Denmark. In 1835 he found time to visit the Swiss Jura and Bonn as well as to be a very capable President of the Geological Society. Some idea of his inherent drive and determination comes from an incident late in his life when, though too enfeebled through ill health to climb on foot, he had himself carried up Vesuvius rather than miss the opportunity of witnessing a unique geologic feature.

In the third volume of *The Principles* (1833) Lyell strongly countered catastrophic reasoning about past processes and seems to have had Conybeare's paper particularly in mind.

> Never was there a dogma more calculated to foster indolence, and to blunt the keen edge of curiosity, than this assumption of the discordance between a former and the existing causes of change. (*Lyell, 1st edn, 1833, III, pp. 2–3*)
>
> We hear of sudden and violent revolutions of the globe, of the instantaneous elevation of mountain chains, of paroxysms of volcanic energy, declining according to some, and according to others increasing in violence, from the earliest to the latest ages. We are also told of general catastrophes and a succession of deluges, of the alternation of periods of repose and disorder, of the refrigeration of the globe, of the sudden annihilation of the whole races of animals and plants, and other hypotheses, in which we see the ancient spirit of speculation revived, and a desire manifested to cut, rather than patiently to untie, the Gordian knot. (*Lyell, 1st edn, 1833, III, p. 7*)
>
> In our attempt to unravel these difficult questions, we shall adopt a different course, restricting ourselves to the known or possible operations of existing causes; feeling assured that we have not yet exhausted the resources which the study of the present course of nature may provide, and therefore that we are not authorized, in the infancy of our science to recur to extraordinary agents. (*Lyell, 1st edn, 1833, III, p. 6*)

The repeated hammer-blows of Lyell's writings and his growing prestige soon began to take effect. Sedgwick, in 1831, was one of the first who was driven to abandon the idea that all so-called diluvial deposits were the result of one flood. He admitted this after a careful study of de Beaumont's work and had the courage to proceed as follows:

> It was indeed a most unwarranted conclusion, when we assumed the contemporaneity of all the superficial gravel on the earth. . . .

To seek the light of physical truth by reasoning of this kind, is, in the language of Bacon, to seek the living among the dead, and will ever end in erroneous induction. Our errors were, however, natural, and after the same kind which led many excellent observers of a former century to refer all the secondary formations of geology to the Noachian deluge. Having been myself a believer, and, to the best of my power, a propagator of what I now regard as a philosophic heresy, and having more than once been quoted for opinions I do not now maintain, I think it right, as one of my last acts before I quit this Chair, thus publicly to read my recantation. (*Quoted by Clark and Hughes, 1890, I, p. 370*)

However, Sedgwick, as we shall see later (p. 171), did not go so far as to reject altogether the agency of catastrophes. Greenough's change of opinion was more fundamental; not only did he give up the idea of the Flood but he crossed to the uniformitarian camp by admitting that Lyell had proved the ability of present processes to erode on the scale required by geologic evidence.

Allow me, in conclusion, to say a few words upon a subject in connection with which my name has of late been brought forward much more prominently than I could have desired; – I mean Diluvial Action.

Some fourteen years ago, (*Greenough, 1819*) I advanced an opinion, founded altogether upon physical and geological consideration, that the entire earth had, at an unknown period, (as far as that word implies any determinate portion of time,) been covered by one general but temporary Deluge. The opinion was not hastily formed. My reasoning rested on the facts which had then come before me. My acquaintance with physical and geological nature is now extended; and that more extended acquaintance would be entirely wasted upon me, if the opinions which it will no longer allow me to retain, it did not also induce me to rectify. New data have flowed in, and with the frankness of one of my predecessors, I also now read my recantation.

The varied and accurate researches which have been instituted of late years throughout and far beyond the limits of Europe, all tend to this conclusion, that the geological schools of Paris, Freyberg and London have been accustomed to rate too low the various forces which are still modifying, and always have modified, the external form of the earth. What the value of those forces may be in each case, or what their relative value, will continue for many years a subject of discussion; but that their aggregate effect greatly surpasses all our early estimates, is I believe incontestably established. To Mr Lyell is eminently due the merit of having awakened us to a sense of our error in this respect. The vast mass of evidence which he has brought together, in illustration of what may be called Diurnal Geology, convinces me that if, five thousand years ago, a Deluge did sweep over the entire globe, its traces can no longer be distinguished from more modern and local disturbances. The first sight of those comparatively recent assemblages of strata, which he designates the Eocene, Miocene and Pleiocene Formations, (unknown but a few years ago, though diffused as extensively as many which were then honoured with the title of universal), shows the extreme difficulty of distinguishing their detritus from what we have been accustomed to esteem Diluvium. The Fossil Contents of these formations strongly confirm this argument. M. Deshayes has shown that they belong to a series unbroken by any great intervals, and that, if they be divided from the secondary strata, the chasm can have no relation to any such event as is called The Flood. (*Greenough, 1833-4, pp. 69-70*)

However, we would be wrong if we looked upon these recantations of former noted diluvialists as ensuring a complete victory for Lyell. Among those who were impressed by his grandeur but did not blindly follow his lead was Henry De la Beche,

> then a handsome and fashionable young man, just beginning to show the quick and shrewed observation of nature, and rare power of philosophical induction, which eventually gave him so honourable a rank in British geology. (*Geikie, 1875, I, p. 116*)

If it had not been for De la Beche's work in the West Country the British Geological Survey would probably not have been formed till much later. It was the obvious utility of his geological map and his enthusiasm that convinced the Government of the advantages of such surveys and in 1832 he was commissioned to prepare a detailed stratigraphical map of south-west England under the nominal authority of the Board of Ordnance. In 1835 he was given a Treasury grant of £300 a year and by 1845 the department of Geological Survey had grown into a thriving organization. Consequently it was transferred from its military jurisdiction and was made a separate branch of the Office of Works with De la Beche as its first Director-General. This development was extremely important to geology as it meant that henceforth Britain would be mapped on a coherent basis instead of being left to the whims of private enthusiasts, and, what was more valuable, that the geologic exploration of the country would be undertaken continuously by a skilled staff. It also meant, as in the case of Andrew Ramsay who joined the staff in 1841, that lack of means need no longer deter anyone really interested in a geological career.

In addition to his survey activities, De la Beche drew the public's attention to contemporary research by a series of publications. His *Manual of Geology* (1831) soon won general renown being translated into French and German as well as being published in the United States. This was followed in the same year by *Researches in Theoretical Geology* and in 1835 by a popular exposition of geological method entitled *How to Observe in Geology*.

A chronological examination of these books shows the slow influence of Lyell's successive publications on the progress of De la Beche's ideas.

> The book that was published earliest contains scarcely any mention of Lyell's name or of his teachings. On the contrary it seems to presume that catastrophes of great extent occurred on earth, causing the extinction of all life. Nor is it strange that he should have held this view; he had collaborated with Conybeare – a die-hard diluvian, . . . upon a study of the newly discovered plesiosaurus. In that paper (*Transactions of the Geological Society*, 1821), which, though written by Conybeare, bears the name of De la Beche

as co-author, the strongest vituperation is poured upon the iniquitous folly of Lamarck and his followers for their theory that living animals, and possibly man himself, are the modified descendants of extinct creatures. De la Beche, when not collaborating, was better balanced, more reasonable, less immoderate. His later works in the order of their appearance show an increasing, though never quite complete, acceptance of the idea expressed by Lyell. (*Drachman, 1930, pp. 52–53*)

Above all, it should be noticed that De la Beche was opposed to Lyell's view that marine currents cut valleys.

When explaining the Formation of Valleys, Mr De la Beche contends that the 'Bursting of Lakes' as it has been termed, could not take place in the way supposed. The Area, comprised within soundings, that is, within the 100 fathom line, round the British Islands, is delineated on a map, in order to show, that within that area at least, no Valleys are produced by Tides and Currents; whence it would follow, that such effects cannot be satisfactorily referred to such cases. (*Greenough, 1834–5, p. 173*)

We shall now see that this question of the potency of marine currents becomes increasingly important as more and more geologists take their lead from Lyell, who begins to favour the sea as a destructive agent more powerful than rivers and so begins to veer from the fluvialistic teachings of Hutton.

CHAPTER TWELVE

Lyell's *Principles*, 1833-75: Uniformitarianism Wavers

'Nothing of him that doth fade,
But doth suffer a sea-change
Into something rich and strange'
(W. Shakespeare: *The Tempest*, I, ii, 394)

'THE PRINCIPLES OF GEOLOGY'

The continued material success of Lyell's masterpiece makes remarkable reading. The work continued to sell rapidly even after its issue as a four-volume third edition in 1834.

In 1838, the portion dealing with historical geology (the fourth volume of the previous editions) was taken out of *The Principles*, re-written and published separately as *Elements of Geology*. This work had gone into six editions by 1865 and was succeeded six years later by a somewhat similar and equally popular *The Student's Elements of Geology*. Even the latter was characterized by the Lyellian method of proceeding from the newest or known geological systems back to the oldest or unknown.

The sixth edition of *The Principles*, thus curtailed, appeared in three volumes in June 1840. The effect of the change was to restrict it mainly

to the physical side of geology – to the subjects connected with the morphological changes which the earth and its inhabitants alike undergo. Thus it made the contents of the book accord more strictly with its title, while the 'Elements' indicated the working out of the aforesaid principles in the past history of the earth and its inhabitants – that is, the latter book deals with the classification of rocks and fossils, or with petrology, and historical geology. (*Bonney, 1895, pp. 74–75*)

The demand then slackened though further three-volume editions were issued in 1847 and 1850, a single-volume edition in 1853 and two-volume editions in 1867–8, 1872 and 1875.

Each edition was revised carefully and insertions added to keep it abreast of new scientific discoveries and fresh shifts of emphasis. Yet from a geomorphological viewpoint it is surprising how much of the original matter remained substantially the same.

The section on the action of ice is so altered and enlarged as to be practically new; for when the first edition of the 'Principles' was published comparatively little was known of the effects of land-ice, and the art of following the trail of vanished glaciers had yet to be learnt. But, with this exception, the part of the book dealing with the action of the forces of Nature – heat and cold, rain, rivers, and sea – remains comparatively unaltered, as do the first five chapters, which give a sketch of the early history of the science of geology. (*Bonney, 1895, pp. 79–80*)

One must compare an early and a late edition, such as the first or third and the tenth or eleventh, in order to realize how great were the changes. . . . New material was incorporated into every part; it makes its appearance sometimes on every page; changes are made in the order of the subjects; many chapters are entirely rewritten; nevertheless, a considerable portion corresponds almost word for word in the two editions; Lyell was no hurried writer. (*Bonney, 1895, p. 78*)

As our concern is with the history of the study of landforms we are deprived of the pleasure of attempting to recount Lyell's other publications. Seventy-six memoirs by him are recorded in the Royal Society's *Catalogue of Scientific Papers*. Nor are we concerned directly with the rapid growth of stratigraphical and palaeontological knowledge although he played a full part in it and it naturally emphasized the growing belief in the influence of structure and faulting on landscapes. Yet it was this general background of scientific investigations as well as the overpowering magnetism of Lyell which now encouraged geomorphic thought to undergo an unfortunate deviation from Huttonian principles. The point of departure concerned the erosion of valleys which Hutton had attributed to the work of the rivers that ran in them. Lyell, though at first adopting this view in Auvergne, gradually became obsessed with the erosive power of the sea and in more and more instances ascribed to it the creation of valleys, possibly subject to minor and later modifications by rivers. He also became increasingly attracted by the idea of local floods caused by the bursting of large continental lakes and by the influence of structure on certain topographical features. These deviations from more traditional uniformitarianism were already strongly evident in the 1833 volume of *The Principles*. In order to appreciate them fully it is helpful to study in detail the keen powers of perception shown by Lyell in dealing with any erosive process.

LYELL'S ACUTENESS AS AN OBSERVER

We shall find no difficulty in illustrating these qualities by extracts on the work of rivers and glaciers taken from the first (1830) and fifth editions (1837) of *The Principles*. In the following description of meandering, for which he owed something to Scrope, the mechanical process is explained down to the smallest detail including the origin of those features which are now geomorphically called oxbow lakes:

By a repetition of these land-slips, the ravine is widened into a small, narrow valley, in which sinuosities are caused by the deflexion of the stream first to one side and then to the other. The unequal hardness of the materials through which the channel is eroded, tends partly to give new directions to the lateral force of excavation. When by these, or by accidental shifting of the alluvial matter in the channel, and numerous other causes, the current is made to cross its general line of descent, it eats out a curve in the opposite bank, or in the side of the hills bounding the valley, from which curve it is turned back again at an equal angle, so that it recrosses the line of descent, and gradually hollows out another curve lower down in the opposite bank, till the whole sides of the valley, or river-bed present a succession of salient and retiring angles. . . .

When the tortuous flexures of a river are extremely great, the aberration from the direct line of descent is often restored by the river cutting through the isthmus which separates two neighbouring curves. Thus, in the annexed diagram, the extreme sinuosity of the river has caused it to return for a brief space in a contrary direction to its main course, so that a peninsula is formed, and the isthmus is consumed on both sides by currents flowing in opposite directions. In this case an island is soon formed – on either side of which a portion of the stream usually remains. (*Lyell, 1st edn, 1830, I, pp. 170–1*)

The transporting power of water is considered with the same meticulous attention to detail and the same ability of dissection or analysis that made Lyell's work stand out so far above that of most other contemporary geologists. In this particular case he has seized on the hydrostatic principle that an object when carried in water loses part of its weight and that the friction at the sides and on the bed of the river slows down movement of the objects carried. He even troubles to calculate the velocity necessary to move the various types of matter normally carried by rivers:

In regard to the transporting power of water, we may often be surprised at the facility with which streams of a small size, and descending a slight declivity, bear along coarse sand and gravel; for we usually estimate the weight of rocks in air, and do not reflect sufficiently on their comparative buoyancy when submerged in a denser fluid. The specific gravity of many rocks is not more than twice that of water, and very rarely more than thrice, so that almost all the fragments propelled by a stream have lost a third, and many of them half of what we usually term their weight.

It has been proved by experiment, in contradiction to the theories of the earlier writers on hydrostatics, to be a universal law, regulating the motion of running water, that the velocity at the bottom of the stream is everywhere less than in any part above it, and is greatest at the surface. Also that the superficial particles in the middle of the stream move swifter than those at the sides. This retardation of the lowest and lateral currents is produced by friction and when the velocity is sufficiently great, the soil composing the sides and bottom gives way. A velocity of three inches per second is ascertained to be sufficient to tear up fine clay, – six inches per second, fine sand, – twelves inches per second, fine gravel, – and three feet per second, stones of the size of an egg. (*Lyell, 1st edn, 1830, I, pp. 172–3*)

His interest was activated in this direction by the findings of early experimenters whose results he cites in full as being illustrative of the immense amount of transportation of which a river is daily or yearly capable:

Quantity of sediment in river water, – Very few satisfactory experiments have as yet been made, to enable us to determine, with any degree of accuracy, the mean quantity of earthy matter discharged annually into the sea by some one of the principal rivers of the earth. Hartsoeker computed the Rhine to contain in suspension, when most flooded, one part in a hundred of mud in volume; but it appears from two sets of experiments, recently made by Mr Leonard Horner, at Bonn, that $\frac{1}{16000}$th would have been a nearer approximation to the truth. Sir George Staunton inferred from several observations, that the water of the Yellow River in China contained earthy matter in the proportion of one part to two hundred, and he calculated that it brought down in a single hour two million cubic feet of earth, or forty-eight million daily; so that, if the Yellow Sea be taken to be 120 feet deep, it would require seventy days for the river to convert an English square mile into firm land, and 24,000 years to turn the whole sea into terra firma, assuming it to be 125,000 square miles in area. Manfredi, the celebrated Italian hydrographer, conceived the average proportion of sediment in all the running water on the globe, which reached the sea, to be $\frac{1}{175}$, and he imagined that it would require 1,000 years for the sediment carried down to raise the general level of the sea about one foot. Some writers, on the contrary, as De Maillet, have declared the most turbid waters to contain far less sediment. One of the most extraordinary statements is that of Major Rennell, in his excellent paper, before referred to, on the delta of the Ganges. 'A glass of water', he says, 'taken out of this river when at its height, yields about one part in four of mud. No wonder, then,' he adds, 'that the subsiding waters should quickly form a stratum of earth, or that the delta should encroach on the sea!' (*Lyell, 5th edn, 1837, I, pp. 229-30*)

Mr Everest found that, in 1831, the number of cubic feet of water discharged by the Ganges per second was, during the

Rains, (4 months)	.	.	.	.	494,208
Winter, (5 months)		.	.	.	71,200
Hot weather, (3 months)		.	.	36,330

so that we may state in round numbers, that 500,000 cubic feet flow down during the four months of the flood season, from June to September and only 100,000 during the remaining eight months. . . .

In computing the quantity of water, Mr Everest made no allowance for the decreased velocity of the stream near the bottom presuming that it is compensated by the increased weight of matter held in suspension there. Probably the amount of sediment is by no means exaggerated by these circumstances: but rather underrated, as the heavier grains of sand, which can never rise into the higher parts of the stream, are drifted along the bottom.

Now the average quantity of solid matter suspended in the water during the rain, was, by weight $\frac{1}{428}$th part; but as the water is about one half of the specific gravity of the dried mud, the solid matter discharged is $\frac{1}{856}$th part in bulk, or 577 cubic feet per second. This gives a total of 6,082,041,600 cubic feet for the discharge in the 122 days of the rain. The proportion of sediment in the waters at other seasons was comparatively insignificant, the total amount during the five winter months being only 247,881,600 cubic feet, and during the three months of hot weather, 38,154,240 cubic feet. The total annual discharge, then would be 6,368,077,440 cubic feet.

In order to give some idea of the magnitude of this result, we will assume that the specific gravity of the dried mud is only one-half that of granite (it would, however, be more); in that case, the earthy matter discharged in a year would equal 3,184,038,720 cubic feet of granite. Now about $12\frac{1}{2}$ cubic ft of granite weigh one ton; and it is computed that the great Pyramid of

Egypt, if it were a solid mass of granite, would weigh about 6,000,000 tons. The mass of matter, therefore, carried down annually would, according to this estimate, more than equal in weight and bulk forty-two of the great pyramids of Egypt, and that borne down in the four months of the rains would equal forty pyramids. (*Lyell, 5th edn, 1837, I, pp. 230–1*)

Then with his mind full of pictures of vast rivers muddy with the sediments of countless square miles of land, Lyell passes on to describe a hydraulic relationship, the reasoning of which in geomorphic circles has only recently been accepted. He calculated that if, below a tributary junction, the river increased its width by an amount proportionate to the width of the tributary, then this process would cause the main river to become so wide and shallow that the current would decline and silting begin. Lyell was quick to realize that the increased discharge was accommodated by an increase in the depth of the main river rather than by a simple increase in the width:

When this mechanical power of running water is considered, we are pre-pared for the transportation of large quantities of gravel, sand, and mud, by the torrents and rivers which descend with great velocity from moun-tainous regions. But a question naturally arises, how the more tranquil rivers of the valleys and plains, flowing on comparatively level ground, can remove the prodigious burden which is discharged into them by their numer-ous tributaries, and by what means they are enabled to convey the whole mass to the sea. If they had not this removing power, their channels would be annually choked up, and the valleys of the lower country, and plains at the base of mountain-chains, would be continually strewed over with fragments of rock and sterile sand. But this evil is prevented by a general law regulating the conduct of running water – that two equal streams do not, when united, occupy a bed of double surface. In other words, when several rivers unite into one, the superficial area of the fluid mass is far less than that previously occupied by the separate streams. The collective waters, instead of spreading themselves out over a larger horizontal space, contract themselves into a column of which the height is greater relatively to its breadth. Hence a smaller proportion of the whole is retarded by friction against the bottom and sides of the channel; and in this manner the main current is often accelerated in the lower country, even where the slope of the river's bed is lessened.

It not unfrequently happens, as will be afterwards demonstrated by examples, that two large rivers, after their junction, have only the surface which one of them had previously; and even in some cases their united waters are confined in a narrower bed than each of them filled before. By this beautiful adjustment, the water which drains the interior country is made continually to occupy less room as it approaches the sea; and thus the most valuable part of our continents, the rich deltas, and great alluvial plains, are prevented from being constantly under water. (*Lyell, 5th edn, 1837, I, pp. 170–1*)

Examples of rivers actually eroding and so affording some measure of the rates of downward cutting are cited from Britain and the volcanic areas of Italy and France, over which Lyell had worked with Murchison:

The power even of a small rivulet, when swoln by rain, in removing heavy bodies, was lately exemplified in the College, a small stream which flows at a moderate declivity from the eastern water-shed of the Cheviot-Hills. Several thousand tons' weight of gravel and sand were transported to the plain of the Till, and a bridge then in progress of building was carried away, some of the archstones of which, weighing from half to three-quarters of a ton each, were propelled two miles down the rivulet. (*Lyell 1st edn, 1830, I, pp. 174–5*)

The rapidity with which even the smallest streams hollow out deep channels in soft and destructible soils is remarkably exemplified in volcanic countries, where the sand and half-consolidated tuffs oppose but a slight resistance to the torrents which descend the mountain side. After the heavy rains which followed the eruption of Vesuvius in 1822, the water flowing from the Atrio del Cavallo cut, in three days, a new chasm through strata of tuff and ejected volcanic matter, to the depth of twenty-five feet. I found the old muleroad, in 1828, intersected by this new ravine. (*Lyell, 1st edn, 1830, I, p. 176*)

The spectacle of Niagara for a moment catches his imagination and he computes how long it will take the falls to cut back to Lake Erie.

If the ratio of recession had never exceeded fifty yards in forty years, it must have required nearly ten thousand years for the excavation of the whole ravine; but no probable conjecture can be offered as to the quantity of time consumed in such an operation, because the retrograde movement may have been much more rapid when the whole current was confined within a space not exceeding a fourth or fifth of that which the falls now occupy. Should the erosive action not be accelerated in future, it will require upwards of thirty thousand years for the falls to reach Lake Erie (twenty-five miles distant), to which they seem destined to arrive in the course of time, unless some earthquake changes the relative levels of the district. (*Lyell, 1st edn, 1830, I, p. 181*)

No process was too insignificant for Lyell's notice if there was a possibility that it might be employed as a uniformitarian advocate. Lyell realised that the ice of glaciers was responsible for the movement and deposit of considerable masses of material. He mentions moraines and obviously understood the mechanics of their formation.

Effects of ice in removing stones. – In mountainous regions and high northern latitudes, the moving of heavy stones by water is greatly assisted by the ice which adheres to them, and which forming together with the rock a mass of less specific gravity, is readily borne along. The snow which falls on the summits of the Alps throughout nine months of the year is drifted into the higher valleys, and being pressed downward by its own weight, forms those masses of ice and snow called glaciers. Large portions of these often descend into the lower valleys, where they are seen in the midst of forests and green pastures. The mean depth of the glaciers descending from Mont Blanc is from 80 to 100 feet, and in some chasms is seen to amount to 600 feet. The surface of the moving mass is usually loaded with sand and large stones, derived from the disintegration of the surrounding rocks acted upon by frost. These transported materials are generally arranged in long ridges or mounds, sometimes thirty or forty feet high. They are often two, three or even more in number, like so many lines of intrenchment, and consist of the debris which have been brought in by lateral glaciers. The whole accumulation is called in Switzerland 'the moraine', which is slowly conveyed to inferior valleys, and left where the snow and ice melt, upon the plain, the larger blocks remaining, and the smaller being swept away by the stream to which the melting of the ice gives rise. This stream flows along the bottom of each glacier issuing from an arch at its lower extremity. (*Lyell, 5th edn, 1837, I, p. 173*)

LYELL'S SWERVE TOWARDS MARINE AND STRUCTURAL INFLUENCES

It seems strange that Lyell with his usual acuteness did not proceed one step farther and realise that glaciers were the cause of isolated boulders (erratics) and 'diluvial' deposits. But close inspection of the above quotation will reveal two distinct ideas: first, the moving by water of stones to which ice had adhered, a phenomenon which occurs even in England in cold winters; and, second, the transportation of debris by land-ice. Lyell preferred to use the flotation idea on a marine scale and, instead of the more correct solution, ascribed erratics and drift deposits to the action of debris-laden icebergs which calved off from a land glacier and drifted across submerged areas dropping their load indiscriminately on the sea-floor (now exposed as land).

> In northern latitudes, where glaciers descend into valleys terminating in the sea, great masses of ice, on arriving at the shore, are occasionally detached and floated off together with their 'moraine'. The currents of the ocean are then often instrumental in transporting them to great distances. (*Lyell, 5th edn, 1837, I, p. 173*)
> Wherever they are dissolved, it is evident that the 'moraine' will fall to the bottom of the sea. In this manner many submarine valleys, mountains, and platforms become strewed over with scattered blocks of foreign rock, of a nature perfectly dissimilar from all in the vicinity, and which may have been transported across unfathomable abysses. (*Lyell, 5th edn, 1837, I, p. 174*)

This preference for marine intervention is interesting. It arose partly because of the great impression that the potent erosive force of waves had made on Lyell's mind. In fact this period was one of intense attraction to the sea for the public, poets and geologists alike and men such as Lyell, were fascinated by the rapid crumbling of sea-cliffs and by the possibility of islands being entirely removed by wave-erosion.

> Although the greater indestructibility of some rocks may enable them to withstand, for a longer time, the action of the elements, yet they cannot permanently resist. There are localities in Shetland, in which rocks of almost every variety of mineral composition are suffering distintegration; thus the sea makes great inroads on the clay slate of Fitfel Head, on the serpentine of the Vord Hill in Fetlar, and on the mica-schist of the Bay of Triesta, on the east coast of the same island, which decomposes into angular blocks. (*Lyell, 5th edn, 1837, I, p. 247*)

> Destruction of Islands. – Such devastation cannot be incessantly committed for thousands of years without dividing islands, until they become at last mere clusters of rocks, the last shreds of masses once continuous. (*Lyell, 5th edn, 1837, I, p. 248*)

FIG. 32. Lyell's illustration of boulders drifted by ice on to the shores of the St. Lawrence River at Richelieu Rapid in the Spring of 1835

The waves constantly undermine the low chalk cliffs, covered with sand and clay, between Weybourne and Sherringham, a certain portion of them being annually removed. At the latter town I ascertained, in 1829, some facts which throw light on the rate at which the sea gains upon the land. It was computed, when the present inn was built, in 1805, that it would require seventy years for the sea to reach the spot: the mean loss of land being calculated, from previous observations, to be somewhat less than one yard annually. (*Lyell, 5th edn, 1837, I, p. 252*)

Lyell, having already used marine work in the dispersion of 'drift' by floating ice and in the destruction of coasts, now applies it to the erosion of valleys, and assumes (wrongly we must notice!) that currents and waves in a shallow sea could easily erode the sea-floor into a topography resembling escarpments and wide vales. He applies the idea to the erosion of the Weald. The means whereby the uplifted central Wealden dome had been eroded into a 'vale' with opposite in-facing scarps had long puzzled geologists. Conybeare, as we have seen, suggested that the strata had been originally deposited in the form of escarpments and that the valleys had later been carved by combined marine and diluvial action. Lyell, who will have nothing to do with the Deluge, postulates a marine and structural origin. He sees the horizontal strata of chalk at sometime after the commencement of the Eocene period slowly heaved up into a dome by the slow action of subterranean forces. As the dome rose from the ocean bed, its surface was subjected to increasing tension until great cracks and rents appeared in the strata. Through these newly made fissures sea-currents would sweep, eroding away the softer strata. In time a wide vale would be cut out of the softer beds while the more resistant rocks bordering the vale would stand up above the sea as marine cliffs.

Suppose the ridge or dome first elevated to have been so rent and shattered on its summit as to give more easy access to the waves, until at length the masses . . . were removed. Two strips of land might then remain on each side of a channel, in the same manner as the opposite coasts of France and England, composed of chalk, present ranges of white cliffs facing each other. A powerful current might then rush, like that which now ebbs and flows through the Straits of Dover, and might scoop out a channel in the gault. We must bear in mind that the intermittent action of earthquakes would accompany this denuding process, fissuring rocks, throwing down cliffs, and bringing up, from time to time, new stratified masses, and thus greatly accelerating the rate of waste. (*Lyell, 1st edn, 1833, III, pp. 294-5*)

Arguing against a diluvial explanation, Lyell points out that if such an event had occurred it would have left behind some evidence of its passage in the form of flints and chalk, whereas in actual practice such residue is never found. The sea, in contrast, by continually washing backwards and forwards over its own bed would tend to clear away fragments breaking off from the cliffs, a cycle of operations which clearly better fits the factual circumstances:

Indeed, if we adopt the diluvial hypothesis of Dr Buckland, we should expect to find vast heaps of broken flints drifted more frequently into the valleys of the Gault and Weald clay, instead of being generally confined to the summit of the chalk downs.

On the other hand, it is quite conceivable that the slow agency of oceanic currents may have cleared away, in the course of ages, the matter which fell into the sea from wasting cliffs. (*Lyell, 1st edn, 1833, III, p. 297*)

But in order that this explanation should be satisfactory, we must suppose that the rise of the land in the south-east of England was very gradual, and the subterranean movements for the most part of moderate intensity. During the last century earthquakes have occasionally thrown down at once whole lines of sea-cliffs, for several miles continuously; but if this had happened repeatedly during the waste of the ancient escarpments of the chalk now encircling the Weald, and if the shocks had been accompanied by the sudden rise and conversion of large districts into the land, the Weald would have been covered with the ruins of those wasted rocks, and the sea could not possibly have had time to clear the whole away. (*Lyell, 5th edn, 1837, II, p. 413*)

Satisfied with his marine explanation of the form of the main vale, Lyell turns to the question of the origin of the transverse valleys and adopts the structuralist view that they follow cracks formed in the chalk dome during its uplift.

He then uses the presence of transverse valleys as an argument in his marine theory, as their existence would have assisted the scouring action of the sea:

In whatever manner these transverse gorges originated, they must evidently have formed ready channels of communication between the submarine longitudinal valleys and those deep parts of the sea wherein we imagine the tertiary strata to have been accumulated. If the strips of land which first rose had been unbroken, and there had been no free passage through the cross fracture, the currents would not so easily have drifted away the materials detached from the wasting cliffs, and it would have been more difficult to understand how the wreck of the denuded strata could have been so entirely swept away from the base of the escarpment. (*Lyell, 1st edn, 1833, III, p. 302*)

In substantiation of his theory of doming and marine denudation, Lyell criticizes the opposing theory of Buckland, who attributed the anticlinal vale of Kingsclere to fracturing and tearing apart that left a central vale, dominated by in-facing scarps and later modified by the Flood (p. 117). By means of true-to-scale diagrams Lyell showed that his entirely erosional version was the more reasonable:

The reader might imagine, on regarding Dr Buckland's section . . . where, for the sake of elucidating the geological phenomena, the heights are exaggerated in proportion to the horizontal extent, that the solution of continuity of the strata bounding the Valley of Kingsclere had been simply due to elevation and fracture, unassisted by aqueous causes; but by reference to the true scale . . . , it will immediately appear, that a considerable mass of chalk must have been removed by denudation. (*Lyell, 1st edn, 1833, III, pp. 306–7*)

We will give only two more quotations to show how far Lyell was beginning to draw away from Hutton, and how in his desire to justify his principle of uniformity he is at times prepared to abandon the erosive action of rivers in favour of marine denudation of a rising land-mass, earthquake fractures and the bursting of lake-dams.

> Excavation of Valleys. – In our attempt to explain the origin of the existing valleys in the South-East of England, it will be seen that we refer their excavation chiefly to the ocean. . . .
> Few of the ordinary causes of change, whether igneous or aqueous, can be observed to act with their full intensity in any one place at the same time; hence it is easy to persuade those who have not reflected long and profoundly on the working of the numerous igneous and aqueous agents, that they are entirely inadequate to bring about any important fluctuations in the configuration of the earth's surface. (*Lyell, 1st edn, 1833, III, pp. 319–20*)
> In speculating on catastrophes by water, we may certainly anticipate great floods in future; and we may therefore presume that they have happened again and again in past times. The existence of enormous seas of fresh water, such as the North American lakes, the surface of the largest of which is elevated more than 600 feet above the level of the ocean, and is in parts 1200 feet deep, is alone sufficient to assure us, that the time may come, however distant, when a deluge may lay waste a considerable part of the American continent. No hypothetical agency is required to cause the sudden escape of the confined waters. Such changes of level, and opening of fissures as have accompanied earthquakes since the commencement of the present century, or such excavation of ravines as the receding cataract of Niagara is now effecting, might breach the barriers. (*Lyell, 5th edn, 1837, I, p. 94*)

THE EFFECT ON LYELL OF CONTEMPORARY CRITICISM, 1835–45

Lyell's swerve towards marine, structural and other rather catastrophic causes, such as dam-bursts, may be attributed partly to the impressions actual observations of these phenomena had made upon his mind. In other words, they were part of the principle of uniformity even if, as we now know, often misapplied by him in detail. But Lyell was also influenced by contemporary criticism which happened to be strongly biased towards catastrophic causes. In this opposition, the fluvialists were in a marked minority and exerted little influence.

Huttonian Criticisms. For the fluvialists, Scrope again appears as a lone but unavailing opponent of such views:

> There does not appear any reason for introducing the sea into the Weald valley, as Mr Lyell does, for the purpose of effecting its denudation, and the removals of the materials of its ancient strata. The agency of rain and rivers acting through an indefinite time may have alone accomplished this. And since no traces of marine deposits of the eocene, or any later age, are met with throughout the whole basin, we must hesitate to believe that the sea covered it during any part of the tertiary period. (*Scrope, 1835, p. 440*)

Of all the geologists of this period, he was the most faithful to Hutton's geomorphic principles, and for this reason his temporary loss to politics is to be regretted.

In 1839 Fitton also took up the cause of the Huttonians and, in his review of

Lyell's *Elements*, suggested that Lyell had not paid sufficient acknowledgement to the earlier work of Hutton:

> he ought to have referred more emphatically to the works of that very remark-
> able man as a source for which his doctrines were first obtained. (*Fitton, 1839*)

The statement seems fair enough because Lyell's proclamation of the uniformity of existing causes owed almost everything to Hutton's earlier theories.
Moreover, in a letter to Fitton, Lyell admits that he clearly understood the
fundamental principles of Hutton's theory but differed in his appreciation of
the merit of Hutton's ideas which he (Lyell) thought were diminished in
value because they came at such a late stage in the development of geologic
thinking. In other words, it seems to us that Lyell did not really understand
the true value of Hutton – that he stood above all for continuity of process;
that he was the first great uniformitarian. Lyell's letter, however, may be
allowed to speak for itself:

> the whole article runs off fluently and in entertaining style, and varied until
> you get into your elaborate disquisition on Hutton, which to all the general
> readers, and to many of the initiated, will, I fear, be somewhat heavy. It has
> been useful to me, as I found it difficult to read and remember Hutton, and
> though I tried, I doubt whether I ever fairly read more than half his writings,
> and skimmed the rest. Considering at how late a period, as compared to
> Steno, Hook, Leibnitz, and Moro, he came into the field, and consequently
> how much greater were his opportunities, I think his knowledge and his
> original views were confined to too small a range of the vast science of
> geology, to entitle him to such marked and almost exclusive pre-eminence
> as you contend for in his behalf. If you had not felt some natural indignation
> at the unpardonable neglect with which the French and Germans have treated
> him, you would not, I think, on reading downwards from the theories of the
> older writers, have considered his merits on the whole as so transcendant.
> Your citation (immediately following the woodcut) is a more perfect enuncia-
> tion of the gradual acquisition of the metamorphic character by successive
> modification than I had remembered to exist in his writings.
> . . . In distinctly alleging that my defective appreciation of Hutton's claims
> was the ground of the pages which you write in his defence and eulogy, a
> point again enforced in the last sentence, you should have been careful to
> distinguish between the total neglect of Cuvier, Von Buch, Humboldt, Boué,
> Brongniart, and others, and the inadequate rank in the relative scale which in
> your judgement I have assigned, as the simple reader of your review must
> confound me with all the rest. As you do not complain that my historical
> sketch was disproportionate in length to the general plan of my 'Principles',
> I must assume that you think Hutton occupies too small a space, yet in my
> first chapter I gave Hutton credit for first separating geology from other
> sciences, and declaring it to have no concern with the origin of things, and
> after rapidly discussing a great number of celebrated writers, I pause to give,
> comparatively speaking, full-length portraits of Werner and Hutton, giving
> the latter the decided palm of theoretical excellence, and alluding to the two
> grand points in which he advanced the science. First, the igneous origin of
> granite, secondly that the so-called primitive rocks were altered strata. I dwelt
> emphatically on the complete revolution brought about by his new views

respecting granite, and entered fully on Playfair's illustrations and defence of Hutton, and he is again put prominently forward in the 'Elements', where no other but Lehmann and Werner are mentioned. The mottos of my first two volumes were especially selected from Playfair's Huttonian Theory, because although I was brought round slowly, against some of my early prejudices, to adopt Playfair's doctrines to the full extent, I was desirous to acknowledge his and Hutton's priority, and I have a letter of Basil Hall's in which after speaking of points in which Hutton approached nearer to my doctrines than his father, Sir James Hall, he comments on the manner in which my very title-page did homage to the Huttonians, and complimented me for thus disavowing all pretensions to be the originator of the theory of the adequacy of modern causes.

Yet, to how many of your readers, who will never see my work, will your elaborate advocacy of Hutton seem to imply that I overlooked, or have been unwilling to acknowledge even in a moderate degree, his just pretensions! It was my business, in tracing the progress of our science to its present state, to estimate the importance of each writer, and adjust the quantity of space due to him in my historical sketch, not simply according to his originality and genius, but partly at least in proportion to his influence; and I still think that Werner's eloquence, popularity, enthusiasm, and position at Freyberg, placed him in this point of view as much above Hutton as I have represented him to fall below him in reference to the truth of his theories. Yet as an admirer of Hutton all I could have wished is, that your panegyric on Hutton had appeared as aiding and seconding my efforts, since I trust that no book has made the claims of Hutton better known on the Continent of late years than mine. (*Lyell, Mrs, 1881, II, pp. 47-50. Letter to Fitton, 1st August 1839*)

Perhaps, however, we should not be too ready to criticize Lyell for his failure to acknowledge Hutton's full worth as *Theory of the Earth* is indeed tedious and difficult to read and as Fitton says

The original work of Hutton is in fact so scarce that . . . no copy exists at present in the libraries of the Royal Society . . . or even of the Geological Society of London. (*Fitton, 1839, p. 455*)

Diluvialist Criticisms. It is not surprising that Lyell was much more influenced by the criticisms of the diluvialists and structuralists which gained in attractiveness as they themselves swerved increasingly towards his own opinions. The diluvial attack was led by Sedgwick and Buckland. We have already noticed (p. 156) how in 1831 Sedgwick frankly recanted many of his views. Yet he still clung to a modified catastrophic idea of Creation and held that hills and valleys were formed beneath the ocean and subsequently, after uplift, were acted upon by the rivers running through them. He also, as did Buckland, retained his belief that geology will ultimately serve to prove the testimony of the Bible. 'Truth cannot oppose truth, but must, in all cases, be coincident' (Sedgwick, 1838, p. 288). However, his religious beliefs did not destroy his scientific sense, and he was the first to attack one of the less enlightened of his brethren, the Very Reverend William Cockburn, D.D., Dean of York, who in a paper to the geological section of the British Association attempted to explain the Mosaic Cosmogony literally (Clark and Hughes, 1890, II, pp. 76–78). Sedgwick in his own ideas on this topic followed the lead of Buckland by the neat expedient of placing the bulk of geologic creation in the period of chaos before the six days commenced.

About this time, William Buckland makes a happy reappearance on the geologic landscape with the publication of his Bridgewater Treatise, *Geology and Mineralogy, considered with reference to natural theology* (1837). By the will of the Earl of Bridgewater, money was left to trustees so that they might arrange the writing by prominent scholars of eight treatises the themes of which should, by the aid of science, prove 'The Power, Wisdom, and Goodness of God as manifested in the Creation'. Buckland had been commissioned as far back as 1830 to write such a Treatise but it took him six years of spasmodic bursts of exertion to complete the work. The theme was similar to that of *Vindiciae Geologicae*; the reconciliation of the discoveries of geology with the precepts of divine revelation. Yet over the passage of time and in the face of Lyell's persuasion, much of his accompanying framework of flood and catastrophe had undergone drastic modification. Buckland's *Treatise* was written during the period of the first publication of Lyell's *Principles* and, where the combination of the logic of Hutton and the soothing plausibility of Playfair had failed to convince him, the masterful opinions of Lyell had almost succeeded in less than a decade. The following passages show the extent of Buckland's conversion to Huttonian notions.

> Geology has already proved by physical evidence, that the surface of the globe has not existed in its actual state from eternity, but has advanced through a series of creative operations, succeeding one another at long and definite intervals of time; that all the actual combinations of matter have had a prior existence in some other state; and that the ultimate atoms of the material elements, through whatever changes they may have passed, are, and ever have been, governed by laws, as regular and uniform, as those which hold the planets in their course. (*Buckland, 1837, I, pp. 10–11*)

He retraces earlier steps when he explains how the early geologists confused themselves by endeavouring to rationalize the period of biblical creation with the facts shown by geological evidence. This inadequacy of time he thinks can be eliminated, not necessarily by considering the biblical 'days' as long periods, but by a new interpretation of the second verse of Genesis. By this means an indefinite pause is created during which the world could assume its present forms and this is followed by a period of chaos which can be taken to refer to the decaying and renewing worlds of Hutton:

> Still there is, I believe, no sound critical, or theological objection to the interpretation of the word 'day', as meaning a long period; but there will be no necessity for such extension, in order to reconcile the text of Genesis with physical appearances, if it can be shown that the time indicated by the phenomena of Geology may be found in the undefined interval, following the announcement of the first verse. (*Buckland, 1837, I, p. 18*)

We have in this second verse, a distinct mention of earth and waters, as already existing, and involved in darkness; their condition also is described as a state of confusion and emptiness, (tohu bohu), words which are usually interpreted by the vague and indefinite Greek term, 'chaos', and which may be geologically considered as designating the wreck and ruins of a former world. (*Buckland, 1837, I, pp. 24–26*)

In speaking of the agents of destruction and construction Buckland's analysis, compared with his earlier catastrophic reasonings, appears drastically modified and is virtually a total acceptance of the main Huttonian principles:

The detritus of the first dry lands, being drifted into the sea, and there spread out into extensive beds of mud and sand and gravel, would for ever have remained beneath the surface of the water, had not other forces been subsequently employed to raise them into dry land; these forces appear to have been the same expansive powers of heat and vapour which, having caused the elevation of the first raised portions of the fundamental crystalline rocks, continued their energies through all succeeding geological epochs, and still exert them in producing the phenomena of active volcanoes; phenomena incomparably the most violent that now appear upon the surface of our planet. (*Buckland, 1837, I, pp. 42–43*)

Within a few years of the Bridgewater Treatise the conversion of Buckland to uniformitarianism was almost complete. The thinker who once waxed eloquent over mighty floods and catastrophes now attributes the erosion of substantial masses of limestone rock to the corrosive action of acidulated rain water. When discussing fissures, caverns and bowl-shaped cavities in limestone areas:

Dr Buckland referred all these phaenomena to the action of acidulated waters, and was of opinion that the frequent volcanic eruptions during the tertiary period might have impregnated the sea in certain regions with carbonic acid, which would account for the extensive corrosion and destruction which the chalk underwent at that epoch, when the gravel-beds of the eocene period were supplied with flints, set free by the solution of the chalk in which they had been formed, and subsequently rounded by the action of the water. He considered also that the carbonic acid contained in rain-water has produced in more recent periods, and still continues to produce similar effects in corroding and forming cavities on the surface of chalk beneath permeable beds of gravel or sand. (*Buckland, 1839, pp. 76–77*)

Soon Buckland is following Lyell in praising the structuralist ideas of Hopkins on the Weald and in ascribing to 'violent marine currents' the conglomerates in the carboniferous strata of the Forest of Dean (Buckland, 1841, p. 493). With such close agreement, the former diluvialists were not likely to check the uniformitarian swerve towards the over-emphasis of marine and structural notions.

Structuralist Criticisms. Very different and more lasting was the influence on Lyell of contemporary structuralists. This school of thought took many forms and seems at times to lack uniformity as well as unity. Some of its adherents were purely 'fracturists' or believers in extensive faulting and rifting; others were 'convulsionists' or believers in spasmodic heavings due to earth-heat; others were enamoured of 'grand ancient metamorphoses' of a more amorphous type.

Among the first group was William Hopkins, a Cambridge mathematician, and friend of Sedgwick. Hopkins in his study of the Weald concentrated his analysis on the stresses and strains produced in a rising land mass. The longitudinal and transverse valleys which he found to be a feature of the area, he supposed owed their existence to the presence of two distinct systems of parallel faults, one longitudinal and the other transversal at right angles to it. He explained this fracturing as due to surface tension, when the strata was forced upwards. The tension first caused surface-fracturing along longitudinal parallel cracks but if the upward pressure continued, the only way in which tension could be released would be by means of transverse cracks at right angles to those already formed. This fault pattern explained the geometrical pattern of the valleys.

> The first effect of our elevatory force, will of course be to raise the mass under which it acts, and to place it in a state of extension, and consequently of tension. The increase of intensity in the elevatory force might be so rapid as to give it the character of an impulsive force, in which case it would be impossible to calculate the dislocating effects of it. This intensity and that of the consequent tensions will therefore be always assumed to increase continuously, till the tension becomes sufficient to rupture the mass, thus producing fissures and dislocations, the nature and position of which it will be the first object of our investigation to determine. (*Hopkins, 1838, p. 11*)

> It is evident, however, that in whatever manner a system of parallel fissures may be produced, that, after their formation the only tension of the mass between them must be in a direction parallel to them. Consequently, should any other system be subsequently formed, it must necessarily be in a direction perpendicular to that of the first system. (*Hopkins, 1838, p. 36*)

Hopkins later repeats this idea but admits that he can find little evidence for displacement or fracturing along the valleys, except the presence of perennial springs, which he assumed to be thrown out at fault-lines.

> Transverse valleys exist in the green sand ridge of this part of the district as well as in that on the southern side. The author also alludes to what he conceives to be incipient valleys of this description, and states his reasons for believing them to be indications of transverse fractures. He conceives this opinion to be strongly corroborated by the existence of the perennial springs by which these valleys are characterized. Several are pointed out, especially in Leith Hill and the Seven Oaks ridge overlooking the valley of the Weald.

> Transverse river-courses through the Chalk escarpment form one of the most striking features in the geology of this district. The analogy which they bear to the transverse valleys across the green sand ridges would seem to leave no doubt of their being referable to the same physical cause; and as there are in many instances direct evidence which renders the origin of these latter valleys in transverse fractures highly probable, the same conclusion

FIG. 33. *William Hopkins*
(*From a portrait in the possession of the Department of Mineralogy and Petrology,*
Cambridge University)

appears almost equally probable with respect to the river-courses through the Chalk. In the evidence of dislocation which the Chalk itself affords, there is nothing, however, very conclusive; but it must be remembered, that the evidence of faults is always difficult to detect in a massive formation like the Chalk, possessing not more than two general divisions which admit of distinct identification.

The central chalk ridge of the Isle of Wight is traversed in like manner by three transverse valleys, two of which are river-courses. The author has pointed out some direct evidence in support of the conclusion, that the central one (that of the Medina) has originated in transverse dislocation. (*Hopkins, 1841, p. 365*)

P. J. Martin in a paper on the chalk formations uses much the same arguments. He starts with a dome-like uplift which brings about geometrical fracturing or cracking, and these cracks are then widened to form the present valleys by the violent aqueous erosion which resulted directly from the disturbance of the waters thrown into motion by the sudden rise of the land. From his mention of drift and the use of the word 'violent' he obviously does not postulate a gradual uplift as did Lyell:

> He begins by an examination into the arrangement of the great chalk dome of Hampshire and Wiltshire, . . . its anticlinal lines of disturbance or upheaval, and their connections with those of the Weald and the smaller western denudations of Pewsey, Wardour and Warminster.
>
> He finds that six great anticlinal lines are the main instruments of the upbearing of this abraded chalk; that the three which characterize the smaller anticlinal western valleys are projected onward, and in a manner decussate three others which emanate from the western extremity of the greater valley of the Weald, the vale of Wolmar Forest . . . ; and that these lines do not inosculate or enter into each other; approximating indeed, but little in any part of their course; severally dying out, and their respective synclinal lines playing off into each other. Their course is rather irregular, and their force exceedingly variable; but their general parallelism is maintained throughout, their progress being East and West, with a point to the North. (*Martin, 1840–1, p. 349*)
>
> The author concludes this paper with some reference to the subject of transverse fractures in these several longitudinal fissures, and the cross drainage, . . . which will be adduced as illustrative of the strong probability, if they do not amount (in connection with the phaenomena of drift) to absolute proof, of the close relation of the acts of upheaval and violent aqueous abrasion. (*Martin, 1840–1, p. 350*)

The next development of structuralist thought which we will discuss was mentioned by Lyell, who saw in it an answer to the arguments that mountains could not have been produced by a slow process. Darwin in 1838 proposed a scheme of slow volcanic pressure, taking the form of repeated strokes, as of a pump—one mass hardening before the onset of the next infusion of volcanic matter.

> At the last meeting of the Geological Society, Darwin read a paper on the Connection of Volcanic Phenomena and Elevation of Mountain Chains, in support of my heretical doctrines; he opened upon De la Beche, Phillips, and others (for Greenough was absent) his whole battery of the earthquakes and volcanoes of the Andes, and argued that spaces at least a thousand miles long were simultaneously subject to earthquakes and volcanic eruptions, and that the elevation of the Pampas, Patagonia, &c., all depended on a common cause; also that the greater the contortions of strata in a mountain chain, the smaller must have been each separate and individual movement of that long series which was necessary to upheave the chain. Had they been more violent, he contended that the subterraneous fluid matter would have gushed out and overflowed, and the strata would have been blown up and annihilated. He therefore introduces a cooling of one small underground injection, and then

the pumping in of other lava, or porphyry, or granite, into the previously consolidated and first-formed mass of igneous rock. When he had done his description of the reiterated strokes of his volcanic pump, De la Beche gave us a long oration about the impossibility of strata in the Alps, &c., remaining flexible for such a time as they must have done, if they were to be tilted, convoluted, or overturned by gradual or small shoves. He never, however, explained his theory of original flexibility, and therefore I am as unable as ever to comprehend why flexibility is a quality so limited in time. (*Lyell, Mrs, 1881, II, pp. 39-40. Letter, 12 March 1838*)

Here, however, we must hasten to point out that as far as geological thought is concerned Lyell had more influence on Darwin than Darwin on Lyell. In fact the great naturalist during this period of his career, seems to have been a remarkably close disciple of Lyell and made excessive use of the Lyellian theory of marine denudation. In the Blue Mountains in Australia, Darwin was impressed by the vast proportions of the valleys cut in the sandstone platforms. Though the sides of the valleys had a parallel sequence of similar strata, thus discounting the possibility of structural influences, he is unable to credit the present rivers with the necessary power to accomplish the denudation. Instead he adopts the agency of the sea, and considers the steep valley sides to be former cliffs. He passes over the difficulty of the occurrence of narrow valleys where one might expect to find broad bay-like recesses or ocean channels by supposing that the embryo pattern was created below the sea at the time of the deposition of the strata:

Valleys in the sandstone platforms. . . . The grand valleys, by which the Blue Mountains and the other sandstone platforms of this part of Australia are penetrated, and which long offered an insuperable obstacle to the attempts of the most enterprising colonist to reach the interior country, form the most striking feature in the geology of New South Wales. They are of grand dimensions, and are bordered by continuous lines of lofty cliffs. (*Darwin, 1844, p. 151*)
The first impression, from seeing the correspondence of the horizontal strata, on each side of these valleys and great amphitheatre-like depressions, is that they have been in chief part hollowed out, like other valleys, by aqueous erosion; but when one reflects on the enormous amount of stone, which on this view must have been removed, in most of the above cases through mere gorges or chasms, one is led to ask whether these spaces may not have subsided. But considering the form of the irregularly branching valleys, and the narrow promontories, projecting into them from the platforms, we are compelled to abandon this notion. To attribute these hollows to alluvial action, would be preposterous; nor does the drainage from the summit-level always fall, as I remarked near the Weatherboard, into the head of these valleys, but into one side of their bay-like recesses. Some of the inhabitants remarked to me, that they never viewed one of these bay-like recesses, with the headlands receding on both hands, without being struck with their resemblance to a bold sea-coast. This is certainly the case; moreover, the numerous fine harbours, with their widely branching arms, on the present coast of New South Wales, which are generally connected with the sea by a

narrow mouth, from one mile to a quarter of a mile in width, passing through the sandstone coast-cliffs, present a likeness, though on a miniature scale, to the great valleys of the interior. But then immediately occurs the startling difficulty, why has the sea worn out these great though circumscribed, depressions on a wide platform, and left mere gorges, through which the whole vast amount of triturated matter must have been carried away? (*Darwin, 1844, pp. 152–3*)

To apply these ideas to the sandstone platforms of New South Wales, I imagine that the strata might have been heaped on an irregular bottom by the action of strong currents, and of the undulations of an open sea; and that the valley-like spaces thus left unfilled might, during a slow elevation of the land, have had their steeply sloping flanks worn into cliffs; the worn-down sandstone being removed, either at the time when the narrow gorges were cut by the retreating sea, or subsequently by alluvial action. (*Darwin, 1844, p. 154*)

Darwin applies similar reasoning to the Chilean Cordillera. The line of the valleys he relates to fractures and structural flexures but the main work of erosion he again explains as being carried out by the sea. The difficulty of accounting for the simultaneous erosion of parallel ridges he evades by supposing that they were elevated at different periods of time. In the concluding paragraphs he shows the typical contemporary inability to appreciate the long-term erosive powers of rivers.

The valleys within this range, often follow anticlinal but rarely synclinal lines; that is, the strata on the two sides more often dip from the line of valley than towards it. On the flanks of the range, the valleys most frequently run neither along anticlinal nor synclinal axes, but along lines of flexure or faults; that is, the strata on both sides dip in the same direction, but with different, though often only slightly different, inclinations. As most of the nearly parallel ridges which together form the Cordillera run approximately north and south, the east and west valleys cross them in zig-zag lines, bursting through the points where the strata have been least inclined. No doubt the greater part of the denudation was effected at the periods when tidal creeks occupied the valleys, and when the outer flanks of the mountains were exposed to the full force of an open ocean. I have already alluded to the power of the tidal action in the channels connecting great bays; and I may here mention that one of the surveying vessels in a channel of this kind, though under sail, was whirled round and round by the force of the current. We shall, hereafter see, that of the two main ridges forming the Chilian Cordillera, the eastern and loftiest one, owes the greater part of its angular upheaval to a period subsequent to the elevation of the western ridge; and it is likewise probable that many of the other parallel ridges have been angularly upheaved at different periods; consequently many parts of the surfaces of these mountains must formerly have been exposed to the full force of the waves, which, if the Cordillera were now sunk into the sea, would be protected by parallel chains of islands. The torrents in the valleys certainly have great power in wearing the rocks; as could be told by the dull rattling sound of the many fragments night and day hurrying downwards; and as was attested by the vast size of certain fragments, which I was assured had been carried onwards during floods; yet we have seen in the lower parts of the valleys, that the torrents have seldom removed all the sea-checked shingle forming the terraces, and have had time since the last elevation in mass only to cut in the under-lying rocks, gorges, deep and narrow, but quite insignificant in dimensions compared with the entire width and depth of the valleys. (*Darwin, 1846, pp. 298–300*)

Criticisms by traditional igneous catastrophists. It is obvious that Darwin, as with the diluvialists, has become so close a disciple of Lyell that his original geomorphic speculations must remain unimportant. Unfortunately the same ultimate geomorphic unimportance can also be attached to the remaining group of contemporary structuralists yet to be discussed – the traditional igneous catastrophists. Whewell, writing in 1839, classified the students of 'geological dynamics' into catastrophists and uniformitarians.

> Both these opinions have received their contributions during the preceding year: Mr Darwin having laid before us his views of the formation of mountain chains and volcanoes, which he conceives to be the effect of a gradual, small and occasional elevation of continental masses of the earth's crust; while Mr Murchison gathers from the researches in which he has been engaged, the belief of a former state of paroxysmal turbulence, of much deeper rooted intensity and wider range than any that are to be found in our own period; and M. de Beaumont, in France, has endeavoured to prove that Etna and many other mountains must have been produced by some gigantic and extraordinary convulsion of the earth. (*Whewell, 1839, p. 92*)

The Murchison mentioned is Roderick Impey (1792–1871), son of an old Scottish family and later Fellow of the Royal Society; president of the Geological Society; holder of the Orders of St Anne and of Stanislaus of the Kingdom of Russia; president of the Geographical Society; Knight; Director-General of the Geological Survey; K.C.B.; baronet; and doctor of Oxford, Cambridge and Dublin. When he retired from the army, Murchison found himself an active man of twenty-three with very little to do. At first he indulged in fieldsports and other light-hearted social activities until his intelligent, cultured wife, with the help of friends, persuaded him 'to pursue science instead of the fox'. Murchison relates his own mobilization into the service of science:

> As time rolled on I got blasé and tired of all fox-hunting life. In the summer following the hunting season of 1822–3, when revisiting my old friend Morritt of Rokeby, I fell in with Sir Humphry Davy, and experienced much gratification in his lively illustrations of great physical truths. As we shot partridges together in the morning, I perceived that a man might pursue philosophy without abandoning field-sports; and Davy, seeing that I had already made observations on the Alps and Apennines, independently of my antiquarian rambles, encouraged me to come to London and set to at science by attending lectures on chemistry, etc. (*Quoted in Geikie, 1875, I, p. 94*)

Having avidly acquired a few scientific principles, Murchison was lucky enough to win the friendship of Lyell and the two worked together in the volcanic areas of France and Italy where, as we have already described, they

FIG. 34. *Roderick Impey Murchison*
(*By Stephen Pearce in 1856. National Portrait Gallery*)

wondered at the river-eroded valleys (Geikie, 1875, I, p. 150 et seq.). These
travels were enlivened by the unison in the person of Murchison of both
hypochondria and indigestion due to constant over-eating (Letter of 11th
June 1828, in Lyell, Mrs, 1881, I, p. 189).

Though Lyell does not always speak highly of Murchison's deductions
or intellect he respected his enthusiasm and no doubt also his vitality and
perseverance. The following almost contemporary portrait seems just:

> He had hardly any imaginative power. . . . He was not a profound thinker,
> but his contemporaries could hardly find a clearer, more keen-eyed, and
> careful observer. He had the shrewdness, too, to know wherein his strength
> lay. Hence he seldom ventured beyond the domain of fact where his first
> successes were won, and in which throughout his long life he worked so hard
> and so well. (*Geikie, 1875, II, p. 346*)

Any hope of Murchison acquiring more advanced geomorphic views was destroyed when he ceased working with Lyell and became a close friend of Sedgwick, with whom over the next twenty years (1830–1850) he elucidated the main intricacies of the Devonian, Silurian (Ordovician) and Cambrian systems. This onerous task drew Murchison away from geomorphic studies and the close study of complex ancient strata, often fantastically contorted, inevitably caused his thoughts to be dominated by the power of underground forces or subterranean agencies. Perhaps, too, Sedgwick's catastrophic leanings also influenced Murchison's outlook. In any event, he became a confirmed catastrophist and in this respect his mind resembled a Silurian matrix, impervious and resistant. His book on *The Silurian System* (1838) did nothing to advance the science of geomorphology and in editions as late as 1854 he still contrasted 'processes now going on in nature on a small scale' with 'those infinitely grander ancient metamorphoses'.

We are not here concerned with the famous quarrel between the two friends which caused such a deep stir in contemporary geologic circles except that it increased popular interest in stratigraphy at the expense of physiography and led to the withdrawal of Sedgwick from the Geological Society, leaving Murchison, apart from Lyell, the most prominent member. It was at this stage, particularly during his presidency, that Murchison's views grew increasingly antagonistic to anything new. By the 1860's he had reached a nadir of reaction.

For some years there had been growing among the younger and more active geologists of Britain a conviction that the old doctrine of Hutton as to the origin of valleys by the erosive action of running water – a doctrine which, in spite of the admirable confirmation of it adduced by Mr Scrope from Central France, had never been generally adopted – was substantially true. Foremost among those who maintained this view, and enforced it by cogent argument and illustration, were the Directors of the Geological Survey of Great Britain and Ireland – Ramsay and Jukes. The former, moreover, started and worked out the remarkable idea that besides the excavation of valleys by river-action and the slow washing of the land by rain, there has been an extensive erosion of hollows and basins by glacier-ice, and that to this process we must attribute the great predominance of rock-basin lakes scattered over those tracts of the northern hemisphere which can be shown to have been buried under land-ice. As may be supposed, the President of the Geological Society regarded these doctrines as rank heresy, not to say sheer nonsense. He opposed them chiefly in his addresses to that Society; but his opposition, though vigorous enough in its language, dealt more in strong denial and protest, with the citation of the crowd of geological authorities who sided with him, than in serious argument. The force of evidence had constrained him to yield somewhat of the old exclusiveness with which he had fought for his icebergs, but having given up some points, and consented to admit the power of glaciers to polish and score the face of a country, and to pile up huge moraine-mounds, he felt himself free to set his foot down firmly and refuse to go a step further in the way of excavation, as his friends, the 'ice-men', would have had him. (*Geikie, 1875, II, pp. 316–17*)

Murchison was still talking in terms of catastrophes in 1864 at a time when the word had almost fallen into geologic disuse:

> In seconding the motion of thanks to Lyell for his address at Bath, I felt bound to say a few words in defence of my opinions as to the grander intensity of causes in old geological times than in the present or Man period; and as Lyell had used the words 'some great convulsion and fracture', to account for the great rent and fault out of which the hot Bath water flows, I said I was happy to receive that indication of the right view, and that I should in future range my friend Sir Charles along with myself among the 'convulsionists'. (*Quoted in Geikie, 1875, II, pp. 318–19*)

Darwin's comment on Murchison may not be far wide of the mark: 'How singular so great a geologist should have so unphilosophical a mind'.

LYELL'S YEARS OF PRE-EMINENCE

By 1845, Lyell was the mirror of all that really mattered in geologic thought and an analysis of his opinions on landscape-evolution will indicate clearly the condition of contemporary geomorphology. Yet before proceeding we must point out that although he was only middle-aged his geomorphological views were already beginning to crystallize and that henceforth, although he greatly advances his knowledge of stratigraphy and palaeontology, he makes relatively little progress in his notions on landforms.

By now he accepts completely Huttonian ideas on the origin of new strata from the breaking down of the old

> The formation of every new deposit by the transport of sediment and pebbles necessarily implies that there has been, somewhere else, a grinding down of rock into rounded fragments, sand, or mud, equal in quantity to the new strata. All deposition, therefore, except in the case of a shower of volcanic ashes, is the sign of superficial waste going on contemporaneously, and to an equal amount elsewhere. (*Lyell, 1852, p. 66*)

Yet he seems to have decided more and more against the ability of rivers to erode on a grand scale. This was undoubtedly because he felt that in many districts the size of the present rivers was not proportionate to the size of the great valleys in which they flowed. The huge valleys of the principal rivers he ascribed to other causes. Some, as in the folded Jura, to underlying structure, but the majority to the action of marine currents sweeping across a rising continent. Erosion would naturally be greatest, he felt, when the land-mass was just below sea level and so within the sphere of wave-action.

> Origin of valleys. . . . Many of the earlier geologists, and Dr Hutton among them, taught that 'rivers have in general hollowed out their valleys'. This is true only of rivulets and torrents which are the feeders of the larger streams, and which, descending over rapid slopes, are most subject to temporary increase and diminution in the volume of their waters. The quantity of mud, sand, and pebbles constituting many a modern delta proves indisputably that no small part of the inequalities now existing on the earth's surface are

due to fluviatile action; but the principal valleys in almost every great hydrographical basin in the world, are of a shape and magnitude which imply that they have been due to other causes besides the mere excavating power of rivers. (*Lyell, 1852, p. 70*)

It has now been ascertained that the rising and sinking of extensive portions of the earth's crust, whether insensibly or by a repetition of sudden shocks, is part of the actual course of nature, and we may easily comprehend how the land may have been exposed during these movements to abrasion by the waves of the sea . . . Under these circumstances, the flowing waters may have power to clear away each stratum of incoherent materials as it rises and approaches the surface, where the waves exert the greatest force; and in this manner a voluminous deposit may be entirely swept away, so that, in the absence of faults, no evidence may remain of the denuding operation. It may indeed be affirmed that the signs of waste will usually be least obvious where the destruction has been most complete; for the annihilation may have proceeded so far, that no ruins are left of the dilapidated rocks. (*Lyell, 1852, p. 70*)

No combination of causes has yet been conceived so capable of producing extensive and gradual denudation, as the action of the waves and currents of the ocean upon land slowly rising out of the deep. (*Lyell, 1852, p. 71*)

Though Lyell chose the sea because he deemed it to be more powerful than the river, he did not visualize it as an agent of planation cutting a platform across the rocks regardless of differences of resistance. This planation idea was the product of Ramsay's mind. Under Lyell's scheme the sea etched out the inequalities provided by the initial fractures and differences in rock hardness. The peculiarity of such reasoning is that Lyell could argue thus with an air of complete conviction when from his own careful observations of the sea-eroded platforms below the cliffs he could plainly see that there was no prolonged marine selection of erosional lines, but unequivocal evidence of the cutting of a universal flat, erosion surface:

> Lastly, it remains only to speak of the terraces, which extend with a gentle slope from the base of almost all the inland cliffs, and are for the most part narrow where the rock is hard, but sometimes half a mile or more in breadth where it is soft. They are the effects of the encroachment of the ancient sea upon the shore at those levels at which the land remained for a long time stationary. The justness of this view is apparent on examining the shape of the modern shore wherever the sea is advancing upon the land, and removing annually small portions of undermined rock. By this agency a submarine platform is produced on which we may walk for some distance from the beach in shallow water, the increase of depth being very gradual, until we reach a point where the bottom plunges down suddenly. This platform is widened with more or less rapidity according to the hardness of the rocks, and when upraised it constitutes an inland terrace. (*Lyell, 1852, p. 74*)

Much of Lyell's geomorphic work is filled with this idea of marine erosion and he still uses it to explain the features of the Weald, where the scarps are considered to be former cliffs and the central vale an eroded marine platform. A new feature, the river terrace, which except for the writings of Hitchcock

(p. 315) had lacked accurate description since the work of Hutton and Playfair, next attracted his attention. Unfortunately, when he saw rivers cutting down through past alluvial deposits he did not assume the rivers were smaller than formerly, for this would have been a denial of uniformitarian principles, but supposed the landmass to have experienced subsidence followed by uplift. During the period of subsidence a wide plain would be deposited through which the river would later cut when the land rose and the river gradient was steepened. This belief in oscillations of the level of the sea and of continents seems a natural outcome of Lyell's detailed studies in marine sedimentation or littoral stratigraphy. He accounted for river-terraces at differing heights by changes in the height of either sea or land but though this was correct for some cases, he was wrong when he concluded that all terraces were produced by intermittent uplift. We shall see later that in many districts they originated from the choking up of stream channels during the Pleistocene:

> It has long been a matter of common observation that most rivers are now cutting their channels through alluvial deposits of greater depth and extent than could ever have been formed by the present streams. From this fact a rash inference has sometimes been drawn, that rivers in general have grown smaller, or become less liable to be flooded than formerly. But such phenomena would be a natural result of any considerable oscillations in the level of the land experienced since the existing valleys originated.
>
> Suppose part of a continent, comprising within it a large hydrographical basin like that of the Mississippi, to subside several inches or feet in a century, as the west coast of Greenland, extending 600 miles north and south, has been sinking for three or four centuries, between the latitudes 60° and 69° N. There might be no encroachment of the sea at the river's mouth in consequence of this change of level but the fall of the waters flowing from the interior being lessened, the main river and its tributaries would have less power to carry down to its delta, and to discharge into the ocean, the sedimentary matter with which they are annually loaded. They would all begin to raise their channels and alluvial plains by depositing in them the heavier sand and pebbles washed down from the upland country, and this operation would take place most effectively if the amount of subsidence in the interior was unequal, and especially if, on the whole, it exceeded that of the region near the sea. If then the same area of land be again upheaved to its former height, the fall, and consequently the velocity, of every river would begin to augment. Each of them would be less given to overflow its alluvial plain; and their power of carrying earthy matter seaward, and of scouring out and deepening their channels, would continue till, after a lapse of many thousand years, each of them would have eroded a new channel or valley through a fluviatile formation of modern date. The surface of what was once the river-plain at the period of greatest depression, would remain fringing the valley sides in the form of a terrace apparently flat, but in reality sloping down with the general inclination of the river. (*Lyell, 1852, pp. 84–85*)

There terraces are seldom continuous for great distances, and their surface slopes downwards, with an inclination similar to that of the river. They are readily explained if we adopt the hypothesis before suggested, of a gradual

FIG. 35. *Sir Charles Lyell*
(*Replica by Lowes Cato Dickinson. National Portrait Gallery*)

rise of the land; especially if, while rivers are shaping out their beds, the upheaving movement be intermittent, so that long pauses shall occur, during
which the stream will have time to encroach upon one of its banks, so as to
clear away and flatten a large space. This operation being afterwards repeated
at lower levels, there will be several successive cliffs and terraces. (*Lyell, 1852,
p. 85*)

We must hasten to point out that what to us now seem defects seemed progress to most of Lyell's contemporaries. He had out-topped contemporary

criticism or had converted most of the critics. Even opposition on moral grounds was now so out of date that rigid adherence to the Bible was almost a matter of ridicule. Scientists from all over Europe corresponded with him and sought his advice on technical details. The Crown sought his views on university reform; he was knighted in 1848 and two years later was chosen as one of the commissioners in charge of the Great Exhibition. In 1853, he went to New York as a commissioner for the International Exhibition. Honours and engagements were thrust upon him from all sides but he managed to visit the Canaries and Madeira in 1854 and the Continent annually from 1855 to 1860.

If Werner had been kingly then Lyell was imperial in his splendour. When dissent from his geomorphic teachings did at last come, the criticisms were treated more as a form of extreme eccentricity than as an argument worthy of serious regard, particularly as their author, Col. G. Greenwood, was virtually unknown in academic circles. Nevertheless Greenwood's *Rain and Rivers; or Hutton and Playfair against Lyell and All Comers* is far too pertinent to be omitted here. He strongly objected (often out of context!) to statements like the following by Lyell, relating to the region of Calabria;

> the formation of valleys by running water can never be understood if we consider the question independently of the agency of earthquakes. (*Lyell, 7th edn, 1847, I, p. 466*)

But he reserved his severest strictures for Lyell's ideas on marine denudation which he set out to prove to be totally incorrect. Marine erosion, Greenwood argued, might produce a flat surface but it was totally impossible for it to create a sloping surface intersected by intricately related systems of hills and valleys. At best, the sea might scoop out a wide depression.

> That the action of waves on land slowly emerging from the deep should have a tendency to wash away soft parts and to leave hard parts, I can conceive; but to attribute the formation of our valleys to this cause, is to suppose that the materials of all the valleys running from the tops of all the heights on the globe were originally softer than the materials of the intervening ridges; but in almost all cases we can see that this is not so, by the corresponding strata on the opposite sides of valleys . . .
>
> In regard to currents: a current might decapitate a continent as it rose, supposing equal softness of materials, or it might scoop a horizontal groove of any size or depth, or (granting lines of hard intervening ridges) many grooves. But they must all be horizontal and in one direction. No marine current could make a single channel sloping from a height to the sea; still less the myriads on myriads of dry upper valleys which ramify in all directions, from all river valleys, through and to all sides of the tops of all elevations, whether high or low. (*Greenwood, 1857, p. 3*)

Almost as a touch of irony Greenwood cites as his best example Lyell's own description of the absence of marine action in the volcanic region of Auvergne, the area where Lyell had worked with Murchison and about which they had so triumphantly announced that they could find no evidence of previous diluvial action.

Unfortunately, Greenwood's call was lost amid the tremendous interest now aroused by the new palaeontological advances.

THE CLOSE OF AN EPOCH

At this time Lyell was sixty years old and his health was showing signs of failing yet he retained his normal mental vitality and eagerly followed new scientific developments. In 1858 he made more ascents of Etna and settled in favour of himself and Scrope their contention that the outward dip of ash and lava from the volcanic plug was due to accumulative deposition and not to the conical upthrust of strata ('crater of elevation') as was postulated by Von Buch and his followers. This, however, was probably his last important contribution to landscape study. He and Sedgwick and others were now distracted from non-palaeontological aspects of geology by the implications of EVOLU-TION, whether man was an advanced species of man-like animal or a divine reproduction of the Godhead. A combined paper on the subject by Darwin and A. Russel Wallace (1858) was published with the help of Lyell and when Darwin's *Origin of Species* was published in the following year there could be no doubt that its public reception owed much to the long acceptance of *The Principles of Geology* and uniformitarianism. Lyell, although formerly not favourable to the transmutation of species, changed his views and made an independent investigation of the evidence. In 1863 he produced *The Antiquity of Man* in which he recognized that the human remains found in some caves and gravels were of great antiquity. The book ran into three editions in a few months.

Some of these findings were incorporated into *The Principles* which were thoroughly revised in 1867–8 and slightly revised in 1872 and 1875. These last editions show clearly enough that Lyell's geomorphic ideas had not developed appreciably since his concentration on Evolutionary topics. In them, with characteristic eloquence, Lyell retraces the old arguments in favour of uniformitarianism as opposed to catastrophism. Vast deluges, he adds, are no longer possible.

> If we restrict ourselves to combinations of causes at present known, it would seem that the two principal sources of extraordinary inundations are, first the escape of the waters of a large lake raised far above the sea; and, secondly, the pouring down of a marine current into lands depressed below the mean level of the ocean. (*Lyell, 1872, I, p. 108*)

He then describes how the levelling of the earth's excrescences and the creation of new strata can be divided between the activities of two sets of agents: the erosive aqueous agents of river, rain and sea; and the igneous agents of volcanoes or earthquakes responsible for the uplift of land. Sometimes, however, it is impossible to distinguish between the two forces because each,

instead of working to undo the operations of the other, may be jointly responsible for producing a feature by a process of mutual adjustment.

It is difficult, in a scientific arrangement, to give an accurate view of the
combined effects of so many forces in simultaneous operation; because when
we consider them separately, we cannot easily estimate either the extent of
their efficacy or the kind of results which they produce. We are in danger,
therefore, when we attempt to examine the influence exerted singly by each,
of overlooking the modifications which they produce on one another; and
these are so complicated, that sometimes the igneous and aqueous forces co-
operate to produce a joint effect, to which neither of them unaided by the
other could give rise, – as when repeated earthquakes unite with running
water to widen a valley; or when a thermal spring rises up from a great depth,
and conveys the mineral ingredients with which it is impregnated from the
interior of the earth to the surface. (*Lyell, 1872, I, pp. 321-2*)

When taking this argument further, his disagreement with Huttonian principles becomes obvious.

It is probable that few great valleys have been excavated in any part of the
world, by rain and running water alone. During some part of their formation,
subterranean movements have lent their aid in accelerating the process of
erosion, such movements being intermittent and often suspended for ages,
and in many cases causing changes of level without any vibratory jar, their
influence may easily be underrated or overlooked by geologists. (*Lyell, 1872,
I, p. 335*)

His convictions about the inadequacy of rivers are in no way shaken when
he examines examples of rapid rain and river erosion.

Earth-pyramids or stone-capped pillars of Botzen in the Tyrol . . . It is
not often that the effects of the denuding action of rain can be studied separately or as distinct from those of running water. There are, however, several
cases in the Alps, and especially in the Tyrol near Botzen, which present
a marked exception to this rule, where columns of indurated mud, varying in
height from twenty to a hundred feet, and usually capped by a single stone,
have been separated by rain from the terrace of which they once formed a
part, and now stand at various levels on the steep slopes bounding narrow
valleys. (*Lyell, 1872, I, pp. 329-30*)

Action of rivers. . . . When travelling in Georgia and Alabama, in 1846, I
saw in both these States the commencement of hundreds of valleys in places
where the native forest had recently been removed. One of these newly
formed gullies or ravines is represented in the annexed woodcut, from a drawing which I made on the spot. It occurs three miles and a half due west of
Milledgeville, the capital of Georgia, and is situated on the farm of Pomona,
on the direct road to Macon.

In 1826, before the land was cleared, it had no existence; but when the
trees of the forest were cut down, cracks three feet deep were caused by the
sun's heat in the clay; and during the rains, a sudden rush of water through
the principal crack deepened it at its lower extremity, from whence the
excavating power worked backwards, till in the course of twenty years, a
chasm measuring no less than 55 feet in depth, 300 yards in length, and

Ravine on the farm of Pomona, near Milledgeville, Georgia, as it appeared
January 1846.

Excavated in twenty years, 55 feet deep, and 180 feet broad.

FIG. 36. *Lyell's illustration of gully erosion on the Georgia piedmont near
Milledgeville, 32 miles north-east of Macon*

varying in width from 20 to 180 feet, was the result. The high road has been
several times turned to avoid this cavity, the enlargement of which is still
proceeding, and the old line of road may be seen to have held its course
directly over what is now the widest part of the ravine. (*Lyell, 1872, I,
pp. 337–9*)

Yet despite this superfluity of evidence for rain and river erosion, Lyell firmly returns to his main thesis that marine currents are the principal instruments in the production of surface topography.

> After these preliminary remarks on the nature and causes of currents, their velocity and direction, we may next consider their action on the solid materials of the earth. We shall find that their efforts are, in many respects, strictly analogous to those of rivers. I have already treated, in the third chapter, of the manner in which currents sometimes combine with ice, in carrying mud, pebbles, and large fragments of rock to great distances. Their operations are more concealed from our view than those of rivers, but extend to wider areas, and are therefore of more geological importance. (*Lyell, 1872, I, p. 507*)

It is a lamentable fact that out of so promising a beginning so little was ultimately achieved in geomorphology. Hutton's thesis of river erosion, which had seemed at first to have a fair chance of being widely accepted, had been pushed into the background and the influence of its true implications was postponed to a much later date. This, the only real blemish to be found in Lyell's achievements, happens to be fundamental. Even today we feel strangely loathe to criticize harshly the great man who once ahd for all assured the acceptance of uniformitarian principles and largely rid geologic thinking of its catastrophic outlook.

His skill in observing details in the field was exemplary. His *Principles* grandly conceived, splendidly written and beautifully illustrated by example and diagram, formed a worthy key-stone to the vast structure of contemporary geology. But here we are concerned only with the history of geomorphology and, while freely admitting the tremendous difference in scale between a water-worn gully and the Wealden vales, we still find it hard to forgive the inadvertence of one who professed a belief in the power of slow processes but refused to attribute great landscape features to river erosion.

Lyell, honoured and revered at home and abroad, died on 22 February 1875 when in his seventy-eighth year and was buried in the nave of Westminster Abbey. Charles Darwin writes of him, 'The science of geology is enormously indebted to Lyell – more so, as I believe, than to any other man who ever lived'.

CHAPTER THIRTEEN

The Glacial Theory

Out of whose womb came the ice?
And the hoary frost of heaven, who hath gendered it?
The waters are hid as with a stone,
And the face of the deep is frozen.

JOB (1520 B.C.)

The problem of what we know today to be glacial phenomena had long puzzled geologists, and such features, because the idea of extensive land glaciers and continental ice-sheets had not yet been conceived, were treated as being similar in origin to mountains and valleys. Geologists had confused themselves by telescoping together many different phenomena within their limited conception of the timespan of earth history, and seeking one explanation to explain them. The ignorance of glacial processes cannot be attributed wholly to any lack of glacial and fluvioglacial landforms although in this respect British geomorphologists were no doubt greatly handicapped by the absence of present glaciers. The more northerly and the mountainous tracts of North America and Eurasia abound in glaciated landscapes but the existing ice-caps and glaciers are restricted to high altitudes and polar latitudes. It was perhaps inevitable that the best descriptions of ice-affected scenery came from Scandinavia and the Alps. Typical were the observations of Nathaniel Wraxall who in his *Tour through . . . the Northern Parts of Europe* describes the profusion of erratic blocks to be found all over the countryside:

> It is difficult to depicture the aspect of the country in which Louisa (Loviisa, in southern Finland) stands. The earth may almost be said to have disappeared from view, so completely was it covered with stones, or rather rocks. It seems as if they had fallen from the sky; and . . . the road, compelled to respect these formidable impediments, performs a thousand tortuous evolutions, in order to avoid them, and serpentines beautifully for many miles. (*Wraxall, 1775, pt. 3, sect. 2, p. 95*)

As we have already mentioned, the presence of these erratics, as well as of the associated 'drift' and rock-striations, was generally explained by the action of a universal flood or oceanic inundation which swept huge boulders and smaller material on to plains and highlands alike.

Not everyone was satisfied with this sea-current theory, which fitted well some of the facts but failed to account for others. Thus De la Beche during his continental travels in 1819–20 saw many of these phenomena and, though he offered no explanation of their origin, he was clearly puzzled as to how a vast current could sweep simultaneously boulders in opposing directions. He noticed that erratic blocks of the same kind of rock occurred on both the northern and southern flanks of the Alps but was at a loss to imagine 'how a grand débâcle from the Alps . . . could operate in two different directions at the same time'. (Quoted in North, 1943, p. 5.) It seemed, he reasoned, that here the passage of the boulders from their places of origin showed a pattern radiating from the mountains and following the line of the present valleys and lakes. Yet De la Beche lacked the vision to connect the features with ice.

EARLY GLACIOLOGISTS

The first real advance in glaciology seems to have been made as far back as 1723 when Johann Jacob Scheuchzer (1672–1733), a Swiss naturalist, proposed a theory to account for the progressive forward movement of glaciers. He envisaged the surface water entering the cracks in the glacier, freezing and expanding, thus causing the whole mass to move downhill in the only direction open to it:

> The cause of the motion (he wrote) is not owing to any miracle, as those ignorant of physics suppose. . . . The water flowing from the sides of the mountain on to the glacier enters its fissures and freezes again, and as it needs more space when thus frozen, as experiments have shown, it causes the glacier to thrust forward and to carry with it sand and stones, some of them of great size. (*North, 1943, p. 8*)

When more was known of glaciers and they were regarded as important erosive agents this became termed the Dilatation Theory. At the time its reasoning passed almost unnoticed and its significance for geomorphology was wholly unappreciated.

In 1744, P. Martel, a Swiss engineer, supported this finding and held that the movement of glaciers was proved by the transport of stones and sand down valleys occupied by these masses of ice.

> It is to be observed that the Glaciere is not level, and all the Ice has a Motion from the higher Parts towards the lower . . . which has been remarked by many Circumstances. First, By great Stones, which have been carried quite into the valley of Chamouny; they showed one of a very large size which several old People assured us that they had seen upon the Ice. (*Martel, 1744, p. 21*)

It will be recalled that some fifty years later in his *Theory of the Earth* Hutton, influenced by De Saussure, specifically accredited glaciers with the power to move the large boulders found in the Alps. In fact he states that this, considering the force required, was the only possible means by which their present position could be explained. Hutton was unable to visit the Alps because of war and had to base his conclusions on the evidence of De Saussure's descriptions of the mountain areas. However, Playfair made a tour of the Alps in 1815 and came to the same conclusions as Hutton.

> For the moving of large masses of rock, the most powerful engines without doubt which nature employs are the glaciers. . . . These great masses are in perpetual motion, . . . impelled down the declivities on which they rest by their own enormous weight, together with that of the innumerable fragments of rock with which they are loaded. These fragments they gradually transport to their utmost boundaries. . . . In this manner, before the valleys were cut out in the form they now are . . . huge fragments of rock may have been carried to a great distance; and it is not wonderful, if these same masses, greatly diminished in size, and reduced to gravel or sand, have reached the shores, or even the bottom, of the ocean. (*Playfair, 1802, pp. 388–9*)

This prescience of Hutton and Playfair should not be overstated. They had not propounded any glacial theory; at its boldest they had said that glaciers could move large rocks and, from the evidence of such rocks in distant places, probably had. The exact means and the extent of such action was left to conjecture. The knot was slowly and much later to be unravelled by the combined endeavour of many persons.

In 1815 a local guide J. P. Perraudin living in the Canton of Valais postulated a former extension of the glaciers, and consequently an increase in their size, as the cause of the boulders lying high up on the valley sides. As he was only a 'habile chasseur de chamois, et amateur de ces sortes d'observations' his views were never known within the main geological circle. In 1821, I. Venetz, a professional Civil Engineer living in the same canton, drew attention to Perraudin's idea in a paper which he read before the Société Helvétique des Sciences Naturelles. While plotting the alpine variations of temperature he noticed that in addition to the morainic debris around the snouts of the glaciers there were similar morainic deposits far beyond their present termination. From this fact he agreed with the conclusion of Perraudin that the glaciers must at some earlier time have covered a greater area. Even this public announcement of the idea received negligible publicity and twelve years passed before the paper was published. Undeterred, Venetz pursued his speculations and became so impressed by similar findings in areas as far away as the Jura, that in 1829 he addressed the society again and explained the presence of all erratic blocks in Switzerland and northern Europe as due to transport by ice. Even so, these grand notions might well have been neglected, if they had not attracted the attention of Jean de Charpentier (1786–1855), Director of Mines for the Canton of Vaud. We stress this because a very similar theory put forward in 1832 by Professor A. Bernhardi of the Academy of Forestry at Dreissigacker received no scientific recognition until 1875; an isolated note of Peter Dobson of Connecticut, published in the United States in 1826, received no mention until Murchison drew attention to it in his anniversary address to the Geological Society in 1842; and in Britain an accurate observation of De la Beche in his *Geological Manual* (2nd edn, 1832) seems to have been utterly ignored. Quotations from these neglected writers are given below.

> The polar ice once reached as far as the southern limit of the district (in Germany) which is still marked by the erratics. This ice, in the course of thousands of years, shrank to its present proportions, and the deposits of erratics must be identified with the walls or mounds of rock fragments which are deposited by glaciers large and small, or in other words, are nothing less than the moraines which this vast sea of ice deposited in its shrinkage and retreat. (*Bernhardi, 1832, Jahrbuch für Mineralogie . . ., III, pp. 258–9*)
>
> I have had occasion to dig up a great number of bowlders, of red sandstone, and of the conglomerate kind, in erecting a cotton manufactory; and it was not uncommon to find them worn smooth on the under-side, as if done by their having been dragged over rocks and gravelly earth, in one steady position. (*Dobson, 1826, p. 217*)
>
> I think we cannot account for these appearances, unless we call in the aid of ice along with water, and that they have been worn by being suspended and carried in ice, over rocks and earth, under water.
>
> It is stated in the *Edinburgh Encyclopedia*, Vol. XIII, p. 426, that 'fields of ice sometimes rise from the bottom, and bring with them masses of rock, of several hundred tons weight. These masses of stone are imbedded in the ice, they are carried along with the ice, and deposited on shores at a great distance from their original situation.' (*Dobson, 1826, p. 218*)
>
> Ice would seem to afford a possible explanation of the transport of many masses (erratic blocks); for the glaciers which descend the valleys of the high northern regions are, like those of the Alps, charged with blocks which have fallen from the heights. (*De la Beche, 1832, p. 173*)

Hence it was fortunate that De Charpentier went to work to examine the mechanics of Venetz's idea. After a close observation of the glacial processes in operation he found sufficient information to enable him to read a paper on the subject at Lucerne in 1834. This was shortly afterwards published under the title *Notice sur la Cause probable de transport des Blocs Érratiques de la Suisse*. In this he contended that the blocks had been transported by ice and not by water, and drew support for this view from the striated surfaces to be found in glacial areas, which he decided were attributable to the action of moving ice and not to the action of boulder- or gravel-laden water.

> He gave credit to Venetz for his pioneer work, claimed that the travelled blocks had not been transported by water but by ice, and, for the first time in the study of the problem (as far as published or known unpublished records go), introduced a reference to the polished and striated surfaces associated with glacier floors – such surfaces had been noticed before . . . but they had been attributed to the action of gravel-laden water. De Charpentier explained that glaciers press themselves into all the holes and hollows in their beds, polishing the surfaces even of overhanging rocks, an effect which running water laden with stones could not produce. (*North, 1943, pp. 17–18*)

Yet even De Charpentier failed to attract the attention of geologists of high standing.

granite blocks

granit Blocks near ṯ Château de l'Hermite on ṯ Salene Jeneva

FIG. 37. *Glacial erratics sketched by:*
Above: *Buckland. Showing granite blocks resting on limestone pedestals.*
Below: *De la Beche. A granite block near Neuchâtel, drawn in 1819.*

a large Block of Granite upon a fine . Neuchâtel ṯ

THE INTERVENTION OF JEAN LOUIS RODOLPHE AGASSIZ

It was at this stage in the glacial theory that Agassiz made his first venture in glaciology. He was born in 1807 of Swiss parents and entered Heidelberg medical school in 1826 where he became interested in fishes and palaeontology. After graduating, he passed on to Paris in 1831 and in the following year was appointed Professor of Natural History at Neuchâtel. Soon, as a specialist in fossil fishes, he achieved considerable distinction and remained in the front rank of palaeontologists until his dramatic switch to glacial geology. By then, most geologists knew Agassiz's work on fossils and in England, Lyell, Buckland and Sedgwick were on terms of appreciative correspondence with him. It is therefore fair to suggest that glaciology benefited quite as much from the use of Agassiz's name as from the application of his ideas.

In 1836 Agassiz and De Charpentier together closely examined the glaciers of Diablerets and Chamonix. Agassiz was immediately convinced of the correctness of his friend's glacial theory and of the possibilities provided by this new erosive and transporting agent. In July 1837, he put his ideas before the Annual Congress of Swiss scientists. In the part devoted to evidence of the former extent of glaciers he closely followed De Charpentier, even as far as imagining that the alpine glaciers extended to the foot of the Jura:

> Finally, the observer, leaving the summit of the Jura in the same direction where the boulders are found, and constantly following their tracks, finds himself at the bottom of the upper valleys of the Alps, and at the glaciers which dominate them, where he sees at last that the deposits have become real moraines. (Quoted by North, 1943, p. 18)

He asserted with the maximum emphasis that glaciers were solely responsible for the production of all the purported 'diluvial' phenomena. In accounting for the movement of the masses of ice involved he resurrected the dilatation theory of Scheuchzer, which had been revived in 1819 by T. de Charpentier, the brother of Jean. Expanding Charpentier's theory, Agassiz postulated that the extension of the glaciers was not a local circumstance but part of a climatic change which affected the whole of Europe. He assumed a general lowering of temperature which ultimately caused the greater part of northern Europe to be covered by vast ice-sheets, leaving only a few isolated mountain peaks protruding above this world of ice. The concept of a continental ice-field stemmed from his association with Karl Schimper, a scientist who, after studying erratics in Bavaria during 1836, had come to the conclusion that they had been precipitated from melting icebergs and by 1837 enlarged his reasoning to include a Europe covered by 'ein grosses Eisfeld'.

> Even the valleys and lakes had been filled with ice (which, he thought, explained why they were not subsequently obliterated by detritus carried by the ice); and that the erratic blocks were carried on the surface of the ice, and so, when it melted, were dropped on mountains and high on the valley sides. (Schimper, quoted by North, 1943, p. 19)

Another departure from De Charpentier's theory was that Agassiz supposed, wrongly as is now known, that this period of glaciation took place before the raising of the Alps to their present altitude. He put the glaciation before the orogenesis because if the Alps were thrust up through the ice-mass, this would cause the ice to shatter and break off in huge icebergs; and the rocks to fracture and roll down over the ice. Thus would be set in motion his vast glacial machine. When the ice melted, the rocks, sand and gravel would be deposited at the points they had then reached.

Thus Agassiz's first theme, apart from a few notable variations, was largely a summary of the notions of little-known men. His prime addition was the prestige of his name.

CONTEMPORARY NON-GLACIAL EXPLANATIONS OF ERRATICS AND
DRIFT DEPOSITS

Before describing the reception of Agassiz's ideas on the Continent and in Britain it seems advisable to summarize the traditional notions on the origin of erratics and drift deposits. We shall thereby discover the strength of the opposition and the extent of possible progress. Although theories based on the Flood or a universal ocean dominated, some other ideas are of interest because of the ingenuity of their mechanisms.

In 1719, Emanuel Swedenborg, better known for his part in the founding of the Church of New Jerusalem, wrote a small memoir in Swedish on certain glacial features using the sea and the waves as the means whereby the deposits were so arranged and laid down:

> ... he drew attention to boulders differing in character from the rocks upon which they rest, and he associated them with the long sinuous ridges of sand and stones known in Sweden as Asar. He noticed that the ridges were more or less parallel to one another: that the stones in them showed signs of having been rubbed and polished; and that the scattered boulders occurred not only on the plains, but also in lakes and even among the mountains.
>
> Swedenborg thought that the smooth stones and the sand were evidence that the sea had been concerned in their disposition, and he invoked a deep universal ocean in which the materials were gathered together, and by the action of which they were distributed and arranged. He supposed that very deep waves tore up the stones and sand and clay from the sea floor, and piled them in heaps and ridges, carrying the fragments even to the tops of the mountains.
>
> To counter objections based upon the great weight of the stones that the sea would have to move, Swedenborg pointed out that while stone is normally about 2½ times as heavy as water, in fresh water the apparent weight is appreciably reduced, and in salt water the stone is relatively lighter still. (*North, 1943, pp. 5–7*)

The next explanation by Daniel Tilas, another Swede, was a little nearer the truth. About 1740 he observed that as the boulders were traced farther away from their source they assumed a more rounded form. Therefore he also supposed the agency of a great flood but to account for the movement of some of the blocks he preferred the explanation of their transport by icebergs.

Jean Guettard in 1762 put forward the novel suggestion that the boulders merely represented the final residue of former granitic mountains which had decomposed during the course of time.

> ... he suggested ... that the granite boulders so abundant in the plains between the Baltic and the Carpathians were the debris of granite mountains that had been destroyed in situ, and he claimed, in support of his suggestion, that the sand associated with the boulders is usually more like that produced by the decomposition of granite than that resulting from the disintegration of sandstone. (*North, 1943, p. 7*)

J. A. de Luc in his *Lettres Physiques et Morales sur les Montagnes* (1778) presented a far more complicated scheme. He adopted the biblical notion of a hollow interior within the earth and he went on to suppose that the boulders and gravels were fired out from this interior during volcanic disturbances:

> He supposed that the irregularities of the earth's surface were due to subsidence of the lower regions, and he postulated great caverns filled with some expansible fluid such as air. The subsidence of large masses of the earth's crust would, he thought, exert considerable pressure upon the imprisoned fluid, the escape of which through cracks would accordingly be violent. This would result in the disruption of the rocks at the bottom of the sea and would account for the grinding together of the fragments to produce rounded pebbles and gravel. (*North, 1943, p. 7*)

Naturally, this was a ready target for the scientific mind and the scheme was strongly attacked by De Saussure in 1787 who demonstrated its implausibility in respect of several points.

> De Saussure asked for examples of explosions that could shoot large blocks of stone through the air for distances of 12 or 15 leagues: he wanted to know how it was that a block shot from the Alps to the Jura country did not, on landing, either smash to pieces or bury itself in the ground. On the contrary, he observed, the boulders simply rested upon the surface and often touched it only at a few points. (*De Saussure, Voyages dans les Alps, I; North, 1943, pp. 7–8*)

From his work in the Alps De Saussure was able to draw a detailed picture of glacial operations, and it was from his descriptions that Hutton and Playfair had been able to make their own assumptions. Like De Charpentier he understood that the position and composition of the moraines was dependent upon the extent, advance and retreat of the valley glaciers. In addition he realized that traces of Alpine fragments could be followed down the Alpine valleys until they came to rest against the sides of the Jura and that their position opposite the alpine valleys of their origin could only point to the conclusion that they must have come by this route. Yet, although he knew the nature of moraines, he failed to appreciate that these erratic boulders also had a glacial origin and he tamely solved the problem of their transport by using the traditional diluvial explanation.

A different theory was put forward by Professor Reade of Berlin, who investigated the distribution of erratic boulders of Scandinavian origin upon the north German plain. Reade placed his faith in transportation by icebergs and his theory was well known but received little credence as long as diluvialism remained in favour with the geologic élite. Von Buch in Germany, a student of Werner, and Sefström in Sweden were among the many powerful propagators of diluvial doctrines. By 1810 Von Buch was explaining German erratics by the usual mammoth flood and he ascribed the scratches to the friction of the boulders against the bottom during their rapid move-

Vüe circulaire des Montagnes
qu'on decouvre ou sommet ou Glacier
ce Buet.

FIG. 38. *A circular view of the mountains surrounding a glacial cirque*
(*From De Saussure, 1779*)

ment from one place to another. Sefström, who was an extremist, would have had the Flood carry the boulders from the far north to the foot of the Alps.

British geologic thinking up to about 1832 did not possess this diversity of opinion. In Britain the only generally credible explanation was that of the Flood, which, as we have already described at some length, had been especially vindicated by the two main publications of Buckland and the combined work of Conybeare and Phillips. Here, however, we ought to point out that the problem was greatly complicated by the frequent occurrence at various levels of marine shells and deposits picked up by former ice-sheets from off the sea-bed around Britain or by the passage over fossiliferous strata.

The diluvialists were quite prepared to accept sea-currents capable of carrying large boulders hundreds of feet uphill and while they were straining to make the facts fit their theories, Lyell in his 1830 edition of *The Principles* suggested an alternative. Though contradictory in some ways of Agassiz's theory, he did choose a glacial agency. In its simplest essentials Lyell conceived all Europe and North America, except the mountain-tops, covered by the sea, and these changes of sea level were accompanied with an expansion and contraction of the polar ice-caps. As the northern hemisphere approached a warmer phase, swarms of icebergs would break away from the polar masses, and carrying on and in them debris, would float unimpeded across the vast sea that was covering Europe. As they melted or became stranded on the isolated peaks they would deposit their rocky burdens. The germ of Lyell's idea probably originated, and certainly drew strength, from the accounts of the Arctic explorers (e.g. Captain Bayfield, 1835–6). In addition his own researches in Scandinavia had confirmed that in many places rocks could be observed either on or in ice. His letter in reply to Sedgwick and the following quotations from *The Principles* make this point:

I have got from Denmark a notice by Dr Pingel, on the gradual subsidence of Greenland. As to erratic blocks, I have no positive general theory, but after seeing Sweden, I cannot for a moment believe they are due to diluvial action. I saw none in Italy, Spain, or Sicily, a few in the Jura and Alps, hundreds in Denmark, thousands everywhere in Sweden, and of stupendous size. It seems to me a northern phenomenon when best developed. Ice does annually carry large stones on Lake Wener, and in the Gulf of Bothnia, not so big as cathedrals I grant, but tolerable sized pebbles, such as many men could not move. (*Lyell, Mrs, 1881, I, p. 459, Letter, 25 Oct. 1835*)

In addition to the facts enumerated in my paper on Sweden in the *Philosophical Transactions* for 1835, in regard to the agency of ice-islands, I may mention a fact observed by Dr Beck on the coast of Jutland. He has ascertained that on the breaking up of the fringe of ice which encircles the coast there during winter, small islands of ice float off and carry with them not only small gravel from the beach but stones four feet in diameter firmly frozen into the solid mass. These ice-floes are sometimes driven eastward in the Cattegat and have been known to stop up the narrow part of the passage of the Great Belt, and to cause new reefs of rocks thus transported on which vessels, and a few years ago a Danish man-of-war, have been stranded. If such power can be exerted by ice-islands, only a few hundred feet in diameter, in latitudes corresponding to those of England, we may be well prepared to find that islands several leagues in circumference may remove blocks of the magnitude of small houses. (*Lyell, 1836, p. 381*)

It may be asked whether I refer all erratics, even those of Switzerland and the Jura, to the carrying power of ice. In regard to those of Switzerland I have elsewhere endeavoured to show that a combination of local causes might have contributed to their transfer; for repeated shocks of earthquakes may have thrown down rocky fragments upon glaciers, causing at the

FIG. 39. *Melting iceberg carrying erratic boulders*
(From Prestwich, 1886)

same time avalanches of snow and ice, by which narrow gorges would be choked up and deep Alpine valleys, such as Chamouni, converted into lakes. In these lakes, portions of the fissured glaciers, with huge incumbent or included rocks might float off, and on the escape of the lake, after the melting of the temporary barrier of snow, they might be swept down into the lower country.

M. Charpentier has lately proposed another theory which he informs us is merely a development of one first advanced by M. Venetz. The Alpine blocks, according to these writers, were not carried by water, for had this been the case the largest would be either in the Alpine valleys or near the base of the great chain, and we should find their size and number diminish as we receded from their original point of departure. But the fact is otherwise, many of the blocks on the Jura, or those farthest removed from the starting-place, being of the largest dimensions. They suppose, therefore, in accordance with the

FIG. 40. *The nineteenth-century concept of the drifting of erratics by icebergs*
(From Prestwich, 1886)

opinion of M. de Beaumont and others, that the elevation of the Alps occurred at a comparatively modern epoch, and that when these mountains were first upheaved they were more lofty than now, and more deeply covered with snow and glaciers. After the principal movement had ceased, a lowering of the Alps took place, the dislocated and shattered beds requiring time to form. According to this hypothesis, therefore, the erratic blocks are monuments of the greater magnitude and extent of the ancient glaciers under a different configuration of the surface. I have not now space for all the ingenious arguments adduced, after a minute examination of the ground by M. Charpentier in support of this theory, but must refer you to the original memoir. (*Lyell 1836, pp. 382–3*)

As the principal prophet declaiming the revelations of uniformitarianism, it was incumbent upon Lyell to discover some natural explanation for glacial phenomena. The submergence of the continental portion of the northern hemisphere was a perfectly natural and frequent occurrence to one skilled in stratigraphy and palaeontology. Yet in many ways Lyell's theory was no more firmly founded on fact than was the theory of the diluvialists and at a later date Agassiz reserved equal criticism for both. It seems strange too that Lyell should adhere so closely to his own iceberg theory when he was so well aware of Dr Charpentier's glacial notions.

This lopsided interpretation of glacial phenomena prevailed until the introduction of Agassiz's theory into Britain. Adherence to the diluvialist or to the iceberg or to a hybrid school of thought continued to confound the basic premises of many descriptions of glacial and fluvioglacial phenomena by British geologists during this period. Mackenzie, for instance, in 1835 when offering a suggested explanation for the parallel roads (terraces) of Glen Roy called into being a complicated combination of subsiding flood and violent wave action. Darwin falls into the same error when he examines the area in more detail. He refutes the notion that a violent flood has been involved because of the undisturbed nature of the stratified deposits. He argued, logically, that because the surface of the land showed no signs of elevation and the rivers lacked the power to remove the rocky barriers that must have existed, the Glen must have been filled by an arm of the sea at a time when the sea level was higher than it is now.

It is admitted by every one, that no other cause, except water acting for some period on the steep side of the mountains, could have traced these (terrace) lines over an extensive district. (*Darwin, 1839, pp. 39–40*)

These irregularly stratified beds, near the mouth of the Spean, attain a thickness of several hundred feet, and they consist of sand and pebbles, many of the latter being perfectly waterworn. Higher up the valley, near the bridge of Roy, the thickness before the central portions were removed appears to have been about sixty feet, but of course the thickness varies according to the original irregularities of the rocky bottom of the valley. Now it may be asked by what agency has this sloping sheet of waterworn materials been deposited along the course of the valley? From the presence of the horizontal shelves we know that there has been no change in the relative level or inclination

FIG. 41. The 'Parallel Roads' (lake shore terraces) of Glen Roy
(Geological Survey of Great Britain)

of the country since this district was last covered with water, and therefore we may argue with safety, that the action of the rivers, as far as it is
determined by their inclination, must have been the same since that period
as it now is, with the exception of that amount of change which they may have
effected in their own beds. . . . Now if we look at any portion of these rivers,
for instance the Roy above its junction with the Spean, we find it has cut a
narrow steep-sided gorge through the solid rock, which is in many parts
between twenty and thirty feet deep, whilst on each side there are remnants,
as above stated, of a continuous bed of gravel, at least sixty feet in thickness.
These beds have certainly been deposited by rapid currents of water, but not
by any overwhelming debacle, as may be inferred from the presence of cross
layers, and the alternate ones of fine and coarse matter. Seeing also the
evident relation of dimension and materials which exist between these
deposits and the valleys in which they occur, it can scarcely be doubted that
the detritus of which they are composed was transported by the existing
rivers. But are we to suppose that the river, as in the case of Roy, first deposited
along its whole course these layers one over another, thus raising its bed sixty
feet above the solid rock, and then suddenly commenced without the smallest
change in the inclination of the country, not only to remove the matter before
deposited, but when having gained its former level, to act in a directly
opposite manner, and to cut a deep channel in the living rock? Assuredly
such a supposition will not be received; and whatever part the river had in
the accumulation of these waterworn materials, from the very moment
(neglecting the annual oscillations of action from the changing seasons) it
ceased to add and began to move, its power must have undergone some
most important modifications. (*Darwin, 1839, pp. 50–51*)

The phenomenon demands an explanation, and the only obvious solution
is . . . that it had been occupied by an expanse of gradually subsiding water,
either of a lake or of an arm of the sea. This conclusion, therefore, may be
urged with only little less certainty regarding many, if not all, of the valleys in
this part of Scotland. . . . But it may be asked, would not the hypothesis of a
succession of lakes explain the appearance, the matter accumulated above
each delta sloping upwards from one level to another. I can only answer this
with respect to those valleys which I have myself seen; in the Spean, Roy,
Tarf Water, and some others, it is easy, as before stated, to replace in imagination the solid rock; and although some small lakes would be thus formed by
the replaced barriers (as probably would be the case in every valley), the
fringe of stratified alluvium we now are speaking of skirts the valley at an
elevation above them. To assume that these rocky barriers were formerly
much higher, and were demolished by some means independent of the action
of the river (for this action tends only to form a narrow wall-sided gorge, as
may be seen in those barriers which certainly did exist), would be so gratuitous
as the imaginary erection of one great barrier across the mouth of the valley,
and would explain, from the continuity of the slope, the appearances far less
perfectly. . . . Therefore it has not been the water of several lakes any more
than of one lake, which slowly retiring from these valleys, determined the
accumulation of the beds, where we now see them. There is, then, as we have
conclusive evidence that an expanse of slowly subsiding water did occupy
these spaces, but one alternative, . . . namely, that the waters of the sea, in
the form of narrow arms or lochs, such as those now deeply penetrating the
western coast, once entered and gradually retired from these several valleys.
(*Darwin, 1839, pp. 55–56*)

We wonder today why Mackenzie (1835) and Darwin did not use ice as the barrier to Glen Roy, but examples of ice-dammed lakes, such as still occur in the great mountain ranges, were apparently not then known to British geologists.

Murchison in his work on Siluria and South Wales, considered the force required to move large erratic boulders and turned for help towards Lyell's iceberg theory. However, his practical knowledge ensured that although he considered an iceberg origin might be applicable in some cases, it could not be applied to others because the rounded appearance of some of the boulders indicated a fairly extensive period of aqueous attrition. This he reasoned would not occur if the boulders were carried in ice as they would then retain their sharp edges.

> After an explanation of the theories hitherto proposed to account for the transport of large boulders to distant points, the author states that the evidences in question seem to him to be subversive of the diluvial hypothesis which imagines that the blocks were carried over the land, it being proved that here, at least, they were accumulated under the sea. He does not think we have yet been furnished with a full explanation of any method by which such blocks can have been transported to distances of 100 miles. (*Murchison, 1835–6, p. 335*)

> It is further submitted that under the physical features of the region when this drift was formed, i.e. when a great arm and strait of the sea separated England from Wales, submarine currents alone could not have been powerful enough to propel these large blocks, though the question is one which ought to be more completely disposed of by those versed in the laws of dynamics. Mr Murchison next takes into consideration the theory of the transport by ice. After allusion to the views of Esmarck, De l'Arriviere, Hausman, &c., it is shown that Mr Lyell has thrown great additional light on this subject by his observations on Sweden and the Alps, by which it really appears that under certain limitations 'ice floes' may have been *verae Causae* in the transport of large blocks, depositing them under seas and lakes at great distances from the source of their origin. In the Salopian case, however, though it is possible such means may also have been employed, there are many arguments which weaken the application of the hypothesis, such as the rounded and worn exterior of the boulders, and their diminution in size and quantity from north to south. (*Murchison, 1835–6, p. 335*)

THE IMMEDIATE RECEPTION OF AGASSIZ'S IDEAS

The glacial theory had to displace the muddled thinking described above before it could hope to win acceptance. On the Continent it gained few adherents because of the great deference paid to the opinions of Von Buch, Humboldt, De Beaumont and other arch-diluvialists including the younger De Luc, a prolific writer who later tried desperately to dissuade Buckland and Murchison from accepting any form of glacial action. Agassiz's address before the Helvetic Society in 1837 received a distinctly hostile reception.

> Von Buch was in the audience; his disapproval, tinged with tolerant contempt, bubbled over when the young President stopped speaking. Humboldt, a

staunch friend and benefactor, expressed his objections by letter. 'Over your and Charpentier's moraines Leopold Von Buch rages as you may already know . . . I, too, though by no means so bitterly opposed to new views, and ready to believe that the boulders have not all been moved by the same means, am yet inclined to think the moraines due to more local causes. . . . Your ice frightens me . . . I am afraid you spread your intellect over too many subjects at once.' (Fenton and Fenton, 1952, p. 117)

In Britain, fortunately for the success of the theory, Agassiz already had good friends. Towards the end of the year in which he delivered his address he tried to enlist the support of Sedgwick. In a letter to him (30 November 1837) he stresses the novelty of his glacial idea and claims that it is the first serious attempt to account for the numerous examples of rock striation:

There is another question on which I would warmly welcome your opinion. It is the problem of the erratic blocks scattered on the flanks of the Jura. None of the explanations already given seems to me to solve the problem. Among the several phenomena there is one in particular which has been omitted in all the theories and which you say occurs in the north of England: I mean the polished surfaces on which in parts the erratic blocks rest. I have tried to relate this phenomenon as well as several others to the transport of the blocks, by supposing that they are the result of masses of ice which, at a certain period, filled the great valley of Switzerland and on which the blocks were carried from the Alpine summits to the flanks of our mountains. I made this topic the subject of my opening lecture at the congress of Swiss Naturalists at Neuchâtel. My opinion was strongly opposed by M. de Buch and Élie de Beaumont. . . . I shall eagerly await your reply, happy if I find my view supported by the distinguished geologist in whose knowledge I have the most confidence.' (Trans. from Clark and Hughes, 1890, I, pp. 505–6)

Sedgwick, in his reply of 5 March 1838, showed sympathy and seemed prepared to go as far as conceding that glaciers might act in certain cases, on the grounds that he felt the iceberg theory inadequate to explain the position of all the blocks as it required icebergs to travel over vast distances to almost every part of the world. In the case of English erratics he is not even prepared to allow the agency of glaciers, but holds that their axial distribution in accordance with the valley lines and their rounded form argues attrition and a distinctly diluvial origin.

Till I see your memoir on the erratic blocks of the Alps I don't know how I can offer any opinion, as I don't at present know exactly what is your hypothesis. Where has it been printed? . . . On the subject of the erratic blocks of Switzerland it strikes me that no one can possibly account for them without the aid of the carrying power of ice. Without knowing what it is, I am, therefore, favourably disposed towards at least part of your hypothesis. A great deal of evidence, both positive and negative, has been advanced in favour of the iceberg theory. For example, Mr Darwin has shown that throughout South America erratic blocks are found within the limits of latitude where glaciers are, or may have been, down to the level of the sea; and that they are wanting in the tropical latitudes, where ice could never have existed near the sea level. In England (where everything is on so small a scale, yet where we have such a fine succession of phenomena, illustrating almost every important point in the geological history of the earth), we have a most interesting series of erratic blocks. I don't think the iceberg theory can be applied to them, because they go in almost all directions, and not towards any prevailing point of the compass, and because they follow the exact line of waterworn detritus and comminuted gravel. Such blocks I attribute to currents produced during periods of elevation and unusual violence. There are many instances of rocks grooved deeply, and partially rubbed down, by the currents of what we formerly called diluvium, a word which is passing in some measure out of use in consequence of the hypothetical abuse of the term by one school of geologists. There are very fine examples of this kind near Edinburgh. Stones transported in this way are always rounded by attrition, and in every question about the origin of erratic blocks we ought to regard their condition (viz. whether rounded or not), as well as their geographical relation to the parent rock. (*Clark and Hughes, 1890, I, pp. 503–4*)

Agassiz was to have more success with Buckland who was already becoming less convinced of diluvialism particularly as Lyell was opposed to it. Among Buckland's manuscripts are numerous jottings which, although disjointed and often undated, seem sufficient to hint at his evaluation of some contemporary theories. He is perhaps worried at the velocity of a current needed to move great blocks.

6. Huttonian idea that vallies were formed by Driblets: Blocks: Lime st. Block, 5 tons, jumped clean over the Breakwater in Nov. storm, 1822, Plymouth. Granite blocks hop about like parched peas, if the moving fluid have only sufficient velocity and bulk. Lake bursting in Val de Bagnes, water ran 28 miles an hour . . .

11. M. v. Buch thinks that ye Blocks scattered over ye Jura and elsewhere have been dispersed at once by a violent shock, and have not been simply rolled upon some inclined plane nor carried along by avalanches, nor projected by explosion of gas.

He finds that ye Blocks of ye Jura must have required an impetus of 357 feet per second to travel in water . . . this is an impetus five times less than that of a Cannon ball, and if ye water, as it must have been, were loaded with sludge, ye weight of Blocks could be then much diminished. He refers for proof of his theory to ye bursting of the Lake de Bagnes, wh. being situated 150 feet above ye valley rushed on with an impetus of 33 feet per second. (*Quoted by North, 1943, pp. 14–15*)

The jottings that follow show his interest in alternative propositions. Sir James Hall's theory of a continuous inclined plane from the Alps to the Jura is examined and rejected, because independent evidence showed that there had been breaks in the supposed plane prior to the movement of the blocks and drift:

7. Sir J. Hall's inclined plane from Mont. Blanc to Jura will not do-for admitting Geneva Lake to be of Diluvian excavation, still ye great valley of Switz existed as such before ye deposition of ye Molasse. It may have been filled by Molasse. If ye lasher at Ifley can excavate ye Hole below, the Diluvian rush from ye Rhône Valley cd. produce ye lake of Geneva. W.B. 1822.

De Luc's explosive theory is treated with a certain amount of ridicule:

> 8. Granit Blocks-De Luc. De Luc supposed Granit Blocks to have been thrown up by the Expansive power of gas, generated at the time of their formation. G.B.G. p. 128 – a libel on gas! (A reference to G. B. Greenough's *First Principles of Geology* (London), 1819, p. 128. De Luc's notion affords an excellent example of the tendency to overdo the application of current discoveries, for, as Greenough observed, 'the hypothesis was constructed at a time when the imaginations of all men were so dazzled by the brilliant discoveries then making in pneumatic chemistry, that it was almost as difficult to speculate without gas as to breath without air)'. (*Quoted by North, 1943, p. 14*)

The iceberg theory is quizzed and he takes a sceptical attitude towards it, being of the opinion that the rounded shape of the blocks rather argued against such a causation.

> 20. Theory of Professor Read (i.e. Wrede) of Berlin. Blocks floated over an Ice (see Bakewell) packed in Ice like salmon from Scotland to Billingsgate – must have floated over or have been rolled across from Norway, before ye existence of ye present Baltic. Rounded form of the Blocks is against their coming over on an iceberg . . .
> 22. Blocks moved by Ice on Shores of Lake Huron – Dr Bigsby. In ye Spring ye Ice occasionally removes fragments of great size. . . . During winter ye Ice forms round the Blocks which are in the shallows, and on being broken up in May by Mild weather and seasonal rise of water it carries them to some other shore. (On envelope postmarked 1823). (*Quoted by North, 1943, p. 15*)

Though undoubtedly Buckland still retained a bias towards the diluvial explanation he was obviously receptive to new theories. The Oxford professor had known Agassiz since 1834 but it was not until 1838 that he paid a visit to Switzerland in order to study the special phenomena for himself. There appears to be little doubt of his almost immediate conversion. Sollas writes that 'a few days' personal investigation sufficed to convince Buckland of the truth of Agassiz's opinions'. (Quoted in McCallien, 1941, p. 318.) The arch-diluvialist himself tells us in more detail:

> Dr Buckland's attention was first directed by Prof. Agassiz in October 1838 to the phaenomena of polished, striated, and furrowed surfaces on the south-east slope of the Jura, near Neuchâtel, as well as to the transport of the erratic boulders on the Jura, as the effects of ice; but it was not until he had devoted some days to the examination of actual glaciers in the Alps, that he acquiesced in the correctness of Prof. Agassiz's theory relative to Switzerland. (*Buckland, 1840–1, p. 332*)

Buckland's defection from the diluvial school brought forth a certain amount of superior amusement from some of his colleagues. The lyrically inclined Duncan even composed a poem to immortalize this geologic incident.

> Buckland, loquitur.
> 'Say when, and whence, and how, huge Mister Boulder,
> And by what wond'rous force hast thou been rolled here?
> Has some strong torrent driven thee from afar,
> Or hast thou ridden on an icy car?

FIG. 42. *Cartoon by Sopwith of Buckland's conversion to the glacial theory*
(From Gordon, 1894)

Which, from its native rock once torn like thee,
Has floundered many a mile throughout the sea,
And stranded thee at last upon this earth,
So distant from thy primal place of birth.'

Boulder, respondit.
'Thou wise Professor, who wert ever curious,
To learn the true, and to reject the spurious,
Know that in ancient days an icy band
Encompassed around this frozen land,
Until a red-hot comet, wandering near
The strong attraction of this rolling sphere,
Struck on the Mountain summit, from whence torn
Was many a vast and massive iceberg borne:
And many a rock, indented with sharp force
And still-seen striae, shows my ancient course:
And if you doubt it, go with friend Agassiz
And view the signs in Scotland and Swiss passes.'
(*Quoted by North, 1943, p. 21*)

AGASSIZ IN BRITAIN

By 1840, Agassiz had virtually eclipsed De Charpentier. His publication
Études sur les Glaciers (1840) had splendidly forestalled De Charpentier's
parallel work *Essai sur les Glaciers* by nearly a year and as De Charpentier's
ideas differed very little from those of Agassiz, all the credit was attracted to
the latter. The sole point of divergence concerned the date of origin of the
Alps, which in contradiction to Agassiz, De Charpentier insisted had preceded the onset of the glaciation. The general acclamation that rather unfairly
went to the pupil instead of the master caused a coldness to develop between
these two old friends. Agassiz went on to triumph while De Charpentier had
to rest content with a few isolated and belated acknowledgements of the part
he had played.

In his book Agassiz set out to combat the rival flood and iceberg theories,
and aimed some shrewd blows at Lyell's reasoning. The point that the perched
blocks rest on a bed of rounded pebbles and sand and this bed in turn often
covers a striated surface was a highly intelligent criticism of the iceberg transportation scheme. The intimate conjunction of all these elements would by
Lyell's scheme have been highly fortuitous but by the glacial theory it was a
natural relationship.

This (ice-rafting) explanation, although very ingenious, is nevertheless not
applicable to the erratic blocks of the Jura for this reason: the erratic blocks
of the Jura do not rest directly on the ground. Wherever the rounded pebbles,
which ordinarily accompany the great blocks, have not been reworked
by later action, we notice that they form a bed several inches and sometimes even many feet thick on which the angular blocks rest. These pebbles are
very much rounded, even polished, and piled up in such a manner that the
greatest are at the surface and the smallest, which frequently grade to a fine

sand, are at the bottom, immediately upon the polished floor. Thus Monsieur Lyell's mode of transportation would well explain why the blocks are not rounded, granting that they were protected by the ice that covered them; but it in no way accounts for the presence of the rounded pebbles found below them nor for the formation of the polished and striated surface on which this pebble bed rests . . . (*Agassiz, 1840A, pp. 283-4*)

Thus Agassiz, an international palaeontologist on intimate terms with most of the best scientific brains of Europe, had already achieved some status as a glaciologist, when, to use his own modest words,

In the early part of the summer of 1840, I started from Switzerland for England with the express object of finding traces of glaciers in Britain. This glacier-hunt was at that time a somewhat perilous undertaking for the reputation of a young naturalist like myself, since some of the greatest names in science were arrayed against the novel glacial theory. (*Quoted by McCallien, 1941, p. 316*)

He joined the September meeting of the British Association at Glasgow and read a paper explaining all aspects of the glacial theory. The formation of various types of moraines was examined and the suggestion made that the general distribution of morainic deposits represents the area formerly covered by ice. The scratches on the rock he attributed not to ice alone but to the scouring movement of hard rock fragments imbedded in the bottom of the ice. Glacial features were such universal phenomena that he was obliged to wave his icy wand far beyond the region of the Alps until he had brought the vast northern countries under a covering of ice comparable in area and thickness to what is now found only in Greenland and Antarctica. When he comes to dispose of all this ice he arrives at the novel conclusion that the mass of water so released would suitably explain the rounded appearance of the stony portions of the drift deposits. This allowance of a combination of causes must have gone a long way to reassure many of Buckland's doubts. It is no longer a problem of preferring ice to water; ice plays its part, so do icebergs and at the end water and ice combine to sort out and re-work the glacial residue. This inspired fluvio-glacial amendment of the theory accommodated all circumstances.

M. Agassiz particularly drew attention to facts relative to the manner of the movements of glaciers, which he attributes to the introduction and freezing of water in all their minutest fissures, whereby the mass of ice is continually expanded. The effects of the movement, produced by this expansion, upon the rocks beneath the ice, are very remarkable. The bases of the glaciers, and the sides of the valleys which contain them, are found to be polished and scratched by stones fixed in the lowest region of the moving ice. The fragments of the rocks that fall upon the glaciers are accumulated in longitudinal ridges on the sides of the ice, forming deposits of stony detritus, which are called lateral moraines. As these descend into lower valleys, they assume a central place on the moving ice, and are called medial moraines. As the glaciers are continually pressed forwards, and often in hot summers melted back at

FIG. 43. *Louis Agassiz in 1844*

their lower extremity, it results that the polished surfaces, occasioned by
friction on the bottom and sides, are left uncovered, and that terminal moraines, or curvilinear ridges of gravel and boulders, remain upon the rocks
formerly covered by the ice. Thus we can discover, by the polished surfaces
and the moraines, the extent to which the glaciers have heretofore existed,
which is much beyond the limits they now occupy in the Alpine valleys.
It is stated to result from similar facts observed by Professor Agassiz,
that enormous masses of ice have, at a former period, covered the great
valley of Switzerland, together with the whole chain of the Jura, the sides of
which, facing the Alps, are also polished, and interspersed with angular
erratic rocks, disposed like boulders in the moraines; but since the masses of
ice were not confined between two sides of a valley, their movements were in
some respects different, and the boulders were not deposited in continuous
ridges, but dispersed singly over the Jura at different levels. Professor Agassiz
proposes the hypothesis that at a certain epoch all the north of Europe, and
also the north of Asia and America, were covered with a mass of ice, in which

the elephants and other mammalia found in the frozen mud and gravel of arctic regions, were imbedded at the time of their destruction. The author thinks that when this immense mass of ice began quickly to melt, the currents of water that resulted have transported and deposited the masses of irregularly rounded boulders and gravel that fill the bottoms of the valleys; innumerable boulders having at the same time been transported, together with mud and gravel, upon the masses of the glaciers then set afloat. Professor Agassiz announced that these facts are explained at length in the work which he has just published, *Études sur les Glaciers de la Suisse*, illustrated by many large and accurate plates, which were laid before the Geological Section. (*Agassiz, 1840B, pp. 113–14*)

After the Glasgow meeting, Agassiz toured Scotland in the company of Buckland and Murchison, searching for glacial evidence. His ideas failed to make any impression on Murchison but Buckland seems to have been fairly thoroughly convinced, and soon induced Lyell to accept the glacial theory in part:

'by showing him a beautiful cluster of moraines within two miles of his father's house', on which Lyell 'instantly accepted it (the Glacial Theory) as solving a host of difficulties that have all his life embarrassed him'. (*North, 1943, p. 22*)

With the aid of such support, the glacial theory could be said to have been firmly established by the November meeting of the Geological Society. At this particular meeting Agassiz read a paper on glaciers and ice-sheets, his main theme being supported by the supplementary papers of Buckland and Lyell on glacial evidence in Scotland and northern England and in Forfar-shire respectively. The theory met with some challenging opposition at the meeting, though the championship of Buckland and Lyell assured it a serious hearing. In fact it received more than this. After 1840 the principle that glaciers accounted for morainic ridges and much of the scattered deposits was never seriously disputed. It was the idea of vast continental ice-sheets that geologists found hard to believe.

In his address Agassiz emphasized points which he thought would most interest British listeners and would best meet their main objections. For instance, he takes pains to explain the origin of the parallel grooves found striating rock surfaces, in order that he may disprove the British idea that they had been caused by water action.

These effects, consisting of surfaces highly polished, and covered with fine scratches, either in straight lines or curvilinear, according to the direction of the movement of the glacier, are constantly found, not only at the lower extremity, where they are exposed by the melting of the glaciers, but also, wherever the subjacent rock is examined, by descending through deep crevices in the ice. Grains of quartz and other fragments of fallen rocks, which compose the moraines that accompany the glaciers, have afforded the material which, moved by the action of the ice, has produced the polish and scratches on the sides and bottom of the Alpine valleys through which the glaciers are continually, but slowly descending. It is impossible to attribute these effects to causes anterior to the formation of the glacier, as they are constantly present and parallel to the direction of the movement of the ice. . . . they are constantly sharp and fresh beneath existing glaciers, but less distinct on surfaces which have for some time been left exposed to atmospheric action by the melting of the ice. . . . The polished surfaces beneath the ice are often salient and in high relief. The sides also of the valleys adjacent to the actual glaciers are frequently polished and scratched at great heights above the ice, in a manner identical with the surface beneath it, but different from the polish of the bed of the torrent. (*Agassiz, 1840C, pp. 321–2*)

To illustrate his arguments in a more topical fashion Agassiz refers to the evidence to be found in Scotland and at the same time appeals to the suscepti-bility of his diluvial audience by carefully reassuring them that in his scheme ice action does not totally replace the agency of water, but in fact the two may often combine to produce any one landscape feature.

To avoid useless discussion, he states, that in attributing to the action of glaciers a considerable portion of the results hitherto ascribed exclusively to that of water, he does not wish to maintain that everything hitherto assigned to the agency of water has been produced by glaciers; he only wishes that a distinction may be made in each locality between the effects of the different agents; and he adds, that long-continued practice has taught him to distin-guish easily, in most cases, the effects produced by ice from those produced by water.

Proceeding to the consideration of facts, he says the distribution of blocks and gravel, as well as the polished and striated surfaces of rocks in situ, do not indicate the action of a mighty current flowing from north-west to south-east, as the blocks and masses of gravel everywhere diverge from the central chains of the country, following the course of the valleys. Thus in the valleys of Loch Lomond, and Loch Long, they range from north to south; in those of Loch Fyne and Loch Awe from east to west; and in the valley of the Forth from north-west to south-east, radiating from the great mountain masses between Ben Nevis and Ben Lomond. (*Agassiz, 1840–1, p. 328*)

The till of Scotland, or the great unstratified accumulation of mud and gravel, containing blocks of different size heaped together without order, and containing no organic remains but bones of Mammalia and insignificant frag-ments of shells, he is of opinion was also not produced by true glaciers, although intimately connected with the phenomena of ice. The polished and striated surfaces of the blocks leave no doubt on M. Agassiz's mind that these masses have been acted upon by ice in the same manner as the blocks which are observed under existing glaciers, and which are more or less rearranged by water derived from the melting of the glaciers. (*Agassiz, 1840–1, p. 329*)

The common origin of moraines, and of accumulations of rounded pebbles and of blocks, M. Agassiz says cannot be doubted. The former are simple ridges formed on glaciers; the latter materials rounded and polished under glaciers, or great masses of ice, and exposed by the melting of the ice, and are disposed by the water thus produced. (*Agassiz, 1840–1, p. 330*)

To emphasize his argument, he obviously felt he must discredit the possi-bility that flood action could have played any part in the whole process that he was trying to explain. The rounded form of pebbles was a feature that British geologists had always associated with a moving mass of water. Agassiz did not deny that water might have contributed to their roundness, but argued instead that the distribution of the pebbles in accordance with the termination of the various glaciers precluded or made unlikely the possibility that they could have been transported by water alone. If water had been the transporting agent the pebbles would have been found disposed in similar deposits all the way along each of the glacial valleys.

> With respect to the valley of the Aar, M. Agassiz says it is easy to prove that the rounded pebbles of Alpine rocks spread along its river, because between the glacier from which it issues and Berne, the flowing of the stream is interrupted by the barrier of Kirchet, the lake of Brienz, and the lake of Thun; and because between these lakes its velocity is so small, that it transports only mud and very fine gravel, and that the pebbles over which the river flows below Thun do not issue from the lake. Supposing that the volume of the Aar was formerly greater, why, asks M. Agassiz, are not the lakes of Brienz and Thun filled in the same manner as the plain of Meiringen and the bottom of the valley which separates the two lakes? All difficulties, however, he is of opinion, vanish, if the pebbles be considered the detritus of retreating glaciers, and the hollows occupied by the lakes of Brienz and Thun were filled with glaciers. (*Agassiz, 1840–1, p. 329*)

His conclusion is grand in the extreme and placed the whole of the British Isles beneath an immense ice-sheet. As perhaps a concession to the popular English theory, Agassiz in the last few lines generously incorporates Lyell's iceberg theory to explain in certain circumstances the presence in England of some Scandinavian rocks.

> If the analogy of the facts which he has observed in Scotland, Ireland, and the north of England, with those in Switzerland, be correct, then it must be admitted, M. Agassiz says, that not only glaciers once existed in the British Islands, but that large sheets (nappes) of ice covered all the surface. (*Agassiz, 1840–1, p. 331*)
>
> It must then be admitted, the author argues, that great sheets of ice, resembling those now existing in Greenland, once covered all the countries in which unstratified gravel is found; that this gravel was in general produced by the trituration of the sheets of ice upon the subjacent surface; that moraines, as before stated, are the effects of the retreat of glaciers; that the angular blocks found on the surface of the rounded materials were left in their present position at the melting of the ice; and that the disappearance of great bodies of ice produced enormous debacles and considerable currents, by which masses of ice were set afloat, and conveyed in diverging directions, the blocks with which they were charged. He believes that the Norwegian blocks found on the coast of England have been correctly assigned by Mr Lyell to a similar origin. (*Agassiz, 1840–1, p. 331*)

This audacious challenge by the young Swiss naturalist, delivered within the London stronghold of some of his most august opponents, received invaluable support from the two accompanying papers by Buckland and Lyell.

The former's acceptance of the glacial theory was not qualified by its restriction to particular areas, as Lyell's was eventually to become. Indeed Buckland's conversion involved a firm disbelief in Lyell's alternative theory and he was quick to say so.

> In his recent examination, in company with Mr McLaren, of the Castle Rock at Edinburgh, Dr Buckland found further proofs of the correctness of the glacial theory, by discovering at points where he anticipated they would occur, namely, on the north-west angle of the rock, distinct striae upon a vertical polished surface; and at its base a nearly horizontal portion of rock, covered with deep striae; also on the south-west angle obscure traces of striae and polished surfaces. Some of these effects may be imagined to have been produced by stones projecting from the sides or bottom of the floating masses of ice; but it is impossible, Dr Buckland observes, to account by such agency for the polish and striae on rocks at Blackford Hill, two miles south of Edinburgh, pointed out to him by Lord Greenock in 1834. On the south face of this hill, at the base of a nearly vertical cliff of trap, is a natural vault, partly filled with gravel and sand, cemented by a recent infiltration of carbonate of lime. The sides and roof of the vault are highly polished, and covered with striae, irregularly arranged with respect to the whole surface, but in parallel groups over limited extents. These striae, Dr Buckland says, cannot be referred to the action of pebbles moved by water; 1st, because fragments of stone set in motion by a fluid cannot produce such continuous parallel lines; and 2ndly, because if they could produce them, the lines would be parallel to the direction of the current; it is impossible, he adds, to refer them to the effects of stones fixed in floating ice, as no such masses could have come in contact with the roof of a low vault. On the contrary, it is easy, he says, to explain the phaenomena of the polish by the long-continued action of fragments of ice forced into the cave laterally from the bottom of a glacier descending the valley, on the margin of which the vault is placed; and the irregular grouping of the parallel striae to the unequal motion of different fragments of ice, charged with particles of stone firmly fixed in them, like the teeth of a file. (*Buckland, 1840–1, pp. 336–7*)

In his paper Buckland also picks out similar glacial proofs which he had observed elsewhere in Scotland and in northern England.

> Sir George Mackenzie pointed out to the author in a valley near the base of Ben Wyvis, a high ridge of gravel, laid obliquely across, in a manner inexplicable by any action of water, but in which, after his examination of the effects of glaciers in Switzerland, he (Buckland) recognizes the form and condition of a moraine. (*Buckland, 1840–1, pp. 332–3*)
>
> The vast congeries of gravel and boulders on the shoulder of the mountain, exactly opposite the gorge of the Tumel, Dr Buckland is of opinion was lodged there by glaciers which descended the lateral valley of the Tumel from the north side of Schiehallion and the adjacent mountains, and were forced across the valley of the Garry, in the same manner as modern glaciers of the Alps (that of the Val de Bagne, for example) descend from the transverse, and extend across the longitudinal valleys. (*Buckland, 1840–1, p. 334*)
>
> Dispersion of Shap Fell Granite by Ice. – The difficulties which had long attended every attempt to explain the phaenomena of the distribution of the Shap Fell boulders, Dr Buckland considers, are entirely removed by the

application of the glacial theory. One of the principal of these difficulties has
been to account for their dispersion by the action of water; northwards along
the valley, descending from Shap Fell to Shap and Penrith; southwards in
the direction of Kendal and Morecombe Bay; and eastward, over the high
table-land of Stainmoor Forest, into the valley of the Tees, as far as Darling-
ton. (*Buckland, 1840–1, p. 348*)

Lyell is at this time equally forthright in his support of the glacial theory
and, as always, his examination of the evidence (in Forfarshire) was conducted with minute exactness and analytical skill. He classified the superficial
deposits and notices that they frequently acted as barriers to lakes and glens
and often surrounded swamps or peat bogs which invariably proved barren
of organic remains. These features which formerly seemed inexplicable now
appeared to be quite natural features of a glaciated landscape.

Three classes of phaenomena connected with the transported superficial
detritus of Forfarshire, Mr Lyell had referred, for several years, to the
action of drifting ice; namely 1st, the occurrence of erratics or vast boulders
on the tops and sides of hills at various heights, as well as in the bottoms of
the valleys, and far from the parent rocks; 2ndly, the want of stratification in
the larger portion of the boulder formation or till; and 3rdly, the curvatures
and contortions of many of the incoherent strata of gravel or of clay resting
upon the unstratified till. When, however, he attempted to apply the theory
of drifting ice over a submerged country to facts with which he had been
long acquainted in Forfarshire, he found great difficulty in accounting for the
constant subterposition of the till with boulders to the stratified deposits of
loam and gravel; for the till ascending to higher levels than the gravel, and
often forming mounds which nearly block up the drainage of certain glens
and straths; for its constituting, with a capping of stratified matter, narrow
ridges, which frequently surround lake-swamps and peat-mosses; and for the
total absence of organic remains in the till.

Since, however, Professor Agassiz's extension to Scotland of the glacial
theory, and its attendant phaenomena, Mr Lyell has re-examined a considerable portion of Forfarshire, and having become convinced that glaciers
existed for a long time in the Grampians, and extended into the low country,
many of his previous difficulties have been removed. (*Lyell, 1840–1,
pp. 337–8*)

With his customary fine precision and scrupulous attention to detail Lyell
also discards his iceberg theory in a locality where he had found on one side
of a loch a mass of boulders clearly derived from the other three sides.

The phaenomena exhibited by the lateral mounds, Mr Lyell states, agree
well with the hypothesis of their being the lateral moraines of glaciers; and he
adds, that he had never been able to reconcile these phaenomena, particularly
the want of stratification, with the theory of the accumulations of the detritus
during submergence, and the removal by denudation of the central portions
of a deposit which had by that means filled the glens. The distribution of an
enormous mass of boulders on the southern side of Loch Brandy, and clearly
derived from the precipices which overhang the Loch on the three other sides,
is advanced as another proof in favour of the glacial theory. It is impossible

to conjecture, Mr Lyell says, how these blocks could have been transported half a mile over a deep lake; but let it be imagined that the Loch was once occupied by a glacier, and the difficulty is removed. (*Lyell, 1840–1, pp. 339–40*)

Faced by this well-ordered and formidable array of facts and examples, the opposition might have been expected to wilt. But geologists are traditionally doughty opponents and like politicians are not easily frightened into capitulation, even by the truth. The glacial theory met with a mixed reception from the members of the London Geological Society. Strong arguments were made against it and at the end we get the impression that most members withheld their final judgement until some future date. Fortunately, notes of the actual discussion that took place after the delivery of Buckland's paper have been preserved and present the hurly-burly at first hand.

Mr Murchison called upon the mathematicians and physical geographers present to speak of the objections to Dr Buckland's glacial hypothesis, he himself should attend only to the facts of the case. Of the scratches and polish on the surface of certain rocks there is no doubt, and 'Are glaciers the cause?' is the question. Could they be done by ice alone? If we apply it to any as the necessary cause, the day will come when we shall apply it to all. Highgate Hill will be regarded as the seat of a glacier, and Hyde Park and Belgrave Square will be the scene of its influence. Dr Buckland has in his paper assumed that all these heaps of diluvium are moraines; but I would rather examine the subject under the old name Diluvium, and with our old ideas of diluvial action, than by using the term moraines, assumed the question proved. On Schiehallion there are . . . rocks. If Schiehallion had been covered with glaciers there ought to be some (indications). If the height be great the result should be proportionate. There ought to be a co-ordinate relation in the phenomena. But in the Highland mountains, not one-third the elevation of the Alps, we have moraines two or three times the magnitude of any known in Switzerland. Formerly, when we found traces of fragmented rocks disposed around a mountain, we attributed them to the successive periods of elevation in that mountain. The parallel roads of Glen Roy were compared to sea-beaches; now all are attributed to the action of ice. And not only these, but Edinburgh and Stirling, and other places equally out of the reach of such actions, did glaciers ever exist in the higher chains, are to be covered with a mass of ice! These grooved and striated surfaces and heaps of boulders are also to be found in Scandinavia, on the east of the Gulf of Bothnia, all proceeding from the north and north-west. Have these crossed the gulf on ice? In Russia, too, we shall find them where there are no mountains. And if we look to the remains of marine shells found in beds elevated, differing in no respect from those in our present seas, except that they are called 'Pleistocene' (by James Smith and Lyell), we have proof of a lower elevation of the very time (the period following upon the more tropical epochs), when these glaciers should be introduced. On these accounts I am still contented to retain our old ideas, that when a mountain was elevated or a body of water passed over a series of elevations, the diluvium would descend with the (streams) and be disposed in mounds and terraces according to the direction of currents, &c.

PROFESSOR AGASSIZ. – Mr Murchison has objected to the glacial theory in the only way in which it could be objected to. He allows that the whole is granted as soon as you grant a little bit. For here, as in other cases, we argue from what is proved, to what is to be proved. In Switzerland the action of glaciers is yearly seen by thousands of foreigners, and of these facts there can be no doubt, (nor as to the former) extent of glaciers. In the Glacier de L'Aar, grooves, &c. are to be found in the valley seven leagues (twenty-two miles) from the end of the present glaciers. Did we find these surfaces only on the hard rocks, we might suppose they were merely uncovered by the action of the glaciers; but on the soft limestone rocks these grooves are only to be seen on the surfaces from which the glacier has just retreated. Many glaciers traverse such rocks only (equivalents of our Lias), and there the grooves are annually renewed in winter, and removed by the atmospheric action in summer. I have been many hundred feet under the glacier of Monte Rosa, and found the quartzose sand forming a bed beneath, and acting like emery upon the rocks. A moraine may be distinguished by certain characters from any other accumulation of fragmented rocks. From the sides of the glaciers moving faster in the middle, there is a continual tendency to throw the fragments into lines at the sides (lateral moraines), and when two glaciers descending from different gorges unite, a medial moraine is formed. The lateral (moraines) are exposed to constant friction with the rocks with which they are brought in contact, and their terminations are passed over by the whole mass of the glacier, so that they become rounded and striated, whilst the medial moraines remaining on the surface, continue angular. When the glacier retreats in the summer, the medial moraine, composed of angular fragments, is spread out over the surface of the lateral and terminal moraines, composed of rounded fragments; and it is by these characters that we have proved the existence of moraines in Scotland, Ireland, and the north of England. There are moraines in the Alps 200 feet wide, composed of boulders several feet in diameter.

MR LYELL spoke of the size of moraines, and the way in which they might, under certain circumstances, attain any magnitude. A glacier had been known to retire half a mile in a single summer, (a number of) moraines have been in succession left, and in severe winters all these might be driven successively into one by the downward motion of a glacier.

MR GREENOUGH spoke of the arguments derivable from analogy, &c. and objected to the mode in which the Geological Society was in the habit of accounting for phenomena. Instances of accumulations of travelled rocks (occur in) North Germany; from a careful comparison some of these must have crossed the Baltic. In the valleys of Switzerland some deposits must have crossed Lake Geneva, and ascended very high mountains. Does Professor Agassiz suppose that the Lake of Geneva was occupied by a glacier 3,000 feet thick? (AGASSIZ. – 'At least!') (Mr Greenough then referred to the) changes of climate necessary to account for these phenomena (and to the) objection from the tropical nature of remains in recent deposits. (He considered it to be the) climax of absurdity in geological opinions. In one period, the Crag, we have three opposite conditions blended: corals, tropical; peat, temperate; shells, pronounced by Dr Beck, arctic!

MR LYELL. – Mr Greenough confuses four distinct epochs under the name of Crag. The first comparatively tropical (Coralline Crag), the others temperate (Red and Norwich Crag), and the period of the peat bogs (Lacustrine deposits) more recent than any.

MR JOHN EDWARD GRAY. – The corals of the Crag appear to me as arctic as the shells. I know no reason for making them tropical.

MR GREENOUGH (remarked) on the size of the blocks on mountains, the agency of floating ice, and on mountains as the physical boundaries of different kinds of diluvium.

DR MITCHELL inquired if Dr Buckland confined the glaciers to the Highlands or whether he made them descend to the Lowlands.

DR BUCKLAND expressed himself ready to answer any question on the subject under discussion, or any involved in his paper, but considered the present question irrelevant.

DR MITCHELL considered his question relevant to the subject.

DR BUCKLAND rose to reply, but MR WHEWELL rose. (Cheers, and 'Mr Whewell!').

MR WHEWELL – At this late hour it is impossible to go into the question of the physical changes necessary to allow of the existence of glaciers in this country. I shall, therefore, confine my remarks to the subject as discussed this evening, and it does appear to me that the way in which Mr Lyell has treated it is not the most fair and legitimate. He says: 'If we do not allow the action of glaciers, how shall we account for these appearances?' This is not the way in which we should be called upon to receive a theory. Now, it is not within our reach at present to refer each set of phenomena in geology to its adequate cause, but that is no reason why we should receive any theory that is offered to account for it. This glacial theory is brought forward to explain what has hitherto, to a great extent, been found inexplicable – the nature and position of diluvial detritus over considerable areas and in widely different climates. So far as it is founded on strict comparison and analogy it is to be received, but we must not over-rate its influence; and it appears to be incomplete in three important particulars: – Firstly, in accounting for such an extent of diluvium over such wide areas, in countries of such opposite physical structure, surface, climate &c. Secondly (from the) marine remains of the glacial period, showing the continents to be submerged. Mr Darwin has described an island capped with snow in the equivalent latitude of Yorkshire, and by supposing an equal extent of water in our Polar regions, we might induce a degree of cold sufficient for that; but these glacial phenomena are found over too wide an extent to allow of that. (MR LYELL. – 'I have attempted to account for that in my paper' – here interrupted. DR BUCKLAND. – 'So have I in a paper which is not yet written!')

MR WHEWELL, continuing. – Our attention to-night is limited to Dr Buckland's paper. Thirdly, the physical conditions under which glaciers now exist. We find them universally stretching out from lofty mountain-chains, which take their rise in *warm* climates, so as to allow of the downward motion and the retiring in summer. Mr Lyell speaks of the prodigiously *rapid* retreat of a glacier which amounted to half a mile in a single summer. But where shall we obtain mountains as *fulcra* for glaciers, stretching many leagues into the plains, producing such results as are ascribed to their action in Scotland?

DR BUCKLAND resigned the chair to Mr Greenough, and argued that *a priori* credit to be attached to his 'narrative', from the circumstance of his having been a 'sturdy' opponent of Professor Agassiz when he first broached the glacial theory, and having set out from Neuchatel with the determination of confounding and ridiculing the professor. But he went and saw all these things and returned converted. And he considered the testimony of four such competent observers as himself, Agassiz, Renouard, and (De Charpentier) who, next to Saussure, had spent more time in the Alps than any other geologist, sufficient to prove to all the truth of their observations and the correctness of their inferences. He referred to Professor Agassiz's book, and condemned the tone in which Mr Murchison had spoken of the 'beautiful' terms employed by the professor to designate the glacial phenomena. That highly expressive phrase *roches moutonnées*, which he had done so well to revive, and that other 'beautiful designation' the *glacier remanié!* remanié! remanié! continued the doctor most impressively, amidst the cheers of the delighted assembly, who were, by this time, elevated by the hopes of soon getting some tea (it was a quarter to twelve P.M.) and excited by the critical acumen and antiquarian allusions and philological lore poured forth by the learned doctor, who, after a lengthened and fearful exposition of the doctrines and discipline of the glacial theory, concluded – not, as we expected, by lowering his voice to a well-bred whisper, 'Now to,' &c. – but with a look and tone of triumph he pronounced upon his opponents who dared to question the orthodoxy of the scratches, and grooves, and polished surfaces of the glacial mountains (when they should come to be d—d) the pains of eternal itch, without the privilege of scratching! (*Woodward, 1907, pp. 138–42*)

GLACIAL RETREAT OR ADVANCE?

After the initial presentation of the glacial theory nothing much seems to have been done in Britain, either to prove or disprove it. Buckland, its most loyal disciple, continued to search out examples of glacial features in this country but unfortunately his geological career were nearing its end and after his appointment in 1845 as Dean of Westminster his official duties, including the installation of a sanitary system, precluded extensive scientific activity. Lyell after his first fervent acclamations of allegiance seems to have drifted back in favour of his iceberg theory. Agassiz himself, the founder, succumbed to the lure of an appointment to the Lawrence Scientific School at Harvard and left Europe in 1847. Hence the glacial idea lingered on without much impetus until from 1850 onwards Ramsay picked up the theme again and finally brought about its acceptance with his detailed description of glacial phenomena in Wales. Although in 1840 there were probably less for than against it, the theory maintained an appreciable popularity partly because of the steady decline of diluvial notions. Geological publications over this period illustrate the instability of the position of the theory in geological circles for between them they contain a mixture of views showing acceptance, part-acceptance and contradiction. Few publications were as pro-glacial as a paper by Byres who warmly accepted Agassiz's ideas.

The slope of the mountain at Porth-Treiddyn is to the north and north-east:
    At the spot which had been cleared, above the quarry, the detritus lying on the surface of the rock was from five to twelve feet thick. It consisted of various soils and gravel, of blocks of rock similar to the flags, and of boulders of porphyry and green-stone, in some of which the felspar crystals were very distinct. Some of these boulders were from three to four feet long, and from two to three feet in diameter.
    It is here especially, that we have presented to us a perfect type of the

glacier action. The surface of the uncovered portion of the rock, where it has not been disturbed by the workmen's tools, is rounded and polished in the most extraordinary manner. The surface is furrowed; and the furrows where the rock is uneven, are from 1 to 2 feet deep, with their edges beautifully rounded off. On the broader slopes, striae are very distinct. With a few exceptions, which I shall presently take occasion to notice, the furrows, striae, scoops, grooves, and undulations, all shape their course, not in the direction in which the mountain slopes, northwards; but in a diagonal or slanting direction, towards the valley of the Glasllyn, that is to say, towards the east or north-east. (Byres, 1844, p. 371)

The same pro-glacial certainty cannot always be attributed even to Buckland who, in his Anniversary Address to the Geological Society, seems to have undergone a distinct backward tendency and to have acquired some reservations in regard to Agassiz's theory. In the absence of the master glaciologist, Buckland's pristine conviction was weakening and doubts were once more entering his mind. We are mildly surprised to find him giving an appreciation of Murchison's version of the iceberg theory.

> Mr Murchison, in an admirable chapter (c. 39) of his *Silurian System*, on the Position and Mode of Transport of Boulders which occur in the Northern Drift, has stated good reasons for believing that such a change in climate may have taken place at the epoch of the transport of erratic blocks as permitted the formation of icebergs on the shores and rivers of Cumberland, Scotland, and Ireland; which being drifted southwards strewed their load of large stones and gravel over the bottoms of then adjacent seas. He also quotes with approbation the ingenious imagination by Mr C. Darwin, of a proportional distribution of the land and water in central and northern Europe, very different from the present, and under which the southern part of Scotland might present an island 'almost wholly covered with everlasting snow', having each bay terminated by ice-cliffs, from which great masses yearly detached would transport fragments of rocks to distant regions; and infers, that as in other parts of the world there are conditions in which ice becomes a motive power, such conditions may also have existed in our latitudes.
>
> Mr Murchison has also proposed to explain the dispersion of erratic blocks now resting on beds of clay and sand containing recent species of arctic shells over large districts in the interior of Russia, by supposing 'that they had been floated in icebergs, which breaking loose from ancient glaciers in Lapland and the adjacent tracts, were drifted southwards into seas which have been since laid dry'. He further suggests, that icebergs loaded with detritus may, by grating upon the bottom of these seas, have produced the parallel striae and polished surfaces on the rocks over which they were drifted; and concludes with admitting so much of the glacial theory as to allow that in former days glaciers probably advanced further to the south, and occupied many insulated tracts, and to a much greater extent than at the present day. (Buckland, 1841, pp. 513–14)

After considering whether the striations and ridges of drift could be claimed as positive evidence by either party Buckland concluded that they could not. He felt that far too little was known of what went on below an iceberg for any definite opinion to be expressed.

With respect to elongated ridges and tumuli of gravel, it remains to discriminate how far they may have been derived from, or modified by, the action of ice under one or more of the three following conditions: 1. Were they lodged by glaciers alone, without the agency of water, in the form of moraines on their flanks and front? 2. Have they been stranded by icebergs loaded with gravel upon the shores of lakes, or estuaries, or seas? 3. Have they been dropped in deep water by floating and melting icebergs, and re-arranged by whirlpools and conflicting currents in the form of oblong reefs and groups of obtuse cones which they actually present? Another large field of inquiry must be forth-with entered upon, in the distinctions we shall have to make between raised sea-beaches and each of the three last-named residuary effects of glacial action.

With respect to scorings also and dressing on the surfaces of rocks, it is very desirable that we should find some criterion whereby to distinguish between the grinding effects of glaciers marching slowly along dry land, and of icebergs dredging the bottom of the sea, and of large stones and gravel drifted simply by water, in producing striae, grooves and furrows, together with rounded and polished surfaces on the rocks over which they respectively advance.

I see not yet by what test we may distinguish these residuary phenomena where they occur in regions now remote from either of the causes most competent to their production, viz. in countries that now enjoy a temperate climate and are in some cases elevated nearly four thousand feet above the level of the sea; for where the supposed agent is ice armed and transfixed with stones projecting like the teeth of a file from its base and sides, the effects of similar instruments on similar materials would probably be the same, by whatever cause a slow progressive motion may have been imparted to them; and whether on dry land or beneath the sea.

It remains, moreover, to ascertain to what extent the sudden elevations of land may have produced great movements of water and diluvial inundations by gigantic waves, analogous to those which are occasioned by modern submarine volcanic action; and to inquire into the effects that may have been produced on the sides and bottoms of valleys of denudation by the drifting of the hard materials that must have been swept through them at and after the time of their excavation. (*Buckland, 1841, p. 515*)

Such moderation may have come with old age for in his final analysis Buckland tries to draw the schools of thought together by suggesting the operation at one and the same time of all three factors, glaciers, icebergs and torrents.

One great cause of the difference of opinion between the diluvialists and the glacialists, is the exclusiveness with which each party would insist upon the agency of the cause which they respectively adopt; the diluvialist apparently errs in refusing to admit the agency of glaciers in mountain valleys that are below the existing limits of ice and snow; whilst Agassiz may have erred in urging too far his theory of expansion as the great locomotive power of glaciers over regions whose surface is too little inclined to admit their progression by the force of gravity; a middle way between these two extreme opinions will probably be found in the hypothesis, that large portions of the northern hemisphere which now enjoy a temperate climate have at no very distant time been so much colder than they are at present, that the mountains of Scotland, Cumberland and North Wales, with great part of Scandinavia and North America, were within the limits of perpetual snow accompanied by glaciers; and that the melting of this ice and snow was accompanied by great debacles and inundations which drifted the glaciers with their load of detritus into warmer regions, where this load was deposited and re-arranged by currents at vast distances from the rocks in which it had its origin. (*Buckland, 1841, p. 516*)

Buckland followed his earlier paper with another in which he produced further examples of glacial evidence in Britain. In it he argued forcibly against the possibility of origin by flood and emphasized that as large boulders were resting on small gravel the conjunction of the two could only be satisfactorily explained by the glacial theory.

Valley of the Llugwy. – A good example of glacial action, the author states, may be seen between the bridge of Pont-y-Gyffyng and Capel Curig, consisting, first, of dome-shaped or rounded masses (roches moutonnées); secondly, of rounded portions of hard rocks, which, where they had been protected from the weather, exhibited polished, fluted and striated surfaces, the lines ranging parallel with the direction of the valley; and thirdly, of a mound of gravel or a moraine close above the dome-shaped hummocks, and immediately in front of the point of confluence of the upper valley of the Llugwy with that of Nant-y-Grryd. (*Buckland, 1841–2, p. 580*)

The base of this moraine is composed of small detritus, but its top is crowded with large blocks, almost in contact with each other . . . , it is difficult to imagine how any current of water possessing sufficient velocity to move them to their present position could have failed to sweep away the mounds of small gravel on which they are lodged, or how drifting icebergs could have failed to drop scarcely one single block along the valley of the Llugwy beyond the limited area where so many thousands are accumulated. (*Buckland, 1841–2, pp. 580–1*)

Then, as if anxious that his dogmatisms should not offend, he again tendered a composite analysis, in which glaciers and icebergs were each accorded a place.

In conclusion, Dr Buckland says, he must refrain from entering on the general subject of Diluvium and Drift, a sufficient number of facts not having been accumulated to admit of final conclusions. (*Buckland, 1841–2, p. 584*)

Darwin, writing about his visit in 1831 to South America as a member of the exploration vessel *Beagle*, still utilized Lyell's iceberg theory because, though he understood something of the action of glaciers, the idea of continental ice-sheets had not yet appealed to him. The absence of a steep slope, the presence of marine remains and the stratified layers of certain deposits, caused him to discard Agassiz's theory for the same reasons that Murchison had done. In addition the supposition of a submersion beneath the sea which went with the iceberg theory fitted in more neatly with Darwin's own idea on the marine denudation of plains and valleys, which has previously been discussed.

Mr Darwin says it is impossible to explain the distribution of boulders without the agency of ice, but he adds, that neither the till of the Straits of Magellan which passes into, and is irregularly interstratified with, a laminated sandstone containing marine remains, nor the stratified gravel of Chiloe, can have been produced like ordinary moraines. The boulders, likewise, on the lower levels at the head of the Santa Cruz river, he considers, could not have been distributed in their present position by glaciers, the surface having been modelled by the action of the sea; . . . the blocks of Tierra del Fuego and Chiloe were certainly transported by floating ice, and most probably those of the low and high plains of Santa Cruz. Finally, he is of opinion, from the general angularity of the blocks, and from the present nature of the climate of the southern parts of America, which favours the descent of glaciers to the sea in latitudes extraordinarily low, that it is more probable that the boulders were transported on the surface of icebergs, detached from glaciers on the coast, than inbedded in masses of ice, produced by the freezing of the sea. (*Darwin, 1841, p. 430*)

The opinion of Hopkins also continued to make a lasting impression, because of the aura of mathematical conclusiveness that surrounded it. When describing the topography of the Lake District and its possible origins, the author postulated a sudden emergence and tension-fracturing of the surface. These fractures were enlarged into valleys by the action of the sea as it drained off the rising land. His theory is quite distinct from that of Lyell, and Hopkins also makes it clear that he totally rejected the glacial theory.

The author conceives the valleys of the district to have been formed during this gradual emergence; the action of denuding causes being facilitated by previous dislocations, the masses, the removal of which formed the valleys, would at the same time be transported and spread over the surrounding country. The formation of the existing lakes must have been one of the most recent events in the geological history of this region. (*Hopkins, 1842, p. 761*)

His principal reason for disregarding the glacial theory was that as an explanation of the movement of erratics it often involved mathematical impossibilities. Hopkins later found his mathematics to be in error but at the time of their publication his pronouncements were eagerly hailed as incontrovertible.

Modes of Transport – Glacial Theory. – This theory, in its application to the transport of blocks across Stainmoor, involves such obvious mechanical absurdities, that the author considers it totally unworthy of the attention of the Society. (*Hopkins, 1842, p. 762*)

Lyell's iceberg theory comes in for severe criticism. Hopkins realizes correctly that icebergs might account for a few erratics but considered that such reasoning became strained when it ascribed the uniform beds of 'diluvium', or deposits showing size-sorting with distance, to a similar process.

Iceberg Theory. – There appears to be no doubt that floating ice may have played an important part in some cases in the transport of large blocks, but the author doubts whether such agency has been at all employed in the case under consideration. In the first place, he cannot but consider it absurd to attribute the formation of a bed of diluvium spread out with approximate uniformity over an extended area to the action of floating ice. Such a distribution of the transported matter is the necessary effect of broad currents of water, while, at most, it is the merely possible effect of floating ice. Secondly, there appears no adequate reason why blocks transported by floating ice should diminish in size as their distance from their original site increases; why the Cumbrian blocks on the eastern coast of Yorkshire should be generally much smaller than those less remote from the place whence they came. Thirdly, the theory in its application to the case before us involves a great physical difficulty – a depression of temperature, for which no adequate cause has yet been assigned. The author does not admit the parallel which has been drawn between this case and that of places in equal latitudes in South America or that of the island of Georgia. (*Hopkins, 1842, pp. 762–3*)

Instead of leaving the deposition of these beds of diluvium to the work of itinerant icebergs, Hopkins substituted the idea that they had been laid down by a current. The radial pattern he explained was due to the action of the flood in following the line of the main valleys when the rock was heaved up in the form of a dome or pyramid. In order to meet possible objections to the ability of the flood to move the large erratic boulders he indulged in an interesting digression on the proportionate ratio between velocity of water and the weight of matter that could be carried. Though his main theory is now known to be wrong the conclusion arrived at in this particular digression is correct: the scientific rule which he then propounded is now termed *Gilbert's 6th Power Law.*

> He accounts for the existence of currents diverging from the centre of the district in question by a repetition of paroxysmal elevations. (*Hopkins, 1842, pp. 762–3*)
>
> With respect to the magnitude of the blocks which might be moved by a current of given velocity, the author remarks, that the facility with which the transport of a block may be effected depends principally on its form. The more it approximates to perfect sphericity, the less, *caeteris paribus*, will be the force necessary to remove it. The author conceives that there is no doubt whatever but that blocks, not more spherical than many rolled blocks are observed to be, of five tons weight and upwards, might be moved under favourable circumstances by a current of ten miles an hour. That the force of a current increases in the ratio of the square of its velocity has been distinctly established by experiment for all velocities up to eleven or twelve miles an hour; nor does there appear to be any reason for doubting that the same law holds for much greater velocities. Assuming this law, the author states it as the result of a simple calculation, that if a certain current be just able to move a block of given weight and form, another current of double the velocity of the former would move a block of a similar form, whose weight should be that of the former in the ratio of $2^6 : 1$, i.e. of 64 to 1. If the velocity of the second current were treble that of the first, the weights of the two similar blocks would be in the ratio $3^6 : 1$, i.e. of 729 to 1, and so on for other velocities. (*Hopkins, 1842, pp. 764–5*)

Of all the British critics of the glacial theory Murchison was probably the most obstinate; he was certainly the most consistent, for he never accepted the theory. Wedded in his early days to the idea of catastrophes he allowed his fidelity to be impaired when Lyell enticed him to accept the iceberg theory. But having travelled so far nothing would shift Murchison further. Only his own physical transport in time to the moment when Britain was covered in ice might have convinced him. In his Anniversary Address to the Society in 1842 he made a lengthy attack on the glacial theory. Along with many other geologists of his time he could only associate glaciers in his mind with high mountain ranges and where these were absent and erratics were present in large numbers, as in Britain and the North Germanic plain, he was convinced the glacial theory was insupportable. In contrast Lyell's iceberg theory met with favour and he seized with enthusiasm on any apparent support which might be provided by the occurrence of marine fossils, terraces and raised beaches in conjunction with the drift deposits. Though not at all apprehensive about sinking all northern Europe below the Arctic Ocean he brought his scepticism to bear upon the rival ideas of a frozen continent or a rising land-mass, propositions which seem to us equally credible.

I must, however, remind M. Necker, that if he assumes that all great erratic blocks are to be referred to some neighbouring chain, now the seat of glaciers, he forgets the cases in Scotland and England, and indeed many others, far removed from mountain ranges, and which must be classed, as I shall presently show, with submarine deposits. (*Murchison, 1842, p. 672*)

Dr Buckland has not confined his views of the action of glaciers to Scotland, but applies them largely to the North of England and to Wales. He has recently endeavoured to satisfy us, that the rocks on the sides of the chief valleys in the latter country which open out from a common centre of elevation are striated, worn and polished in the direction of the present watercourses, and these he conceives to be evidences of former glaciers, which filled up all the valleys radiating from Snowdon to a distance of many miles from a common centre. I confess I see almost insurmountable objections to this view. Apart from other evidence, the very physical geography of this tract is at variance with the construction of such an hypothesis. In the Alps, and indeed in every other part of the world in which they have been observed, the length of glaciers is in ratio to the height of the mountains from which they advance, or, to use the words of Agassiz, from which they expand. Now whilst in the present days, a small glacier hangs to the sides of a mighty giant like Mont Blanc, having the altitude of 15,000 feet, our Welsh hills, having a height only of 4,000 feet, had glaciers, by the showing of Dr Buckland, of a length of many miles. Again, in the same memoir, which fills so large a portion of the Principality with glaciers, the author comments upon certain facts already well known to us, viz. the existence upon Moel Tryfane and the adjacent Welsh mountains of sea shells of existing species, at heights of 1500 and 1700 feet above the sea, where they are associated with mixed detritus of rocks transported from afar, all of which have travelled from the North, the hard chalk and flints of the North of Ireland being included. How are we to reconcile these facts with the theory that the greater part of the country in question was frozen up under the atmosphere in some part of the same modern period? Unable otherwise to explain how marine shells should be found on mountains which are supposed to have been previously and during the same great period occupied by terrestrial glaciers and accumulation of ages, Dr Buckland invokes anew the aid of the old hypothesis of a great wave. ... At one moment the argument used is, that scratches and polishings of rock must have been done by ice, because in existing nature it has been found that ice can produce such effects; and in the same breath we are told that beds of shells have been placed on a mountain by an agency which is truly supernatural.

In fact, the 'glacier' theory, as extended by its author in proving too much, may be said to destroy itself. (*Murchison, 1842, pp. 676–7*)

My own belief, Gentlemen, as you know, has been, that by far the greatest quantity of boulders, gravel, and clay distributed over our plains and occupying the sides of our estuaries and river banks, was accumulated beneath the waters of former days. (*Murchison, 1842, p. 679*)

To justify his support for Lyell's iceberg theory Murchison cited the discovery in January 1840 of Antarctica by an American squadron and unhappily suggested, for again he was wrong, that what they thought to be dry land was no more than a moving ice pack.

The Antarctic expedition, under the distinguished navigator Captain James Ross, has, as might have been expected, thrown considerable light upon the glacial theory. A few years only have passed since the existence of an enormous mass of ice-clad land in the antarctic region, was announced by an American squadron of geographical research. This great icy tract, which was described as exhibiting hills and valleys, and even rocks upon its surface, has entirely disappeared in the short intervening time; for Captain Ross has sailed completely through the parallels of latitude and in the same longitude which it was said to occupy. As we cannot suppose that the American navigators were deceived by atmospheric phaenomena, so must we believe that what they took for solid land, was one of the enormous accumulations of ice called 'packs', the great source of those numerous ice islands, which periodically encumber the Southern Seas. (*Murchison, 1842, pp. 681–2*)

The example of striation, which Agassiz had laboured so hard to emphasize as one of the principal trade marks of glacial movement, Murchison dismissed glibly as being mere structural phenomena.

My conviction was that these grooves, though then attributed by Dr Buckland to glacial action, are due neither to that agency, nor to any rush of waters, but are simply the result of the changes which the mass of the rock underwent, when it passed from its former molten or pasty condition into a solid state. (*Murchison, 1842, p. 675*)

Ultimately, like Buckland, Murchison tries to end on a note of reconciliation only this time the attempt seems rather transparent.

Thus far had I written, Gentlemen, – in short I had, as I thought, exhausted the glacial subject at all events for this year, – when two most important documents were put into my hands. The first of these is the discourse of my predecessor, who has so modified his first views, that I cannot but heartily congratulate the Society on the results at which he has now arrived. I rejoice in the prudence of my friend, who has not permitted the arguments of the able advocate to appear as the sober judgment of so distinguished a President of the Geological Society. In fact, it is now plain that Dr Buckland abandons, to a great extent, the theory of Agassiz, and admits fully the effects of water as well as of ice, to account for many of the long-disputed phaenomena. Whilst this admission involves the concession for which we have been contending, viz. that the great surfaces of our continents were immersed, and not above the waters when by far the greater number of the phaenomena on the surface of rocks was produced, I reject for those who entertain the same opinions as myself, the simple division into 'glacialists' and 'diluvialists', into which Dr Buckland has divided the combatants on this question; for to whatever extent the former title has been won by Agassiz and himself, we who have contended for the submarine action of ice in former times, analogous to that which we believe is going on at present, can never be merged with those who, under the name of diluvialists, have contended for the rush of mighty waves and waters over continents. Besides glacialists and diluvialists, my friend must therefore permit me to call for a third class, the designation of which I leave to him, in which some of us desire to be enrolled who have advocated that modified view to which the general opinion is now tending. (*Murchison, 1842, pp. 685–6*)

Reinforced in his opinions by the recent views of the opponents of Agassiz, Murchison makes the subject of the following year's Anniversary Address an exposure of the glacial theory and in his anxiety to discredit 'the ice-men' used any arguments antagonistic to them. In the work of Hopkins and Sedgwick he found a kindred dislike of glacial theories. Both men dispensed with glacial and iceberg action and attempted to explain 'drift' as the result of deposition by violent waves impelled outwards from a rising land-mass. They appreciated that this could not be done by ordinary waves which normally are quite shallow. Instead they substituted 'waves of translation' which according to them had the ability to move material far below the surface of the sea. Thus they conjured up an hitherto unknown mechanism which could propel huge boulders along the sea-bed.

I have reason to think that Professor Sedgwick entirely concurs with Mr Hopkins, in the necessity of calling into play enormous aqueous currents (which followed elevations) to account for much of the coarse drift which has been poured off upon the flanks of this mountain chain (the Lake District). Unwilling to call in the assistance of icebergs, and opposing the hypothesis of depression of temperature, Mr Hopkins shows, that the propelling forces of water alone are fully adequate to produce the transport of all the blocks and gravel around the Cumbrian chain, and to spread them out in the great masses of drift which encumber the adjacent northern centres, accounting for the existence of diverging currents by a number of paroxysmal elevations. Assuming moderate upheavals of certain areas of land, to height of 50 feet each, from beneath an ocean having a depth of 300 and 400 feet only, he informs us that a number of great divergent waves would be the consequence. . . .

In describing the motion of such masses of water, he invokes the aid of those waves of 'translation' whose properties have been reduced to laws by the ingenious and valuable researches of Mr Scott Russell and who, giving us measures of their relative velocity and power, has brought forward exact proofs of the transference by them of solid bodies immersed in water. Such waves have in fact been generated by the experiments of Mr Scott Russell, exactly in the same way as Mr Hopkins supposes waves to have originated on the great geological scale. These experiments prove, that a sudden elevation of solid mass from beneath the water, causes a corresponding elevation of the surface of the fluid, which infallibly produces a wave of translation of the first order. Now this wave is termed one of translation because it is found not to rise and fall like common waves, but wholly to rise and maintain itself above the level of the water. Arguing that this wave is propagated with a velocity which varies with the square root of the depth of the ocean, Mr Russell determines the velocity of wave transmission: but what is of most importance to the geologist is, that the old idea of the agitation and power of waves extending a little way down only in the sea, is found to be not true as touching waves of translation; for Mr Scott Russell has ascertained that when they are in action, the motion of the particles of the water is nearly as great at the bottom as at the top. He further shows, that the body moved at the bottom, is not rolled backwards and forwards as by a common surface-wave, but has a continuous forward motion during the whole transit of the wave's length. A complete transposition does therefore result from the wave-transit; and the wave of translation, says Mr Scott Russell may be regarded as mechanical agent for the transmission of power as complete and perfect, as the lever or the inclined plane.

Arguing from these remarkable data of Mr Scott Russell, and applying them to our geological phenomena, Mr Hopkins states, that currents of twenty-five and thirty miles an hour may be easily accounted for if repetitions of elevations of from 160 to 200 feet be granted; and with motive powers producing a repetition of such waves, our author has no difficulty in transporting to great distances masses of rock of larger dimensions than any boulders in the North of England. (*Murchison, 1843, pp. 91–92*)

At this stage Murchison's argument becomes more than a little inconsistent. Having given a certain credence to this diluvial type of explanation, he casts it aside when he comes to consider the north of England because in his opinion it is inadequate to account for the presence of erratics whose origin was obviously from far-distant foreign parts.

Admitting with Mr Hopkins, that nearly all the great boulders at a distance from the mountains from whence they were derived were accumulated upon the bottom of the sea, which bottom was subsequently elevated, I regret, that whilst he allows the possibility of floating ice having played a part in some cases in the transport of large blocks, he should doubt its agency in the country under consideration; for the North of England is full of examples of far-borne detritus, the position of which seems to me inexplicable by calling in the power of water alone.

Many practical geologists must, I think, admit that in the desiccation of the former bed of the ocean, and in its conversion into our present lands, the submarine outline of hill and dale has been, to a great extent, preserved; and if this be granted, no waves of translation nor any force of water can have hurled blocks across high ridges and deep valleys which are transverse to the direction which the erratics have taken, and of such relations England offers numerous examples. (*Murchison, 1843, p. 92*)

In almost the same breath he restores Hopkins to favour on the strength of his (incorrect!) mathematical assertion that it was impossible for ice to move across a flat surface or a gentle inclination.

> The glacial theory, as first propounded, has now, I apprehend, very few supporters, but to any such I recommend the perusal of the theoretical investigation of Mr Hopkins, 'On the motion of Glaciers' for he has shown, by clear mathematical analysis, that the locomotion of such bodies over large and flat continents is a theory founded in mechanical error, and involving conclusions irreconcilable with the deductions of collateral branches of physical science. (*Murchison, 1843, p. 93*)

Murchison concludes by professing – or should we say pretending? – to maintain an open mind and then imputes to so-called glacial features a submarine origin, coupled in some cases with a rising land-surface. In the last instance, he unconsciously makes his first really accurate observation on this topic, as many shore terraces were in fact marine beaches elevated above sea level.

> Ready as I have always been to admit the transport of blocks in ice from any former seat of congelation, and particularly in tracts where marine shells of arctic characters are associated with the surrounding drift, I am here, however, bound in candour to state, that in my desire to see established as a fundamental point of reasoning, 'the submarine condition of by far the largest parts of the surface of Northern Europe when erratics were disturbed', I have been carried too far in my Discourse of last year, in the endeavour to limit the number of centres from which icebergs may have been detached. Whilst, therefore, I cannot bring my mind to advocate that extent of glacial action in which my friend Dr Buckland believes, and have failed, in recent excursion, to observe any proofs of the existence of 'moraines' on the flanks of Snowdon, I allow that, looking to the sea-shells which lie around it at different altitudes, the summits of Snowdon may, like those of Spitzbergen, have constituted ice peaks in the midst of an ocean, others being at the same time in existence in the mountains of Cumberland. This is, I am persuaded, the full extent to which we can admit the application of the glacial theory in England. (*Murchison, 1843, pp. 93–94*)

It is perhaps surprising to find that there were at this time anti-glaciologists fiercer than Murchison. Probably the most disbelieving was Macintosh who made the wildest assertions in denying the glacial theory including the following which as an example of scientific thought is likely to remain uncopied and without parallel.

> The surface of rocks of various kinds and at very different elevations has frequently been observed to exhibit a polish, which has usually been attributed of late to the action of ice. The author considers that some of these cases are doubtful; that the polish is not of long endurance when the surface is exposed; and that in some, at least, of the examples in North Wales, the polish has been produced by the sliding of the boys on the smooth rock. (*Macintosh, 1845, p. 599*)

Yet we must add that Macintosh's paper called forth one other quite extraordinary comment. At the end of it, the youthful Ramsay, later to become the chief British exponent of glaciology, was said to have exclaimed gleefully, 'Jolly night at the Geological. Buckland's glaciers smashed!'

THE MID-NINETEENTH CENTURY-INTERGLACIAL

The state of the glacial theory about 1850 was more that of an interglacial period than of an ice advance. The older geologists as a whole, with the exception of Buckland, still seemed to have very little use for the glacial theory and where they had accepted it did so only with great reluctance and with many reservations. In 1848 Sedgwick is still talking (to Hugh Miller) about waves of translation:

> I believe our Till (if we may so call it) was formed exclusively by water – by waves of translation – and not by ice; and I by no means agree with what Lyell has written about it. (*Clark and Hughes, 1890, II, pp. 148–9*)

By 1851, when dealing with the Pennines, his views seemed to have modified but he will only allow part of the glacial theory, and could not altogether accept an 'enormous extension of land-ice'.

John Phillips as late as 1853 was still speaking of the boulders near Horton in Ribblesdale, Yorkshire, as follows:

> The blocks . . . may be regarded as uplifted and floated by ice, and dropped on surfaces which had been swept by currents clear of other loose matter. (*Phillips, 1853, pp. 111–12*)

Not even Lyell can be exempted from this class of die-hards or reactionaries. In the 1852 edition of his *Manual* he still gives excessive prominence to his iceberg theory.

Yet Lyell's ideas appear to fluctuate for in a letter of 1857, written shortly after he had revisited Switzerland and France, he accepts the inapplicability of his own iceberg theory to the Alps, the Jura and parts of Scotland at least.

> In order to escape from the necessity of appealing to such a gigantic mound of ice, I ventured, you may remember, to suggest that the sea may have floated the Alpine erratics to the Jura, . . . But the entire absence of marine remains in the associated gravel, mud and moraine, whether here or anywhere in Switzerland, the conformity of the distribution of the travelled blocks here with the shape of so many valleys, and above all, the sight of the Alpine snows at Berne and elsewhere, has made me strongly incline, with Charpentier, Agassiz and others, to embrace (as James Forbes did) the theory of a terrestrial glacier . . . at Berne in the region of the supposed ancient glacier of the Aar, I had a grand day with Escher von der Linth, and went over all the arguments for and against the land- and the sea-ice theory. . . . Escher pointed out to me that on the right side of the moraine of the Aar, you have fragments of the rocks which, far above the Lakes of Thun and Brientz, belong to the right side of the valley of the Aar, and on the left side those derived

from very different formations occurring on the left side. It is evident therefore that the two lakes were then full of ice. Another interesting point is this.
It is clear that a greater glacier like that of the Rhône coming down the Vallais
– filling the lake of Geneva – rising as they assume 3,000 or more feet above
its level, and crossing to the Jura, must have blocked up the mouths of the
minor and lateral valleys. Places were also pointed out to me, one of them near
Vevay, where the old colossal masses of ice of the Rhône glacier blocked up
tributaries which in summer brought down pebbles at points far above the
level of the lake of Geneva, 1,000 feet or more. At these points of junction a
mixture of stratified alluvium proper to the said tributary torrent, and of unstratified mud and blocks from the Vallais, are observed in spots on which it
seems most unnatural for any such accumulations to have taken place, unless
one admits Charpentier's theory or some modification of it. Now, just such
old moraines of mixed character, called here 'diluvium glaciare,' are observable in Forfarshire at the openings of lateral valleys into the larger one of
Clova, or Water Esk, only explicable by imagining the deeper glen to have
been once filled with ice. Indeed, if the hypothesis now generally adopted
here to account for the drift and erratics of Switzerland, the Jura, and the
Alps be not all a dream, we must apply the same to Scotland, or to the parts
of it I know best. All that I said in May 1841 on the old glaciers of Forfarshire (see *Proceedings of the Geological Society* for that year) I must reaffirm, . . .
There can in the first place be no doubt that ice was the carrying power, and
the distance travelled by such blocks and others is the same whether our
hypothesis employs floating ice or a land glacier. Escher has pointed out to
me that in several cases where the valleys bend at sharp angles, as in the case
of the Rhine above the Lake of Constance, the floating ice is out of the question. He has established this by the aid of a very peculiar rock called the
granite of Pontelyas, near Trons. Fortunately this granite is exceedingly
unlike any other in Switzerland or the Alps, large regular crystals of common
felspar in a base of green felspar, with another dark mineral (I forget the
name) dispersed like mica. As fragments are traced from its starting point we
have positive proof of its origin. (*Lyell, Mrs, 1881, II, pp. 249–52, Letter
16 Aug. 1857*)

We shall show in chapter eighteen how, unfortunately, Lyell's return to the
glacial theory came too late and was too incomplete to have an appreciable
effect on contemporary thought. In Britain the older geologists were either
like Murchison, too firmly tied to old-fashioned theories to accord unbiased
consideration of anything new or, like Lyell, too closely concerned with the
implications of the new organic evolutionary theories to pursue diligently the
problems of glaciated landscapes. Even the researches by Forbes and Tyndall
on glacier mechanics had to wait many years before their significance was
fully grasped and the detailed investigation of glacial problems in the British
Isles was delayed until the rise of the younger generation, in the persons of
Ramsay and Geikie.

Yet despite the slow advance of British glacial studies, in retrospect
Agassiz's visit to Britain may be considered the turning point in the growth of
glacial theories. After about 1840, glacial geology, freed increasingly from confusion with a universal Flood, grew slowly as a separate branch of the science;

FIG. 44. *Caricature of Thomas Henry Huxley drawn about 1861 by Agassiz, who believed in the 'special creation' theory of Cuvier. The rhyme reads:*

Idée solitaire!!
Idée unique!!
École Polytechnique
FRS; ASS!

in the same way, geomorphology, happily freed from the shackles of diluvialism and gradually enriched by the acceptance of glacial action, returned more safely to the uniformitarian study of fluvial and marine mechanisms as the prime creators of landforms.

CHAPTER FOURTEEN

The Beginning of American Geomorphology

THE AMERICAN SCENE

So far the monopoly of geomorphological ideas had rested with the Old World, and so it was to remain for some years longer. When the United States did finally enter into the geological arena it was to dominate, but that moment was not to come until the nation had grown to the stature of a major power. By 1845, although she had outgrown colonial ties and had been involved in consequence in two wars with Britain, she was not yet fully developed nationally. Up to the middle of the eighteenth century the Appalachians had still marked her western boundary; even in 1820 the large-scale European immigration had scarcely begun; and in 1845 the landscapes of the centre and west of the continent were known only in the broadest outline. Hence, in this chapter, we are concerned with a people living mainly in the eastern United States, where the physical setting differs markedly from that of western Europe.

In western Europe, and in Britain particularly, the complex variety of the rock series presented at the surface by the often complicated and small-scale structures, coupled with the knowledge acquired by mining, was the main reason for the focusing of attention upon stratigraphy and structure. In America, though complex sedimentary rocks did occur in exposed positions, many large areas like the atlantic coastal plain were of a relatively simple structure and the superficial agencies were consequently made to appear more important. Whereas in Britain the obvious structural elements tended to obscure the not so obvious glacial phenomena, in America the relative simplicity of the structure underlined the magnitude and the extent of the 'drift' features. What is more, examples of the work done by ice were more numerous in America and on a more gigantic scale. The unique features included the Great Lakes and Niagara which inevitably figure frequently in early geological discussions.

But this variety of landforms had not as yet evoked a corresponding variety in geological theories. The human setting was quite stereotyped. What geological learning the Americans possessed had European roots and was based on European experiences and prejudices; knowledge of their own country was too inadequate to allow American geologists to form indigenous theories. In addition, geology was a science followed only by the amateur enthusiast and comparison of notes and results was hindered by the absence of any major journal. Nor were most of the main towns large enough to constitute the nuclei of popular scientific meeting-places. In consequence, American science had to take the lead from European spokesmen and was often woefully backward in its theoretical outlook. There was the same tendency to confuse glacial features with the results of a flood, and the same history of a lingering allegiance to Werner, ultimately to be abandoned on the appearance of the American edition of Lyell's *Principles* in 1837, and his first visit in 1841.

TRANS-ATLANTIC NEPTUNISTS

In the late eighteenth and early nineteenth centuries the influence of Werner dominated geologic thought in America. The advocated catastrophes may not always have been entirely Wernerian but they seldom departed far from Neptunism proper.

Yet we must begin somewhat earlier than Werner* and notice the remarkable contribution to American geological knowledge made by Lewis Evans, who was born near Pwllheli in North Wales in 1700 and, after emigrating to Pennsylvania, became a noted surveyor, draughtsman and map-maker (G. W. White, 1951; 1956). In 1743 he wrote a journal describing the country between Philadelphia and Lake Ontario and during the next twelve years published several annotated maps and a pamphlet of 36 pages entitled *An analysis of a general map of the Middle British Colonies*. The journal was not printed until 1776 when his patron, Governor Thomas Pownall, issued parts of it and the *Analysis* from a London publisher. Thus although Evans himself escaped the teachings of Werner, his ideas were fated to appear at the height of Neptunism. In spite of this, his writings and excellent maps had a considerable influence on early American geologists and, as Professor George White observed, it would be interesting to know if Hutton ever saw them.

The following passages show how advanced, for America and even for Britain, were some of Evans's geomorphological observations. River-erosion and deposition are taken for granted.

> Various Systems and Theories of the present Earth have been devised in order to account for this Phaenominon (fossils in rocks). One System supposes that the Whole of this Continent, the highest Mountains themselves, as they now appear, were formerly but one large Plain, inclining with a considerable Slant towards the Sea; that this has been worn into its present Appearance of Ridges, with Vales between them, by the Rains of the Heavens

---

* The first written recognition of valley-erosion by rivers in America appears to be the remarkable statement by John Josselyn in his *New-Englands Rarites discovered* . . . (1672, p. 3) that 'the *White Mountains* . . . are inaccessible but by the Gullies which the dissolved Snow hath made;'. But this incidental observation had no influence on later geomorphological studies either in London, where it was published, or in America. It should be emphasized that Lewis Evans's maps and *Analysis* were published in Philadelphia and that the Welsh-born geographical genius was the real father of American geomorphology.

and Waters of the Earth washing away the Soil from the upper Parts, and carrying it down to Seawards. That the Soil thus carried down and lodged in various Places hath in a Series of Ages formed the lower Plains of the Jerseys, Pennsylvania, Maryland, Virginia, and the Carolinas . . . The Downfall of Waters from the melting of the Snow, the Rains, and the swollen Springs is such amongst the Mountains, and the Discharge from thence so great, that the Freshes on the Susquehanna River, where it is a Mile broad, rise 20 feet, although they are discharged with a violent and precipitate Current. These Freshes carry down with them immense Quantities of Soil which they begin to drop as the Velocity of their Course slackens in gliding over the lower Plains, and which they finally lodge in Bars and Islands at the Mouths of the Rivers where they meet the Sea. Thus have been many very extensive Countries formed at the Mouths of all the great Rivers in the World. (*Quoted in Pownall, 1776, pp. 29–30*)

In attempting to explain the presence of shells on mountain-tops, Evans refers, probably for the first time in American literature, to the idea of isostasy.

But we must have recourse to some other Explanation in order to account for the Situation of the Shells on the Tops of the Mountains.

It is easy to shew the Earth and Sea may assume one another's Places, but positively to assert how that hath actually happened in Times past, is hazardous; we know what an immense Body of Water is contained in the great Lakes at the Top of the Country, and that this is damm'd and held up by Ridges of Rocks: Let us suppose these Ridges broken down by any natural Accident, or that in a long Course of Ages a Passage may be worn through them, the Space occupied by the Water would be drained: This Part of America, disburthened of such a Load of Waters, would of course rise, as the immediate Effect of the shifting of the Centre of Gravity in the Globe at once or by Degrees, much or little, accordingly as the Operation of such Event had Effect on that Centre. . . .

Some such Changes may have come gradually and advanced by such slow Degrees, as that in a Period of a few Ages would not be perceptible; History therefore could take no Notice of them. (*Quoted in Pownall, 1776, p. 30*)

Among the numerous annotations in the blank spaces of Evans's excellent maps of 1749 and 1752 are remarks on the Endless (Appalachian) Mountains. The notes on the 'long uniform ridges' of these mountains become decidedly Huttonian.

They furnish endless Funds for Systems and Theories of the World; but the most obvious to me was, That this Earth was made of the Ruins of another.

Evans, however, had not yet freed himself from the Deluge although he had progressed far beyond the teachings of John Woodward.

These Mountains (Appalachians) existed in their present elevated Heights before the Deluge, but not so bare of Soil as now. . . . Their Height rendered them no doubt less exposed to that general Devastation, and preserv'd them unhurt, while the Soil and the loose Parts of the lower Hills and Vallies, agitated by a greater Weight of Water, were borne away, suspended in the dashing waves, and thrown downwards in Stratas of different kinds, as the Billows roll'd from different Parts; . . . But the Power he [Dr Woodward] ascribes to the Water of the Deluge is too much of a Miracle to obtain Belief. We have glaring Marks of a Deluge of far more recent Date, in which the Compass of Britain might not perhaps have furnished the Dr with.

Evans's ideas, probably because of his indispensable maps, did not suffer the same neglect in America as did those of Hutton in Britain or of Lamarck in France. Yet his statements on river-erosion were overlooked in favour of his more catastrophic suggestions.

Evans supplied geological information to Peter Kalm, the Scandinavian scientist whose *Travels in North America . . .*, first published in Swedish in 1753 and later in several other European languages, contained many geological observations including the suggestion that Ireland and North America were once either joined together or else linked by a chain of islands. Kalm was probably the first to describe American erratics and till (drift) but did not speculate on their origin. In 1783, a notable soldier, François Jean, Marquis de Chastellux, also described erratics in his *Travels in North America . . .* but his hypothesis on their origin failed to satisfy even himself. By this date it was almost inevitable that the great abundance of erratics and drift would be first convincingly explained by catastrophes, particularly the bursting of lake-dams. In 1793, B. de Witt attributed the sixty-four different varieties of rock found deposited along the shores of Lake Superior to catastrophic convulsions.

Now, it is almost impossible to believe that so great a variety of stones should be naturally formed in one place. . . . They must, therefore, have been conveyed there by some extraordinary means. I am inclined to believe that this may have been effected by some mighty convulsion of nature, such as an earthquake or eruption, and perhaps this vast lake may be considered as one of those great 'fountains of the deep' which were broken up when the earth was deluged with water, thereby producing that confusion and disorder in the composition of its surface which evidently seems to exist. (*Quoted in Merrill, 1906, p. 300; 1924, p. 12*)

Some fifteen years later, S. Akerly modified Evans's ideas in order to account for erratics and 'drift' in New York state. He conjured up the mechanism of a collection of large inland lakes, like the present Great Lakes, held in check by a north-easterly range of hills. At some period, he thought, these hills were broken up by underground disturbances and the dammed waters released to rush down the present Hudson valley, strewing the unconsolidated debris indiscriminately in their path:

After the waters of the Deluge had retired from this Continent, they left a vast chain of lakes, some of which are still confined within their rocky barriers; others have since broken their bounds, and united with the ocean. The highlands of New-York was the southern boundary of a huge collection of water, which was confined on the west by the Shawangunk and Kaats-kill mountains. The hills on the east of the Hudson confined it there. When the hills were cleft and the mountains torn asunder, the water found vent and over-flowed the country to the south. . . . The earth, sand, stones and rocks brought down by this torrent were deposited in various places: as on this (New-York) island, Long Island, Staten Island and the Jerseys'. (*Akerly, 'American Mineralogical Journal'*, *1*, *1814*, *p. 193*)

The authors of the official textbooks were no more advanced in their ideas. Maclure's *Observations on the Geology of the United States* published in 1809, and re-published in 1817, closely followed Werner's classification of rocks, and made no mention of Hutton. Cleaveland's *Elementary Treatise on Mineralogy and Geology* (1816 and 1822) showed that the author had adopted a combination of Cuvier's theory of the earth and, with reservations, Werner's Neptune theory, of which he says

> Though its general outlines may be correct, we are yet unable to give its details. It seems, however, to be rather incumbered with difficulties, than absolutely confronted with existing facts. It is obliged to admit the existence of certain operations, which cannot be repeated even on a small scale, and whose processes cannot be described. (*Cleaveland, 1816, p. 593*)

Despite this, he too adheres to Werner's classification and theory of the origin of rocks.

S. L. Mitchill in his *Observations on the Geology of North America* (1818), also showed ideas on the formation of the earth that had been influenced by Jameson's translation of Cuvier's essay. When he explained the features that were to be found in his own country he appeared to have been struck by the significance of the Appalachian water gaps, imagining, like Evans and Akerly, that they had been caused by the escape of lake waters. Mitchill postulated that the Great Lakes were shrunken remnants of a salt sea which broke through the Appalachian barrier as a flood, draining away to the east, forming and enlarging the Appalachian water gaps (much in the same way as Akerly had pictured the formation of the Hudson valley) and depositing the debris as unconsolidated drift (Merrill, 1906 and 1924).

We must interpolate here an interesting French incursion into American geomorphology by Count Volney who was a friend and associate of Mitchill and of Lamarck. Volney travelled widely in the eastern states and in his *View of the Soil and Climate of the United States*, made many acute first-hand observations on their physical features. In his description of the terraces of the Ohio, he suggested that they had been formed when the river flowed at higher levels and he even drew columnar sections of the terrace materials. He also gave a long account of the Niagara Falls and its retreat. The fossils he collected near Cincinnati were studied and reported on in detail by Lamarck. Fortunately, translations of Volney's work were published in London and in Philadelphia and were widely read.

Another writer of note during this early period was Amos Eaton (1776–1842) whose first few years of manhood were most unpromising and would

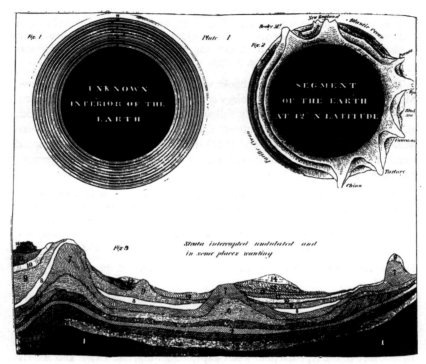

FIG. 45. *Amos Eaton's (1820) idea of the internal structure of the earth*

undoubtedly have blighted the spirit of a less resilient man (McAllister, 1941).
On coming of age he took up the law as a career but this plan was ruined in
1811 when he was tried for fraud, found guilty and sentenced to hard labour
for life. Despite such a calamity, by 1814 his exemplary conduct had gained
him the privileges of the 'good behaviour' convict and somehow he had also
managed to prepare a manual of surveying. Apparently he absorbed many of
his early ideas from a reading of Kirwan's work on mineralogy. By 1815 the
ethics of his original trial had been questioned and as a result he was pardoned.
On his release he went at once to study under Benjamin Silliman at Yale, and
from this moment his rise in the geological world was so rapid that within
two years he had been appointed Lecturer in Mineralogy at Williams College.
By 1818 he had published his first book, *Index to the Geology of the Northern
States,* which proved strictly Wernerian in outlook and strongly tinged with
Eaton's own deep religious conviction that religion and science were reconcil-
able. Yet when Eaton refers to the occurrence of boulders at high altitudes
he had some doubts even of the power of the Flood to roll stones uphill. He
preferred in this instance to use Conybeare's idea of filling up the present

valley in order to make the whole into an inclined plane down which the boulders rolled to their present resting places. Writing of the granite erratics of the Connecticut valley, Eaton asked:

> What force can have brought these masses from the western hills, across a deep valley seven hundred feet lower than their present situation? Are we not compelled to say that this valley was once filled up so as to make a gradual descent from the Chesterfield range of granite, syenite, etc., to the top of Mount Tom? Then it would be easy to conceive of their being rolled down to the top of the greenstone where we now find them. (*Quoted in Merrill, 1906, p. 302*)

Eaton's Wernerian bias is shown by the absence of any mention of the rival ideas of Hutton and Playfair unless, of course, he was unaware of them. In 1822 he helped in a survey of the Erie Canal and was responsible for the publication of part of the subsequent report. By now he was completely sure of his own abilities and was in the front rank of American geologists.

> Eaton was cockily proud of his book, saying that it contained more geological facts than all other American works put together. (*Fenton and Fenton, 1952, p. 145*)

In 1830 he published a *Geological Textbook*, which ran to two editions. In this he admitted that the Lake Ontario basin had been eroded but concluded that the Hudson valley had been produced by a fusion of the structural elements at a time of extreme volcanic heat. Elsewhere his writing is preoccupied with primitive religious beliefs of the fierce wrath of God as witnessed by the infliction of the Flood upon the inhabitants of the earth:

> the geological records of divine wrath poured out upon the rebellious inhabitants of the earth at that awful period, can never be effaced or changed. These latter records add, to the Mosaic account, that even the antediluvian beasts of the forest and fens partook of the ferocious nature and giant strength of an antediluvian man. (*Eaton, 1832, p. 10*)

The whole history of Eaton represents the triumph of determination over adversity but unhappily for American geomorphology his superhuman energies were misdirected into the sterile channels of Wernerism and religious fanaticism. He died in 1842, some seventeen years after founding Rensselaer College.

With his death the first period of isolated endeavours and of Wernerian attitudes in America really ends.

THE FIRST AMERICAN GEOLOGICAL ORGANIZATIONS

The works quoted above demonstrate that although in America there was no intellectual dearth, there was, however, a marked lack of any organization which would provide intercommunication between the various writers. As it was they were often widely scattered and not unnaturally tended to lean towards Europe and European doctrines.

FIG. 46. *Benjamin Silliman*

This vacuum was filled in 1818 by the inspiration of Benjamin Silliman, who in his youth had visited England, Scotland and Holland in search of geological instruction and had met such famous figures as Sir James Hall, Jameson and Playfair. When he returned to America in 1806 he accepted an appointment as Professor to the chemical and geological department at Yale College. Fired by the continental example, he had brought back with him sufficient specimens to build up a useful fossil and mineral collection. He continued to augment this throughout the period of his professorship until eventually it was built up to a size and quality that became of world-wide renown.

The founding of a scientific magazine in 1818 was his second venture. The earlier editions appeared without the aid of a professional publisher, Silliman being responsible both for the editing and financing of the magazine. During this time sales often barely covered the costs of production. The magazine started as *Silliman's Journal* and its contents were unrestricted; articles of an artistic as well as of a scientific nature being accepted. Later it was called the *American Journal of Science and Arts*, and finally, when the artistic element was dropped, the *American Journal of Science*. Though the appearance of this scientific publication did not cause any immediate change in American geological ideas, the dependency on European changes of opinion still remaining, it did prove the means whereby American scientists could pool their observations, form native schools of thought and in time advance to a position where the New World would be able to teach the Old. The magazine still flourishes.

In the following year (1819) the American Geological Society was established.

THE RISE TO PREDOMINANCE OF DILUVIALISM

The period that followed the founding of *Silliman's Journal* was transitional between Neptunism and uniformitarianism but was dominated by diluvial theorists. It corresponded with the period in Britain between the publication of Playfair's *Illustrations* and of Lyell's *Principles*.

We will begin with an all too brief account of authors not absolutely controlled by diluvialism and who, in fact, use fluvialistic or uniformitarian suggestions. First, a work by Gilmer in 1818 explaining the formation of the Natural Bridge of Virginia (*Fig. 47*) contains a revolutionary statement of natural causes. The writer argues along Huttonian lines that natural causes should be sought before assistance is obtained from super-natural ones.

> In the present state of geology, the phenomenon (the Natural Bridge) does not require us to resort to the operation of the unknown, or even of doubtful agents. And instead of its being the effect of a sudden convulsion, or an extraordinary deviation from the ordinary laws of nature, it will be found to have been produced by the very slow operation of causes which have always, and must ever continue, to act in the same manner. (*Gilmer, 1818, p. 190*)
>
> It is probable, then, that the water of Cedar Creek originally found a subterranean passage beneath the arch of the present bridge, then only the continuation of the transverse ridge of hills. The stream has gradually widened, and deepened this ravine to its present situation. Fragments of its sides also yielding to the expansion and contraction of heat and cold, tumbled down even above the height of the water. (*Gilmer, 1818, p. 190*)

Second, a work by Hitchcock in 1819 on a glacial lake in Connecticut, which, although not nearly so advanced in conception, does provide some evidence that stream cutting was understood. Here, however, stream-erosion formed a small part of the larger idea that the landscape had been produced by a series of lake-bursts sweeping across the surface of the land.

In contrast to these two forward-looking writers there was no dearth of authors whose work reflects either diluvialism or older traditional notions. In 1820 H. H. Hayden published his *Geological Essays*. His profession as a dentist practising in Baltimore and formerly as an architect did not deter him

FIG. 47. *Natural Bridge, Rockbridge County, Virginia from the south-east side*
(Photograph by C. D. Walcott, U.S. Geological Survey, in 1891).

from having decided geologic views, which were dominated by the idea of
the Flood, to which he attributed the creation of all alluvial deposits including those of the Mississippi delta. He believed that all the water from the
Flood came from the rapid melting of the polar ice-caps and, in this way, by
postulating ice-rafting, he was able to explain the position of erratics. Referring to the unconsolidated Tertiary and Recent deposits of the Atlantic coastal
plain, he said:

> Viewing the subject in all its bearings, there is no circumstance that affords
> so strong an evidence of the cause of its formation (Atlantic Plain) as that of its

having been deposited by a *general current* which, at some unknown period, flowed impetuously across the whole continent of America; and that from north east to south west . . . the course of this current . . . depended on that of the general current of the Atlantic Ocean. That from some unknown cause, its waters rose . . . to that degree, that it overran its ancient limits, and spread desolation on its adjacent shores. (*Hayden, 1820, pp. 40–41*)

His whole work was an unhappy mixture of the intelligent and the dogmatic; it was, as a modern critic says,

> verbose in the extreme . . . and more argumentative than logical. (*Merrill, 1924, p. 81*)

Yet Hayden's book must have been considered important by his contemporaries for Silliman reviewed it at length in 1821. One of the purposes of the book was to prove that the unconsolidated and alluvial deposits of the Atlantic coastal plain had been produced by the action of Noah's Flood. Though Silliman did not completely accept this explanation, for he felt that a longer period was necessary to account for the marks of attrition than was afforded by the biblical catastrophe, he accepted Hayden's main thesis that the position of these heterogeneous deposits far removed from their places of origin and bearing no relation to the present sources of water could only be explained satisfactorily by the action of a flood or floods.

> he has endeavoured to adduce facts sufficiently numerous and strong, to prove that the whole region, (the Atlantic coastal plain) with the attendant phenomena, is the result of the operation of currents, that flowed from the northeast to the south-west; or from the north to the south over the whole continent of America. (*Silliman, 1821, pp. 48–49*)
>
> The almost universal existence of rolled pebbles, and boulders of rock, not only on the margin of the oceans, seas, lakes and rivers; but their existence, often in enormous quantities, in situations quite removed from large waters; inland, . . . in high banks, imbedded in strata, or scattered occasionally, in profusion, on the face of almost every region, and sometimes on the tops and declivities of mountains, as well as the vallies between them; their entire difference, in many cases, from the rocks in the country where they lie . . . rounded masses and pebbles of primitive rocks being deposited in secondary and alluvial regions, and vice versa; these and a multitude of similar facts have ever struck us as being among the most interesting of geological occurrences, and as being very inadequately accounted for by existing theories. . . .
>
> The attrition of the common waters of the earth, and even that exerted during the comparatively short period of the prevalence of the deluge of Noah, would do very little towards producing so mighty a result, and we must assign this operation to the more recent periods of the prevalence of the great chaotic deluge, whose existence is distinctly recorded in the first chapter of Genesis, and equally admitted by all geologists. (*Silliman, 1821, pp. 49–50*)

Throughout his review Silliman's own ideas have a habit of intruding upon those of the writer he is discussing. From a few sentences it is possible to discern that Silliman's thinking, though advanced enough to discount the possibility that the flood alone can have been responsible, is still captivated by medieval and mystical fantasies; ideas of caverns within the earth from which at any time a flood of waters can rush forth at God's command, a notion which he might well have inherited from Werner who adopted it as part of one of his schemes.

> If, however, in our turn, we might be indulged in stating an hypothesis, we would beg leave to suggest the following as a cause which may have aided in deluging the earth, and which, were there occasion, might do it again.
> The existence of enormous caverns in the bowels of the earth, (so often imagined by authors) appears to be no very extravagant assumption. It is true it cannot be proved, but in a sphere of eight thousand miles in diameter, it would appear in no way extraordinary, that many cavities might exist, which collectively, or even singly, might well contain much more than all our oceans, seas, and other superficial waters, none of which are probably more than a few miles in depth. If these cavities communicate in any manner with the oceans, and are (as if they exist at all, they probably are,) filled with water, there exist, we conceive, agents very competent to expel the water of these cavities, and thus to deluge, at any time, the dry land. (*Silliman, 1821, pp. 51–52*)
> It is very possible, that anterior to the deluge of Noah, and to the peopling of the globe by rational beings, and during the gradual draining of the earth from the grand chaotic deluge, several floods more or less partial or extensive, may have taken place, – thus accounting for partial formations, as the parasitical trap rocks, etc. (*Silliman, 1821, p. 53*)

Some further idea of Silliman's knowledge regarding geological processes and the source of his belief can be traced in the following extracts from his review of Conybeare's and Phillip's textbook (pp. 78–79). Though the article was entered anonymously it is believed from the style and content that he was the author. In it he dwells upon the recent rapid rise of geology as a science and gives a list of the chief contemporary figures responsible for this progress. The absence of Hutton's name among them is a notable exception. Silliman praises the textbook for its excellence of style, an accomplishment upon which most nineteenth-century writers prided themselves. Indeed the maintenance of stylistic standards was a vital part of every educated scientist's training, and any writer aspiring to attract notice had to attain a high literary standard.

> It was not long however, before the present constellation of European geologists arose. It was a new and interesting field, and many a gentleman of distinguished talents and science was allured into it. It is sufficient to mention in Great Britain, the names of Jameson, Playfair, MacCulloch, Greenough, Webster, Conybeare, Buckland, Phillips, Aikin, Weaver, Seymour, Griffith, Farey, Bakewell, Parkinson, Sowerby, and Miller; and on the continent of Europe, those of Cuvier, Brongniart, Daubuisson, Humboldt, Von Buch, Brocci, De Luc, Brochant, and Delametherie. In the hands of such men, geology, for the last twenty years, has outstripped even chemistry in its progress. Instead of a few unassisted individuals, struggling with almost every

difficulty, and producing as the result of a life of laborious, self denying effort, a single volume, which, in the opinion of the world, placed its author among the candidates for Bedlam; we now find in the 'crystal hunter', many a Reverend, and many a Right Honourable: while numerous respectable societies, pour forth their splendid annual, and semi-annual quartos, rich with subterranean intelligence. (*Anon.* (*Silliman?*), *1824A, pp. 204–5*)

With the purity and neatness, not to say occasional elegance of style, in which this work is mostly written, we are much gratified: especially when we recollect, how often works of merit, in this department of knowledge, give pain to the classical man, by the homely and uncouth dress in which they are presented. (*Anon.* (*Silliman?*), *1824A, pp. 206–7*)

On geology Silliman has no original contribution to offer here. He is clearly obsessed by diluvial notions, and when he talks of the distinction between alluvium and diluvium he is merely repeating what Conybeare and Phillips said in their textbook:

A fuller and very satisfactory account of the diluvial formation is given in the Introduction: . . . And although it occurs in every part of the globe, covering probably more surface than any other; yet has it hitherto been passed, by geologists, in silence; or scarcely noticed; or blended with alluvium. It is the 'geest' of Kirwan and Jameson, and the 'diluvian detritus' of Buckland; . . . which, however, he has put down under the general name of alluvium. It must not be supposed, however, that all beds of loose sand and gravel are diluvial. Such beds often occur interstratified with regular strata of clay, and even consolidated rocks; as will be seen in the sequel. Deposits of sand, gravel, and boulder stones are not diluvial, unless they occur above all regularly stratified beds, and the ingredients are confusedly mingled together. This formation exists abundantly, in the United States. (*Anon.* (*Silliman?*), *1824A, p. 211*)

This clinging to old ideas was not entirely induced by a lack of the necessary imagination to invent new ones. In America as in Europe scientists were constrained within limits by religious beliefs and any who denied the Flood were considered both ungodly and unreasonable. Most of the upholders of the Flood depended no longer on mysticism and dogma but on intelligent scientific explanations with a religious background. But the example of Werner and the medievalists had taught the danger of too much hypothesis and 'facts' were now valued as much as clever literary arguments. In a contemporary review of *Reliquiae Diluvianae*, Catcott and Penn were both criticized because they failed to observe this factual rule.

Catcott was one of these disciples, and his work on the deluge is probably the best of that school. He has treated the subject of diluvial currents with great ability. But he was seduced by the extravagances of hypothesis, and inserted in his work a plate exhibiting 'the internal structure of the terraqueous globe, from the centre to the circumference', and with great seriousness, advises his readers to make themselves well acquainted with this, as rendering plain and clear the philosophical explanations of the Flood. In explaining this plate, he remarks 'that the opinion of the ancients concerning the earth's resembling

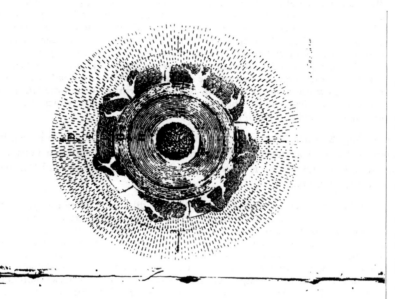

An Explanation of the COPPER-PLATE,

REPRESENTING

The internal structure of the terraqueous Globe, from the Center to the Circumference, and the Air around it.

D. The *outward Expanse* or *the eyes Firmament of Heaven*.

E. A *circular Space* filled with water during the height of the Deluge, but now with the Air that came from the central Hollow of the earth; and at present constitutes what we call our *Atmosphere*.

F. The *shell of the earth* broken into innumerable *apertures* and *fissures*, of various shapes and sizes; the *larger* of which, f.f.f.f.f. being filled with the water that descended from the surface of the earth, form *Seas* and *Lakes*; the *lesser* (which branch from the former, or pass immediately from the under-part of the shell of the earth to the top of the highest mountains) serve as canals for the water which supplies *Springs* and *Rivers* to run in; the *least* of all (denoted by the *irregular black strokes* in the solid shell of the earth) represent the cracks thro' which *vapours* principally ascend.

G. H. The *Great Abyss* of water within the earth; with which all Seas, Lakes, Rivers, &c. communicate; and from whence they receive their supplies. G. H. are divided from each other by a dotted circle, because *one of them* represents the water that, during the Deluge, covered the whole surface of the earth, but which was afterwards forced down, thro' the above-mentioned larger apertures and fissures, to its original place, as the inward Air was forced out thro' the lesser and oblique fissures: and the *other of them* represents that part of the Abyss which, during the Deluge, remained beneath the earth.

I. A *solid Ball* or *Nucleus* of terrestrial matter, formed from what the water in its descent from the surface, tore off, and passage through the strata of the earth, tore off, and carried down with it into the Abyss, and reposited at the lowest place, the center of the earth.

☞ So that the Opinion of the Ancients concerning the Earth's resembling an *Egg* has great propriety in it: for the Central Nucleus, (I.) by its innermost situation and shape, may well represent the *Yolk*; the *Abyss* of water, (G. H.) which surrounds it, and in a middle position, may stand for the *clear Fluid* of the *White*; the *Crust of the Earth* (F) (allowing only for its breaks and cracks) by its roundness, hardness, uppermost situation, and little inequalities on its surface, is justly analogous to the *Shell*. And on this account the term *the shell of the earth* is frequently used in this treatise.

FIG. 48. *Alexander Catcott's idea (1768) of the internal structure of the earth*

an egg, has great propriety in it; for the central nucleus, by its innermost situation and shape, may well represent the yolk; the abyss of water, which surrounds it, and is in a middle position, may stand for the clear fluid of the white; the crust of the earth (allowing only for its breaks and cracks), by its roundness, hardness, uppermost situation, and little inequalities on its surface, is justly analogous to the shell'. (*Anon, 1824B, pp. 153–4*)

The recent *Comparative Estimate of the Mineral and Mosaical Geologies*, by Granville Penn, Esq. . . . . exhibits him in the character of a good scholar, who is well versed in philology, and who has read most of the modern treatises on geology, but really we do not fear to hazard the assertion, that he has not seen much of rocks in their native beds. Yet he has made a vehement attack upon modern geology. He assumes in the first place, that the mineral and Mosaical geologies are directly opposed to each other, and absolutely irreconcilable; so that if the one be true, the other is false . . . In this discussion, he brings forward not a single fact in proof of his position, and notices but few of the facts which geologists adduce to support the contrary; but relies on abstract reasoning. (*Anon, 1824B, pp. 154–5*)

In the actual discussion on geology that followed, the reviewer revealed himself as an admiring apostle of the diluvial school but, like De la Beche and certain other British geologists, he is circumspect enough not to link the origin of all valleys with the Flood and adopts for valley-formation a variety of causes which may act singly or in combination. He is typically European in believing that rivers erode to a very limited extent in mountain areas or in a particular instance like the bursting of a lake but that the tremendous features of valley-erosion were due to either the intervention of a flood or structural forces.

It is quite obvious that the immense quantity of rounded pebbles and boulders, scattered over the earth's surface, must have been derived from the solid strata; and it is quite as obvious that the process by which this diluvial detritus was brought into its present form, must have produced vallies. Indeed, every body who looks at the innumerable vallies existing on the earth's surface imputes them at once to the action of running water. But the general belief is, that existing streams, avalanches and lakes, bursting their barriers, are sufficient to account for all their phenomena, and not a few geologists, especially those of the Huttonian school, at whose head is Professor Playfair, have till recently been of this opinion. So long ago, however, as the time of Catcott, this subject was ably handled, although his views have been much neglected till of late. But it is now very clear to almost every man, who impartially examines the facts in regard to existing vallies, that the causes now in action, mentioned above, are altogether inadequate to their production; nay, that such a supposition would involve a physical impossibility. We do not believe that one thousandth part of our present vallies were excavated by the power of existing streams. We are aware that some mountain torrents do exert, within narrow limits, a powerful agency. We have seen hard, unstratified, quartzose masses, of several tons in weight, torn out of their beds by a mountain torrent, and removed a considerable distance. But in level countries, and where the stream has no great descent, it is found that rivers have not power to move except in a few extraordinary instances, even small pebbles. In very many cases of large rivers, it is found, that so far from having formed their own beds, they are actually in a gradual manner filling them up. . . .

Again; how happens it that the source of a river is frequently below the head of a valley, if the river excavated that valley? Rivers also sometimes change their beds; but if they excavated their own beds, how could they change them? And to suppose that rivers formed their own banks, is to suppose they were once without banks. The most powerful argument, however, in our opinion, against the supposition we are combating, is the phenomena of transverse and longitudinal valleys; both of which could not possibly have been formed by existing streams. . . .

We must not, however, attribute the origin of all vallies to diluvial action. In primitive and mountainous districts especially, 'the original form in which the strata were deposited, the subsequent convulsions to which they have been exposed, and the fractures, elevations and subsidences which have affected them, have contributed to produce vallies of various kinds on the surface of the earth, before it was submitted to that last catastrophe of an universal deluge which has finally modified them all'.

Existing vallies, then, have been produced by three distinct classes of agencies. 1. By the present streams, the bursting of lakes, etc. 2. By the last universal deluge. 3. By the original construction of the strata, and diluvial actions previous to the last. It may be difficult, in all cases, to refer the origin of particular vallies to its proper period. It is sufficient however, for the purpose of this argument, to show, that there exist cases clearly referable to all the agencies above mentioned. When, for instance, we find on the margin of a valley, diluvial pebbles and bowlders, evidently torn out of that valley, we can have no hesitation in ascribing its excavations to the last universal deluge.

Excavations formed by that catastrophe are called 'valleys of denudation'. (*Anon., 1824B, pp. 332–4*)

It would not be difficult to find a dozen American geological writers in later years who adopted a more diluvial or more catastrophic approach than Silliman's. J. Finch (1824) used a simple diluvial explanation for the nature of the superficial deposits on the Atlantic coastal plain; another author in 1826 (Anon.) explained the rush of waters by a sudden halt in the rotation of the earth; J. Geddes in a paper on the Ontario valley (1826) triumphantly concluded that the existence of large valleys beyond the present heads of the rivers proved that the valleys had been formed before the advent of the rivers. He was even prepared to deny 'that the cataract of Niagara had in time travelled from near Lewiston to its present site' (Geddes, 1826, p. 213). As late as about 1840, C. T. Jackson in describing the geology of Maine (1839) and Rhode Island (1840), and A. Gesner in writing on Nova Scotia (1840), draw exclusively upon the diluvial theory.

It happened that in America diluvialism was encouraged by the presence of the Great Lakes, particularly as for each great lake there were hundreds of minor lakes, all seemingly pointing towards a former general inundation of the continental interior. It is not therefore surprising that many of the early American geological descriptions incorporated the idea of a flood or of a

FIG. 49. *Edward Hitchcock*

galaxy of major lakes. In his sketch of the Connecticut river region E. Hitchcock includes in his hypothesis the existence of former lakes and the occurrence of some great diluvial debacle. This was an acute observation, for today these lake beds are referred to drained pro-glacial lakes. Hitchcock was not prepared to rely on the lake alone to account for the present features, for he was fairly sure that the rush of the escaping lake waters would be insufficient to roll the larger erratics some hundreds of feet uphill. For this reason he concluded that an earlier flood must have occurred and chronologically he considered that the Noachian deluge was the requisite agent. A little further on he suggested an alternative means of moving these erratics – by floating ice.

Any one who examines the passage of the Connecticut and many of its tributaries, through several mountains will be led, I think, to the conclusion that the waters of this river once flowed over the great valley along its banks, forming an extensive lake: and also, that when this began to subside, by the wearing away of the outer barriers, other barriers would appear and produce other lakes of inferior extent.

It is no argument, as some have thought, in favour of such a supposition, that so much rock occurs in this basin which is evidently a recomposition of the detritus of older formations; and that organic remains are found in these rocks. For every geologist knows that all this must be referred to a period anterior to that, in which the last grand diluvian catastrophe happened to the globe and left our continents in their present form. Nor is the mere occurrence of masses of stone, evidently rounded by the attrition of running water, any evidence in favour of this hypothesis; for we must look for the cause of this also, as far back at least as the Noachic deluge. – No current of water with which we are acquainted is sufficient to transport such masses of rock in the situations in which we find them: 'for though we can readily conceive how the agency of violent currents may have driven these blocks down an inclined plane, or, if the *vis a tergo* were sufficient, along a level surface, or even up a very slight and gradual acclivity, it is impossible to ascribe to them the Sisyphean labour of rolling rocky masses, sometimes of many tons in weight, up the face of abrupt and high escarpments'. Rounded masses of rock may however occur under such circumstances as to show them to have been removed by currents posterior to the deluge. (*Hitchcock, 1824, pp. 16–17*)

At the outlet of the Connecticut through the mountains below Middletown, a little south of the Chatham cobalt mine, and six or seven hundred feet above the present bed of the river, I saw rounded masses of old red sandstone, several inches in diameter, mixed with the fragments of the rocks in place. Such a fact I never noticed at any other place in the primitive region along the river; certainly not on the east side of it. And I was led irresistibly to the conclusion, that they were conveyed thither by the ice of the ancient lake, which would be floated to the ocean through this outlet. (*Hitchcock, 1824, p. 18*)

This whole sketch shows that Hitchcock possessed good powers of observation and an ability to appreciate the smaller details of the main phenomena. But his final analysis of their causes presents a confused picture of lake bursts, floating ice and biblical floods.

W. Keating, who in 1825 went on Long's expedition to the source of St Peter's River in the region west of Lake Superior, revived this popular idea of lake bursts. He regarded the deposits of what had been glacial Lake Agassiz as being of lacustrine origin and explained the draining of the lake by its subsequently bursting through its former barriers.

The whole region comprising the headwaters of the Winnipeek River was looked upon as having been at a comparatively recent period an immense lake interspersed with innumerable barren, rocky islands, which had been drained by the bursting of the barriers which tided back the waters. The innumerable bowlders which he found covering the valley were regarded as due to the flood of waters caused by the bursting of these natural dams. (*Merrill, 1924, p. 104*)

THE WANING OF DILUVIALISM

Between about 1830 and the visit of Lyell in 1841, American geology remained in this state of confusion. Diluvialism weakened and changed but Huttonian theories made little progress and uniformitarianism was only just making headway.

It is tempting to discuss the Huttonians first, if only for their boldness. Whereas in Britain fluvial ideas had now begun to attract a few adherents, in America the picture remained totally unbalanced with only one side really represented. In 1829 Silliman edited the first American edition of Bakewell's *Introduction to Geology*. In 1833 Hitchcock, writing of the geology of Massachusetts in the first geological report of that state, said that most valleys were of primary origin and were modified to a small extent only by deluges and other abrading agencies. Yet only in a few rare cases where a stream had cut into a particularly soft stratum was he prepared to concede that the origin might have been secondary, and in discussing Cape Ann he wrote:

> It cannot fail to impress every reasoning mind with the conviction, that a deluge of tremendous power must have rushed over this cape. Nothing but a substratum of unyielding syenite could have stood before its devastating energy. (*Hitchcock, 1833, p. 148*)

The year 1833 brought a small glimmer of light to the general diluvial gloom when W. W. Mather, in his *Elements of Geology*, became the first American writer to give a favourable description of Hutton's views. This new attitude may have been prompted by reading Lyell's works. Unhappily it was no more than a passing flicker and went unnoticed by most American geologists. Yet a British visitor, G. Fairholme (1834), let his imagination speculate on the time necessary for the Niagara Falls to cut back to their present position. The whole implication of Fairholme's argument is that present processes differ very little from those that have acted in the past.

> It is perhaps not so well known as it ought to be, that almost the whole continent of North America consists of vast plains, composed of secondary strata of various kinds, and that calcareous formations in a horizontal stratification form the leading characteristics in the geology of that country. (*Fairholme, 1834, p. 12*)
> It is to one of these interruptions in the general level of that part of the world that the cataract of Niagara owes its origin. Two vast plains, or steppes (as they are termed in the North of Europe), extend themselves in different directions. One is spread over Upper Canada and New York towards the north, while the other embraces the shores of Lake Erie and its surrounding States towards the south-west. Between these great plains there is a considerable difference of level; and as the former is lower than the latter, all the waters which are drained from the one must experience a fall, more or

less violent according to the nature of the line of demarcation over which they must pass, before they can subside into the general level of the other. (*Fairholme, 1834, p. 12*)

This fall of 1 foot in 160 would not be thought remarkable in any inland stream of common size; but in the case of such a vast body of water as the Niagara contains, (which forms a large proportion of the whole fresh water of the earth, and has been calculated at one hundred millions of tons per hour,) the action, when once set in motion, must be enormous; nor can we feel surprise at the powerful effects which it produces, and on which I am now about to remark. (*Fairholme, 1834, p. 13*)

It is clear, therefore, that the waters of the whole of the upper lakes, in seeking their level in the ocean, have to descend from the higher to the lower plain; and as this descent does not take place by a wide valley, such as forms the usual channel for most other rivers, it is equally obvious that a period once existed, when these waters first began to overflow, and when they must have made their way over the upper surface of the country in the form of a great rapid, and that the violence of this superficial action on secondary strata of a horizontal form has gradually occasioned a cataract, which would naturally commence near the base, and which has in the course of many ages gradually worked its position backwards, until we now find it nearly at the greatest possible height which the nature of the ground will admit of. If this point be admitted, it is equally obvious that a continuance of the action must occasion a continuance of the effect, and that a time must consequently arrive when the whole barrier between the two lakes will be intersected. (*Fairholme, 1834, p. 14*)

It has been already stated that the distance from the fall to Queenston Ferry, where the river has resumed its more tranquil and navigable course, is about seven miles, or 37,000 feet; and if we divide this distance according to the data already stated, or at the rate of 4 feet per annum, we find that a little more than 9000 years would be necessary to complete the section as we now find it. It is scarcely necessary, however, to point out that such calculation would not be fairly stating the facts of the case. As we now see the fall, we find that it has attained within 50 perpendicular feet of the summit level and that it is now, consequently, acting upon a vastly greater resisting body than was at first opposed to it, in the commencement of its labours. (*Fairholme, 1834, p. 17*)

Fairholme even goes so far as to deny that the American continent was above sea level before the Flood; this, he felt, could not have been so because otherwise such rivers as the Niagara would during the passage of time have completely worn the continent away.

Yet with all this appreciation of the power of present processes Fairholme retains, as of necessity, the operation of the Flood, whose waves, tides and currents would have acted upon the surface in the same way, he thought, as they do now on the floors of shallow seas.

Let us first firmly establish the fact of a universal deluge about the very period denoted by Scriptural chronology. When this great point is admitted, and proved beyond cavil (which is not supposing too much), let us candidly apply to such diluvial water, the full power and action of the tides, and more especially of the currents, by which our seas are kept in such continual circulation and which are a natural and necessary consequence of the rotary motion of a solid globe covered by a fluid ocean. (*Fairholme, 1834, p. 22*)

That this confused work, which is somewhat reminiscent of the marine denudation theory of Lyell, represented a British viewpoint rather than a change in American traditional geologic thinking may be seen from the writings of H. D. Rogers, afterwards famous as a structural geologist. Rogers believed that the Niagara Falls might have enlarged and maintained the present chasm but that the original valley and falls were produced by diluvial denudation.

It is a very generally received opinion and may so far as present evidence extends, be taken for granted, that the country adjacent to Niagara and the lakes was originally covered with a vast lake, or rather inland sea, which some change in the configuration of the region contracted to the still very extensive masses of fresh water now remaining. The passage of such a body of water over the surface would deeply indent all the exposed portions of the land. Rushing in its descent from Lake Erie to Lake Ontario, from a higher to a lower plain, and across a slope like that at Queenstown, it would inevitably leave a deep and long ravine. But further, the whole of this region has been grooved and scarified by the same far sweeping currents which denuded the entire surface of North America, and strewed its plains and mountains with boulders, gravel and soil from the north. Such a diluvial valley, of greater or less length and depth was, I cannot help believing, probably the commencement of the present remarkable trough below the Falls. (*Rogers, 1835, p. 329*)

It is an observation of Professor Sedgwick, that the existing vallies of any country are generally the result of the joint agency of many causes and the remarkable valley of the Niagara river, notwithstanding the simplicity of its present features, may exemplify this principle. (*Rogers, 1835, p. 330*)

It will easily be imagined that what the Great Lakes were to the diluvialists, the Niagara Falls were to the fluvialists. Such a force constantly turned the mind towards fluvial erosion. Hence it is not surprising that in the next year J. B. Gibson contradicts Rogers' idea that the Falls were caused by diluvial action.

Rogers ingeniously maintains in the late January number of this Journal, that the cataract had not its commencement at Queenstown Heights at all. . . . When we see the river working in the rock like an endless saw, it is difficult to think that it did not make the groove in which we find it. If this groove were originally but a valley of denudation, why are its sides perpendicular even at the brink, and why is the original inclination of its slope broken by a cataract now? In the opinion of Professor Rogers and many others, an inland sea, vastly more immense than the present fresh water lakes, sent a current along the course of the Niagara river, tearing up the exposed portion of the land, and imperfectly excavating the rough and unshapen trough below the falls. The traces of an overwhelming current are doubtless every where visible; and it is reasonable to suppose that, seeking the lowest part of the barrier, it would gradually narrow and confine its action at that point, at least sufficiently to mark out the course of the subsequently diminished stream. But we are unable to imagine how a widespread torrent could have spent its entire action on a strip six hundred yards in breadth, giving to the sides of the gutter made by it, the character and appearance of perpendicular walls. No such walls are found in the water gaps of the Alleghany mountains. (*Gibson, 1836, pp. 204–5*)

Having argued so ably against the older ideas, Gibson then spoils this initial impression by reverting to the traditional obsession with the biblical Flood. His reason for believing this was quite plain. To him the widespread glacial deposits showed the extent of the flood waters while the over-deepened basins of the Great Lakes showed that they could not have been denuded by the existing erosional agents. He believed therefore that they must have been enlarged by some supernatural factor. Unfortunately, like most of his colleagues, he chose the flood and not ice-sheets and glaciers.

> The present basins are but the remaining, as they were the deepest, parts of the ancient channels. . . . They evidently owe their excavation to some other power than that of the waters which now possess them and to what other are they so likely to owe it as to that of the great denuding current of which we every where see such convincing proofs? To account for their origin at all, we must believe that they were scooped out by it, the abrasions of the lias being but partial in that quarter, and entire further on where the force of the current was increased by concentration in narrower channels. (*Gibson, 1836, pp. 208–9*)

How the flood acted and where it came from becomes clearer as we read on. Gibson imagined it starting from the cold ocean areas of the Pacific or more northerly Bering Strait, penetrating the Rocky Mountains, sweeping eastwards across the flat intercontinental area and eventually finding outlets through either the Saint Lawrence or the Hudson valley or the Gulf of Mexico. Lyell's icebergs were employed to transport the diluvium. When Gibson is combating the hypothesis of Rogers one finds him giving credence to a considerable amount of erosional activity on the part of the Niagara. Yet when he contemplates the water gaps of the Appalachians and the valleys of the Hudson and Delaware rivers he propounds the retrograde view that even a flood could not have produced the great erosion which is visible and therefore he supposes the land must have been below the sea and the gaps must have been worn by marine erosion.

> A current setting in the direction of this line, from the Pacific a little south of Behrings Straights, would sweep over the primitive region at the northern extremity of the Rocky Mountains, and may well have supplied the fragments in question from that point. There is certainly no proof to assign them an origin elsewhere; nor can we otherwise account for their existence where we find them. The angular masses occasionally, though rarely, found even within the region of frost, have doubtless been borne on icebergs; but that they, as well as the spheroids came from that quarter, there is not a rational ground to dispute. It is evident from the greater degree of denudation to the northeast, that the main body of the current discharged itself into the gulf of St Lawrence; but that diverging portions of it found their way into the gulf of Mexico, is equally evident. Such a division of it could not but be produced

by its efforts to clear the obstruction presented to it by the Alleghany Mountains; to effect which it would necessarily pass round both ends of the chain. The proof of it however rests not merely on the configuration of the continent. (*Gibson, 1836, p. 209*)

But these two estuaries, though the principal, were not the only ones. The rolled pebbles found along the water courses leading to them, are found also along the shores of every river which flows through the passes of the Alleghany from the table land of the lakes. They are found on the Hudson, the Delaware, and the Susquehannah; and they are not found on any river which descends from either flank of that mountain. Portions of the great diluvial current must, therefore, have cleft it at the places occupied by those rivers. No feebler power could have cut their channels out. They could not themselves have done it. The disuniting force must have been applied at vastly higher levels than they could have attained by filling up the vallies; and the evaporation from the expanse of surfaces thus produced, would have nearly equalled the quantity of water returned to them in rain, so that there would have been but little comparatively to tumble over the ridges. But what power could there have been to sustain a mass of water at an elevation sufficient to cause an overfall at any point of the principal or dividing ridge, without which no fresh water stream could make its way through that barrier or produce the stupendous results observable in every member of the chain? An ocean beating against its side might perforate it; but it is infinitely more probable that the breaches were effected even before the mountains had raised their heads above the deep. (*Gibson, 1836, p. 210*)

The whole of Gibson's work makes a curious assemblage of advanced and reactionary thinking. He and most other contemporary American geological writers, apart from fundamentalists and confirmed diluvialists, borrowed parts of other people's ideas and combined them into notions that fitted New World phenomena. Shades of Werner, Cuvier, Buckland and even Desmarest bob up here and there, but no writer discards these outworn ideas and attempts to devise a scheme strictly applicable to the geologic features of his homeland. Many workers, such as Thompson in New York State (1833), continued to be diluvialists confused by the problem of glacial phenomena. The coming of Lyell's *Principles* was to be a boon to such thinkers who unfortunately, as happened to Gibson, soon adopted the iceberg and other less desirable theories of the uniformitarians.

G. E. Hayes was one of the first American writers to exhibit strong symptoms of this indoctrination. Unhappily for him the principle which he chose for adoption included Lyell's misguided theory of marine denudation. In an article on the geology of western New York his opening general description places emphasis on the influence of structure, and he decries the possibility that the streams could have been responsible for the form of the topography. The presence of over-deepened lakes also convinced him that a more powerful agent than the streams must be sought.

Superimposed on the mountain limestone, we have a series of shales and slaty sandstones of great aggregate thickness, dipping, as do the formations already noticed, in a southerly direction, but less able to resist the powerful, degrading action to which all have evidently been exposed. The deep valleys, which penetrate this formation in a southerly direction form the great limestone terrace; the dividing ridges, also, which have their northern terminations on the same terrace, becoming more rugged and mountainous as they approach the Pennsylvania state line, with their sides deeply furrowed by precipitous gullies and ravines, are sufficient proofs that other causes of denudation than the insignificant streams which traverse these valleys, have been in operation.

One peculiar feature, which adds greatly to the picturesque scenery of Western New York, arises from the fact that many of these valleys have been excavated to a level below the general escarpment of the limestone terrace, which consequently forms a barrier at their mouths, and gives rise to most of those beautiful sheets of water so justly admired by the lovers of fine scenery. This feature will again be alluded to further on. (*Hayes, 1839, pp. 86–87*)

Arguing along uniformitarian lines he states that it is needless and probably groundless to suppose the existence of a flood, when the action of the sea, a known and existing agent, could as effectively and more naturally have formed the present landscape.

Why not then lay aside the fashion of attempting to explain such phenomena by invoking the assistance of the Noachian Deluge, or of tremendous inundations, sweeping over the tops of the highest mountains, produced 'by the flux and reflux of mighty deluges, caused by the sudden elevation of mountain chains in various parts of the globe'? Sound philosophy forbids these violent presumptions, particularly when the facts admit of explanations more consonant with the natural order of events.

The condition of a continent, gradually elevated from the ocean, whether by volcanic action, or by the expansive force of crystallization, or by any other cause whatever, would be such as to account for all the geological phenomena hitherto attributed to the mechanical action of water. Every portion of a continent thus reclaimed, must, in succession, have been the bed, and then the beach of an ocean. Every portion must have been subjected to the action of the waves and the tides, when lashed into fury by the raging storm; and for a period of time only limited by the greater or less rapidity of the elevatory process.

When any considerable portion had become permanently elevated above tide water, it would form a water shed, collecting the rain into rivulets, which, finding their way to the ocean, would cut out narrow channels for their beds. But the effect of these streams in the formation of valleys, by denuding and tearing up the rocky strata, would be insignificant in comparison with the action of the surge at those points where their waters were disembogued. As each portion of such channels would successively be exposed to their combined action, and must successively form the bed of an estuary at the valley's mouth, we can readily account for their excavation, to a greater or less extent, in proportion to the hardness of the rocky bed, to the violence of the waves and tides, and the duration of their action. In these estuaries, the comminuted materials would assume nearly a horizontal position, and when left dry, would resemble the alluvial plains or 'bottoms', which border most of our rivers. Should a sudden rise of a few feet take place, the water would at first recede; but by the action of the waves and tides on this alluvial mud, they would soon regain possession of that part of their former bed, bordering the stream to a greater or less extent. The centre of the valley would thereby be lowered; and this process being repeated, a series of terraces, or steps, would result, precisely similar to those in the valley of the Connecticut river, which Prof. Hitchcock attributes to the fluviatile action of existing streams. Valleys could thus be formed where streams of no great magnitude ever flowed, and where currents, except the ordinary ones of the ocean, never existed. (*Hayes, 1839, pp. 88–90*)

To justify what he had just put forward he examined the diluvial evidence and compared the plausibility of his explanation with those that had gone before.

Every inch of surface has been subjected to the denuding agent; the tops of the highest hills, no less than the limestone platform, bear the scars and scratches of the contending elements. The surface, except on the steep escarpments, is every where covered with a thick coat of diluvium, composed of water-worn pebbles, boulders, sand, &c. The valleys are often deeply filled with these materials, more or less comminuted; and sometimes they contain large quantities of detrital matter, little worn, evidently derived from strata similar to those of the adjoining hills.

The condition of an ancient inland lake which has burst its barriers and disappeared, could not account for these things; nor could its drainage from a higher to a lower plain, as suggested by Prof. Rogers, excavate the deep and long ravine through which the Niagara now flows. It is equally idle to suppose, that the existing streams have excavated the valleys through which they flow; much less could they have effected the comminution and uniform distribution of the coat of diluvium. And as for a sudden inundation, deluge, or any succession of them, (aside from the improbability of nature stepping so far out of her ordinary track,) had they been sufficiently powerful to tear up the strata, and lay bare so large a district of the limestone rocks, we should hardly expect to find the work so systematically accomplished. A great deluge, it is true, may account for the uncovering of the limestone and by sweeping heavy boulders over its surface, might have produced the 'diluvial scratches'. But portions of this rock are highly polished, and indicate a much longer continuance of the watery friction than is consistent with the notion of a deluge. The systematic and parallel arrangement of the long sloping ridges, composed of shale and sandstone, no better adapted to resist a sudden and overwhelming inundation than those portions which have been removed from the intermediate valleys, could hardly have resulted from any sudden irruption of water. (*Hayes, 1839, pp. 92–93*)

Convinced of the marine theory, Hayes employed it exclusively to solve the problems of the landscape under discussion. For instance he makes mountain peaks originate in the following manner:

Suppose this dividing ridge to have attained an elevation above tide water. The southern slope would present to the waves the smooth surface of the strata; whereas their basseting edges would be exposed on the northern declivity. Deep notches would soon be worn into it from both sides, which would occasionally interlock, and sometimes meet; thereby cutting the ridge into a series of islands, with transverse passes between them. These islands may form the highest peaks of the range; and the passes correspond to the elevated valleys, in which the principal streams take their rise. (*Hayes, 1839, p. 93*)

The Niagara Falls he envisaged as a great reef, or limestone ridge, which suddenly appeared above the sea. In time tidal action would breach the barrier and as the reef rose higher the scouring force of the sea acting like a vast tidal bore would cut out the present gorge and, as the reef continued to tilt, so the Falls would gradually appear. The theoretical standpoint is pure Lyell and is a forerunner of Darwin's (1846) early descriptions of southern Chile and of Australia's Blue Mountains discussed on pp. 177–9.

THE AMERICAN VISITS OF CHARLES LYELL

During the winter of 1841–2 Lyell brought to the American scene the same proselytizing spirit that had already captivated British geologists. He was received in America as a figure worthy of admiration and the audience at a course of lectures that he gave at Lowell Institute, Boston is said to have averaged three thousand. Yet the great man soon showed that he had come to be informed as well as to inform. He wished above all to test the theories he had worked out in Europe. His geomorphic conceptions were, as we have seen, far from faultless and his obsession with marine denudation was eventually to prove most unfortunate, but these blemishes should not be allowed to hide the true value of his contribution to American geology. The principles of uniformitarianism *were* basically correct as Lyell taught them, and it was a technique of reasoning essential to the progress of American geologists.

Lyell, as was his custom, visited as many places as possible and exchanged ideas with numerous local experts. When it became known that he was going to publish the results of his visit many American geologists, who had freely discussed their ideas with him, were troubled by the fear that he would utilize material gathered from them. When James Hall discovered that Lyell was writing his *Travels*, he wrote in 1842:

> I have condemned unqualifiedly such a course in Mr Lyell – a course which I should not have anticipated and which from my intercourse I thought him incapable of. (*Quoted in Merrill, 1924, p. 668*)

However, these fears were groundless, for Lyell's *Travels in North America* were concerned with general principles and gave full credit to his American informants.

These published conclusions on his visit indicate that the new features had at least stimulated his imagination even if they did not produce any radical alteration in his basic ideas. Throughout, it is obvious that, as regards geomorphology, he came searching for evidence of marine denudation and this supplied the key to the majority of his explanations. For example, he explains how the Niagara Falls were formed first by marine erosion and then by the action of the rivers when the land was upraised; the original escarpment which marked the starting place of the Falls he believed had been produced by the action of the sea; the ravine on the other hand he thought had been cut by the river itself, though it might have been guided initially by the direction of a shallow gully left by the sea.

Again, in his description of the Great Lakes area, Lyell's belief in marine action blurs his vision and deters him from using his knowledge of glacial action. The overdeepened basins of the Great Lakes he supposed to have been first scooped out by the scouring action of estuarine currents and then when the whole area was again depressed the deposits and the present raised beaches were formed. He does all this when he must have known that the glacial theory offered a far simpler solution.

The author next describes the ridges of sand and gravel surrounding the great lakes, which are regarded by many as upraised beaches. He examined, in company with Mr Hall, the 'Lake ridge', as it is called, on the southern shore of Lake Ontario, and other similar ridges north of Toronto, which were formerly explored by Mr Roy, and which preserve a general parallelism to each other and to the neighbouring coast. Some of these have been traced for more than 100 miles continuously. They vary in height from ten to seventy feet, are often very narrow at their summit, and from fifty to 200 yards broad at their base. Cross stratification is very commonly visible in the sand; they usually rest on clay of the boulder formation. and blocks of granite and other rocks from the north are occasionally lodged upon them. They are steeper on the side towards the lakes, and they usually have swamps and ponds on their inland side; they are higher for the most part and of larger dimensions than modern beaches. Several ridges, east and west of Cleveland in Ohio, on the southern shore of Lake Erie, were ascertained to have precisely the same characters. Mr Lyell compares them all to the osars in Sweden, and conceives that, like them, they are not simply beaches which have been entirely thrown up by the waves above water, but that many of them have had their foundation in banks or bars of sand, such as those observed by Capt. Grey running parallel to the west coast of Australia, Lat. 24° S., and by Mr Darwin off Bahia Blanca and Pernambuco in Brazil, and by Mr Whittlesey near Cleveland in Lake Erie. They are supposed to have been formed and upraised in succession, and to have become beaches as they emerged, and sometimes cliffs undermined by the waves. The transverse and oblique ramifications of some ridges are referred to the meeting of different currents and do not resemble simple beaches. (*Lyell, 1842–3, pp. 20–21*)

The author then endeavours to trace the series of changes which have taken place in the region of Lake Erie and Ontario, referring first to a period of emergence when lines of escarpment like that of Queenstown, and when valleys like that of St Davids were excavated; secondly, to a period of submergence when those valleys and when the cavities of the present lake-basins were wholly or partially filled up with the marine boulder formation; and lastly to the re-emergence of the land, during which rise the ridges before alluded to were produced, and the boulder formation partially denuded. He also endeavours to show, how during this last upheaval the different lakes may have been formed in succession, and that a channel of the sea must first have occupied the original valley of the Niagara which was gradually converted into an estuary and then river. The great Falls, when they first displayed themselves near Queenstown must have been of moderate height and receded rapidly, because the limestone overlying the Niagara shale was of slight thickness at its northern termination. On the further retreat of the sea a second fall would be established over lower beds of hard limestone and sandstone previously protected by the water; and finally, a third fall would be caused over the ledge of hard quartzose sandstone which rests on the soft red marl, seen at the base of the river-cliff at Lewistown. These several falls would each recede further back than the other in proportion to the greater lapse of time during which the higher rocks were exposed before the successive emergence of the lower ones. Three falls of this kind are now seen descending, a continuation of the same rocks on the Genese River at Rochester. Their union, in the case of the Niagara into a single fall may have been brought about in the manner suggested by Mr Hall, by the increasing retardation of the highest cataract in proportion as the uppermost limestone thickened in its prolongation southwards, the lower falls meanwhile continuing to recede at an undiminished pace, having the same resistance to overcome as at first. (*Lyell, 1842–3, pp. 21–22*)

When a fuller description of Lyell's American visit was put forth in two volumes in 1845 it could hardly fail to influence and interest Americans, though geomorphologically a great deal of it was spoilt by over-preoccupation with his pet theories of marine-denudation and iceberg-drifting. Many novel speculations by Americans must have been aborted by wrong conclusions arising out of a belief in these two theories. In the first volume, Lyell strikes hard at the contention that the Niagara Falls had a diluvial origin and it is refreshing to read of how many examples he gives of the river's erosive power. He had no doubt at all that it owed its origin to erosion without the intervention of any force other than that of falling water. The only qualification he makes is that the slope of the river may have been guided by the course of an original channel present before the Falls ever came into existence. Towards the end he again confuses glacial evidence, for although he correctly appreciated that the Falls were cutting down through drift deposits, he wrongly identified these as marine in origin.

In like manner, the Falls of Niagara teach us not merely to appreciate the power of moving water, but furnish us at the same time with data for estimating the enormous lapse of ages during which that force has operated. A deep and long ravine has been excavated, and the river has required ages to accomplish the task, yet the same region affords evidence that the sum of these ages is as nothing, and as the work of yesterday, when compared to the antecedent periods, of which there are monuments in the same district. (*Lyell, 1845, I, p. 23*)

The sudden descent of huge rocky fragments of the undermined limestone at the Horseshoe Fall, in 1828, and another at the American Fall, in 1818, are said to have shaken the adjacent country like an earthquake. According to the statement of our guide in 1841, Samuel Hooker, an indentation of about

FIG. 50. *Lyell's bird's-eye view of the Niagara Falls and gorge*

forty feet has been produced in the middle of the ledge of limestone at the lesser fall since the year 1815, so that it has begun to assume the shape of a crescent, while within the same period the Horseshoe Fall has been altered so as less to deserve its name. Goat Island has lost several acres in area in the last four years, and I have no doubt that this waste neither is, nor has been a mere temporary accident, since I found that the same recession was in progress in various other waterfalls which I visited with Mr Hall, in the State of New York. (*Lyell, 1845, I, p. 27*)

The supposed original channel, through which the waters flowed from Lake Erie to Queenston or Lewiston, was excavated chiefly, but not entirely, in the superficial drift, and the old river-banks cut in this drift are still to be seen facing each other, on both sides of the ravine, for many miles below the Falls. A section of Goat Island from south to north, or parallel to the course of the Niagara shows that the limestone had been greatly denuded before the fluviatile beds were accumulated, and consequently when the Falls were still several miles below their present site. From this fact I infer that the slope of the river at the rapids was principally due to the original shape of the old channel, and not, as some have conjectured, to modern erosions on the approach of the Falls to the spot. (*Lyell, 1845, I, p. 31*)

Among the other regions visited by Lyell was the Appalachian Range. In this case when he explains the system of parallel ridges he does not employ his marine theory but uses the far simpler explanation of control by the underlying structure; he made the hills mark the outcrops of hard rock, whereas the water gaps coincided with faults.

No traveller can fail to remark the long and uniform parallel ridges, with intervening valleys, like so many gigantic wrinkles and furrows, which mark the geographical outline of this region; and these external features are found by the geologist to be intimately connected with the internal arrangement of the stratified rocks. The long and narrow ridges, rarely rising more than 2000 feet above the valleys, and usually not more than half that height, are broken here and there by transverse fissures, which give passage to rivers, and by one of which the Schuylkill flows out at Reading. The strata are most disturbed on the south-eastern flank of the mountain chain, where we first entered, and they become less and less broken and inclined as they extend westward. (*Lyell, 1845, I, pp. 66–67*)

Farther on there reappears an attempted justification of his iceberg and marine theories. On the Delaware River he supposed that the scraping of trees by ice and the filling of a canal with gravel and sand were proof of the carrying power of ice. The sand and gravel could equally well, of course, have been borne by the water which carried the ice.

On our way, we heard much of a disastrous flood which occurred last spring on the melting of the snow, and swept away several bridges, causing the loss of many lives. I observed the trees on the right bank of the Delaware at an elevation of about twenty-four feet above the present surface of the river, with their bark worn through by the sheets of ice which had been driven against them. The canal was entirely filled up with gravel and large stones to the level of the towing path, twenty feet above the present level of

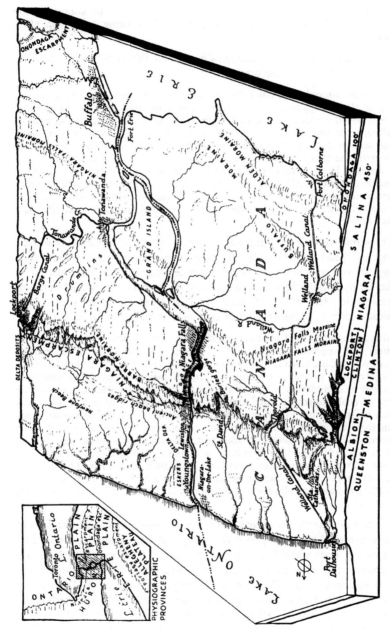

FIG. 51. Block diagram of the Niagara Falls region
(Drawn by Erwin Raisz)

the stream, which appeared to me to be only explicable by supposing the stones to have been frozen into and carried by the floating ice. (*Lyell, 1845, I, pp. 82–83*)

In Brooklyn he sought corroboration of his marine theory by supposing once again that drift had a marine origin.

> At South Brooklyn, I saw a fine example of stratified drift, consisting of beds of clay, sand, and gravel, which were contorted and folded as if by violent lateral pressure, while beds below of similar composition and equally flexible, remained horizontal. These appearances, which exactly agree with those seen in the drift of Scotland or the North of Europe, generally accord well with the theory which attributes the pressure to the stranding of ice islands, which, when they run aground, are known to push before them large mounds of shingle and sand, and must often alter greatly the arrangement of strata forming the upper part of shoals, or mud-banks and sand-banks in the sea, while the inferior portions of the same remain unmoved. (*Lyell, 1845, I, pp. 190–1*)

Lyell's second volume on America is more remarkable for its geomorphic conclusions. Amid the inevitable repetition of the marine and iceberg theories he suggests one avenue of research which Ramsay was to make use of in Britain. By hypothetically filling in the void spaces between parallel outcrops of strata which had obviously been severed, Lyell found he could more graphically emphasize the amount of erosion that had been carried out.

> In discussing with Dr Locke the probability of the former continuity of the Illinois and Appalachian coal-fields . . . over that flat dome on the middle part of which Cincinnati is built, we endeavoured to calculate the height which the central area would have attained, if the formations supposed to have been removed by denudation were again restored. In that case the thickness of the strata of coal, subjacent conglomerate, Devonian and Upper Silurian beds, which must have been carried away, could not, if we estimate their development from the mean of their aggregate dimensions on the east and west of Cincinnati, have been less than 2000 feet. The tops of the hills near Cincinnati, composed of the blue limestone, are about 1400 feet above the level of the sea. If, then, the formations presumed to have been destroyed by denudation were replaced, the height of the dome would be about 3500 feet, or exceeding the average elevation of the Alleghany Mountains. (*Lyell, 1845, II, pp. 41–42*)

Another useful observation occurs when a description of the river terraces of the Ohio caused Lyell to suppose that they had been caused by an initial sinking of the land; when this sinking took place it made the river flood and deposit where formerly it had eroded, to be followed by renewed uplift and renewed stream-cutting. Here we may glimpse the effect of Hitchcock's thought on Lyell. The account was used again almost word for word in the passages in *The Principles* that we have already quoted (pp. 184–5).

Lyell was at his weakest when describing glacial phenomena. Instead of the simpler land-ice theory of Agassiz he constantly confronts the reader with a vision of a land half sunk below Arctic seas and of hundreds of earth-covered icebergs floating, bumping and scratching over and against the continental surfaces. He himself seems to have sensed nothing peculiar in the different directions taken by the glacial grooves. Writing of New York state he stated:

First, the country . . . acquired its present geographical configuration, so far as relates to the outline of the older rocks, under the joint influence of elevatory and denuding operations. Secondly; a gradual submergence then took place, bringing down each part of the land successively to the level of the waters, and then to a moderate depth below them. Large islands and bergs of floating ice came from the north, which as they grounded on the coast and on shoals, pushed along all loose materials of sand and pebbles, broke off all angular and projecting points of rock, and when fragments of hard stone were frozen into their lower surfaces, scooped out grooves in the subjacent solid strata. The sloping beach, as well as the level bottom of the sea, and even occasionally the face of a steep cliff, might all be polished and grooved by this machinery; but no flood of water, however violent, or however great the quantity of detritus, or size of the rocky fragments swept along by it, could produce straight, parallel furrows, such as are everywhere visible in the district under consideration. (Lyell, 1845, II, pp. 82–83)

Thirdly, after the surface of the rocks had been smoothed and grated upon by the passage of innumerable icebergs, the clay, gravel and sand of the drift were deposited, and occasionally fragments of rock, both large and small, which had been frozen into glaciers, or taken up by coast ice, were dropped here and there at random over the bottom of the ocean, wherever they happened to be detached from the melting ice. During this period of submergence, the valleys in the ancient rocks were filled up with drift, with which the whole surface of the country was over-spread. Finally; the period of re-elevation arrived, or of that intermittent upward movement, when the ridges to be described in the next chapter were formed in succession, and, when valleys, like that of St David's which had been filled up, were partially re-excavated. (Lyell, 1845, II, p. 84)

Lyell re-visited America in 1845 and stayed there till the autumn of 1846 but his observations on this protracted visit showed little geological advance. He was impressed by the rate of soil erosion and gullying in certain parts of Georgia where the process had been greatly accelerated by deforestation (Lyell, 1849, II, pp. 28–30). He also correctly describes the origin of oxbow lakes, realizing that they are abandoned portions of the former river channel.

The great number of crescent-shaped lakes to the westward of the Mississippi, which formerly constituted bends in its ancient channel, are also monuments of the antiquity of the great plain over which the river has been wandering. (Lyell, 1849, II, p. 192)

He again reverts, in more detail this time, to his conclusion that river terraces are the result of the river trenching through its own alluvial deposits, the change in operation arising from a variation in the level of the land. When Lyell endeavours to calculate the age of the Mississippi delta his reasoning assumes a distinctly modern outlook, rarely found in his contemporaries. In

Lyell's case it assisted him to appreciate the relationship between, on the one hand, the velocity, width and depth of a stream and, on the other, the amount of matter carried in suspension.

> Finding it impossible to calculate the age of the delta, from the observed rate of the advance of the land on the Gulf in each century, I endeavoured to approximate, by a different method, to a minimum of the time required for bringing down from the upper country that large quantity of earthy matter which is now deposited within the area of the delta. Dr Riddell communicated to me, at New Orleans, the result of a series of experiments which he had made, to ascertain the proportion of sediment contained in the waters of the Mississippi. He concluded that the mean annual amount of solid matter was to the water as 1/1245 in weight, or about 1/3000 in volume. Since that period, he has made another series of experiments, and his tables show that the quantity of mud held in suspension, increases regularly with the increased height and velocity of the stream. On the whole, comparing the flood season with that of clearest water, his experiments, continued down to 1849, give an average annual quantity of solid matter somewhat less than his first estimate, but not varying materially from it. From these observations, and those of Dr Carpenter and Mr Forshey (an eminent engineer, to whom I have before alluded), on the average width, depth, and velocity of the Mississippi, the mean annual discharge of water and sediment was deduced. I then assumed 528 feet, or the tenth of a mile, as the probable thickness of the deposit of mud and sand in the delta; founding my conjecture chiefly on the depth of the Gulf of Mexico between the southern point of Florida and the Balize, which equals, on an average, 100 fathoms, and partly on some borings, 600 feet deep, in the delta near Lake Pontchartrain, north of New Orleans, in which the bottom of the alluvial matter is said not to have been reached. The area of the delta being about 13,600 square statute miles, and the quantity of solid matter annually brought down by the rivers 3,702,758,400 cubic feet, it must have taken 67,000 years for the formation of the whole; and if the alluvial matter of the plain above be 264 feet deep, or half that of the delta, it must have required 33,500 more years for its accumulation, even if its area be estimated as only equal to that of the delta, whereas it is in fact larger. (*Lyell, 1849, II, pp. 187–8*)

But for all this astuteness the same faults and the same blind spots are still present. He continues to believe in his world of icebergs:

> after reading the accounts given by Sir James Ross and Captain Wilkes, of the transfer of erratics by ice, from one point to another of the southern seas, these travelled boulders begin to be regarded quite as vulgar phenomena, or matters of every-day occurrence. (*Lyell, 1849, I, pp. 36–37*)

and he gives great prominence to the sighting of a few small rocks on an iceberg during his homeward sea journey.

> We sailed within half a mile of several bergs, which were 250 feet, and within a quarter of a mile of one 150 feet in height, . . . I was most anxious to ascertain whether there was any mud, stones, or fragments of rock on any one of these floating masses, but after examining about forty of them without perceiving any signs of foreign matter, I left the deck when it was growing dusk. . . . I

had scarcely gone below ten minutes, when one of the passengers came to tell me that the captain had seen a black mass as large as a boat on an iceberg, about 150 feet high, which was very near. By aid of a glass, it was made out distinctly to be a space about nine feet square covered with black stones. The base of the berg on the side towards the steamer was 600 feet long, and from the dark spot to the water's edge, there was a stripe of soiled ice, as if the water streaming down a slope, as the ice melted, had carried mud suspended in it. In the soiled channel were seen two blocks, each about the size of a man's head. Although I returned instantly to the deck when the berg was still in sight, such was then the haziness of the air, and the rapidity of our motion, that the dark spot was no longer discernible. (*Lyell, 1849, II, pp. 272-3*)

Similarly he is just as pleased when he recognizes the Pleistocene raised beaches of the southern Atlantic states, for superficially they seemed to fit very well into this theory of marine denudation; as the land rose from the sea by stages each stage would be marked by its separate beaches and cliffs.

Going inland twenty miles, we come to the termination of this lower terrace, and ascend abruptly to an upper platform, seventy feet above the lower one, the strata composing which belong to the Eocene period. This upper terrace also runs back about twenty miles to the abrupt termination of a third table-land, which is also about seventy feet higher, and consists of Eocene strata by the denudation of which all these terraces and escarpments (or ancient sea-cliffs) have been formed. Bartram has, with his usual accuracy, alluded to these steps, or succession of terraces, as an important geographical feature of the country, each of them being marked by its own botanical characters, the prevailing forest-trees, as well as the smaller plants, being different in each. (*Lyell, 1849, I, p. 257*)

Lyell paid further visits to the United States in 1852, when he again lectured at Boston, and in 1853. He had a profound effect on the future of American geology but as we will now see, his glacial theories inevitably became the targets for progressive critics.

AGASSIZ'S ARRIVAL AND THE RISE OF NEW WORLD GLACIOLOGY

As with Lyell, the arrival of the ideas of Agassiz preceded his migration to the United States. If we neglect such ignored writers as Dobson (p. 194), the first American author with marked glaciological leanings was probably T. A. Conrad, whose arguments largely repeat those of Agassiz, apart from the use of frozen lakes, a misguided intrusion which seems pardonable in one well acquainted with that phenomenon on a vast scale. Conrad stresses the superimposition of giant boulders upon fine sand and gravel as evidence of the work of ice.

The fall of temperature (so happily illustrated by the genius of Agassiz) which occurred at the commencement of the 'Diluvial epoch' is so well supported by all the known facts, that we feel no hesitation in applying the theory to all the inferior grand formations; . . .
The boulders (erratics) rest usually on sand, gravel, or the natural soil,

which would necessarily have been swept away, had currents transported these huge fragments, leaving them in every instance reposing on indurated strata. The hypothesis of icefloes bringing them from the north, floating on the waters of an ocean, and depositing them where they are now found, has been supported by some of the geologists of the present day; but this was in direct opposition to another theory of these same geologists, that a higher mean temperature prevailed over the northern regions at that period, than now reigns in temperate climes. This would not have been the case, all other things being equal, if the northern half of the continent had been nearly all formed by the ocean, notwithstanding the mean temperature is greatly modified in the same parallel of latitude, by the presence or absence of large bodies of water, rising with the former and falling with the latter physical condition of the globe. Whence then this immense body of ice, which has scattered boulders over so vast a tract of country, appearing too at an epoch subsequent to the extinction of the mastodon and other mammalia, which evidently lived in this region and enjoyed an equatorial climate anterior to the icy period? Nothing can reconcile this apparent contradiction, but the admission of a fall of temperature far below that which prevails in our day, freezing the enormous lakes of that period, and converting them into immense glaciers, which probably continued undiminished during a long series of years. At the same time, elevations and depressions of the earth's surface were in progress, giving various degrees of inclination to the frozen surfaces of the lakes, down which boulders, sand and gravel would be impelled to great distances from the points of their origin. (*Conrad, 1839, pp. 240–1*)

Conrad, however, seems considerably ahead of his time as even in Europe Agassiz's theory was only just receiving its initial publicity and there and in the New World Lyell's iceberg theory was achieving a notable popularity. The confusion of glacial ideas in America may be judged from the opinions of the American Geological Survey which was founded in 1836. W. W. Mather believed that drift was in part transported by currents of water and in part by ice drifted by currents (Merrill, 1906, p. 311). E. Emmons, on the other hand, thought that erratics were due to iceberg floating but that striations could not be made on the sea-floor because it is not bare rock, nor would icebergs move in straight lines as the striae indicate, although he allowed that river icebergs could produce furrows and striations. James Hall, while prepared to believe that blocks carried in the ice might have cut the grooves in the rock, was however inclined to restrict the occurrence of such a contingency to a limited area.

A notable move away from the iceberg theory and towards that of normal glacial action was taken when Hitchcock gave the first anniversary address to the Boston meeting of the Association of American Geologists (published 1841). In his opening sentences Hitchcock compared the glacial phenomena of Europe and America and concluded that both had such a close similarity that their origins must stem from a common source. There follows a masterly analysis of glacial features, written with such lucidity that it may be left entirely to speak for itself.

Until recently, I confess, I have doubted whether some of the most striking of these phenomena were not much more fully developed here than in most countries of Europe. I refer particularly to the smoothing, polishing, scratching and furrowing of the rocks in place, and to those accumulations of gravel, bowlders, and sand, which form conical and oblong tumuli, with tortuous ridges of the same, and which abound in the northern part of the country, from Nova Scotia to the Rocky Mountains. But the recent investigations and accurate descriptions by Agassiz, Buckland, Lyell, Sefstroom, and others, have satisfied me of the almost exact identity of the facts in relation to drift on the two continents. The resemblance, however, seems to be most complete in this respect between Scandinavia and this country. Except in Sweden, I have not yet seen evidence that the scarification of the rocks is as common in Europe as in New England, where if they were denuded of soil it seems to me, one third of the surface would be found smoothed and furrowed. But it is now found to be very common in Scotland, England, and especially in Switzerland. It appears too, that those countries abound in those peculiar accumulations of gravel and bowlders to which I have referred, and which are now regarded as ancient moraines. Bowlders, also, appear to have been dispersed in a similar manner on both continents. (*Hitchcock, 1841, pp. 247–8*)

Now if I have not mistaken the recent descriptions of European drift, its composition and arrangement correspond with those of the drift of this country; and scarcely any thing seems wanting to make out a complete identity.

It is well known that the theory of drift has for some years been the most unsettled part of geology. The mass of geologists have, indeed, admitted that in some way or other, currents of water have been the principal agency employed, because they witness somewhat analogous effects from aqueous action; and, until recently, no other power of adequate energy and extent has been known to exist. Hence they have been willing to retain the term diluvial, as a generic expression, implying simply aqueous agency in general. Yet so many difficulties attend any theory of mere currents, that many geologists have become sceptical in regard to every particular theory that has been proposed. I confess myself to have been long of that number. Yet it has seemed to me of useful tendency to make isolated inferences from the facts developed and although they may seem to favour rival hypotheses, and will need modification, as new light falls on the subject, yet they will form the elements out of which a legitimate theory will ultimately spring. Allow me to present for your consideration, a summary of the most important of these inferences, as they had been developed to my own mind in examining the diluvial phenomena of this country.

In the first place, these phenomena must have been the result of some very general force, or forces, operating in the same general direction; that is, southerly or southeasterly. For in a southerly direction has the drift been so uniformly carried, and the furrows and scratches on the rocks so generally point southerly, that the force which produced these effects must have tended thither. Our valleys have, indeed, considerably modified the course of the drift; but not enough to contradict the general statement. It would be strange if careful examination should not discover here, as in the Alps and in Great Britain, that the moving force had sometimes been exerted outwardly from the axes of high mountains. But I am not aware that as yet any facts of importance in favor of such an opinion, have been brought to light. At any rate, the evidence of a force urging detritus and bowlders in a southerly, or more strictly in a southeasterly direction, is too marked, and has been noticed by too many independent observers, over a breadth of nearly two thousand miles, to be doubted; even though local exceptions should be discovered; . . . and such a uniformity of direction over so vast an area, indicates a very general agency.

Secondly, this agency has operated at all altitudes, from the present sea level, and probably beneath it, to the height of three thousand or four thousand feet. In New England, most of our hills and mountains, not excepting insulated peaks not higher than three thousand feet, are distinctly smoothed and furrowed on their tops and northern slopes, and upon their east and west flanks to the bottom of the lowest valleys. Dr Jackson supposes he has found transported detritus on Mount Katahdin, four thousand feet high. But he could discover no marks of this action at the summit of the White Mountains of New Hampshire, which are six thousand two hundred and thirty four feet high; although the nature of the rock there, is most unfavorable for preserving furrows and markings.

Thirdly, the smoothing and furrowing of the rocks exhibits almost equal freshness at all altitudes, which indicates an approach to synchronism in the producing cause.

Fourthly, the almost perfect parallelism preserved by the grooves and scratches over wide regions, shows that they were made by the projecting angles of very large and heavy masses of great extent, moving over the surface with almost irresistible force, by water or some other mighty agent. There is sometimes more than one set of scratches, which intersect one another at a small angle, as has been shown by Prof. Locke to occur in Ohio, but each set preserves its parallelism most perfectly. Even where they pass over high and precipitous ridges, they are rarely turned out of their course.

Fifthly, this agency appears to have been less and less powerful as we go southerly. We have as yet, indeed, had but few trusty reports on this subject from the southern portions of North America; but had the phenomena of drift been as striking there as in New England, New York and Canada, they would certainly, ere this, have been described. It ought not to be forgotten, however, that De la Beche has described the drift of Jamaica as very similar to that of New England.

Sixthly, the relative levels of the surface have not been essentially changed by vertical movements, since the epoch in which this agency was exerted. They could not have been much changed without disturbing the detritus, often fancifully arranged in the valleys and on the flanks of the hills; not without sometimes breaking up the smoothed and furrowed surfaces of the rocks along their joints or planes of stratification. But such a disturbance I have never witnessed.

Seventhly, the North American continent must have attained essentially, its present height above the ocean, previous to the exertion of the agency. For all our formations, as high at least as the eocene tertiary, are covered with drift; and I know of no evidence of any important uplift subsequent to that which has tilted up our tertiary strata. This work, therefore, could not have been accomplished while the continent was beneath the ocean. Other evidence of this position might be adduced, did time permit.

Eighthly, water must have been one of the forces employed in this agency. The regular deposits of clay and sand which form the upper part of the diluvial deposit, must surely have been accumulated at the bottom of bodies of water, which have subsequently been drained off. Much, also, of the finer part of our drift is more or less stratified, and exhibits that oblique lamination which is peculiar to aqueous deposits. Nor can I conceive of any other mode in which detritus has been transported hundreds of miles, as ours has been, but by the aid of water; although this alone could not do it. In New England, we have been able to trace erratic blocks not more than one hundred or two hundred miles, because we then reach the ocean. But in the central parts of the country, I am informed by Prof. Mather that the primary bowlders from Canada and the western part of Michigan, are found as far south as the river Ohio; which would make their maximum transit from four hundred to five hundred miles; about the same distance as the bowlders from Scandinavia have been carried into Germany. What agency but water could have effected such a transportation?

It is very natural, also, to ascribe the smoothness and furrowing of the rocks to the action of water. But I have in vain examined the beds of our mountain torrents and the shores of the Atlantic, where the rocks have been exposed to the unshielded and everlasting concussion of the breakers, and can find no attrition that will compare at all with that connected with drift; and I am satisfied that to explain it we must resort to some other agency.

Ninthly, ice must have been another agent employed to produce the phenomena of drift. What else could have transported large blocks and gravel over such a wide space as has been mentioned, and have lodged them too, upon the crests of narrow and precipitous ridges; and especially, what other agent could have produced those singular mounds and peculiar ridges of gravel and bowlders that meet us in so many places?

Tenthly, this agency must have been exerted previously to the existence of man upon this continent, and have been of such a nature as to destroy organic life almost entirely. For the remains of man and other existing animals have not been found in drift; but those occurring there belong chiefly to extinct species, while the deposits of clay and sand made during the same period, scarcely contain a species of animal or plant.

Yet eleventhly, this agency must have been comparatively recent. For the disintegration of the surface of the smoothed and furrowed rocks to the depth of half an inch, would usually obliterate all traces of their erosion. Yet in how many places does this effect appear as distinct as if produced during the present century!

Finally, this agency must have been far more powerful than any now operating upon the globe. In the language of Prof. John Phillips, which he applies to the phenomena of drift in general, 'such effects are not at this day in progress, nor can we conceive the possibility of their being produced by the operation of existing agencies operating with their present intensities and in their present direction'. . . . *Treatise on Geology*, Vol. I, p. 296. (*Hitchcock, 1841, pp. 248–52*)

All these observations Hitchcock had apparently inferred for himself and they had received a welcome stimulation and clarification with the publishing of Agassiz's theory. As he admitted, Agassiz's ice mechanism afforded the first really satisfactory explanation of such apparently dissimilar features as morainic accumulations, furrowing of rocks, and erratics and perched blocks occurring on top of gravel deposits. In the last few lines of the next passages quoted he even appears to accept the idea of ice-sheets.

Beyond such independent inferences as these, I confess I have been of late years unwilling to go; and have regarded the numerous theories of diluvial action, which have recently appeared, only as ingenious hypotheses. But it is well known that the Glacier Theory, originally suggested by M. Venetz, and subsequently adopted by M. Charpentier, and more fully developed of late by Agassiz, is now exciting great interest in Europe. To say nothing of geologists in this country who have expressed themselves favorably towards it; it is surely enough to recommend it to a careful examination, to learn that such men as Agassiz, Buckland, Lyell and Murchison, after long examination, have more or less fully adopted it; although on the other hand, it ought to be mentioned, that such geologists as Beaumont, Sedgwick, Whewell, Mantell and others, still hesitate to receive it.

In a country like ours, where no glaciers exist except in very high latitudes, and with the very defective accounts which have hitherto been given to those in the Alps, it is not strange that the attempt to explain the vast phenomena of diluvial action by such an agency, should appear at first view, fanciful, and even puerile. But the recent work of Agassiz, entitled *Études sur les Glaciers*, gives a new aspect to the subject. It is the result of observations made during five summers in the Alps, especially upon the glaciers; about which so much has been said, but concerning which so little of geological importance has been known. Henceforth, however, glacial action must form an important chapter in geology. While reading this work and the abstracts of some papers by Agassiz, Buckland and Lyell, on the evidence of ancient glaciers in Scotland and England, I seemed to be acquiring a new geological sense; and I look upon our smoothed and striated rocks, our accumulations of gravel, and the tout ensemble of diluvial phenomena, with new eyes. The fact is, that the history of glaciers is the history of diluvial agency in miniature. The object of Agassiz is, first to describe the miniature, and then to enlarge the picture till it reaches around the globe. (*Hitchcock, 1841, pp. 252-3*)

In the first place, [The Glacial Theory] explains satisfactorily, the origin of those singular accumulations of gravel and bowlders, which we meet with, almost everywhere, in the northern parts of our country. I cannot doubt that these are ancient moraines; just such as exist in Scotland and England. Were this the proper place, I could point out a multitude of localities of these, most of which have been a good deal modified by subsequent aqueous agency; but some of them retain the very contour which they had as the ice melted away. The lateral moraines are perhaps most common, especially if, with Dr Buckland, we regard our terraced valleys as modifications of these; but I am confident that in our mountain valleys, the terminal and the medial moraine are not infrequent. I have long been convinced that the agency of ice was essential to explain these accumulations; but I was not aware that their antitypes existed in the moraines of the Alps.

In the second place, this theory explains in a most satisfactory manner, the smoothing, polishing and furrowing of the rocks at different altitudes. All these effects are perfectly produced beneath the glaciers in the Alps, nor can I conceive of any other agent by which the work could be executed. It certainly was not done by currents of water alone. One has only to cast his eye upon the splendid plates by Agassiz, of the polish and striae produced by the glaciers, to be satisfied that the multitudes of examples of analogous phenomena in New England, and in New York and Ohio, as described by Profs. Dewey, Emmons, and Locke and Dr Hayes, are precisely identical with those in the Alps.

In the third place, it explains the transportation of bowlders, and their lodgment upon the crests and narrow summits of mountains, and that often without having their angles rounded.

In the fourth place, it accounts for the occurrence of deposits of clay and sand above the drift. For it furnishes the requisite quantity of water to fill the valleys, and the means of damming up their outlets for a season.

In the fifth place, it shows us why these deposits of clay and sand are almost completely destitute of organic remains, either of animals or plants, although probably centuries must have been consumed in their formation.

In the sixth place, it accounts for some rare and peculiar phenomena connected with diluvial action, which seem to me inexplicable on any other known principle. I shall name only two. The first is, that the northern slopes of some of the mountains of New England, although quite steep, and their summits rounded, exhibit striae and furrows which commence several hundred feet below their tops, and pass over them without losing their parallelism; and yet the situation of the drift shows that these markings were made by an ascending and not a descending body. Such might be the effect if the whole surface of the country were covered by a thick sheet of ice expanding in a southerly direction. (*Hitchcock, 1841, pp. 254–6*)

Yet a little farther on it is clear that he relates ice to mountains only, for he seems to feel that the occurrence of morainic ridges far away from any mountains partly negatived the glacial theory.

Another difficulty results from the fact that some of the most remarkable of our moraines are found, not in valleys, but on the sea coast, some of them fifty and others one hundred miles distant from any mountains much higher than themselves. I refer to those remarkable conical and oblong tumuli of drift, sometimes more than two hundred feet high, which occur on Plymouth and Barnstable counties in Massachusetts. I see nothing in this theory that will explain such astonishing accumulations in such circumstances; and yet their existence may not militate against its truth. For even the present mighty glaciers of the Alps, may give us but a faint idea of the effects of the advance and retreat of a sheet of ice thousands of feet thick. We have no evidence in this country, that any of our mountains have been elevated since the glacial epoch; as seems to be proved to have been the case with the Alps, and this circumstance may have produced a considerable modification of glacial action on this continent. (*Hitchcock, 1841, pp. 257–8*)

This paper is a landmark in the American acceptance of the glacial theory. Hitchcock, if not completely satisfied on all points, recommended the main principles of Agassiz's theory. He objected to Lyell's iceberg theory and also to Agassiz's notion of the polar ice accumulations spreading south and later melting during a warmer period, to produce floods of water by means of which the icebergs would be transported. Though convinced of ice action in some form, Hitchcock is doubtful and hesitant about accepting the full implications of either of the rival glacial theories.

This qualified adoption by Hitchcock was echoed later in a paper by Maclaren who acknowledged that the theory resolved many of his doubts about moraines, striations, roches mountonnées, erratic blocks and fluvioglacial

deposits. Although he believed it also provided the answer to the origin
of the newer alluviums, he was still prepared to accept that the older deposits
had been laid down during the flood.

> Glaciers are properly long narrow masses of ice filling the bottom of Alpine
> valleys, but M. Agassiz thinks that sheets of ice, such as are met with in
> Greenland, covered the whole surface of Europe, and all Northern Asia as far
> as the Caspian Sea. This conclusion, which has been adopted in whole or in
> part by Professor Buckland, Mr Lyell, and other eminent geologists, has been
> deduced from a careful study of the phenomena attending glaciers, some of
> which are of so marked and peculiar a kind, as to afford satisfactory evidence
> of their ancient existence in situations where none are now seen. (*Maclaren,
> 1842, pp. 346–7*)
>
> Even though M. Agassiz's opinions should not be fully established, they
> still afford us a new geological agent of great power and widely applicable,
> which may help us to an explanation of some phenomena very difficult to
> account for with our existing means of information. (*Maclaren, 1842, p. 347*)
>
> These very original and ingenious speculations of Professor Agassiz . . .
> cannot be considered as fully established till they have been brought to the
> test of observation in distant parts of the world, and under a great variety of
> circumstances. Supposing the theory to be substantially sound, the magnitude
> of the consequences it involves will undoubtedly bring objections to light,
> which may render modifications necessary, both in its principles and its
> details. In the mean time, it assists us in resolving some difficulties and it
> promises to throw light on what is at present a very obscure subject, the origin
> of the older and newer alluvium. (*Maclaren, 1842, p. 365*)

As in Europe where Sedgwick and Cuvier may be regarded as intelligent
opponents of the glacial theory, so in America many geologists with catastrophic or modified diluvialist or typical uniformitarian views were not satisfied with the authenticity of the new theory. H. D. Rogers, the structural
geologist, exemplifies well this intelligent reactionary approach in the United
States. He was far from unobservant but he viewed the evidence in a light
unfavourable to glacial theories. He commenced, as did Hitchcock, by defining and analysing the main features of the drift topography. One important
difference which he discovered between the two continents was that in Europe
much of the glaciation has been produced by glaciers moving outward from
the principal mountain valleys in a radial pattern while in America the action,
by whatever force was responsible, had moved in one general direction. This
afforded the diluvialists the argument that though the glacial explanation
might be applicable to Europe the circumstances in America were dissimilar
and another cause must be sought.

> 1st, The smoothed and furrowed surface is coextensive or nearly so with the
> drift stratum, and it occurs at all altitudes, from the summits of the loftiest
> mountains of New England and New York, to the beds of the valleys, and
> over the whole broad plain of the lakes and the western states. In the moun-
> tainous and hilly tracts, the northern and northwestern brow and flank of
> each eminence, are much more smoothed and striated than the opposite.

The scratches do not radiate from the high mountain summits, but in the vast plains and prairies of the west, among the confused hills of New England, or on the transverse mountain crests of northern Pennsylvania, and western Vermont and Massachusetts, they maintain invariably in all the higher levels a general southeasterly direction. In lower situations, however, on the slopes of the great drainage valleys, their course is diverted to conform more nearly and sometimes with exact parallelism to the direction of the natural barriers and channels. (*Rogers, 1844, p. 264*)

2nd. Respecting the drift itself, the following appear to be the principal phenomena:

Throughout all the northern tracts of the United States and the adjoining districts of the British provinces, the surface is covered with a loose stratum composed of sand, clay, gravel and boulders of all sizes, variously mingled and locally stratified.

The stratification is characterized by plains of inclined and confused deposition, denoting turbulent currents.

The materials invariably belong to formations lying north or northwest of their present positions, and great spaces occupied by broad plains, wide belts of hills and even mountains, deep valleys and vast sheets of water, intervene. The bowlders have evidently not radiated from any local centres of dispersion. (*Rogers, 1844, p. 265*)

3rd. The third class of facts connected with the drift, relate to the proofs of a lower level in the land at the epoch of its production. In describing the post pleiocene blue clay of Lake Champlain and other northern valleys, I have already cited the proofs that at one period at least in the general era of the drift, the surface of the country in the region of New York and the St Lawrence was lower in level than it now is by as much as perhaps five hundred feet. (*Rogers, 1844, p. 266*)

When Rogers had to suggest a mechanism which had brought into being all the various phenomena, it becomes plain that he was adequately acquainted with the main modern theories, but, as a structural geologist, naturally adopted the hypotheses of Sedgwick and Hopkins, together with all their jumbled apparatus of catastrophes and waves of translation.

The principal hypotheses proposed for explaining the detrital phenomena, are –

First, the theory which attributes the scratches on the rocky floor of the drift, and dispersion of the far-carried fragmentary materials, to the agency of ice, creeping forward with a slow velocity but an enormous momentum, like the glaciers of the Alps, grinding down and finely grooving the jagged asperities of the surface, and bearing on its back the collected rubbish in the mountain slopes, and strewing this still further by a rapid thaw:

Secondly, the theory which imputes the whole to icebergs, loaded with detrital matter, and floating southward, until stranded on the surface of the submerged land, which the ice-fields are conceived to have smoothed and scored through the agency of innumerable fragments frozen into their lower surfaces:

Thirdly, the theory which supposes no general permanent submersion of the land, but imagines one or more paroxysmal movements of the earth's crust in the higher northern latitudes to have sent a portion of the contents of the Arctic seas – water, ice, and fragmentary rock – in a succession of tremendous deluges southward across the continent. (*Rogers, 1844, p. 268*)

My brother and myself entertaining very similar objections to the explanation of the phenomena by icebergs, have ventured farther, and perceiving no necessity for supposing that the cutting fragments and particles were ever pressed upon by ice, have appealed to the enormous erosive power which a thick and ponderous sheet of angular fragmentary rock would possess, if driven forward at a high velocity under the waters of a deep and general inundation, excited and kept in motion by an energetic upheaval and undulation of the earth's crust during an era of earthquake commotion. (*Rogers, 1844, p. 269*)

Let us now give our attention for a moment to the paroxysmal theory which I cannot but think will be found, on careful examination, to be more in agreement with admitted laws of physical dynamics than either of the more popular hypotheses of the day. This doctrine . . . merely supposes that at the epoch of the drift, the polar half of the northern hemisphere was the theatre of violent and perhaps frequently repeated movements of the earth's crust, each particular disturbance emanating probably from a different local region. These disturbances, which are conceived by Von Buch, De Beaumont, Hopkins, De la Beche, Sedgwick, Phillips, and other distinguished geologists, to have been of the nature of simple paroxysmal elevation, and by my brother and myself to have consisted in an energetic and extensive undulation of the crust of the earth accompanying each sudden rise, are deemed sufficient to have caused a rush of the northern waters over all the higher latitudes of Europe and North America, covering the surface with an almost continuous sheet of gravel and bowlders, and polishing and scoring the whole rocky floor. (*Rogers, 1844, pp. 273–4*)

It has been shown by Mr Hopkins, of Cambridge, reasoning from the experimental deductions of Mr Scott Russell upon the properties of waves, that 'there is no difficulty in accounting for a current twenty five or thirty miles an hour, if we allow of paroxysmal elevation of from one hundred to two hundred feet', and he further proves that a current of twenty miles an hour ought to move a block of three hundred and twenty tons, and since the force of the current increases in the ratio of the square of the velocity, a very moderate addition to this speed is compatible with the transportation of the very largest erratics any where to be met with, either in America or Europe.

Holding in view these demonstrable conclusions, let us consider the far more enormous velocity which a broad general current would derive from that mode of paroxysmal action, earthquake undulation, which constitutes, as my brother and myself have endeavoured to show, an essential feature in all movements of elevation. (*Rogers, 1844, pp. 274–5*)

Rogers' attitude to the glacial theory was typical of that of the majority of American geologists in the early 1840's. What was needed most in America to convert the waverers to the new theory was the stimulus of the glaciological master himself. Agassiz had already had a striking personal influence on some British diluvialists when in 1846 he came to Boston to lecture at the Lowell Institute. Finally he made America his home.

American geology now possessed all the accoutrements requisite to a successful and progressive development. The publication of *Silliman's*

Journal in 1818; the founding of a Geological Society in 1819; the establishment of a Geological Survey in 1836; the visits of Lyell and Agassiz, what better auguries could be asked of the future? Even if Hutton was still almost ignored, the appearance of native geologists of the calibre of Hitchcock was full of hope. American geology was destined to lag behind that of Europe for a few more decades but the seeds of change had already been sown and the true beginnings of the later remarkable spurt to world predominance may be traced back to these years.

CHAPTER FIFTEEN

Early Quantitative Geomorphology

During the first half of the nineteenth century many new methods of examining physical features were attempted, such as Robert Everest's measurements of the Ganges' alluvium. These calculations were less important for their findings than for the example they set in showing the necessity for 'accurate' observation and measurement. Such scientific assessments were applied, as today, both to surface forms and to physical processes, for the use of accurate ground survey is as vital as the quantitive measurements of processes. Efforts in these methods were relatively uncommon yet they commanded much attention and reverence partly because they remained unassailable until tested by other measurements. Uniformitarianism was in principle founded on – and sometimes misguided by – the study of actual processes.

Here we can do little more than exemplify the kind of progress made in quantitative geomorphology between 1800 and 1845. There is perhaps no need to emphasize that the authors themselves did not always appreciate the future importance of their studies and that people such as the Rev. R. Everest put forward the results of their inquiries merely as simple scientific recordings. The theorists used these to reach their own generalizations, especially the uniformitarians, who were particularly keen on existing rates of erosion. Lyell's use of the calculation that the chalk cliffs of eastern England were retreating one yard annually (p. 167) fitted in well with his emphasis on marine erosion. He also discusses the velocity of water necessary to move fine clay, sand and stones and quotes liberally of statistics relating to river volume and amount of debris carried by the Rhine, Ganges and Yellow River (pp. 162–3). Yet he never fully appreciated the potency of river erosion. Conybeare, a diluvialist, actually used the slow rates of surface degradation, which scarcely affected the ramparts of prehistoric encampments in 1700 years, to disprove fluvial theories (p. 153). Hopkins, a non-glacialist, was attributing the topography of the Lake District to fracturing and to the cutting of valleys by marine floods when he propounded the law that 'the force of a current increases in the ratio of the square of its velocity' (p. 226).

But the fact remains that a careful measurement is in itself a worthwhile acquisition and that any advances in quantitative geomorphology are of considerable importance. The following calculations by Everest show the importance of the annual amount of material carried by a river; the extent of seasonal and daily fluctuations in the river's carrying power; and the differences in the river's velocity at its centre and near its banks. Such information is essential to a clear understanding of fluvial processes and forms.

In the course of last summer, I made some attempts to ascertain the weight of solid matter contained in a given quantity of Ganges water, both in the dry and rainy season, but I found the weight so variable on different days (when little difference might have been expected) that I can hardly consider the observations numerous enough to give a correct average. Such as they are, however, they may not be without interest in the absence of other information on the subject . . .

1. A quantity of Ganges water taken 27th May, 1831, gave when evaporated, a solid residuum of 1·084 grains per wine quart.

2. July 21st. There has been little rain for some days, and the river was low for the season: a wine quart contained of soluble matter 2·0 grains; of insoluble 16·2; – total 18·2.

5. August 20th. The water had hitherto been taken from the side, but as it was evident that the quantity of matter held in suspension in the middle of the current was much greater than towards the bank, where the water was nearly still, I took two separate portions as before, and obtained, from the middle, 40 grains of insoluble residuum; from the side 20 grains ditto: add for soluble matter, suppose two grains to each, the middle gives 42, the side 22 grs. The river today was at the same height as on the 13th (the maximum). (*Rev. R. Everest, 1832, p. 238*)

34 grains per wine quart was found to be the average for the rains. Now as the wine quart of water weighs 14544 grains, we have about 1/428th part of solid matter by weight. But as the specific gravity of this cannot be stated at less than 2, we have 1/856th part in bulk for the solid matter discharged, or 577 cubic feet per second. This gives a total of 6,082,041,600 cubic feet for the discharge in the 122 days of the rains:—7·8 grains per wine quart was the weight determined for the five winter months or 1/1838th part in weight, and 1/3676th part in bulk, which gives 19 cubic feet per second, or a total of 247,881,600 cubic feet for the whole 151 days of that period:—3·8 grains per wine quart was the weight allowed for the three hot months, which gives a 1/3827th part by weight, and a 1/7654th part by bulk, or about 4·8 cubic feet per second for the discharge of solid matter, and a total of 38,154,240 cubic feet for the discharge during the 92 days. The total annual discharge then would be 6,368,077,440 cubic feet. (*Rev. R. Everest, 1832, p. 241*)

Leonard Horner, Lyell's father-in-law, conducted similar experiments at Bonn on the Rhine which produced equivalent results.

I made two sets of observations, the one in the month of August, and the other in November. The apparatus I used was very simple, but answered the purpose perfectly. . . . It consisted of a stone bottle, capable of containing about a gallon, furnished with a cork covered with leather, and greased; a weight of about 10 lb. was attached to the bottom of the bottle by a rope, of such a length, that, when the weight touched the ground, the mouth of the bottle might be at the desired distance from the bottom of the river. A rope was attached to the ear or handle of the bottle, by which it was let down, and a string was fastened to the cork. As soon as the bottle had reached its destined position, the cork was withdrawn by means of the string, the bottle became filled with the water at that particular depth, and was then instantly drawn up. The water, as soon as drawn up, was emptied into glass jars, on which I had previously marked a certain measure. The quantity of water on which I intended to operate, was a cubic foot, or 1000 ounces, and I collected it at different times; for instance, after one-third of a cubic foot had stood in the jars for some days, I drew off the clear water with a syphon, and another third of water, fresh taken from the river, was added to the sediment left at the bottom of the jars from the first; that was allowed to stand, the clear water was again drawn off, and the last third was added in the same way. When this had stood a sufficient length of time, the accumulated sediment was removed to an evaporating dish (a common saucer will do quite well), and carefully dried in a gentle heat. The dried mass was the amount of solid matter held in suspension in a cubic foot of water, and now in the state of indurated mud. (*Horner, 1834–5, p. 103*)

(First observation – August, 165′ from the left bank, 6′ from the river bottom.) A cubic foot of distilled water weighs 437500 grains, therefore the solid matter amounted to $\frac{1}{20734}$ part of the cubic foot of water. (*Horner, 1834–5, p. 104*)

(Second observation – November, Middle of river, 1′ below the surface.) The residuum, when dried in the same manner, weighed 35 grains, which is $\frac{1}{12500}$ part of solid matter in one cubic foot of the water. (*Horner, 1834–5, p. 104*)

The above experiments shew, that the quantity of solid matter suspended in water, which, in the mass, has a turbid appearance, may be very trifling. But the extent of waste of the land, and of the solid materials carried to the sea, which even such minute qualities indicate, is far greater than we might be led to imagine possible from such fractions. It is only when we take into account the great volume constantly rolling along, and the prodigious multiplying power of time, that we are able to discover the magnitude of the operations of this silent but unceasing agency. In the absence of more accurate data for my calculations, for the sake of shewing how large an extent of waste is indicated by water holding no more solid matter in suspension than is sufficient to disturb its transparency, I shall assume that the Rhine at Bonn has a mean annual breadth of 1200 feet, a mean depth throughout the year of 15 feet, and that the mean velocity of all parts of the stream is two miles and a half per hour. These assumptions are probably not far distant from the truth. I shall take the average amount of solid matter in suspension to be 28 grains in every cubic foot of the water. (*Horner, 1834–5, pp. 104–5*)

We have 145,980 cubic feet of stone carried down by the Rhine past the imaginary line every twenty-four hours, . . . a mass greater in bulk than a solid tower of masonry sixty feet square, and forty feet in height. If we multiply 145,980 by 365, we have 1,973,433 cubic yards carried down in the year, and if this process has been going on at the same rate for the past two thousand years, – and there is no evidence that the river has undergone any material change during that period, – then the Rhine must in that time have carried down materials sufficient to form a stratum of stone of a yard thick, extending over an area more than thirty-six miles square. (*Horner, 1834–5, p. 106*)

Besides these lone precursors of the quantitative school many useful discoveries were made about the processes concerned in the shaping of river beds. J. Yates in an informative paper to the Geological Society distinguishes the various sources of the alluvial deposits and traces how they reach their present positions and assume their present shapes. Weathering he recognized to be an active agent of distintegration which supplied most of the material transported by the rivers. He also appreciated that the rivers in turn sorted this material out, carrying the finer particles in suspension; rounding and rolling the larger fragments along the bottom. Another aspect which he noted was the effect of changes in the level of the river bed and to this cause he attributed the occurrence of alluvial cones and shallows wherever a tributary at a higher level joined the principal stream.

I. – He considers first those processes of disintegration, not dependent upon the action of running water, by which materials are supplied for the formation of alluvium. These are of two kinds.

1. – Earthquakes and landslips, by which large masses are detached suddenly from the mountains, and fall occasionally with so great an impetus as to extend across valleys.

2. – Other processes, such as frost and oxidation, which are far more important in their effects. The agents of this class always divide rocks according to their natural structure of separation, so that every fragment of the debris is bounded by the plane of its cleavage. (*Yates, 1830–1, p. 237*)

II. – The materials thus furnished are distributed by streams, which round off their angles by continual friction, so as to convert them into pebbles, sand and mud. The hard and heavy fragments driven along by streams, also wear down the rocks in place, the latter being acted upon according to their degrees of softness and their proneness to disintegration.

When the detritus thus produced is discharged from a lateral into a principal ravine, or valley, the divergence of the stream gives it the form of a cone; but as the force of running water carried loose materials much further than they would fall by their own weight, the form thus produced is not an acute but an obtuse cone. (*Yates, 1830–1, p. 238*)

III. – Whenever detritus is conveyed by running into standing water, a separation takes place between those finer particles which are held in suspension, and those which it only rolls along the bottom. (*Yates, 1830–1, p. 238*)

IV. – When two streams meet, they neutralize each other's motion, and a deposition takes place at the point of quiescence.

Peculiar appearances ensue, when streams meet at different levels. If a lateral stream brings down a disproportionate quantity of detritus, its bed is raised, but is abruptly terminated by the action of the principal stream. Hence the valleys of mountainous regions exhibit not only level terraces formed in lakes, but others the edge of which have a steep declivity. (*Yates, 1830–1, p. 239*)

A more detailed work by the French civil engineer Alexandre Surell takes Yates' point about irregularities in the surface several stages further. He in fact gives a clear description of the longitudinal profile (or thalwegs) of mountain streams and explains their formation by what the modern geologist now knows as the principle of 'grade'. His study of torrents taught him that streams tend to attain a profile which geometrically can be represented by a critical curve made up of a succession of slopes whose aggregate provides the most efficacious discharge. To this end he believed that streams are constantly eroding their higher reaches and depositing upon their lower courses. He appreciated that where river beds differed in the form of their longitudinal profile this was because, although in each case the process was the same, the stage reached represented a different point in the river's history. This principle that a river is continually working towards an adjusted slope is important because it demonstrates that a river is not merely an instrument of either erosion or of deposition but an active organism which can bring about changes in the composition of its member parts. For instance by a change in the river bed it may at one moment erode through its own deposits and at another time lay down deposits across a denuded area of its own eroding.

> If ascending the course of a torrent we continued the levelling measurements from the alluvial cone into the gorge . . . to the source and traced the complete curve we would verify the fact . . . that streams beds form a curve, convex towards the centre of the earth and increasing gradually in curvature upstream.
>
> At the same time as we traced the stream profile up the gorge, we could measure the height of its steep banks; we could also measure the level of the plain in the midst of which the alluvial cone is spread. This method supplies all the elements for a new curve, which gives the relief of the terrain crossed by the torrent. If the two curves (the stream profile and the terrain profile) are superimposed, they provide a diagram from which we can see perfectly clearly the nature of the effect that torrents have upon the ground, throughout their entire length. In the upper course the curve of the terrain is above that of the thalweg; in the lower course, it is below it. Consequently the two curves must cross each other, and this point of intersection marks the transition from washing away to building up. It is at the downstream end of the gorge and at the top of the alluvial fan (*Fig. 52*).
>
> We see that the water, compelled at first to follow the relief of an uneven terrain, has gradually destroyed the irregularities of the slopes. It has lowered certain points and raised others. Here it has eroded; there it has deposited. The obtuse (*rentrant*) angle formed by the mountain slope and the level plain has been smoothed out as the deposition proceeded and the water has substituted in this stretch a curved line for a stepped line. The result of all these actions has been to form a new stream-bed profile which is better suited to the flow of the water than was the original profile of the terrain.
>
> We should note particularly that it is not only a question of the removal of various roughnesses, planed off by the friction of the water. . . . The steep banks of the gorges (at the exit from the catchment basin) are excavated sometimes to a depth of 100 metres; the alluvial deposits form hills with summits often more than 70 metres above the adjacent plain. It is on these statistics that we must judge the enormous changes that torrents can produce in their longitudinal profiles.
>
> Considered from this point of view, torrents form a subject for useful comparisons. It is impossible to doubt that the creation of their longitudinal profile is entirely their own work. Therefore they allow us to understand a general phenomenon, which in other streams is difficult to grasp but which with them is obvious. . . .
>
> Water flows in the bed of a torrent according to the same laws as in the bed of larger rivers. The longitudinal profile of a torrent is no different from that of a river or of any other stream, except that we would have to reduce the

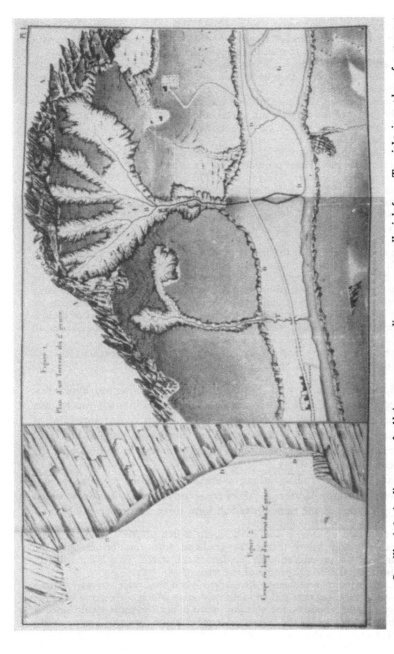

FIG. 52. *Surell's (1841) diagrams of Alpine torrents grading out on to alluvial fans. To right is a plan of a torrent (not yet graded). To left, orientated at right angles to the plan, cross-sections of the topography and stream bed are superimposed to show incipient torrent traversing slope of scree debris (AB); main torrent flowing in gorge in bedrock (BD); and main torrent crossing alluvial fan deposited on floor of main valley (DD)*

horizontal scale while retaining the vertical scale unaltered. It is the relation between the abscissa and the ordinate which varies but the characteristic properties of the curve remain the same. (*Surell, 1841, p. 21*)

An area which we must specially study in the thalwegs of torrents is that where raising by deposition begins, or the interesection of the terrain and stream-bed curves.

In some torrents, the smoothness of the curve is unbroken in this transition and the slopes of the deposits are joined at a tangent with those of the gorge; so well that nothing on the curve, considered alone, would show the point where the building up begins.

In others, on the contrary, the profile curve is broken there in a more or less abrupt manner. These provide us with an example of a bed with a curve that has not yet been completely formed. The gradients are imperfect and the process unfinished. We immediately conclude that in such torrent's building up must be proceeding very energetically, whereas that process, in the former, has already been virtually completed, being no longer stimulated by the same causes and having no longer the same end to attain. This in fact is what actual observation shows.

We also notice that in the first kind of torrents the slope of the deposits is such that materials brought there would be swept along to the river, if the water did not disperse them in its wandering. We can be sure of this fact by comparing that slope with those of other torrents which roll along the same kind of materials but no longer deposit them solely because they can no longer wander either as a result of artificial works or of the effect of natural circumstances. Moreover, we believe that a similar slope must exist for all kinds of materials; we will call this the 'limiting slope'.

In the second type of torrent, the 'slope limit' is not yet reached; it is still being developed. (*Surell, 1841, pp. 22–23*)

Surell, on formulae derived from contemporary treatises on hydraulics, computed the approximate speeds reached by torrents in spate (up to over 14 m/sec) and the size of the blocks that such torrents could move. One of the Alpine torrents had in a recent flood left about one hundred blocks of 30 cubic metres and some blocks up to 60 cubic metres in volume upon its alluvial cone.

However, his general observations on stream profiles are more important. In all rivers, he suggests, there are three consecutive periods or stages, which succeed each other in the same order and form three distinct epochs.

1st. A period of erosion and building-up which prepares the bed of the thalweg and everywhere brings the gradients into equilibrium with the resistance of the ground and the friction of the water. This is intended to establish the longitudinal profile of the stream.

2nd. A period of wandering (*divagation*) when the water strives to attain the shape of cross-section and windings of course which correspond with the greatest stability (because the straight curve is not the most stable since it does not necessarily lead the current over parts where the bank is most solid). Here, the action of the water is restricted to moving the ill-defined thalweg to and fro on the same plane, without sweeping away nor building up notably its bed; it is the liquid mass which is displaced rather than the ground. The

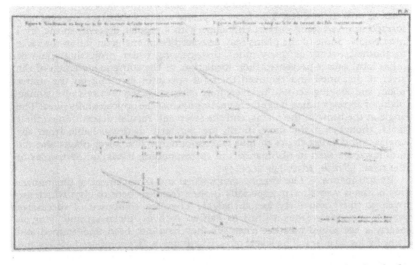

FIG. 53. *Surell's longitudinal profiles of recent Alpine torrents, showing broken thalwegs (lower lines)*

result of this period is to fix the alinement of the course, or, in other words, to determine the plan.

3rd. Finally, a period of grade (*régime*) when the water overflows and, subsides again into an unchangeable bed. (*Surell, 1841, pp. 126–7*)

Rather different quantitative methods of (crude) survey and the analysis of the layer-bands of deposits were made by A. Smith, an American writer who contributed some useful observations on the formation of terraces. He noticed in one area of the Connecticut river that several of these occurred one above the other and the similarity of their curvature with that of the present river led him to conclude, rightly as it happens, that they were produced and then abandoned by the progressive lowering of the river. He also noted that many of these terraces were made up of layers of sand and clay, each layer being distinguishable by an accurately sorted gradation from granular sand to the finest clay. Smith, who writes before the general acceptance of the glacial theory, suggested the (varved) deposits were laid down in lakes held in by rocky barriers. With the exception of the supposition of rocky barriers this was correct in all details. He accounted for the release of the water by supposing that the barriers subsided and were breached by the pent-up waters.

Arranged in terraces. – The plains and meadows are generally so level, that their declination is not perceived by the eye. Three distinct terraces of plain and meadow, in some portions of the valley four, rise to various heights, from fifteen or twenty to two hundred feet, above the surface of the Connecticut.

The upper terrace of plains extends to the hills whether near or distant, and rests against their sides. The faces of plains towards the river are generally formed into sweeping curves, and slope with a regular descent, to the level next below. Some of the plains may be traced from one great fall in the river to another, twenty or thirty miles, disappearing where projecting spurs or other local causes prevented their formation, or the subsequent action of the river undermined and removed them. A spectator standing on the upper plain, and looking across the river, may commonly see portions of a similar plain, of corresponding height terminating against the opposite hills sides. The angle at the brow of the plains, and the steep and regular descent down their faces, though not the same in all, are perfectly distinguishable from the angles and curving slopes of hills and mountains. The lower plains and the meadows, are seen in like manner, in corresponding levels, on both sides of the river. (*Smith, 1832, pp. 214–15*)

By subsidence. – The interior composition and arrangement of the numerous plains is very similar, especially in their frequent beds of clay, which are twenty, thirty and forty feet in height. They are composed of successive layers, commonly from a third to half an inch in thickness, and lying in nearly a horizontal position wherever they have not been undermined and bent by local causes. The layers of clay beds are in fact composed of clay and quick-sand. The order in which the materials of each layer are arranged is invariable, and may be most distinctly seen wherever the clay is highly colored. At the bottom of the layer is the coarsest, heaviest, and least colored portion of the sand. In the center is the finer sand, with an intermixture of clay and deeper color. The top of the layer consists of the finest clay, and is the most highly colored. Beds of clay, consisting of fifty, a hundred, and even more such layers, are found in plains, and sometimes passing under the river, in extensive portions of the valley. (*Smith, 1832, p. 215*)

In ancient lakes. – The facts which have been stated must lead the way in accounting for the formation of the plains arranged in layers, situated so far above the highest inundations of the river. No one, who considers the power of currents and falling water, will maintain that the Connecticut river has not lowered its bed, during a lapse of several thousand years. At some former period, a chain of lakes must have possessed the centre of the valley, connected by streams falling, as now at the rapids of the river, or more abruptly, from the level of higher lakes, to the level of others below. The elevation of the lakes was at least as great, in different divisions of the valley, as that of the existing plains in the same divisions. The hills which press upon the water's edge, on either side of the Connecticut, at White river falls, were probably, at some former period, united by an unbroken ledge of rocks, which lying across the present channel, formed the lower brim of a long basin, and sustained the waters of a lake reaching to the Fifteen mile falls of varying width, corresponding with the distance of the hills which formed the boundary on opposite sides. At Bellows falls was, probably, another rocky barrier, sustaining a lake thirty-five miles long, extending to White river falls. Similar barriers existed below, probably one near Brattleborough, and others in succession, as far down as Middletown, below which the higher plains disappear from the course of the river. (*Smith, 1832, p. 216*)

The lakes become a river. – It may be inferred, from the level surface of the plains, that the lakes in which they were formed were filled with earth before, perhaps long before, the barriers gave way . . . that the lakes disappeared, and a river flowed through the full plains, wearing away one side, and casting up the earth in eddies on the other, as at this day among the meadows . . . that what is now the plain was once meadow. Hence the sweeping curves and regular slopes of the faces of the plains. When a barrier gave way, the river would deepen its channel throughout the whole extent of such a plain. There is no such thing in nature as a rapid stream, running upon a bed of clay or sand. Currents seize the finer earth with which they come in contact, bearing it along, until thrown aside into some eddy, or meeting with other tranquil waters, it is again deposited. Hence when, by the breach of a barrier, the river took possession of a deeper bed at the outlet of the basin, its channel through the whole extent of alluvial plain, would be deepened in like proportion. Thus the lakes disappeared, leaving Connecticut river in their stead. (*Smith, 1832, p. 217*)

There are many physical processes in landscape formation which are notoriously difficult to assess without undertaking a long series of elaborate observations. Among these is weathering, the importance of which was already beginning to be realised in the early nineteenth century. Both Sedgwick and Phillips wrote papers emphasising how this process could produce changes not only in building surfaces but also in the natural topography.

The columnar structure of basalt he rightly referred to shrinkage, and showed that some of the curious forms produced among the Granite Tors should be referred, not to concretionary action, but to the peeling off of the exposed surfaces, urging in illustration the fact that 'ancient pillars of granite have been known to exfoliate in cylindrical crusts, parallel to the axes of the pillars; and even pillars of oolitic limestone, which unquestionably have no spheroidal structure, sometimes exfoliate (e.g. in the second court of Trinity College, Cambridge) in crusts parallel to the axes of the several pillars'. (*Clark and Hughes, 1890, II, p. 100*)

The rapid waste occasioned by fluctuations of heat and moisture is next examined; and it is shown that the south and west fronts of buildings suffer most by these variations; that when the composition of the stone is unequal, the waste of its surface corresponds in general to the nature and arrangement of the particles; but that also there are cases when the atmospheric influences cause an exfoliation of the surface, without reference to the internal arrangement of the particles. (*Phillips, 1831, p. 323*)

The power of frost in connection with other agents is then noticed as very important in producing the fall of mountain precipices; and the author concludes his paper with a description of some remarkable excavating effects of rain on the surfaces of ancient monumental stones and bare limestone rocks. He endeavours 'to show, that within the historic aera hard and durable stones have been greatly furrowed by the rain, and that in more ancient periods the precipitations from above have carved themselves channels of various kinds, and sometimes occasioned real though miniature valleys of great length and continuity'. (*Phillips, 1831, p. 323*)

Finally, we may draw attention to the rare attempts to visualize wide-sweeping summit-levels and so to reconstruct the surfaces of landscapes older than the present. Certainly one of the most striking suggestions of this nature was that made by S. P. Hildreth in 1836 when speaking of the plateau of western Pennsylvania:

> In any part of this region the view from the highest hills, presents one vast plain filled with hollows, and affording no spot much, if any more elevated, than the one on which the spectator stands, – bringing forcibly to the mind, the reflection, that this now hilly and broken region was once, at some remote period, a level, and nearly horizontal plain. (*Hildreth, 1836, p. 7*)

Hildreth was the first to notice the Schooley erosion surface, though quite unconscious that his discovery bore within it the seed of Davis's uplifted peneplain theory. Percival made a similar observation regarding the upland areas bordering the Connecticut valley some six years later:

> The eastern and western primary may both be regarded as extensive plateaus, usually terminated abruptly toward the larger secondary basin, but sinking more gradually towards the south, on the sound. These plateaus present, when viewed from an elevated point on their surface, the appearance of a general level, with a rolling or undulating outline . . . interrupted only by isolated summits or ridges, usually of small extent. (*Percival, 1842, p. 477*)

All these ideas which at the time of their publication were very much on the fringe of the main geologic current are interesting because they show how the beginnings of many important theories take root in ideas conceived, and considered insignificant, many decades before their fruition. In the case of the previous writers, each adopted an approach to the subject which, though not recognized in his time, was to provide a principal stepping stone to the advancement of a new branch of geomorphology.

References: Part Two

ANON. (1820) 'Reflections on the Noachian deluge' *Phil. Mag.*, Vol. 56, pp. 10–14.
— (1824*A*) Review of the 'Outlines of the Geology of England and Wales' by W. D. Conybeare and W. Phillips. *Amer. Jour. Sci.*, Vol. 7, pp. 203–40.
— (1824*B*) Review of the 'Reliquiae Diluvianae' by the Rev. William Buckland. *Amer. Jour. Sci.*, Vol. 8, pp. 150–68 and 317–38.
— (1826) 'Proofs that general and powerful currents have swept and worn the surface of the earth' *Amer. Jour. Sci.*, Vol. 11, pp. 100–4.
— (1834) 'Remarks on the connection between the Mosaic history of the Creation and the discoveries of geology' *Amer. Jour. Sci.*, Vol. 25, pp. 26–41.
AGASSIZ, L. J. (1840*A*) 'Études sur les Glaciers' (Neuchâtel), 346 pp.
— (1840*B*) 'On glaciers and boulders in Switzerland' *Rept. Brit. Ass.*, 10th Meeting (Glasgow), Sect. 2, pp. 113–14.
— (1840*C*) 'On the polished and striated surfaces of the rocks which form the beds of the glaciers in the Alps' *Proc. Geol. Soc.*, Vol. 3, pp. 321–2.
— (1840–1) 'Glaciers, and the evidence of their having once existed in Scotland, Ireland and England' *Proc. Geol. Soc.*, Vol. 3, pp. 327–32.
ARCHIAC, É. J. A. (1847–60) 'Histoire des progrès de la Géologie, 1834–59' (Paris), 8 large vols.
BAILEY, E. B. (1962) 'Charles Lyell' (Edinburgh), 214 pp.
BAILEY, T. (1841) 'On the gravel deposits in the neighbourhood of Basford' *Proc. Geol. Soc.*, Vol. 3, pp. 411–13.
BAIRD SMITH, R. (1842–3) 'On the structure of the delta of the Ganges' *Proc. Geol. Soc.*, Vol. 4, pp. 4–6.
BAYFIELD, CAPT. (1835–6) 'A notice on the transportation of rocks by ice' *Proc. Geol. Soc.*, Vol. 2, p. 223.
BONNEY, T. G. (1895) 'Charles Lyell and Modern Geology' (Macmillan), 224 pp.
BUCKLAND, W. (1820) 'Vindiciae Geologicae' (Oxford), 38 pp.
— (1823) 'Reliquiae Diluvianae' (London).
— (1824) 'On the excavation of valleys by diluvial action' *Trans. Geol. Soc.*, 2nd Ser., Vol. 1, pp. 95–102.
— (1829) 'On the formation of the valley of Kingsclere' *Trans. Geol. Soc.*, 2nd Ser., Vol. 2, pp. 119–30.
— (1837) 'Geology and Mineralogy, considered with reference to Natural Theology', Treatise 4 of the Bridgewater Treatises (London).
— (1839) 'On the action of acidulated waters on the surface of the chalk near Gravesend' *Rept. Brit. Assn.*, 9th Meeting (Birmingham), Section 2, pp. 76–77.
— (1840–1) 'On the evidences of glaciers in Scotland and the north of England', Pt. 1, *Proc. Geol. Soc.*, Vol. 3, pp. 332–7; Pt. 2, Vol. 3, pp. 345–8.
— (1841) 'Anniversary Address' *Proc. Geol. Soc.*, Vol. 3, pp. 469–540.
— (1841–2) 'On the glacia-diluvial phaenomena in Snowdon' *Proc. Geol. Soc.*, Vol. 3, pp. 579–84.

BYRES, R. W. (1844) 'On the traces of the action of glaciers at Porth-Treiddyn, Carnarvonshire' *Proc. Geol. Soc.*, Vol. 4, pp. 370–2.

CHASTELLUX, F. J. DE (1787) 'Travels in North America . . .' (London), 2 vols. French Edn 1786.

CLARK, J. W. and HUGHES, T. M. (1890) 'The Life and Letters of the Rev. Adam Sedgwick' (Cambridge), 2 vols.

CLARKE, W. B. (1837) 'On the geological structure and phenomena of Suffolk' *Proc. Geol. Soc.*, Vol. 2, No. 50, pp. 528–34.

CLEAVELAND, P. (1816) 'An Elementary Treatise on Mineralogy and Geology' (2nd Edn, 2 vols., Boston, 1822).

CONRAD, T. A. (1839) 'Notes on American Geology' *Amer. Jour. Sci.*, Vol. 35, pp. 237–51.

CONYBEARE, F. C. (1905) 'Letters and Exercises of the Elizabethan School-master John Conybeare', with notes and a fragment of Autobiography by the Very Rev. William Daniel Conybeare (Frowde, London), 159 pp.

CONYBEARE, W. D. (1829) 'On the hydrographic basin of the Thames' *Proc. Geol. Soc.*, Vol. 1, pp. 145–9.

— (1830) 'Letter on Mr Lyell's "Principles of Geology" ' *Phil. Mag.*, Vol. 8, pp. 215–19.

— (1830–1) 'Examination of those phenomena of geology which seem to bear most directly on theoretical speculations' *Phil. Mag.*, Vol. 8, N.S., pp. 359–62 and 401–6; Vol. 9, N.S., pp. 19–23, 111–17, 188–97 and 258–70.

— (1832) 'Report on the progress, actual state, and ulterior prospects of geological science' *Rept. Brit. Assn.*, pp. 365–414.

DARWIN, C. (1839) 'Observations on the Parallel Roads of Glen Roy' *Phil. Trans. Roy. Soc.*, Pt. 1, pp. 39–81.

— (1841) 'On the distribution of erratic boulders, and on the contemporaneous unstratified deposits of South America' *Proc. Geol. Soc.*, Vol. 3, pp. 425–30.

— (1844) 'Geological Observations on the Volcanic Islands' (London), Pt. 2. Page numbers refer to 3rd Edn, New York, 1897.

— (1846) 'Observations on South America' (London). Page numbers refer to 'Geological Observations', 3rd Edn, New York, 1897.

DE BEAUMONT, ÉLIE. (1830) 'Recherches sur quelques-unes des Revolutions de la Surface du globe' *Revue Française*, No. XV.

— (1831) 'Researches on some revolutions which have taken place on the surface of the globe' *Phil. Mag.*, Vol. 10, N.S., p. 241.

DE CHARPENTIER, J. (1835) 'Notice sur la Cause probable de transport des Blocs Érratiques de la Suisse' *Annales des Mines*, Paris, Vol. 8.

— (1841) 'Essai sur les Glaciers et le Terrain Érratique du Bassin du Rhône' (Lausanne).

DE LA BECHE, H. T. (1829) 'Notice on the excavation of valleys' *Phil. Mag.*, Vol. 6, N.S., pp. 241–8.

— (1831) 'A Geological Manual' (London), 535 pp.

— (1832) 'A Geological Manual' (London), 2nd Edn.

— (1834) 'Researches in Theoretical Geology' (Knight, London), 408 pp.

— (1835) 'How to Observe: Geology' (Knight, London), 312 pp.

DE SAUSSURE, H. B. (1779–96) 'Voyages dans les Alpes' 4, vols. (Neuchâtel).

DOBSON, P. (1826) 'Remarks on bowlders' *Amer. Jour. Sci.*, Vol. 10, pp. 217–18.

D'OMALIUS D'HALLOY, J. J. (1830) 'Observations sur l'origine des vallées' *Jour. de Géologie*, Tome 2, pp. 399–407.

DRACHMAN, J. M. (1930) 'Studies in the Literature of Natural Science' (Macmillan, N.Y.), 487 pp.

EATON, A. (1818) 'Index to the Geology of the Northern States' (Albany) (2nd Edn, Troy, N.Y., 1820)

— (1830) 'Geological Text Book'.

EVEREST, REV. R. (1832) 'Some observations on the quantity of earthy matter brought down by the Ganges River' *Jour. Asiatic Soc. of Bengal*, Vol. 1, pp. 238–42.

EYLES, V. A. (1947) 'James Hutton and Sir Charles Lyell' *Nature*, Vol. 160, pp. 694–5.

FABER, G. (1954) 'Oxford Apostles' (Penguin), 442 pp.

FAIRHOLME, G. (1834) 'On the falls of Niagara' *Phil. Mag.*, Vol. 5, N.S., pp. 11–25.

— (1837) 'New and Conclusive Physical Demonstrations . . . of the Fact and Period of the Mosaic Deluge' (London).

FEATHERSTONHAUGH, G. W. (1844) 'On the excavation of the rocky channels of rivers by the recession of their cataracts' *Rept. Brit. Assn.*, 14th Meeting (York), Section 2, pp. 45–46.

FENTON, C. L. and FENTON, M. A. (1952) 'Giants of Geology' (Doubleday, N.Y.).

FINCH, J. (1824) 'Geological Essay on the Tertiary formations in America' *Amer. Jour. Sci.*, Vol. 7, pp. 31–43.

FITTON, W. H. (1832–3) 'Notes on the history of English geology' *Phil. Mag.*, Vol. 1, N.S., pp. 147–60, 268–75 and 442–50; Vol. 2, N.S., pp. 37–57.

— (1839) Review of Lyell's 'Elements of Geology' *Edin. Rev.*, No. 140, pp. 217–49.

FULTON, J. F. and THOMSON, E. H. (1947) 'Benjamin Silliman' (New York).

GEDDES, J. (1826) 'Observations on the geological features of the south side of the Ontario valley' *Amer. Jour. Sci.*, Vol. 11, pp. 213–18.

GEIKIE, A. (1875) 'Memoir of Sir R. I. Murchison' (London), 2 vols.

— (1895) 'Memoir of Sir A. C. Ramsay' (London), 397 pp.

— (1905) 'Founders of Geology' (London), 2nd Edn, 486 pp.

GIBSON, J. B. (1836) 'Remarks on the geology of the lakes and the valley of the Mississippi' *Amer. Jour. Sci.*, Vol. 29, pp. 201–13.

GILMER, F. W. (1818) 'On the geological formation of the Natural Bridge of Virginia' *Trans. Amer. Phil. Soc.*, Vol. 1, N.S., pp. 187–92.

GORDON, MRS. (1894) 'The Life and Correspondence of William Buckland' (London).

GREENOUGH, G. B. (1819) 'A Critical examination of the first principles of geology' *Thompson's Ann. Phil.*, Vol. 14, pp. 365–73 and 456–64.

— (1833–4) 'Anniversary Address' *Proc. Geol. Soc.*, Vol. 2, No. 35, pp. 42–70.

— (1834–5) 'Anniversary Address' *Proc. Geol. Soc.*, Vol. 2, No. 39, pp. 145–175.

GREENWOOD, G. (1857) Rain and Rivers' (London), 195 pp.

GREGORY, H. E. (1918) 'Steps of progress in the interpretation of landforms' *Amer. Jour. Sci.*, Vol. 46, 4th ser., pp. 104–32.

HAYDEN, H. (1820) 'Geological Essays' 412 pp.

HAYES, G. E. (1839) 'Remarks on the geology and topography of western New York' *Amer. Jour. Sci.*, Vol. 35, pp. 86–105.

HILDRETH, S. P. (1836) 'Observations on the bituminous coal deposits of the valley of the Ohio' *Amer. Jour. Sci.*, Vol. 29, pp. 1–148.

HITCHCOCK, E. (1819) 'Remarks on the geology and mineralogy of a section of Massachusetts on the Connecticut River' *Amer. Jour. Sci.*, Vol. 1, pp. 105–16.

HITCHCOCK, E. (1824) 'A sketch of the geology, mineralogy and scenery of the regions contiguous to the river Connecticut '*Amer. Jour. Sci.*, Vol. 7, pp. 1–30.

— (1833) 'Report on the Geology . . . of Massachusetts' (Amherst), 700 pp.

— (1841) 'First Anniversary Address before the Association of American Geologists' *Amer. Jour. Sci.*, Vol. 41, pp. 232–75.

HOGARD, H. (1851) 'Terrain Érratique des Vosges' (Épinal), 139 pp.; with Atlas of 32 plates.

HOPKINS, W. (1838) 'Researches in Physical Geology' *Trans. Camb. Phil. Soc.*, Vol. 6, pp. 1–84.

— (1841) 'On the geological structure of the Wealden district, and of the Bas Boulonnais' *Proc. Geol. Soc.*, Vol. 3, pp. 363–6.

— (1842) 'On the elevation and denudation of the district of the lakes of Cumberland and Westmorland' *Proc. Geol. Soc.*, Vol. 3, pp. 757–66.

HORNER, L. (1834–5) 'On the quantity of solid matter suspended in the water of the Rhine' *Edin. New Phil. Jour.*, Vol. 18, pp. 102–6.

HOWORTH, H. H. (1893) 'The glacial nightmare and the Flood' (London), 2 vols.

HUGI, F. J. (1843) 'Die Gletscher und die erratischen Blöcke' (Solothurn), 256 pp.

JACKSON, C. T. (1839) 'Reports on the geology of the State of Maine' *Amer. Jour. Sci.*, Vol. 36, pp. 143–56.

— (1840) 'Report on the Geological and Agricultural Survey of the State of Rhode Island' (Providence), 312 pp.

JUDD, J. W. (1911) 'The Student's Lyell' (Murray, London), 2nd Edn, 645 pp.

KALM, P. (1753) 'Peter Kalm's Travels in North America' (New York), ed. by A. B. Benson, 1937.

LONG, G. (1839) 'On the occurrence of numerous swallow holes near Farnham' *Proc. Geol. Soc.*, Vol. 3, pp. 101–2.

LURIE, E. (1960) 'Louis Agassiz' (Univ. of Chicago), 449 pp.

LYELL, C. and MURCHISON, R. I. (1828–9) 'On the excavation of valleys' *Proc. Geol. Soc.*, Vol. 1, pp. 89–91.

— (1829) 'On the excavation of valleys' *Edin. New Phil. Jour.*, Vol. 7, pp. 15–48.

LYELL, C. (1830) 'Principles of Geology', 1st Edn, Vol. 1.

— (1833) 'Principles of Geology', 1st Edn, Vol. 3.

— (1836) 'Anniversary Address' *Proc. Geol. Soc.*, Vol. 2, pp. 357–90.

— (1837) 'Principles of Geology', 5th Edn. Page numbers refer to the 1st American Edition (2 vols.) which was taken from the British 5th Edn.

— (1840–1) 'On the geological evidence of the former existence of glaciers in Forfarshire' *Proc. Geol. Soc.*, Vol. 3, pp. 337–45.

— (1842) 'On the recession of the falls of Niagara' *Proc. Geol. Soc.*, Vol. 3, pp. 595–602.

— (1842–3) 'On the ridges, elevated beaches, inland cliffs and boulder formations of the Canadian Lakes and the valley of St Lawrence' *Proc. Geol. Soc.*, Vol. 4, pp. 19–22.

— (1845) 'Travels in North America' (New York), 2 vols.

— (1847) 'Principles of Geology', 7th Edn, Vol. I, 810 pp.

— (1849) 'A Second Visit to the United States of North America' (New York), 2 vols.

— (1852) 'A Manual of Elementary Geology', 4th Edn (London).

— (1872) 'Principles of Geology', 11th Edn (London), 2 vols.

LYELL, MRS K. M. (1881) 'Life, Letters and Journals of Sir Charles Lyell' (London), 2 vols.

MACINTOSH, A. F. (1845) 'On the supposed evidences of the former existence of glaciers in North Wales' *Proc. Geol. Soc.*, Vol. 4, pp. 594–601.

MACKENZIE, G. S. (1835) 'On the theory of the Parallel Roads of Glen Roy' *Phil. Mag.*, Vol. 7, N.S., pp. 433–6.

MACLAREN, C. (1842) 'The glacial theory of Professor Agassiz' *Amer. Jour. Sci.*, Vol. 42, pp. 346–65.

MACLURE, W. (1809) 'Observations on the Geology of the United States' *Amer. Phil. Soc. Trans.*, Vol. VI, pp. 411–28; 2nd Edn as book in 1817 (Philadelphia).

MANTELL, G. (1832) 'A notice on the geology of the environs of Tonbridge Wells', in 'Descriptive Sketches of Tonbridge Wells' by J. Britton (London), 148 pp.

— (1833) 'The Geology of the South-East of England' (London), 415 pp.

MARTEL, P. (1744) 'An account of the Glacieres or Ice Alps in Savoy' (London).

MARTIN, P. J. (1828) 'A Geological Memoir on a Part of Western Sussex' (London).

— (1840–1) 'On the relative connection of the eastern and western chalk denudations' *Proc. Geol. Soc.*, Vol. 3, pp. 349–51.

MATHER, K. F. and MASON, S. L. (1939) 'A Source Book in Geology' (McGraw-Hill, N.Y.), 702 pp.

MATHER, W. W. (1833) 'Elements of Geology'.

MCALLISTER, E. M. (1941) 'Amos Eaton: Scientist and Educator, 1776–1842' (Univ. of Pennsylvania Press, Philadelphia), 587 pp.

MCCALLIEN, W. J. (1941) 'The birth of glacial geology' *Nature*, Vol. 147, pp. 316–18.

MERRILL, G. P. (1906) 'The development of the glacial hypothesis in America' *Pop. Sci. Monthly*, Vol. 68, pp. 300–22.

— (1906A) 'Contributions to the History of American Geology' (U.S. National Museum, Washington), pp. 189–733. Largely reprinted in 1924.

— (1924) 'The First Hundred Years of American Geology' (New Haven), 773 pp.

MITCHELL, J. (1838) 'On the drift from the chalk' *Proc. Geol. Soc.*, Vol. 3, pp. 3–5.

MITCHILL, S. L. (1818) 'Observations on the Geology of North America'; in Cuvier's 'Essay on the Theory of the Earth' (New York), pp. 321–428.

MURCHISON, R. I. (1835–6) 'The gravel and alluvia of South Wales and Siluria' *Proc. Geol. Soc.*, Vol. 2, pp. 230–336.

— (1836) 'On the ancient and modern hydrography of the River Severn' *Rept. Br. Assn.*, Pt. 2, p. 88.

— (1842) 'Anniversary Address' *Proc. Geol. Soc.*, Vol. 3, pp. 637–87.

— (1843) 'Anniversary Address' *Proc. Geol. Soc.*, Vol. 4, pp. 65–151.

NORTH, F. J. (1943) 'Centenary of the glacial theory' *Proc. Geol. Assn.*, Vol. 54, pp. 1–28.

OSPOVAT, A. M. (1960) 'Werner's influence on American geology', *Proc. Oklahoma Acad. Sci.*, Vol. 40, pp. 98–103.

PERCIVAL, J. G. (1842) 'Report on the Geology of the State of Connecticut.'

PHILLIPS, J. (1828) 'Remarks on the geology on the north side of the Vale of Pickering' *Phil. Mag.*, Vol. 3, N.S., pp. 243–9.

— (1829) 'Illustrations on the Geology of Yorkshire' (York), 192 pp.

— (1831) 'On some effects of the atmosphere in wasting the surfaces of buildings and rocks' *Proc. Geol. Soc.*, Vol. 1, pp. 323–4.

— (1853) 'The Rivers, Mountains, and Sea-Coast of Yorkshire' (London), 291 pp.

POWNALL, T. (1776) 'A Topographical Description of . . . North America' (London).

ROGERS, H. D. (1835) 'On the falls of Niagara and the reasonings of some authors respecting them' *Amer. Jour. Sci.*, Vol. 27, pp. 326–35.

— (1844) 'Annual Address at the Meeting of the Association of American Geologists and Naturalists' *Amer. Jour. Sci.*, Vol. 47, pp. 137–60 and 247–78.

SCHEUCHZER, J. J. (1723) 'Itinera per Helvetiae Alpinas Regiones facta annis 1702–11', Collected Edn (Leyden).

SCROPE, G. P. (1825) 'Considerations on Volcanos' (London), 270 pp.

— (1827) 'Memoir on the Geology of Central France' (London), 182 pp.

— (1829–30) 'On the gradual excavation of the valleys in which the Meuse, the Moselle, and some other rivers flow' *Proc. Geol. Soc.*, Vol. 1, pp. 170–1.

— (1830) Review of the 1st Edn of Lyell's 'Principles of Geology' *Quart. Rev.*, Vol. 43, pp. 411–69.

— (1835) Review of the 3rd Edn of Lyell's 'Principles of Geology' *Quart. Rev.*, Vol. 53, pp. 406–48.

SEDGWICK, A. (1825) 'On the origin of alluvial and diluvial formations' *Annals of Phil.*, N.S., Vol. 9, pp. 241–57; Vol. 10, pp. 18–37.

— (1830) 'Anniversary Address' *Proc. Geol. Soc.*, Vol. 1, No. 15, pp. 187–212.

— (1838) in 'Report of the 7th Annual Meeting of the Brit. Assn.' *Amer. Jour. Sci.*, Vol. 33, pp. 265–96.

SILLIMAN, B. (1821) Notice of 'Geological Essays . . .' by Horace H. Hayden. *Amer. Jour. Sci.*, Vol. 3, pp. 47–57.

SMITH, A. (1832) 'On the water courses, and the alluvial and rock formations of the Connecticut River valley' *Amer. Jour. Sci.*, Vol. 22, pp. 205–31.

SMITH, S. (1929) 'The Letters of Peter Plymley' (Dent, London), 296 pp.

SPENCER, E. (1835) 'Observations on the Diluvium of the vicinity of Finchley, Middlesex' *Proc. Geol. Soc.*, Vol. 2, p. 181.

SURELL, A. (1841) 'Étude sur les torrents des Hautes-Alpes' (Paris), 1st Edn.

SWEDENBORG, E. (1719) 'Om wattrens höjd och förra werldens starka ebb och flod. Bevis utur Sverige' (Stockholm).

THOMPSON, W. A. (1833) 'Facts relating to diluvial action' *Amer. Jour. Sci.*, Vol. 23, pp. 243–9.

TRIMMER, J. (1832) 'On the diluvial deposits of Caernarvonshire' *Proc. Geol. Soc.*, Vol. 1, No. 22, pp. 331–2.

TUCKWELL, W. (1900) 'Reminiscences of Oxford' (London), 288 pp.

VOLNEY, C. F. C. DE (1804) 'A View of the Soil and Climate of the United States of America . . .' (London and Philadelphia). Trans. by C. B. Brown of Paris Edn. 1803.

VON HOFF, K. E. A. (1822–34) 'Geschichte du durch Überlieferung nachgewiesenen natürlichen Veränderungen der Erdoberfläche' (Gotha), 3 vols.

VON ZITTEL, K. (1901) 'History of Geology and Palaeontology' (London), 562 pp.

WALLACE, A. R. (1898) 'The Wonderful Century' (New York).

WHEWELL, W. (1839) 'Anniversary Address' *Proc. Geol. Soc.*, Vol. 3, pp. 61–98.

WHITE, G. W. (1951) 'Lewis Evans' Contributions to Early American Geology' *Trans. Illinois Acad. Sci.*, Vol. 44, pp. 152–8.

— (1953) 'Early American Geology' *The Scientific Monthly*, LXXVI, pp. 134–141.

— (1956) 'Lewis Evans . . .' *Nature*, Vol. 177, pp. 1055–6.

— (1956) 'John Josselyn's Geological Observations' *Trans. Illinois Acad. Sci.*, Vol. 48, pp. 173–82.

WHITE, H. D. (1920) 'History of the Warfare of Science with Theology in Christendom' (New York), 2nd Edn.

WOODWARD, H. B. (1907) 'A History of the Geological Society of London' (London), 336 pp.

— (1911) 'History of Geology' (London), 204 pp.

WOODWARD, J. (1695) 'Essay Towards a Natural History of the Earth'.

YATES, J. (1830–1) 'On the formation of alluvial deposits' *Proc. Geol. Soc.*, Vol. 1, pp. 237–9.

PART THREE

Marine *versus* Subaerial Erosionists
1846-75

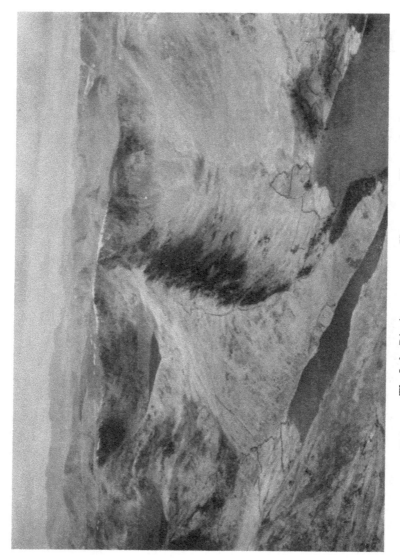

FIG. 54. *The Lake District scenery near Haweswater, Westmorland*
(Aerofilms and Aero Pictorial Ltd.)

CHAPTER SIXTEEN

The Advent of Marine Planation

INTRODUCTION

The three decades we are about to discuss coincided with either the establishment or the enlargement of geological surveys in most civilized states. The prime purpose of the official surveys was the accurate mapping of exploitable areas in order to determine what rocks lay beneath the surface, but this naturally led to the inquiry as to how they had come to assume their present position and composition. The work necessarily involved paying attention to processes still active on the surface and the elaborate geologic memoirs published contained details notably of the sequence and thickness of strata and of the amount of strata removed by erosion, but also of the methods and rates of denudation and deposition. Imaginative inferences were gradually replaced by factual deductions and many theories acquired a sound factual basis. Not that in the past theories had been without facts, but about the mid-nineteenth century geological observations began to multiply at an ever-increasing rate. Geology was moving inexorably towards exactitude. Gone was the age of the speculator dreaming visions of cosmogonical chaos and creation from the security of a study chair.

In 1845, when the period under discussion begins, the dominant ideas were those already described in Part Two of this volume. The most powerful group comprised the numerous supporters of the marine erosion theory of Lyell, who still considered themselves true uniformitarians. Against them were: the structuralists and semi-catastrophists for whom Hopkins was the prime propagandist; and the fluvialists or the small minority who still followed Hutton in believing that rain and rivers had mainly denuded the present landscape.

To these rival schools of thought was added, right at the beginning of the period, another – the followers of Ramsay's idea of marine planation.

A. C. RAMSAY AND THE MARINE PLANATIONISTS

Andrew Crombie Ramsay was born on 31 January 1814 at Glasgow, where his father had a dye business. By all accounts his was a well-educated family and French was the language of the breakfast table one morning a week, while on other mornings a passage was read from the work of some famous author. Because he was delicate Ramsay was sent to a school at Saltcoats, Ayrshire, as the sea air, it was thought, would be beneficial to his health. He actually finished his education at Glasgow Grammar School, where unfortunately his schooling was cut short by his father's death when Ramsay was only thirteen.

His mother was left in straitened circumstances and Ramsay was sent to work in a counting house and later with a firm of linen merchants. Here, in his spare time, he wrote and edited a small journal of poems, caricatures and articles of a light nature, which was circulated among a few of his friends. In addition to this outlet for his literary bent, he was well acquainted with some of the teachers and students of Glasgow University: Lord Playfair was one of his mother's lodgers and Professor Nichol advised Ramsay on his geological reading (Geikie, 1895).

The next change in his life occurred when he entered into partnership in a cloth and calico dealing business, which soon failed. This failure, and his general dissatisfaction with a business way of life, prompted him to look for a post more suited to his tastes. The opportunity occurred in 1841. In the year previous the British Association had met at Glasgow, when Ramsay, probably through his association with Nichol, had been elected to one of the sub-committees. The method employed at such meetings was to divide the scientists up into specialized sub-committees, and to allot to each group a particular task. The sub-committee of which Ramsay was a member had to prepare a map of, and to gather a representative collection of the geological specimens of, the Isle of Arran, an area that Ramsay knew particularly well. The value of his contribution was so obvious that he was asked to read a paper before the Association on his discoveries in Arran. Among his audience were geologists like Greenough, Buckland, Phillips, and Murchison of the older school, and Lyell, De la Beche, Forbes and Agassiz of the newer generation. On these, and on the audience generally, he made such a favourable impression that he was asked to act as guide to a section of the Association which was to visit Arran. Unfortunately Ramsay worked on his notes so conscientiously the night before that the following morning he overslept and missed the boat taking the party to the island! Despite losing this chance of further glory, he was not forgotten. Professor Nichol encouraged him to keep up his geological studies and prevailed upon him to collate his findings on the Isle of Arran and to publish the whole as a book. Ramsay was lucky enough to find a Glasgow publisher who undertook to print the work and eventually paid him £21 for it. His success did not stop there, for in the spring of 1841 Murchison proposed that Ramsay should accompany him on a geological expedition to America. This offer proved irresistible to Ramsay and his mother reluctantly gave her blessing to the venture with the admonition that he should request a salary and also place on a legal basis the arrangement for the payment of his travelling expenses. He had already arrived in London when Murchison suddenly found the plan impracticable and abandoned it. Seeing how disappointed Ramsay was at this turn of events, Murchison approached De la Beche and arranged for Ramsay's appointment as an assistant geologist in the newly-formed Ordnance Survey.

There were four other assistants when Ramsay joined, and each was paid

FIG. 55. *Andrew Crombie Ramsay*

9s. a day for a six-day week. Being part of the Ordnance Department and under military jurisdiction they had to wear a dark-blue uniform and a tightly-fitting, buttoned frockcoat, a most unsuitable garment in which to perform the goat-like antics demanded of geologists. Later they viewed with marked relief their transfer to a civilian status, and with it permission to choose their own clothes, on which, as a link with their military past, they retained the gilt buttons embossed with the crown and crossed hammers.

In his first years with the Survey Ramsay was assigned to South Wales and worked-over thoroughly the counties of Pembroke, Carmarthen, Brecknock, Cardigan, Montgomery and Radnor. The duties of a geologist in those early years were hard. He worked on his own in the open from early morning to nightfall in all kinds of weather which did not actually make surveying impossible. He had to find his own food and accommodation, and in his letters Ramsay records how after a day's climbing he used to return weary to the small country inns, make a heavy meal of the plain fare provided and then check the day's work before going to sleep. On occasions, when in more populated areas, he received invitations of hospitality from some of the wealthier of the local families. With his early training in literature, and a natural liking for verse and song, he was a welcome addition to any such social circle and in the sparser-populated areas where educated young men were scarce was, no doubt, much in demand by families with unmarried daughters. During the winter months when geology in the field became impracticable the assistants went to London where they mapped the work of the previous year.

As a geological assistant Ramsay must have impressed his leader, De la Beche, who in 1845, when the Survey became a branch of the Office of Works, remained in supreme control with the title of Director-General. The reorganization involved the appointment of two new Directors, one responsible for England, Wales and Scotland, and the other for Ireland. Ramsay was appointed Director for the first region. This new post brought with it also responsibility for the co-ordination of Surveys within Britain and for the preparation of the quarterly financial accounts. These latter sometimes occupied more of his time than that taken up by all the rest of his duties put together. Fitting this into, or rather on to, his work in the field must have made a severe call on his already heavily taxed resources of energy. In the credit balance the necessity of overlooking all the geological work carried out within his region must have caused him to reflect upon, and view in the abstract, the various geological problems that were being uncovered.

It is always difficult to discover what precise examples or earlier advice prompts a scientist to arrive at his final synthesis. Ramsay was probably affected by several trends, as well as maintaining a healthy independence of his own. In his youth his first guiding influence was Lyell whose *Principles* he was given as a prize for chemistry at Glasgow Grammar School. Later, the constant collaboration, as part of his duties, with De la Beche had some effect upon him. It does not appear that De la Beche forced his opinions on his staff. He himself belonged neither to the old catastrophist school nor to the radical fluvialists but had a balanced appreciation of what physical processes might perform provided they were allowed sufficient time and were related to the initial factor of structure. As an older man his guidance in matters of theory, if neglected more and more as Ramsay developed his own ideas, undoubtedly had a significant formative influence on the young geologist. Thus Ramsay grew up believing in Lyell's principles of uniformitarianism, though in other respects he was more backward in his outlook. When it came to accepting or rejecting Agassiz's theory Ramsay joined those who derided its credibility. In judging theories concerned with facts outside his own field-work he was inclined to accept the more orthodox and traditional views; when a judgment involved facts with which he was acquainted he made his own decision. It is against these origins that the value of his essay on the *Denudation of South Wales* should be assessed.

The essay was one of a collection of memoirs, intended to give prominence to the work of the Survey, and written by the senior members on the topics or regions with which they had principally been concerned. De la Beche wrote 'On the Formation of the Rocks of South Wales and South West England'; and Forbes 'On the Connexion between the Distribution of the Existing Fauna and Flora of the British Isles and the Geological changes which have affected their Area, especially during the Epoch of the Northern Drift'. Ramsay's contribution is the only one that remains important, the reason being that in it he forcefully brought to the forefront two quite distinct ways of thinking about erosion, though in his presentation he actually joined the ideas together. They may be summarized as: (1) the visual reconstruction of former folded surfaces, and (2) marine planation.

THE VISUAL RECONSTRUCTION OF FORMER FOLDED SURFACES

As a practical investigator of the earth's structural contortions, Ramsay had begun to realize that, although the newer strata often rests conformably upon the older, in other instances there were definite signs of unconformity. Often, again, the strata had been forced into giant folds, which in places gave way to precipices looking as if they had been severed from top to bottom. If he carried his investigations farther he often found the other limb of the same fold some distance away with such an exact correspondence of alignment that it was quite impossible to resist the conclusion that the portion in between must at some time have existed and have since been removed by erosion. While he did not believe that the whole world was covered by the same succession of strata, he thought that this was true for small regions.

Therefore, where a lateral or vertical break in the strata occurred this possibly indicated that the missing portion had been eroded away. He contended that the strata had not been laid down in their present fragmentary pattern and that many of the disconnected strata had been originally of much greater unbroken extent. To obtain an accurate picture of the original stratigraphical pattern, the folded limbs which were broken should be joined, and the missing portions would represent fairly accurately the amount of rock that had been eroded away. In the same way an unconformable surface would provide some indication of the measure of erosion that had taken place before the laying down of the overlying strata.

All this in itself was not new, as Hutton had long ago enunciated the main principles and Lyell and many others had said the same thing, especially about horizontal strata cut by rivers. What gave it greater appeal was the factual detail that accompanied Ramsay's essay; he supplied graphical support with figures of the amount of rock involved and diagrams showing how the various strata could be joined and continued. He made it abundantly clear that strata of a thickness of many thousands of feet had been removed.

Ramsay proceeded to seek out a way whereby the amount of denudation could be accurately calculated and he found the solution in a reconstruction of the surface as it was assumed to be before the denudation took place:

> But to whatever causes the existing features of any country may be assigned, it is evident that without data by means of which we may form true conceptions of that form, it is impossible to reason correctly either of the manner of action of these bygone operations, or of the magnitude of their effects. Such data are to a certain extent supplied by the construction of geological sections on a true scale, vertically and horizontally. Having for their base line the level of the sea. (*Ramsay, 1846, p. 298*)

His work in the field had provided him with an instinctive sense of order. Where others had viewed the area as a chaotic jumble and had based their explanations for its appearance on floods or earthquakes, Ramsay believed he could see the original pattern as it must have been. The broken folds and outliers were related to each other; they represented the result of the action of denudation upon structures going on over millions of years, interrupted at

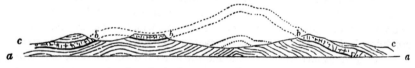

FIG. 56. *A geological reconstruction by Ramsay (1846)*

intervals by changes in sea level. That the discontinuous fragments could be legitimately joined he considered reasonable because he felt it more unreasonable to suppose that they could have been laid down in their present fragmentary distribution, necessitating, as it would, the slow deposit of strata thousands of feet in thickness ending in abrupt precipices:

> What first strikes the eye on examining certain of these sections, is the remarkable curvature and distortion to which the strata composing all the formations, from the top of the Coal Measures downwards, have been subjected. Following these breaks and curves, the same series of rocks are seen repeatedly to rise to the surface, sometimes in rapid, sometimes in widespreading undulations. When, in accordance with the curves indicated by the surface dips, vast masses of rock are carried in these sections deep down into the earth, far below our actual cognizance, it is yet impossible to doubt their underground continuity, when we find again and again, the same set of beds diving downwards in one district, and (perhaps somewhat modified) again outcropping to the surface in remote parts of the country. The abuse of this fact, now familiar to every geologist, in earlier times led to the hypothesis of the original continuity of all strata over the entire circumference of the globe. But if the inference now drawn be legitimate, a little reflection will show, that in the case of curved and conformable strata, the same arguments that apply to the continuity of rocks below the surface, may often safely be employed to prove the original connexion of contorted strata, the upturned edges of which may frequently be far apart. Attention being given to the physical

relations of all the rocks in any country, such restoration of masses of rock to the form they once possessed, is within the limits of safe inference. And if, in the cases above noticed, this original continuity of distant masses, and their spreading over tracts where they have left not a trace, be once granted, then the vast amount of matter we shall be able to show has been removed from such tracts, may well make us cautious in disbelieving the probable or possible destruction of other masses, once resting above the rocks that compose the present surface, but of the former existence of which above that surface, we have at first sight no direct evidence. Outliers, cape-like projections, and anticlinals of various strata, so common on our maps of geological England, sufficiently illustrate the first proposition; and the frequent occurrence of vast thicknesses of strata, disposed vertically or at high angles, afford perfect evidence that such strata were not originally discontinued at their present outcrop, since such supposition would involve the necessity of asserting, that the rocks in question were deposited in successive layers, forming together at their extreme edge, a wall or highly inclined plane, often many thousand feet in height. (*Ramsay, 1846, pp. 298–9*)

Ramsay, indeed, postulated that the strata all over South Wales had been truncated and, by means of his geological restoration, estimated the removal of at least 3500 feet of strata in the Woolhope district.

MARINE PLANATION

The vast quantity of rock removed caused Ramsay to turn to the sea. There is little doubt that his marine planation idea received direct inspiration from Lyell's own marine erosion theory. Lyell goes so far as to suggest this:

> . . . your ideas of denudation acting contemporaneously with subterranean movements, whether of upheaval or depression, agree with those which I published in 1831 in my *Principles* (and, by the way, before that time it was thought a triumphant argument against what were called 'modern causes' to prove that a river could not denude the rocks); the gradual action of the ocean acting concurrently with movements of the land, as exemplified in my denudation of the Wealden, had not, so far as I know, been fully set forth in any geological work, with due allowance also for the resistance of the harder and yielding of the softer rocks. (*Lyell, letter 8 Oct. 1846; quoted by Geikie, 1895, p. 86*)
>
> When I inferred that the denuded dome of the Wealden had lost some 2000 feet and upwards of thickness of strata removed, I also assumed that it was shaved off by the ocean when rising and had never constituted hills 4000 to 5000 feet high. So I think of your denuded tracts. They were never suffered to attain an Alpine elevation. (*Lyell, letter 8 Oct. 1846; quoted by Geikie, 1895, p. 88*)

Ramsay admits as much in his reply:

> Nothing could be further from my wish than to assume as my own any idea started by another, especially by one whose *Principles of Geology* strongly tended to make my geological mind such as it is by first directing its inquiries into proper channels . . . (*Ramsay, letter 19 Oct. 1846; quoted by Geikie, 1895, p. 92*)

If he had stopped short at Lyell's idea that the sea etched out the main topographical features while the land was slowly rising from beneath the waters, then his theory would have been of little interest. But he modified Lyell's theory in several important ways. Instead of imagining that the sea always performed the work of erosion in a *rising* landmass, Ramsay considered the effect of the normal processes of marine erosion acting on a *stationary* or *subsiding* landmass. In this way the sea instead of, as Lyell supposed, carving out inequalities in the surface, would during its transgression slice off an almost level platform or marine plain. Inequalities on the surface would arise from the natural differences in hardness or in the attitude of the strata. Like Lyell he saw the present cliffs, promontories and islands as the future precipices, ridges and hills of a newly-risen landmass. In essence his idea was not superficially very different from Lyell's, except that he had reversed the mechanism. Ramsay laid stress on marine platforms and benches; Lyell emphasized (wrongly) marine dissection. This idea of planation was a radical suggestion, the implications of which were to assume far greater importance than was ever recognized during Ramsay's lifetime. Even Geikie, his contemporary, in an estimate of his life's achievements places far more emphasis on his theme of denudation than on the planation theory:

> In physiography Ramsay's work was abundant, as well as remarkably original and important. It may be grouped in three subdivisions: (1) Denudation in general; (2) The history of river valleys; and (3) The results of the operations of ice.
> (1) The early paper on the Denudation of South Wales, published in 1846 . . . was undoubtedly the most important essay on the subject which up to that time had appeared. Much had previously been written on the question of denudation, but it was of the vaguest nature. It was Ramsay's merit that he based his discussion upon the results of careful surveying. He had traced out the structure of a complicated geological region, and was able to show what should have been the form of the surface had it depended entirely on geological structure. He was thus in a position to demonstrate how much material had been removed by denudation, and how far the process of removal had been guided by geological structure. It is true, as he himself afterwards confessed, that at the time he assigned too much power to the sea, and too little to the subaerial agents, in the lowering of a mass of land. But this exposition of the old base-level of ancient erosion, or 'plain of marine denudation', as he called it, will ever be a classical study in geological literature. (*Geikie, 1895, pp. 358–9*)

Ramsay's idea of the existence of wide uplifted marine plains was based on his own field-work. In America the accordance of summit heights had been already noted by two writers but its true significance had been totally neglected by all geologists. Ramsay repeated the point that there was a general line of slope from the mountain top to sea level which applied as much to the hills as to the valleys; in other words, as in the case of an ordinary plain, there was a general uniformity of height with a gentle slope towards the sea. Ramsay had

FIG. 57. *View of the Cardiganshire coast near Aber Arth, showing a planation surface dissected by rivers*
(Photograph by J. K. St. Joseph)

observed that the heights of the hilltops were similar at corresponding distances from the coast, and also that they presented a regular pattern of gentle descent to the coast; as if they too represented the fragmental remains of an eroded marine plain. This regularity of slope and coincidence of height between neighbouring hilltops made him think that the slope was the sole remaining relic of the original plain; a plain slowly eroded away by an encroaching sea, the subsequent inequalities having been produced mainly by subaerial erosion.

If later Ramsay had to modify his theory and admit that many of his marine plains were really the result of uninterrupted subaerial erosion, it cannot alter the importance of his planation theory for future generations. Hutton had already suggested such plains, only in his case produced by subaerial forces and here again Ramsay was drawing attention to a feature that had long been overlooked. In the hallowed era of W. M. Davis the significance of what Ramsay said will be obvious, as the plain, or peneplain, becomes the cornerstone of the whole theoretical structure. Viewed against the contemporary background Ramsay's theory is not more important than many others; set against the developments that were to come it is an outstanding landmark.

There are several aspects of the new planation theory which amply repay detailed study. The immense quantity of denudation that had taken place induced Ramsay, like Lyell, to discount the action of subaerial agencies except as modifying forces which came into operation after the main work of erosion had been completed. He, for all his acuteness concerning problems of stratigraphy, initially held the popular contemporary view that rivers were incapable of cutting the large valleys in which they now ran. What is more, he still considered that what we know to be glacial and fluvio-glacial drift was a relic of 'the old sea bottoms'.

> As the land sank, the waves, which had not touched these shores for countless ages, once more began the work of destruction on hills long placed high above their reach . . . [Subsequent uplift occurred to the present elevation during which the sea acted especially on the soft rocks] . . . the restless action of the sea gave to our hills and valleys the normal outlines of their present forms, since but slightly modified by atmospheric agencies . . . (*Ramsay, pp. 326–7*)
>
> The power of running water has also considerably modified the surface, but the part it has played is trifling compared with the effects that have sometimes been attributed to its agency. How small a part has been acted by rivers . . . , is shown by the fact that in the lowlands they have rarely succeeded in cutting through the superficial drift, or relics of the old sea bottoms that have remained on the land during its latest elevation. (*Ramsay, 1846, p. 332*)

Although Ramsay used the sea as an instrument of denudation he followed Hutton in supposing that the material denuded went to form new deposits below the sea. At the same time as the sea drove its way across the wasting landmass, it laid down between sea level and the ocean bottom a new series of strata. This process was repeated many times as the land either rose above or sank beneath the waves and in between these phases of upheaval or depression the processes of degradation acted on the exposed surfaces. The extent of degradation, Ramsay said, could be assessed from examining the lines of unconformity, which marked the division between the erosion surface of the older continent and the newer marine deposits superimposed upon it. The thickness, position, attitudes and absence of the various strata could all be interpreted to show from what direction the sea had encroached, the height sea level had reached and the amount of material denuded.

> There is every probability, that this older land was destroyed and denuded back by the elder sea, and a gradual depression ensuing southward, permitted the further encroachment of the later sea, so that its deposits *overlap* the lower members of the same series and rest directly on the ancient contorted strata. (*Ramsay, 1846, p. 319*)

There was nothing violent or catastrophic about Ramsay's theory. He was prepared to allow millions of years in which the sea would eat away the coast, a few feet annually. In his diary, when recounting an anecdote of his admired Director-General, he provides a hint of what must have been his own belief:

> One thing he said to-day amused me much. We were sitting on the sea-beach, eating mutton sandwiches, and watching the action of the waves on the pebbles, when Sir H. said 'I'll tell you what the old gentleman is saying: he's saying: "Only give me plenty of time ha! ha! ha!" ' (*Geikie, 1895, p. 191*)

Ramsay foresaw many objections to his marine planation theory. Some fluvialists averred that the sea was a very slow erosive agent as it often impeded its destructive work by casting up beaches, bars and sandbanks. He countered this argument by suggesting that the accretions were temporary features which were eventually removed. He further suggested that gradual depression of the land might occur, bringing with it accelerated or renewed erosion as the depositional features became more readily subject to wave action.

> When the high ground constituting the restored portions of the sections was, in the progress of geological events, removed, the land again uprose to attain its present elevation; and if, during the progress of this gradual upheaval, occasional oscillations of level occurred (and this seems to have been the case), then the destroying agency of the waves would act, in favourable conditions, with increased power. And thus, in the long lapse of geological time, as the land slowly reached its existing height, the restless action of the sea gave to our hills and valleys the normal outlines of their present forms, since but slightly modified by atmospheric agencies, the loose drifty deposit that covers the hills and valleys so formed, being, as it were, but the dregs of the matter removed from the rising land during its last elevation . . .
>   The line of greatest waste on any coast, is the average level of the breakers. The effect of such waste is obviously to wear back the coast, the line of denudation being a level corresponding to the average height of the sea. Taking unlimited time into account, we can conceive that any extent of land might be

so destroyed, for though shingle beaches, and other coast formations will apparently for almost any ordinary length of time protect the country from the further encroachments of the sea, yet the protections to such beaches being at last themselves worn away, the beaches are in the course of time destroyed, and so, unless checked by elevation, the waste being carried on for ever, a whole country might gradually disappear.

If to this be added an exceedingly slow depression of the land and sea bottom, the wasting process would be materially assisted by this depression, bringing the land more uniformly within the reach of the sea, and enabling the latter more rapidly to overcome obstacles to further encroachments, created by itself in the shape of beaches. By further increasing the depth of the surrounding water, ample space would also be afforded for the out-spreading of the denuded matter. To such combined forces, namely, the shaving away of the coasts by the sea, and the spreading abroad of the material thus obtained, the great plain of shallow sounding which generally surrounds our islands is in all probability attributable . . . (*Ramsay, 1846, pp. 326–8*)

He was also able to explain the irregularities which occurred on land above sea level. The sea, he said, naturally picked out the weaker portions of the coast and removed them first, leaving the more resistant sections as islands, promontories or cliffs. When the sea subsided and the land became dry, these islands, promontories or cliffs formed mountain tops or escarpments. In this way he avoided one of the objections which had been levelled at the marine theory of Lyell, who imagined that the waves eroded and etched out the inequalities while the land remained *below* sea level. To this it had been objected that the sea eroded uniformly and on a plane surface irrespective of rock composition; that it acted more as a levelling agent than as an instrument of dissection. Ramsay combined both ideas. The sea, he said, planed away that portion it had passed over and, in newly attacked coastal areas, infiltrated by probing along and taking advantage of weaknesses in rock:

The waves acting equally on cliffs of unequal hardness soon produce great irregularities of outline, the harder rocks standing out in bold promontories, while the softer materials yielding more rapidly to the shock of the breakers, slowly originate bays, creeks and arms of the sea. If to this we add the influence of exposure to prevalent winds, the indented character of a coast often becomes very remarkable, the general depth of the plain, so to speak, formed by denudation continuing the same, unless varied by oscillations of level. Besides these peculiarities of outline dependent on unequal hardness of rocks, the form of the escarpment bounding the surface beneath the sea, would of course be much modified by the dip and geological position of its component rocks, so that it sometimes happens that a soft material on account of the sea-ward slope of its beds, resists the power of the breakers longer than a harder substance placed under less favourable conditions. But as the chances are equal to these varieties being thrown into favourable or unfavourable positions, it will be found that in a great majority of cases, the promontories on a coast are formed on the less yielding material.

. . . By modifications of these causes islands are formed, which afterwards by further upheaval become the tops of hills. The flat sea bottoms surrounding such newly upheaved islands, are again subjected to denudations, and again unequal hardness and exposure produce further inequalities.

Thus by endless oscillations of level, the contour of a country assumes its varied outline. It is remarkable how frequently the hilltops and the higher lands of Wales are more or less tipped with rock of superior hardness, such land having been saved or upheld, as it were by reason of that superior hardness in the midst of the surrounding denudations. From the same operations most of the greater hollows have been scooped out by the destruction of softer materials. (*Ramsay, 1846, pp. 328–9*)

Several actual examples from Wales are cited as stages in this process of events; the flat-topped Monmouthshire hills were considered as first planed by the sea and the valleys being subsequently etched out by the sea during uplift; in a similar manner, the Towy valley was held to have been excavated along the faulted crest of a Silurian anticline; in Central Wales the gradual regularity of slope from mountain to sea was evidence of a former marine plain.

As the sea has encroached not once but several times, Ramsay was prepared to visualize several such plains, one above the other. When therefore he came across marked breaks in slope he saw them as old sea cliffs which denoted the boundary between a lower and upper plain of marine denudation.

From being just another promising young geologist Ramsay had now taken on stature and rank equivalent to that of Sedgwick and Murchison. His work belonged to the modern uniformitarian school, being opposed to catastrophic principles. If its main idea flowed on from Lyell's marine erosion theory, it also contained original features. His contemporaries praised him for his graphic exposition of the amount of past denudation whereas the historian, with the advantage of time, realizes that his most important contribution lay in his theory of marine planation. In the past, Lyell's opinions had almost the force of law; in the years to come they were to be subject to increasing criticism. When twenty years later the question of marine against fluvial erosion was again revived as a vital geological issue, Ramsay's planation theory became the principal support of the marine faction. Even when the idea of universal marine erosion had been discredited, the planation part of the theory lived on in Davis's cycle of erosion and in the writings of mid-twentieth-century geomorphologists.

CHAPTER SEVENTEEN

The Persistency of Lyell's Marine Dissection Theory

About the mid-nineteenth century it is sometimes difficult to decide whether a geologist belongs to the school of Ramsay, Lyell, or Hopkins. The sea formed the basis of all three theories and the sole distinguishing principle concerned the manner in which the sea acted or had acted upon the land. It was easy for geologists to put forward theories which embraced the ideas of all three 'marine' schools. The view of Hopkins and of the structuralists generally differed fundamentally from that of the other two schools in that it considered the present topography to be an event of the past, over and done with, whereas both Ramsay and Lyell followed Hutton in viewing topography as a temporary stage in a series of stages. Ramsay and the marine planationists in turn differed from Lyell and the marine dissectionists in some essentials. Ramsay's physical machine, as we have just discussed, involved a stable or sinking landmass across which the main body of the sea was massively and slowly slicing off an almost level plane, leaving protuberances upstanding where islands had never been submerged beneath the waves. In Lyell's conceptions, the sea actually cut out the inequalities or patterns in the rocks by the usual process of eroding cliffs and inlets and islands, and these as the landmass was raised above sea level stood out as escarpments, valleys and hills.

THE CONTINUED POPULARITY OF LYELL'S MARINE THEORY: 1846–60

The idea of the marine dissection of a rising landmass into valleys and hills and escarpments remained popular throughout Lyell's lifetime. The theory had been adopted by Darwin in his explanation of the mountain landscape features of Australia and Chile (pp. 177–9), and in Europe it was all too easy for observers to mistake the chalk escarpments, with their steep slopes and uniform faces over long distances, for marine-eroded cliffs. The theory was especially popular in the early 1850's when Lyell, its fountainhead, had in no way modified his ideas. In an article on volcanic cones during that year he ascribed calderas as being due to the marine erosion of cones while they were submerged although he had seen one such caldera in La Palma (the Canaries) breached by a great river ravine. The majority of Lyell's contemporaries accepted his marine theory absolutely or in a slightly modified form. The idea of changes of sea level, and particularly of a rising landmass, was used readily in the explanation of river-terraces, raised marine beaches, and of drift or superficial deposits sometimes with and often without the aid of icebergs.

It is easy to find a dozen authors who reflect utter or partial allegiance to Lyell. E. Hitchcock, an American geologist some of whose work has already been discussed, diverged slightly from Lyellian uniformitarianism. In an account of the river-terraces of the Connecticut valley (1849) he sub-divides the terraces into three kinds according to their origin. The first type was considered to be the relics of the ancient sea-margins or a form of raised-beach. This was a direct legacy from Lyell's marine theory. The other two types were related to the margins of fresh-water lakes and of rivers respectively, which demonstrated that Hitchcock also accepted the idea that many erosional features were the result of either stream or glacial action. Generally speaking, he approves of fluvial erosion, although he still suggests that the presence of gorges proved that much of the American continent was formerly under the ocean.

Another paper by Hitchcock in 1856, on erosion and deposition since Tertiary times, showed the persistency of marine theories in a well-read mind. He asserted that 'alluvial formation' was due to causes still in operation and classified deposits as

1. *Drift Unmodified:* produced by glaciers, icebergs, landslips, and waves of translation; and
2. *Drift Modified:* beaches, submarine ridges, sea bottoms, osars, dunes, terraces, deltas and moraines.

The glacial outwash deposits of New England he considered to be old sea bottoms.

Robert Chambers, the journalist, popular educator and editor of Chambers' Journal, also wrote on the subject of terraces (Chambers, 1848). In 1844 he had published anonymously his *Vestiges of Creation* which contained the boldest and most comprehensive statement of organic evolution since Lamarck and in which he applied a modified uniformitarianism to the concept of organic development. The *Vestiges,* with which his name did not become formally associated for some forty years, was produced after Chambers had spent only two years in mastering several scientific disciplines and appeared anonymously because

> . . . he realized that his name would not lend authority to it, and because he feared that the abuse it received would only bring discredit upon the other enterprises with which he was associated. (*Himmelfarb, 1959, p. 179*)

In his account of terraces (1848) he suggested two means by which they might be formed. For either case, he presupposed a rise in sea level. In the first instance he discusses the origin of those terraces that are to be found in river valleys and which, in certain respects, seemed to accord with the general alignment of the valleys. To account for the fact that the river has

FIG. 58. *Chambers' view from the Links, St. Andrews, looking south-eastward, showing a supposed succession of marine benches*

SECTION OF VALLEY AT DUNKELD.

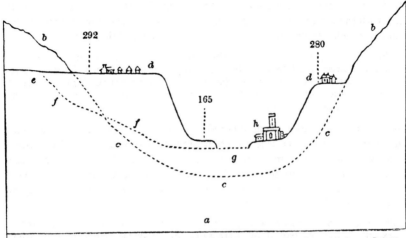

a, present sea-level. b, hills forming the valley. c, their ideal continuation. d d, the terraces. e, f, f, course of the Bran. g, the Tay. h, Dunkeld.

SAME VALLEY IN A SUPPOSED PRIOR STATE.

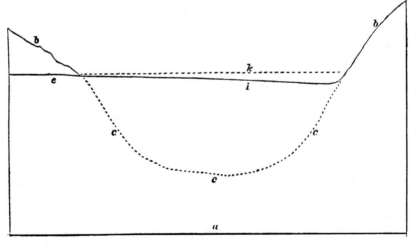

a b c, as above. e, embouchure of the Bran. i, unbroken gravel formation k, surface of the sea.

FIG. 59. *Chambers' sections illustrating his idea of the evolution of the Tay valley at Dunkeld*

actually cut down through its own deposits he supposes that a lake became filled with sediment, then spilt over, the overspill in time widening its outlet until with the change in sea level it was eventually able to cut back through the sediments. The explanation differs little from that of Hitchcock.

> These terraces, which rise above the reach of the present river, yet slope in the same direction. The immediate hollow or trough in which the river runs, has evidently been cut out of what was at first an entire sloping sheet of detrital matter, filling the valley from side to side up to a certain height. (*Chambers, 1848–9, p. 189*)

The second and more usual method by which terraces might originate, Chambers stated was as beaches left behind by a subsiding sea. Chambers' book called forth an important criticism from James Dana, the American fluvialist. There was no necessity, Dana felt, to envisage a rise in sea level; the same terrace formation could just as easily be produced by a lowering of the sea level or by no change at all.

> The beds of rivers being raised, as well as the rest of the land, the amount of descent to the sea would be increased; and in consequence of this, their waters would run more violently, the excavating force would be augmented, and they would go on with a process of rapid degradation, until the former rate of descent was reached. (*Dana, 1849B, pp. 8–9*)
>
> If a slow gradual rise should commence at any time, the river, through its increased excavating power, would begin to sink between its banks. . . . The terrace slope would also show its first beginnings, as an outline of the river flats. Finally . . . it would have been more or less broad flats, which . . . would be commonly bounded by a slope; and, if the former alluvium remained, there would be a shelf or terrace above. (*Dana, 1849B, p. 9*)
>
> Besides depressing their beds, the rivers would act laterally, and carry off the alluvium of their banks . . . until finally, a broad flat or 'bottom land' . . . [would be formed] . . . bounded by a steep slope rising to an upper level. (*Dana, 1849B, p. 9*)

Probably Dana's most telling stroke against Chambers' theory was the statement that any change in sea level would produce terraces not merely in a few places but would extend throughout every part of the river's course.

> [If any change in sea level takes place] . . . there must be river terraces of contemporaneous origin even over the higher lands of the country; every river will tell the tale as well as the beach along the coast. (*Dana, 1849B, p. 11*)

Chambers, however, continued his work and a year later (*Chambers, 1849–50*) he writes of finding erosional terraces in Scandinavia at 522 feet and marine depositional terraces elsewhere at over 2000 feet. His correlation Tables, shown on p. 319, are remarkably modern in lay-out.

If some of the previous writers have seemed somewhat confused in their allegiance, John Phillips makes his belief abundantly clear. His whole mode of thinking shows a close adaptation to Lyell's theory; the land is elevated and broken apart, then wasted by the sea and finally, in an eroded state, raised as

Ridges, &c. on Lake Ontario.	Shell deposit on Hill at Montreal.	Scottish Terraces.	Norway.	France.
996		996		
914		914		
		709		
		687	656	
654		656		
		628-30		629¼
		599	597	599
576		576		576
		562		563¼
542	540	545		545
514		520	515	531
		504		

500 Feet above Level of Sec.

Ontario.	Britain.	France.	Norway.	Spitsbergen.
442	461-6	462-5		
	438-42	442-3		
392	392-4	392		
342	342-8	347		
	279...	281		
242-7	243	208¾		
	202-17	186-92	206	
	186-93	167	147	
	165-70			128
	146	126		
	126-8			
	70 63	69	60	
	40-4			
	10 or 11			

Level of Sea.

Chambers' Tables, showing the correlation of erosional terraces

dry land. In the first note he makes mention of the guiding influence of rock structure and mineral composition from which it is clear that he understood the part played by these factors in fashioning the upper surface. However, much of the later discussion of marine erosion tends to hide this structural emphasis.

> The great features of the earth's surface, the ranges of mountains, the extended plains, the long promontories and retiring bays, depend mainly on the position of the subjacent mineral masses; while the minuter physiognomy of hills and valleys, the sinuosity of rivers, the character of water-falls, and the inequality of caverns, have a further and very important dependence on the internal structure and degree of consolidation of rocks. (*Phillips, 1853, p. 5*)
>
> We shall now suppose these strata to be upheaved, by a force acting . . . (vertically) . . . so as to come within the power of the waves and currents of the sea, and finally to appear above it. By this elevating force the parts over the . . . (greatest uplift) . . . will be first brought under the corroding agency of the waves; the cracks being partially opened by the pressure upon the strata, the continuity of these masses of matter is broken, and their power of uniform resistance to the water is destroyed; the weakest parts yield most, and thus before the strata reach the level of the water, their surface is channelled, and the land as it emerges above the sea exhibits, . . . a broken ridge . . . , between the points of which the strongest currents flow . . .
>
> It is needless to follow further the stages by which, under the same conditions of gradual rise of the land and continual battery of the sea on the parts as they successively come near to and reach the surface, the original islands, . . . become united to slopes and ridges, until they constitute merely the culminating points of the country, . . . On a surface thus constituted, the atmosphere produces further waste, carbonic acid eats away the limestone, moisture softens and crumbles the shale, rains wash away the loosened grains of every kind, rills collect, and rivers carry away the accumulated detritus, and the rough old sea channels, in some places filled up by these deposits, and in others worn still deeper, are changed into those smooth dales or picturesque glens, which are the boast and charm of the North. Rivers run in valleys which the sea made for them. (*Phillips, 1853, pp. 8–9*)

The scarp topography is fitted into this scheme by stressing how sea erosion differentiates between the varying degrees of resistance of the rocks. Yet for all his emphasis on marine action, it is obvious that Phillips is well aware of the slow but constant wearing action of the other elements. He has absorbed a modicum of true uniformitarianism.

In his 1855 publication, *A Manual of Geology*, he accepts the fact that throughout geologic time each type of process has continued acting in a uniform way, but insists that the balance is every so often upset giving one type of process an advantage over the others. This he says shows itself in the more extreme forms of topography. One such concession he makes is the admission that streams can cut their own channels but he qualifies this admission by restricting it to cases where the channels are small.

> . . . the greater action of existing streams has been sufficient to work out their actual channels, though the excavation of the broad valleys in which they

run may have been accomplished by more violent and voluminous waters, flowing in directions predetermined by ancient subterranean movements. (*Phillips, 1855, p. 474*)

Yet Phillips is really recapitulating his old argument, set out in Chapter 10, that a river's main function is to transport and not to erode; that it will more often be seen filling in a valley with deposits than eroding it.

> Wherever the valley originally presented great inequalities, these are constantly diminished by the upfilling of the hollows, and at length the originally rugged chasm is changed by additions and upfillings into the smooth, evenly declining hollow, which, because of that smoothness and uniform declination, is supposed by many to be entirely a valley of denudation. (*Phillips, 1855, p. 479*)

Phillips' work, a very popular textbook in its day, is as clear an exposition of the marine erosion theory as is available. It shows how much reliance was placed on the action of the sea and how little on the action of the rain and rivers. By means of the marine dissection theory, interpretation of surface topography was easy. Wherever the landforms were varied or of great magnitude, geologists merely had to call upon a willing sea to have carved out such features. As all the work of erosion was carried out beneath the ocean, about whose processes little or nothing was known, they could theorize with easy mind. Thus such a facile and plausible theory enjoyed a long reign simply because it did not have to undergo a scientific examination. Moreover, few or no geologists now disputed that continents had risen up above the seas and that seas had previously invaded the lands. For example, the Huttonians agreed with this although they insisted on a slow, imperceptible process as distinct from a violent or catastrophic change.

The popularity of the marine dissection theory is also clearly illustrated in the writings of Edward Hull.

> Scattered at intervals over the Gloucester plain are several outliers of the higher formations, forming isolated hills, which when viewed from a commanding position call forcibly to the imagination a state of physical conditions formerly in existence, when an arm of the sea overspread the plain, bathing on the east the flanks of the Cotteswolds, and on the west those of the Malvern and May Hills, and from the level of which the present outstanding hills arose in the form of islands. (*Hull, 1855, p. 478*)

Hull, like Phillips, had a keen appreciation of the influence of structure, and he added one or two refinements to the marine theory to explain the occurrence of particular features. For instance he drew attention to the fact that an uplifted anticlinal arrangement of strata was more favourable to the formation of a valley by an encroaching sea than was a synclinal arrangement.

> . . . when a slight elevation or anticlinal arrangement of the beds has been produced along a line of country, the axis will create a line of least resistance, and . . . a valley will be formed when the area is subjected to the action of the

Crown Copyright Reserved

FIG. 60. *The Oolite escarpment of the Cotswolds, looking north-east at Cleeve Common near Cheltenham, Glos. An outlier appears in the top left (Photograph by J. K. St. Joseph)*

sea; and conversely, . . . a corresponding synclinal axis will produce a line of greatest resistance, tending to the formation of a headland under the same circumstances. (*Hull, 1855, p. 482*)

D. Sharpe was another English writer who was dazzled by the image of the sea as a giant and inexhaustible force. Not content with attributing to it the scarp topography of England, he extended its potency also to the major relief forms of the Alps. Some parts of his reasoning show the influence of Ramsay's marine planation theory, for as the Alps do not present a regular gradient, but contain many gentle portions finishing in marked steepenings of slope, he assumed that the gentle slopes were marine plains and that the steep slopes marked the maximum limit of marine erosion. To illustrate this graphically he even drew the Matterhorn rising steeply from a marked erosion surface. Another feature showing the former presence of the sea, he believed, existed in the valley terraces, which he considered marked a coastal margin. Whenever a stream entered the sea it would deposit its load, and these deposits, when raised as dry land, would stand out as terraces along the sides of the valley which formerly had been a coastal inlet.

FIG. 61. *Sharpe's (1856) ideas of platforms of marine erosion (EE) in the Alps*
Above: *The Mont Blanc massif, viewed from the Chamonix valley.*
Below: *The Matterhorn rising above an assumed marine surface.*

In the following pages I have attempted to show, that, after the Alps had assumed their present form, and when they already stood as much above the surrounding lowlands as at present, they must have been nearly submerged beneath the sea, out of which their rise must have been, by a series of steps or starts of unequal amount, separated by long intervals of time. The evidence on which these views rest is derived from three sources; 1st, from the traces of erosion on the sides of the mountains, ending upwards in lines of uniform level; 2ndly, from the levels to which the valleys have been excavated; 3rdly, from the elevation of the terraces of alluvium in the valleys . . . (*Sharpe, 1856, p. 102*)
. . . the upper limit of the . . . marine . . . abrasion is usually well marked by a change in the outline of the hill-sides from a moderate to a steeper slope, similar to the indentation produced on a coast by the waves beating against a cliff. The nearly uniform height of this re-entering angle produces a tolerable level line round the mountains, which I shall speak of as the line of erosion. (*Sharpe, 1856, p. 106*)

His reasoning leads him to divide the Alpine slopes into three main erosional levels – at 9000′, 7500′ and 4800′, each level being based upon a topographic step. These main and several minor breaks-of-slope are expressed in the following statistical table.

Lines of Erosions.	Heads of Valleys.	Terraces.
English feet. 9000 7500......	7660	
	6593..................................	6100, 6190
	5756, 5835, 5895	5900, 5930
		5650 ?
	5410, 5455	5464
	{ 5280*, 5257*, 5315*, 5315, 5365, 5370 .... }	5200 ?
	4877, 4910	5000 ?
4800......	{ 4626*, 4642, 4642*, 4730, 4763, 4776 .... }	4720, 4770, 4780, 4782
		4597
	4400...........................	4422
	4305*, 4316*	4343, 4343, 4350
	4240, 4274	4230, 4233
	4100, 4100, 4117, 4143, 4152 ...	4150
	4058..........................	4060
	3936, 3953, 3954, 3985	3970, 4000 ?
	3822, 3828, 3867	{ 3730, 3770, 3770, 3815, 3825, 3865
	3597..........................	3567
		3415, 3444, 3465
		3254, 3300, 3300, 3316
		3100, 3113, 3113, 3120, 3175
	2905..........................	2930, 2930, 2930, 2940
		2770, 2770, 2821, 2840
	2600 ?	2690
	2500..........................	2490
		2330, 2350, 2350
		2200, 2200
		2100 ? 2145, 2145
		1650 ? 1700
		1234, 1300
		981, 1000 ?

The numbers marked ? rest on doubtful measurements.
The numbers marked * are below the truth.
The altitudes placed on the same line are supposed to belong to one water-level.

The small measure of erosion allowed to natural processes, as well as the ignorance of glacial erosion, is clearly shown by Sharpe's insistence that cirques are a special form of scooped-out cliffs.

> These deep indentations at the heads of valleys are not due to the streams which flow in them, for their action tends to equalize the slope of their beds: they must owe their origin to the action of waves beating for a long period against the rock. (*Sharpe, 1856, p. 111*)

When Sharpe attempts to explain the origin of valleys he follows the theory of Lyell and Hopkins in supposing that the sea excavated them by working along original cracks, faults, or fractures in the surface.

THE LYELLIAN DIE-HARDS, 1860-75

The marine dissection theory continued to find stout adherents throughout Lyell's old age. Some, including Scrope, John Phillips and Hull, we have already mentioned; others appear on the scene for the first time. Most are utterly or unmistakably Lyellian but a few seem uncertain whether to accept the teachings of Lyell or Ramsay. Those who openly preferred marine planation we shall discuss elsewhere.

An interesting modification of the marine dissection theory was propounded by S. J. Mackie in relation to the English Weald. He was most concerned with the problem of explaining the erosion of the wide longitudinal vales. To produce the requisite machinery for this, he assumed that England was land-locked with north-east France and that the flood tide, which normally rushed through the Strait of Dover into the North Sea, was diverted north-westward towards the Weald, where it scoured away the weaker strata. He devised the following sequence of events: the Weald rose slowly; cracks appeared in its surface; these were rapidly widened by a flood tide.

> Sweeping round their eddyings as restrained by the ridge of land that formed the barrier to their onward passage, these pent-up tides scoured the Wealden plains, and strewed the bases of the downs with half-worn flints and gravel . . . until the time of ebb, when it would flow out with a strong current, carrying with it the debris, and thus the cause of the clean surface of the Weald and the general absence in the few superficial deposits of any organic remains, the denudation being a tidal one. (*Mackie, 1860, pp. 202-3*)

If, however, we wanted confirmation that the heads of the marine erosionists were still unbowed, Mackintosh would provide it better than Mackie. When others began to modify or waver, Mackintosh grew more violent in his advocacy of marine erosion, and over the next few years he constantly appears as the evergreen, ever-ready opponent of the fluvialists. His standpoint is quite simple: waves, tides and currents are the important denuding agents; rain and frost exert a small influence on the shape of the surface and only produce minor inequalities; they are quite incapable of producing such features as cliffs and pillars, which are the result of what he called lateral erosion.

> Meteoric agents . . . operate from above, and their mechanical effects at least are confined to a decrease of level, or the formation of slight inequalities. (*Mackintosh, 1865A, p. 154*)

Rivers are considered to have slightly modified the area by forming small gullies and by the 'roughening of the surface of rocks'. (*Mackintosh, 1865B, p. 300*)

> The streams which traverse most of the valleys of the Lake District are too insignificant to justify the supposition that these valleys have been scooped out by running water since the last emergence of the land. (*Mackintosh, 1865B, p. 303*)

Mackintosh's method of reasoning revolved around what he referred to as lateral erosion. Rain and rivers, he argued, eroded downwards and because many rivers flowed in flat valleys he presumed that the rate of erosion must be very small. On the other hand the sea advanced by planing off the land beneath the wave-surface in a horizontal fashion, leaving the land out of reach of the waves upstanding as cliff-bound eminences. As much of the land surface was in fact flat and the plains bounded by steep, scree-footed precipices, he believed that such a landscape was a product of the sea, and was content to explain all such topographic features by reference to a marine origin.

> The only difference between many mountain-passes and a sea-strait consists in this, that the strait is still washed by the sea – the pass has long since been forsaken by the billows. (*Mackintosh, 1865B, p. 302*)

In the following years Mackintosh launched many vigorous counter-attacks upon the fluvialists, who assailed him right and left. Scrope replied in a manner reminiscent of Mackintosh's own technique; he studied the terraces on the hillsides of Wiltshire and Dorset which Mackintosh had put forward as marine-eroded beaches and then set out to prove that they were fashioned by other means.

> . . . I have no hesitation in declaring them without exception of artificial origin, worn by the plough, at a time when these slopes were, if they are not still, under arable cultivation. (*Scrope, 1866C, p. 293*)
>
> . . . the ridge of soil raised by the mould-board of the plough has everywhere a tendency, through the action of gravity upon it, to fall down-hill, never upwards. This down-hill tendency of the disturbed soil is greatly assisted by the wash of heavy rains upon the sloping surface, and culture is, slowly, indeed, but surely, travelling downwards, until it is stopped by some hedge, or wall, or bank, . . . (*Scrope, 1866C, p. 294*)
>
> . . . in fact, these terraces brought forward by Mr Mackintosh as 'proofs' of the impotence of rain in moulding the earth's surface, afford on the contrary very pregnant and convincing evidence of its power in altering the configuration of our hill-slopes within very recent and limited times. (*Scrope, 1866C, p. 296*)

Even as late as 1869 Mackintosh remained incorrigible. In that year in his book on the scenery of England and Wales he still relied heavily on marine denudation, particularly by tidal currents and waves. Six chapters were devoted to the marine origin of inland escarpments while one chapter only was given to describing the erosive power of streams, rain and ice.

In contrast to Mackintosh, Edward Hull has altered his views more in favour of fluvial erosion. In an article (1866A) on the denudation of valleys he combines fluvial and marine theories. Rivers in steep-sided valleys in Lancashire and Yorkshire carried, he observed, much material; but, on the other hand, some valleys, such as the Vale of Todmorden, continued as marked features right across the watersheds where no streams existed. These high-level troughs (now known to be glacial melt-water overflow-channels) and the presence of post-Pliocene drift-deposits caused Hull to conclude that the whole area must have been recently submerged and subjected to marine erosion.

Similarly, A. B. Wynne cannot fully retain yet cannot dispense with Lyell's marine dissection theory. Wynne suggests that the chief factors that ought to be studied in the problem of denudation are

(1) the antiquity of the eroding agents;
(2) the influence of structure; and
(3) the frequent submergence of land beneath the sea in the past.

The emphasis on the last-named factor arose partly from contemporary researches into Pleistocene changes of sea level.

> most of the land has been frequently beneath the ocean, many inland cliffs have stood within reach of its denuding influence. (*Wynne*, *1867*, *p. 4*)

Despite this marine bias and a doubt about wide longitudinal vales being of fluvial origin, Wynne thought that some inland cliffs could be produced by subaerial denudation where vertical jointing controlled the pattern of the erosion, and he was far from subscribing fully to Lyell's ideas.

> Although the sea coast has a certain relation to the form of the land, it may be doubted whether it has any to show that the surface of the latter has depended for its shape upon marine denudation. To remove such doubt it must be proved that all parts of the existing surface have been successively acted upon by the sea, and consequently that atmospheric agencies have not since materially altered its configuration; but this is just the point in dispute among the advocates of marine versus subaerial denudation, and is certainly far from being proved. (*Wynne*, *1867*, *p. 7*)

It might be thought that once someone proved conclusively that marine erosion was relatively ineffective compared with subaerial denudation, the marine theory of Lyell would rapidly become unimportant. In fact this was done in 1868 by Archibald Geikie in his famous article on denudation which we discuss in detail in Chapter 23. Geikie, by using known rates of denudation achieved by river-systems and by wave-erosion, showed clearly that 'before the sea could pare off more than a mere marginal strip of land between 70 and

80 miles in breadth, the whole land would be washed into the ocean by atmospheric denudation'.

This was the most effective quantitative demolition of the marine erosion theory ever published and its results must have been singularly disconcerting to the marine erosionists. For years most of them had doubted or ridiculed the ability of rivers to erode deeply the land surface, and here Geikie had proved not only that rivers could erode but that they could demolish at a much faster rate than the sea. If, however, any reader should think that Geikie's argument caused the sudden extinction of Lyell's marine theory, he seriously under-rates the strength of tradition, especially among older geologists. Lyellian die-hards remained very much alive.

Wood was one of these echoes of the past. In an article of 1871 on the Weald he purports to act as a bridge between old and new ideas. He outlines the various theories to account for the form of the Weald: how the theories of the structuralists and marine erosionists have 'of late years . . . met with partial dissent'; how Foster, Topley and Ramsay assume the present features to be entirely due to rivers acting on a plain of marine denudation, whereas Martin and Lyell uphold the marine erosion theory. Then Wood offers his own explanation, that we discover depends almost entirely upon erosion by the sea, to the exclusion of all but a very small accompanying erosion by subaerial forces. The width of the central Wealden area seems to have be-mused Wood, and he, as with other marine erosionists, failed to understand the ability of rivers to excavate wide valleys.

> The form and character of the great Wealden denudation area (or major valley), as distinguished from the valleys proper of the Wealden rivers (or minor valleys), is diametrically opposite to any that can result from river-action, because, however great we concede the power of that action to be, any excavation resulting from it must be coterminous with the excavating agent itself (the river and its tributaries), since every stream, large or small, can only deepen its own proper valley, and the result cannot be any such excavation as the major valley of the Weald, with its well-known contour and escarpments, but only a series of valleys, or minor excavations, ramifying in the directions in which the stream extends, and in some degree at least coinciding with them; and the longer this action is continued, the deeper and more distinct must these features become. (*Wood, 1871, p. 11*)

The escarpments, according to Wood, were not sea cliffs but steep foreshores. Marine erosion took place during uplift and the rivers occupied the elevated channels formed by marine scour. Progressive uplift produced an arm of the sea extending into the Weald, with the central area as an island – the whole area progressively rising. The isthmus of Dover was sufficient, he thought, to produce in the Wealden area an erosive tidal scour which carved out not only the major valley, but the minor ones as well.

In the same year, J. Phillips, in his *Geology of Oxford*, also showed himself to be very much in line with Lyell's marine erosionist theory (*Phillips, 1871:*

Anon, 1872). He had not really embraced the new ideas of marine planation, nor the version which combined it with subsequent subaerial erosion. In his scheme he simply submerges the land 250 feet, then 500 feet and finally 1000 feet beneath the present ocean level and leaves the formation of the present topography to the gradual processes of waves, tides and currents.

> . . . the Thames Valley would be . . . a vast estuary with a sea-loch up to the Kennet Vale; straits between the chalk hills of Chiltern and Lambourne. (*Phillips, 1871, p. 44*)

At the third level of submergence only the higher peaks near Cheltenham, Evesham and the Malvern Hills were above water, and this was followed by emergence.

> . . . and thus our land-surface, as we see it, exhibits in every part the modifications produced by what may be called the 'ordinary action' of daily causes, these being superimposed on broader and greater features generated by elevation and depression on a grand scale, accompanied by powerful waves and strong currents of the sea. (*Phillips, 1871, p. 46*)

Not all marine erosionists were quite as traditional as Phillips. In 1872–3 T. M. Reade described the drift-filled buried channel of the River Mersey and ascribed it to the erosion of fluvio-glacial agents prior to its infilling with drift. Yet in the next year he confused the argument by introducing the idea of marine erosion. By using Beardmore's formula for the surface, middle and bottom velocities of streams (pp. 428–9), he calculated the depth at which marine currents could act. The resultant calculation seemed to show that marine currents were capable of cutting submarine relief on the sea bed by constant erosion due to the movement of water and debris, in exactly the same way as a river erodes its own bed.

When it is remembered that this mathematical application was made only one year before Lyell's death and the final edition of his *Principles*, the persistency of the marine dissection theory seems amazing. We shall show later that its survival depended partly on the stubborn unwillingness of many geologists to accept the full implications of the Huttonian fluvial theory. But may not some of its longevity be due to environmental influences? The British Isles stand on an exceptionally wide continental shelf where violent seas, high tides and racing currents give an overpowering feeling of omnipotence. The sea was Britain's heritage and thousands of Victorian homes echoed the verses of the Romantic and Nature poets.

> Roll on, thou deep and dark blue Ocean – roll!
> . . . .
> Thou glorious mirror, where the Almighty's form
> Glasses itself in tempests; in all time, –
> Calm or convulsed, in breeze, or gale, or storm,
> Icing the pole, or in the torrid clime
> Dark-heaving – boundless, endless and sublime,
> The image of eternity, . . .

CHAPTER EIGHTEEN

The Melting of Lyell's Iceberg Theory

THE WEAKENING OF LYELL'S GLACIAL MARINE THEORY

Whereas Lyell's marine dissection theory was still remarkably unimpaired by criticisms in the mid-nineteenth century, his ideas on iceberg deposition were showing signs of being undermined by the glacial theory. It is true that the widespread occurrence of glacial deposits or drift tended to support Lyell's contention of a recent marine submergence but the distribution of large erratic boulders was often less happily explained by marine or iceberg agencies. Because of Agassiz's writings and lectures, the action of mountain glaciers in grinding their way slowly down valleys and heaping up morainic mounds at their snouts, was well understood, but few geologists were also willing to accept the necessary extension of the glacial theory – that vast ice-sheets had once spread over the lowlands of much of Europe and North America. The solution of the problem was hindered by the fact that, as Agassiz had pointed out, over vast areas, the marshy terrains, gravel deposits and winding ridges of poorly stratified material all bore the signs of having had some connexion with water. Much glaciated topography was in fact fluvio-glacial or the work of ice and of melt-water. This and Lyell's omnipotence largely explain why marine explanations were still so widely followed.

Darwin continued to be a confirmed disciple of the iceberg theory.

Those who believe in the powerful agency of ice in moving boulders will probably conclude that icebergs have in some manner transported them from a lower to a higher level. But the most obvious method by which fragments of rock can get on icebergs is by their having first fallen from the surrounding precipices on glaciers entering the sea, and therefore they must have come from a higher to a lower level. (*Darwin, 1848, p. 317*)

To take the case of North America: Mr Lyell has shown, from an independent train of reasoning, that this country during the glacial period was submerged to a considerable amount: several American geologists have come to the same conclusion, and they believe that the subsidence amounted to two or three thousand feet, or even more . . . (In the estuary of the St Lawrence) annually an enormous number of boulders, both on and near the coast, are frozen into the coast-ice and transported to shorter or greater distances; can we doubt, that if during the year the land sunk a few inches or feet, the boulders, whilst actually frozen in or when refrozen during the ensuing winter, would be lifted up and landed so many inches or feet higher up the coast? (*Darwin, 1848, p. 319*)

Murchison, as the quintessence of all conservative thought on geology, upholds the same principle two years later. He correlates the Alpine with the

Scandinavian erratics and ascribes the position of both to floating ice at a time when much of Europe was beneath the ocean. (*Murchison, 1850*)

R. A. C. Austen, a confirmed anti-fluvialist, supported Lyell in a more indirect way by postulating numerous changes in sea level during the Pleistocene period.

> That, with respect to movements of the earth's crust in this region, during a period which Geologists have agreed to consider as one and indivisible . . . [the Pleistocene] . . . the oscillations (of sea level) . . . have been great, both of depression and of elevation, and that there has been at several distinct periods a constant return to a level, very near the present one . . . These former levels are to a great extent independent of the present relation of land and sea: the external configuration of the country had obviously been acquired prior to any of the successive conditions which have been here described. (*Austen, 1851, p. 136*)

Hopkins, the propounder of the waves of translation theory, is prepared to compromise, allowing that erratics may have been transported by any one of three means, different processes having operated in different areas:

> I consider the distinct recognition of these three agencies of transport–glaciers, floating ice, and currents – as essential to the final establishment of sound theoretical views on this subject . . . (*Hopkins, 1852, p. xxvii*)

This hesitant position was adopted by many contemporary geologists. In the Alps the evidence pointed all one way and Agassiz's arguments on his home ground were too cogent to be ignored but they felt that to extend the arguments to lowland or coastal areas was quite another matter and called for other hypotheses. Therefore there is nothing contradictory in Hopkins admitting the action of glaciers and at the same time stressing his own elevation theory.

> It appears to me, therefore, that we are driven to the alternative either of rejecting all theories on the subject, or of adopting that which would attribute these currents to waves of elevation, resulting from frequent, sudden, but not extensive vertical movements of the central range of elevated land; movements which we may conceive to have thus repeated while the mean movement of the whole region was one either of gradual depression or of elevation . . . we may conceive the centres of the elevatory movements to have been different at different times . . . as in fact they would appear to have been from the different directions in which the transported matter has been driven from the same original site. But movements which could send forth the greater quantity of floating ice would be those which more immediately affected the line of coast; and the coast being deeply indented . . . by the present river-valleys when submerged, torrents would be simultaneously discharged from their mouths which would determine in a material degree the resulting current in the open sea. . . . (*Hopkins, 1852, p. xxxi*)

None of these contributions can be called significant and made no appreciable progress on the much-revered statements by Lyell in the many editions of *The Principles*, which gave, as we have noticed (pp. 200–1), excessive prominence to his iceberg theory.

Strangely enough, it was Ramsay, the youthful derider of glacier notions, who was to make the next real onslaught upon the Lyellian icebergs.

In Chapter 16 we left Ramsay at the time when he had just published his treatise on the *Denudation of South Wales*, which contained the rather startling notion of marine planation. Since then his progress up the scientific scale had been deservedly steady and he was now De la Beche's right-hand man. His ideas as well as his fortunes were beginning to advance. In 1847 he accepted the Professorship of Geology at University College, London, delivering lectures during January, February and March, the only months when he could be spared by the Ordnance Survey. The first set of lectures was a sketch of the progress of geologic thought from Strabo down to Ramsay's own time. It was in the preparation of this series that Ramsay in his diary provides the first hint of a radical change in his beliefs:

> Stuck at Hutton's *Theory of the Earth* and Playfair's *Illustrations* all day (Sunday), and before night read all, and made a complete abstract of the latter . . . Wrote a good bit of my lecture at night. Hutton every day strikes me with astonishment. Lyell does not do him half justice. (*Quoted in Geikie, 1895, pp. 116–17*)

Unfortunately he does not say exactly what impressed him about Hutton's ideas.

In 1848 we read of Ramsay acting as a voluntary special constable during the Chartist meetings in London. His diary captures the feeling of tension at the time.

> Grand Row expected today. Forbes called, and we went down to the Museum before ten; met Playfair. Sir Henry at the Museum very active and mysterious, passing through holes in the back stables of the Scotland Yard Police Office, and bringing out armfuls of cutlasses. Streets full of Special Constables. Chartists afraid, and cowed, all passing off quietly. No procession took place. However, we had a jolly dinner in Sir Henry's room for fourteen, and cigars and coffee in the laboratory afterwards. This was the hardest duty we had to perform. On public grounds, our men were well pleased that things went off quietly; but as private individuals, many seemed rather disappointed that there was no scrimmage, especially Bone and J. A. Phillips, who were very bloodily inclined. Salter was evidently in a funk, and kept up his spirits all day by whistling psalm tunes. (*Quoted in Geikie, 1895, p. 129*)

He still retained his interest in British literature and read widely and quoted frequently—some of his colleagues thought *too* frequently! – especially from Shakespeare, Scott and Keats (*Geikie, 1895, p. 246*). Like many other literary-minded men he was not above indulging in occasional flights of poetry of a somewhat rumbustious vein. Perhaps one of his happiest efforts was that delivered at an anniversary dinner, which captures in a few lines life in the early Survey:

> I joined the Chief in Tenby Bay
> and shillings I caught nine
> 'twas three for breeks, and three for beer
> and shillings three to dine.

When first I left the Land o' Cakes
and took to wearing breeches,
I little thocht that I should join
this corps o' De la Beche's.

There's Forbes' men that work within
and our field-working laddies,
including Jukes, that shaved his chin★
to please the Irish paddies.

When age has put our auld pipes out
by precept and harangue,
new lads will rise without a doubt,
will gan the hammers bang.

With such an imaginative and responsive mind it is not surprising that Ramsay's scientific outlook should be flexible to recent changes. In 1850 in a paper entitled *The Geological Phenomena that have produced or modified the scenery of North Wales*, he acknowledged that glaciers had played some part in the shaping of the landscape. Gone now was the ridicule of the glacial theory. Henceforth he was to be the leading British advocate of the past action of ice. In 1852 he married and spent his honeymoon, his first departure from this country, in Switzerland where he was greatly impressed with the daily demonstrations of the natural processes of denudation.

> The scenery is so large and grand, the cliffs so great, the strikes, dips, and contortions of the great masses of strata so enormous and so grandly exposed, and the immense slopes of talus below, scarred with frequent torrents, give such overwhelming ideas of the incessant effects of atmospheric disintegration. (*Quoted by Geikie, 1895, p. 199*)

The visit also fostered his growing concern for glaciology. On his return he went first to Scotland, to organize the survey of that country, and took a growing interest in its glaciated landscapes, though as yet he still held the view that drift deposits had been dropped by icebergs. When he had completed his work in Scotland he moved to the Midlands and here he purported to find traces of glacial deposits in the Permian beds. In an article of 1852 he cites slight evidence of two glaciations, each of which had preceded the deposition of the Drift, and each of which was imposed on an already-formed topography.

> It is evident that before this glacier period the land had already received its grand contour. (*Ramsay, 1852, p. 374*)

This supposed Permian glaciation was a completely new suggestion. Whereas Agassiz's theory of an ice age applied to relatively modern times, Ramsay's statement implied that there had been more than one glaciation,

★ Jukes was at one time a candidate for the Chair of Geology at Trinity College, Dublin, and to assist his application he shaved off a beard which he had grown. On failing to gain the appointment he allowed it to grow again.

for the Permian system evolved at a relatively early period in geologic history and was removed by many million years from Agassiz's ice age. To confirm his ideas Ramsay visited comparable Permian formations in Germany. By this time De la Beche's health was failing and more and more of the administrative work devolved upon Ramsay, now the father of three children. In April 1855 De la Beche died. Ramsay had hoped to succeed him but when it became apparent that this was not to be, he loyally supported the appointment of Murchison as Director-General.

In the same year Ramsay published a more detailed article in which he positively asserted that the Permian beds possess abundant evidence of an ice age during their formation. The rocks and deposits that are everywhere found in them bear far too close a resemblance to their glacial counterparts in the Pleistocene areas for any other conclusion to be possible. The source of these deposits is also obvious, though Ramsay still retains his old prejudice for icebergs as the means of their transport.

> In the summer of 1852 I traced the boundaries of the Permian breccias that run between the Bromsgrove Lickey and the Clent Hills, having previously visited similar rocks on the flanks of the Abberley and Malvern range. Though much struck with the size and angularity of the fragments, and with the marly paste in which they are imbedded, I did not then venture to propose to myself the solution of these and other peculiarities, at which I have since arrived, viz. that they are chiefly formed of the moraine matter of glaciers, drifted and scattered in the Permian sea by the agency of icebergs. (*Ramsay, 1855, p. 186*)
>
> The lithologic nature of the imbedded fragments has already been described. Everywhere in spite of exceptional fragments in the Malvern district, they seem to have been derived from one set of rocks; they are all enclosed in the same red marly paste, and they are mostly angular or subangular. A well-rounded waterworn pebble is, in places, of rare occurrence. The surfaces of a great majority of the pebbles are much flattened, numbers are highly polished, and, when searched for, many of them are observed to be distinctly grooved and finely striated. (*Ramsay, 1855, pp. 196–7*)

Ramsay must have realized that his conclusion was very radical, considering that most geologists had yet to accept Agassiz's theory which only involved one ice age and he was asking them to accept two. To forestall much of the obvious scepticism he argued his case in some detail. As the distance between the known source and the present position of the deposits was quite considerable, he thought that if the transport had been done by sea or river most of the material would have been rounded by attrition. Yet the materials were either angular or angular on one side and smooth and scratched on the other. He discounted the theory of Hopkins, that the material might have been scattered by waves generated by earth-movements and suggested that transport by water in any form, waves, tidal currents and rivers, could not be reconciled with the angularity of the debris. The glacier was a proved agent for the transport of angular, striated rock fragments.

They were therefore deposited in water with considerable regularity, and, as we have seen, over a large area. It is altogether unlikely that the stones were poured into the sea by rivers in the manner in which some conglomerates are formed on steep coasts, where mountain-ridges nearly approach the shore, 1st, because the fragments, being derived almost exclusively from the Longmynd country, if the sea then washed its old shores, no river-currents passing out to sea could carry such large fragments from thirty to fifty miles beyond their mouths and scatter them promiscuously along an ordinary sea bottom; and, 2ndly, if the rivers merely passed from the Longmynd across a lower land to the sea, transporting stones and blocks of various size, these would have been waterworn in their passage seaward after the manner of all far-transported river-gravels, whereas many of the stones are somewhat flat, like slabs, and most of them have their edges but little rounded. Neither could ordinary marine currents move and widely distribute fragments so large that some of them truly deserve the name of boulders; and except in the case of earthquake waves, which here and there occasionally produce an occasional debacle on a shore, I have no faith in violent currents of sea water (such as have been sometimes assumed to result from imagined sudden upheavals of land) washing across hundreds or thousands of square miles, and bearing along and scattering vast accumulations of debris far from the parent rocks. This is an assumption without proof. . . .

If then, they were not distributed by any of these agents, there remains but one other means of transport and distribution – the agency of ice.

1st. There is proof, the great size of many of the fragments, the largest observed weighing (by rough estimate) from a half to three-quarters of a ton.

2nd. Their forms. Rounded pebbles are exceedingly rare. They are angular or subangular, and have those flattened sides so peculiarly characteristic of Pleistocene drifts, and the moraine matter of the Welsh, Highland, Irish and Vosges glaciers.

3rd. Many of them are highly polished and others are grooved and finely striated like the stones of existing Alpine glaciers and like those of the ancient glaciers of the Vosges, Wales, Ireland and the Highlands of Scotland; or like many stones in the Pleistocene drift. (*Ramsay, 1855, pp. 198–9*)

If this conclusion be correct, and if the parent rocks whence the stones were derived be properly identified, then it follows that the ancient territory of the Longmynd and the ancient Lower Silurian rocks, having undergone many mutations, at length gave birth to the glaciers, which flowing down some old system of valleys reached the level of the sea, and, breaking off into bergs, floated away to the east and south-east and deposited their freights of mud, stones and boulders in the neighbouring Permian seas. (*Ramsay, 1855, p. 200*)

Irrespective of the possible geological inaccuracy of Ramsay's conclusion (as he and his colleagues were unaware of coarse deposition by flash-floods in arid regions, which we describe in Vol. II), for glaciology this article was a great advance, the first real step forward since the advent of Agassiz's theory. It also showed the direction that the major part of Ramsay's studies were to take. From now on he became the chief English investigator of glaciated landscapes and during the next few years he gradually abandoned the iceberg idea in favour of the full implications of Agassiz's teaching.

THE MELTING OF LYELL'S ICEBERGS

After 1855, Lyell's ideas on iceberg deposition began to be assailed from both sides of the Atlantic. In America, Dana when describing *American Geological History* (1856) expressed disbelief in the iceberg theory because, he said, there was no evidence of submergence in the southern parts of North America during the Ice Age; if anything, the southern states had been elevated during this period.

We have already noticed (pp. 232–3) how Lyell himself, fresh from another visit to Alpine glaciers in 1857, had accepted the inapplicability of iceberg transport in the Alps, the Jura and at least parts of Scotland. But he seems to have been far too busy with other aspects of geology to correct his bias towards iceberg deposition elsewhere.

The reduction of icebergs from major to minor influences on the formation of landscapes was primarily the work of Ramsay, who now gradually veered towards the full implications of the glacial theory. His claim that there had been glacial conditions during Permian times was met by the customary amount of scepticism, but his conviction was such that at the 1856 Anniversary Dinner of the Geological Survey he defended his views in verse.

> Few, few believe what I have told,
> Men say that I am overbold,
> What then? they sneered that Welshmen's tails
> Had polished Buckland's rocks in Wales.
>
> And when I'm dead, and these poor bones
> Lie underneath the turf and stones,
> The home of worms and churchyard mice,
> Men then will swallow Permian ice.
> (*Quoted in Geikie, 1895, p. 241*)

In 1858, both his convictions and his knowledge of glaciology were strengthened by a summer stay in Switzerland with John Tyndall, who was one of Europe's leading glaciologists and had continued the work of Agassiz, Forbes and others on the mechanics of glacier movement. After this visit Ramsay seems to have had no reservations about Agassiz's version of the glacial theory. From his own knowledge of Ramsay's opinions at this time, Geikie is able to say:

> During one of these rambles with me in Fife our conversation turned on the Boulder clay and the mysteries of its origin. We both felt how unsatisfactory was the received explanation of iceberg action and submergence. I was thus led to study this deposit, and to reach the conclusion, at which Ramsay also simultaneously and independently arrived from a consideration of other evidence, that the great glaciation was the work of land-ice. This change of (Ramsay's) view was completed before the summer of 1861. (*Geikie, 1895, p. 261*)

Ramsay's conversion to the glacial theory had another effect; it seems to have caused him to forsake the theory of marine dissection, even as an explanation of valley erosion; and by abandoning this theory Ramsay made his biggest break with Lyell. He wrote to Geikie in 1863:

> By the way, I think I have given up the marine denudation of the Weald. Atmosphere, rain and rivers must ha' done it. (*Quoted by Geikie, 1895, p. 280*)

Conviction led to more detailed observation, and this in turn led him to formulate his own conclusions, which appeared in the form of three articles on the Alpine valleys and lakes. Few of Ramsay's conclusions were new, but they were nevertheless extremely important, because they showed unequivocally that he, a leading geologist, had abandoned the iceberg theory and in its place had substituted the original theory of Agassiz.

> I am therefore constrained to return . . . to the theory many years ago advocated by Agassiz, that, in the period of extremist cold of the Glacial epoch, a great part of North America, the north of the continent of Europe, a great part of Britain, Ireland, and the Western Isles, were covered with sheets of true glacier-ice in motion, which moulded the whole surface of the country, and in favourable places scooped out depressions that subsequently became lakes. (*Ramsay, 1862B, p. 204*)

Ramsay's change of views was fundamental; he had not only lost faith in the iceberg theory – he had vigorously criticized its deficiencies. In later papers (1862A and 1864) Ramsay added a few details to Agassiz's early ideas. Thus, he noted that the present Alpine valleys were created by both rivers and glaciers. After post-Miocene or pre-glacial stream-erosion, the valleys were deepened or widened by ice during the glacial period; and finally they have been further modified by rivers since the ice-retreat.

Ramsay's contemporaries were especially excited by his 1862 article in which he stated that many large lake basins had been scooped out by glaciers. Bonney wrote in 1890 that 'few scientific papers have ever excited more interest or more controversy. The latter is not yet decided.' Today we have no doubts, nor do we doubt the destructive effect that Ramsay had upon Lyell's glacial marine theory.

We must, however, notice that Ramsay received some help from others. Indirect assistance came from fluvialists such as Prestwich (p. 448) who decided that many gravel deposits could not be of marine origin, as Hopkins, Wood and Mackintosh supposed, but that they were of local origin, derived from the native rocks, and were a natural feature of the life-cycle of rivers. He felt that the employment of a marine theory of any kind was unnecessary to explain the presence and situation of gravels such as occurred in the Weald and north-eastern France. Here the correct explanation for the existence of gravels at higher levels lay in the occurrence of a severer climate at the time of their formation. Only the larger erratics were connected with ice-transport – in this case by river ice-floes.

> . . . that the excavation of the valleys, is wholly due to river-action; the gravel represents old river-beds, and the brick-earth or loess the deposits left by the

rivers in times of flood. The climate being probably much more severe at the time when the higher gravels were deposited than now, the melting of snow in the spring would raise the river to levels such as they cannot possibly reach now, and the flood deposits would be proportionately increased. The rivers would also be laden with ice, and in this manner Mr Prestwich accounts for the large blocks of stones sometimes found in the gravel. (*Discussed by Topley*, *1875, p. 29*)

Lyell himself commented favourably on Prestwich's work and in his eleventh edition of *The Principles* (1872) writes:

> The beds of gravel often called drift, which contain antiquities of this (Palaeolithic) age, may be said to have been deposited by the existing rivers, when these ran in the same direction as at present, and drained the same areas, but before the valleys had been scooped out to their present depth. The height above the present alluvial plains at which the old drift occurs is often no more than 20 to 30 feet, but sometimes 100 or even 200 feet. (*Lyell, 1872*, *II, p. 566*)

The inevitable conclusion is that less and less of the 'drift' was being attributed to iceberg deposition, and that gradually the recognition of the glacial origin of the unstratified 'drift', which obviously could not be imputed to fluvial action, was removing the need for a universal inundation.

In this connexion Ramsay had a useful helper in T. F. Jamieson who in 1860 published a promising article on the drifted and rolled gravel of North Scotland. Jamieson argued at length that the recent bursting of a valley-dam here had not caused either 'the striae or the deposits associated with glacial action'.

> the well-rounded aspect of the pebbles, indicating long continued rolling by water, forbids the supposition of one catastrophe doing the whole by the sweep of a sudden wave or deluge passing over the country. And further, as this gravel is found along not a few rivers, but along all that I have hitherto examined . . . I think the first of the suppositions enumerated becomes highly improbable for it can scarcely be thought that there has been a bursting of lakes in all the water-courses . . . (*Jamieson, 1860, p. 353*)

He concluded that the debris he was examining had been carried by streams whose discharge was by some cause increased during the Pleistocene period. He went on to assert that the position and pattern of erratic boulders elsewhere in the Scottish Highlands were all in favour of an origin by glaciers or ice-sheets. Jamieson wrote an article of equal importance in 1863 on the origin of the parallel *roads* of Glen Roy, a problem which had attracted the attention of Darwin and many others. His conclusions virtually settled the question once and for all. According to his theory, the 'roads' were really the remnants of beaches that lined freshwater lakes, impounded by the reversal of the drainage which occurred with the accumulation of ice during the Ice Age.

Jamieson's papers are important partly because their certainty hastened Archibald Geikie's entry into the field of glaciology. But to us their date is also of significance. Here was a geologist in the 1860's writing of landscape problems that might have been taken as settled twenty years earlier. The fact seems to be that despite the abundance of eminent geologists in Britain when Agassiz presented his glacial theory, the field for exploration which it offered has been largely ignored, except by Ramsay. How much of this neglect was due to Lyell's marine ascendancy will never be known. Thus when Geikie thinks as follows we can see the shadow of the master uniformitarianist clearly imprinted on the page:

> When such current-driven masses (icebergs) grate along the sea bottom they must tear up the ooze and break down and scratch the rocks. In the course of long ages a submerged hill or ridge may get its crest and sides much bruised, shorn, and striated, and the sea-bed generally may be similarly grooved and polished, the direction of the striation being more or less north and south according to the prevalent trend of the drifting ice. (*Geikie, 1865, p. 84*)

But Lyell was not entirely to blame. The British were seriously handicapped by a lack of national glaciers and no doubt an insular people have a mental affinity with the sea. Consequently present students of geomorphology will see in the iceberg theory a minor issue mistakenly raised to universal stature and in the glacial theory a universal theme battered temporarily into insignificance. However, each *was* an outcome of uniformitarianism and few would fully agree with the suggestion that, in this respect, Lyell, like Britannia in the music-hall skit, 'ruled the waves and waived the rules'.

We cannot disagree – except as to quantities – with what Geikie wrote in 1865.

> Land-ice is thus a most powerful geological agent in new-modelling the surface of the earth. But in northern latitudes the same ice which plays so important a part on the land descends to the sea-level, and breaking off there into icebergs, performs a new series of feats in the great deep. These ice-islands carry with them any soil or rock rubbish that may have fallen upon them from inland cliffs, while they formed part of the ice-sheet of the country. The debris so borne off is, of course, thrown down upon the sea-bottom as each berg melts away after a voyage of perhaps several thousand miles. Year by year whole fleets of these bergs are sent southwards in the arctic regions, so that the bed of these northern seas must be plentifully strewed with earth and boulders . . . (*Geikie, 1865, pp. 82–83*)

But geomorphologists are primarily concerned with the *visible* landscape and a secret of their success is the ability to assess the *relative* effect of processes. None could have blamed Adam Sedgwick, now a grand old man of eighty, if in the lectures he was still giving annually at Cambridge, he had often quoted his warning to Lyell in 1835.

> Your ice theory will, I think, only let you slip into the water, and give you a good ducking.

CHAPTER NINETEEN

The Structuralists:
Catastrophic and Stratigraphic

We have already discussed in Chapter Ten the ideas of the structural or neo-catastrophists, who believed that catastrophic or structural upheavals provided the main key to the problem of the formation of topographic features. As their following remained quite large throughout Lyell's lifetime, it is necessary here to summarize and extend their arguments. Their main strength lay in the obvious necessity of having to postulate frequent changes of sea level, or upheavals of the landmasses, in order to explain both the normal geological succession of ordinary rocks and the many breaks or unconformities in the stratigraphical succession. But their arguments were equally necessary to explain the presence of mountains, of folded structures, of fractures and other features associated with earth-movements and orogenesis. We shall also describe in the following pages how these catastrophic structuralists were now joined by a growing school of erosional structuralists who believed that the varying resistance and disposition of rocks largely control the present topography.

THE CATASTROPHIC STRUCTURALISTS, 1845–61

As already noticed (pp. 173–6 and 225–6) P. J. Martin and W. Hopkins in the 1830's and early 1840's were strongly supporting Buckland's idea (1829) of the formation of anticlinal valleys by upheaval, fracturing and subsequent modification by 'violent aqueous abrasion' along the lines of fractures. Martin extended the fracturing to include numerous transverse rifts which resulted in a rectilinear pattern of valleys. Hopkins, a Cambridge mathematician and close friend of Sedgwick, provided a wider and more scientific basis for the theory. His earlier arguments (pp. 225–6) may be summarized as follows: convulsive emergence of the land, with elaborate fracturing, threw the overlying sea into a turmoil of tidal waves, or 'waves of translation'. These waves because they had been set in motion by earth-movements or volcanic eruptions were not like ordinary oscillatory waves, which only represent a gentle rise-and-fall motion of the surface water without any major progressive movement of the particles themselves. They moved forward as a whole mass and were therefore far more capable of transporting large blocks of rock and of causing destruction to the areas over which they swept. Whewell in 1847, when using this theory to account for the distribution of the European drift deposits, considered one wave inadequate, for it would either deposit all the materials

FIG. 62. *William Whewell*

in one place or would have insufficient force to carry the load. He suggested that the way out of this difficulty was for a succession of waves to be produced by a series of elevatory jerks.

> The great wave is solitary: the fluid before and behind it is at rest; and the particles move only while the wave is passing over them. Therefore the effect of such a wave upon loose materials immersed in the fluid would be only one of two:—either it would carry a single mass along with it, giving to it its own velocity, – or it would give a transient motion to a series of masses in succession, as it passed over each, moving each but a small distance. A single wave of translation cannot explain the situation of a long line of masses each of which has been moved through a great distance.
>
> If indeed we suppose a series of waves of translation each produced by a sudden elevation, or by some other paroxysmal action, we may obtain a greater effect. (*Whewell, 1847, p. 228*)

Making a calculation on the basis of the drift of Scandinavia and northern Russia, Whewell estimated that:

> We shall have the requisite force for instance, if we suppose this area to be elevated by ten jerks of 50 feet each, fifty jerks of 10 feet each or by the same 500 feet any how divided into sudden movements. (*Whewell, 1847, p. 231*)

Hopkins, in 1848–9, wrote further papers on the complex surface features of the Lake District. He retains the mechanism of a sudden elevation of land setting off a succession of tumultuous tidal waves but meets Whewell's criticisms by stressing that a number of jerking movements would be necessary to produce the multiplicity of waves required.

> If a considerable area at the bottom of the sea were suddenly elevated, nearly the whole superincumbent mass of water would be elevated in nearly the same degree, and a great wave, which has been called a wave of translation, would diverge in all directions from the central disturbance, and would be accompanied by a current diverging in like manner, the velocity of which would depend principally on the depth of the sea, the height of the elevation of the water, and the distance to which the wave had been propagated. (*Hopkins, 1848, p. 90*)
>
> . . . the transport of large blocks to considerable distances would require many repetitions of the action of these currents, and therefore many distinct movements of elevation, since each such movement would only produce its single wave of translation. (*Hopkins, 1848, p. 91*)

Unlike some geologists, Hopkins did not think that the boulders were actually carried in the waves but believed that they were propelled along the sea-floor by alternate sliding and rolling. He realized that the power required to move some of the largest boulders would have to be very great but in substantiation of his beliefs he again put foward his theory, the 'sixth power law of traction', to show that the actual force was not so great as might have been expected. The theory itself simply stated that as the velocity of a current of water increased, its ability to transport material increased not by an arithmetical ratio but by a geometrical one (p. 226).

> A curious consequence results from this law, when we estimate the force of the current by weight of the largest block of a given form which it is capable of transporting. Thus estimated, the force varies as the sixth power of the velocity of the current. (*Hopkins, 1848, p. 92*)

Or put more exactly:

> Therefore the moving force of a current estimated by the volume or weight of the mass of any proposed form which it is just capable of moving, varies as the sixth power of the velocity. (*Hopkins, 1849, p. 233*)

Hopkins was aware that his theory was criticized in many quarters because if a current had transported all the drift and boulders to their present positions, then in many cases this would involve the material not only travelling uphill but crossing hills at relatively high altitudes. It was felt that not even 'waves of

FIG. 63. *The Lake District valley of Thirlmere, Cumberland, looking south*
(Photograph by J. K. St. Joseph)

translation' could do this. To meet the objection Hopkins showed that by his scheme few inequalities would exist to impede the distribution of material.

The difficulty in this theory arising from the presumed inequalities of the surface over which the blocks must have been transported, has been, I conceive, in many instances, far too much insisted on; for it has been made to rest on the assumption that the inequalities of surface between the present and original sites of erratic blocks were the same, or nearly so, at the time of transport as at present; an assumption which I regard as totally untenable. There are three obvious causes of inequality of surface – elevation and disruption, denudation during gradual emergence from beneath the ocean, and erosion after emergence. (*Hopkins, 1849, p. 237*)

. . . such great inequalities as are presented by the oolitic and chalk escarpments, have doubtless been due in great measure to denudation, during the period of gradual emergence of the land, the higher levels being raised above the sphere of denuding action, while the lower beds remained exposed to it. (*Hopkins, 1849, p. 237*)

Though basically Hopkins' theory differed from Lyell's because the former made use of catastrophic causes while the latter supposed the slow action of a natural agency, they were close enough in concept not to be contradictory. On the other hand, as we have already noticed (pp. 225–6), Hopkins was completely opposed to the glacial theory, for he held that his wave of translation theory could adequately account for the diffusion of drift material. A series of uplifts of the order of 100 or 150 feet could produce waves capable of moving drift and erratics. Final uplift would allow wave erosion and current action to operate along faults to produce the present topographic inequalities. As the sea retreated from the rising landmass and a fault was progressively exposed, so the waves would excavate a valley from the head down towards its eventual mouth.

If the catastrophic background is ignored, that is very much what Lyell was postulating. Therefore Hopkins is important for the way in which his theory bridged the old gap between the catastrophists and the uniformitarians. They have now virtually joined forces on the issue in competition with the heresies of the fluvial and the glacial schools.

H. D. Rogers, one of America's senior geologists, continued to find much to admire in Hopkins' theory as also did C. H. Weston (1850). Murchison, as might have been expected, still approved warmly of structural catastrophic ideas while, as already noticed, P. J. Martin still possessed virtually the same opinions that he had expressed twenty years earlier.

If I have succeeded in enabling my readers to realize in their minds the picture of the conjoint action of earthquake and flood which I have in my own, they will be able to understand the confused flux and reflux, and the clash of opposing torrents which must necessarily follow in the train of so extensive a displacement of solid matter; whether the convulsion took place at the bottom of the sea or in the open air; and whether or not it was prolonged by a continuous heaving and falling (terrene undulations of incalculable violence) of some continuance. We have no means of following the great bulk of the displaced materials. Much of it was doubtless cast off over the great synclinals on each side. . . . we are soon convinced that the drift I have attempted to describe, is only the remnants of those materials, and the last leavings of the retiring waters. (*Martin, 1851, p. 370*)

Thus, this form of catastrophic structuralism enjoyed a healthy longevity. Fisher in 1861 repeated its main essentials. After examining the tidal estuaries in Essex, he decided that such long inlets could not have been formed by the action of normal waves and tidal currents, which would produce a slow,

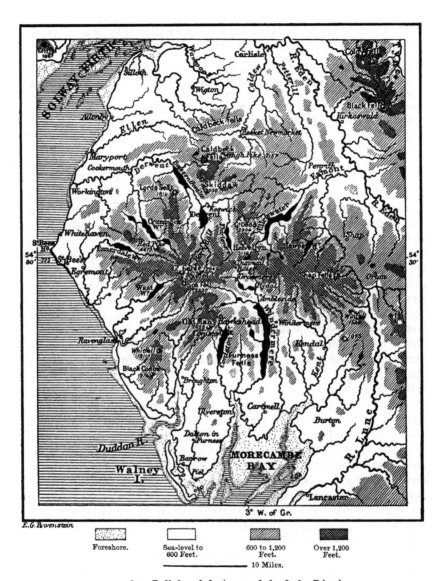

F.G. Ravenstein

Foreshore.   Sea-level to 600 Feet.   600 to 1,200 Feet.   Over 1,200 Feet.

————— 10 Miles.

FIG. 64. *Relief and drainage of the Lake District*

uniform inland retreat of the coastline but were the marks of a current or rush of water descending the slopes of the hills.

> Mr Fisher does not see any other way for such a form of surface as obtains in this district, in many others composed of yielding strata, than by a super-incumbent mass of water rapidly draining off from a flat or slightly dome-shaped area. Slight depressions, cracks, or lines of readily yielding materials would first determine the course of the streams of drainage; and these would cut channels which would be more or less completely scoured out according to the velocity of the water. (*Fisher, 1861, p. 2*)

THE STRATIGRAPHIC OR EROSIONAL STRUCTURALISTS

About the mid-nineteenth century the more catastrophic form of structuralism underwent an important change. The convulsionists and other catastrophic structuralists had always stressed the dominance of violent earth-movements and had usually imputed the newer or minor landscape features to rushing water, impelled by earth-upheavals and guided by rifts or earth-fractures. On the other hand, there were numerous geologists who considered that, irrespective of such catastrophes, the landscape was much affected by litho-logy, or the resistance and disposition of the surface rocks. It was, however, not until the 1850's, with the advance of geological mapping, that topo-graphers began to impute almost complete dominance to the control of litho-logy on topography and to rank it far above surface erosional processes in the formation of landforms. In this respect Peter Lesley must be given pride of place.

Born in Philadelphia of a Scottish Presbyterian family, Lesley graduated from the University of Pennsylvania in 1838, and then, partly because of his frail health, undertook three years' work for the Pennsylvania Geological Survey in the Anthracite Region under H. D. Rogers (Davis, 1915; Cate, 1959). On this first assignment he took with him 'volumes by Cowper and Carlyle, and a flute', describing himself in after life as 'never anything but an amateur, except in topographical geology' (Cate, 1959, p. 18). In 1841 the Survey suffered a contraction and Lesley spent three years at the Princeton Theological Seminary, before continuing his theological studies in 1844 in Europe, where he was impressed with the topographic similarities of the Jura Mountains and the Appalachians. In the following year he became a preacher in Boston and there helped Rogers with the preparation of the Pennsylvania Survey *Reports*. In 1851 he returned to full-time employment with the Pennsylvania Geological Survey, although his cataclysmic ideas on erosion continued to show the effects of his theological training. In 1856 Rogers' geological map of Pennsylvania appeared and Lesley, annoyed because he considered that insufficient credit had been given to the contribution of Rogers' associates, wrote his *Manual of Coal and its Topography* in only six weeks (Stephenson, 1903).

FIG. 65. *J. Peter Lesley*
(*From a portrait by Mrs. L. Bush-Brown*)

Lesley's ideas on erosion were akin to those of Hopkins; that a sudden mountain uplift was followed by waves of translation. However, when Lesley confines himself to the definition and topographic significance of structural sequences in the Appalachians he became far more original and truly significant. In his *Manual* on the North American coal measures he argues very forcibly that structure determines the shape of every surface feature whether it is large or very small. He makes his position crystal clear in the first few lines below:

The Science of Topography, like every other science, proceeds to deduce from a few elementary laws an infinity of forms, by which these few laws may or actually do express themselves upon the surface of the earth. The possible is always an infinite series, the actual a limited and fortuitous selection from it – not always, therefore, the most striking, perfect, and complete. (*Lesley, 1856, p. 121*)

He is careful to emphasize that the region being discussed is in no way unique or peculiar, but has within its boundary all the normal erosional forms.

> In the coal measures we have, on the contrary, long and regular mountain crests vanishing horizontally in parallel lines into the horizon, broken at intervals with gaps and curving in and out in zigzags including inner sets of zigzags, system within system, down through the series of rocks. We have all the phenomena of drift and denudation, diluvial scratches, eddy hills and terraces, in great perfection. (*Lesley, 1856, p. 122*)

The argument opens with an echo of Lamarck's assertion that mountains were the remains of the earth as it was originally laid down, while the valleys represented what had been taken away. Lesley makes the additional assertion that because the mountains are examples of the original structure it is logical that they should be examined to uncover the key to the present surface forms.

> All topography resolves itself into a discussion of the Mountain. The surface of the earth may be considered as a congeries of mountains (and hills are but smaller mountains) touching each other at their bases. Valleys have only a negative topographical existence and represent the absence or removal of mountain land. Mountains are solid portions of the earth's crust, while valleys are but vacua in it, taken possession of by the water and the air. The propriety of this view, opposed though it be to the economical history of the human race, will become more evident when the denudation of valleys is discussed. In geology it is important to true views to consider all changes to be mountainous, and valleys to be an incidental consequence.
>
> A Mountain has three Elements, top, side, and end, and the primary discussion of a mountain is that of its slopes, its crest-line, and its termini. Gaps are irregularities in its crest-line, as terraces are in its slopes. The cross-section of a mountain shows why its slopes, and usually also why its crest-line and its termini are not only what they are, but could not be different. It is a deep set feeling among men, that if there be accidental forms upon earth they are to be found in mountains. There could not be a greater mistake; for if there be natural forms unalterably predestined by the direction and intensity of natural forces they are those of mountains. Not a wrinkle on the side, not a notch on the crest, not a flexure in the trend of a mountain or a hill but is an evidence of laws which have operated upon it with the nicest precision. Not a ravine, not a rod of cliff, not a waterfall, but exists in the immediate vicinity of its own explanations. The place where a stream breaks through, however apparently accidental, was determined by positive relationships between the rocks of that locality; nor can any investigation be more exciting than that which rewards itself with perpetual discoveries of cause and effect in a wilderness of apparent lawlessness and unexplained confusion. (*Lesley, 1856, pp. 125–7*)

Having stated the general principles he examines the various structural forms that actually occur. Commencing with unfolded strata, he observes that: where strata are vertical, the slopes are sharp and ragged; where they are tilted, a precipice occurs at the exposed face and a gentler slope down the dip of the strata; where they are horizontal, a tableland is created and if the rock is resistant this will represent a feature whose shape is nearest to its original form and upon which the denudational processes have had the least effect.

> The Mountain Slope. To illustrate the influence which its interior structure has upon the form of a mountain, the accompanying series of cross-sections are introduced. The first set show how a stratum of sand-rock or other hard material enclosed in softer stuff, arranges the height and slopes of its mountain to suit its own dip. Where it is vertical the mountain is low, sharp, and symmetrical; at 60 degrees, it is higher with a front side long; at 30 degrees higher still, with a long, gently sloping back, and short steep front, covered with angular fragments from a range of cliffs above; when horizontal, the mountain is at its maximum height, forming a tableland with precipices and steep slopes in front. It is needless to suggest the infinite variations of this simplest law of mountain form.
>
> When two such sand-rocks lie in neighbourhood they of course form a double mountain, subject to the same vicissitudes of external aspect in view of similar changes of internal structure. (*Lesley, 1856, p. 127*)

However, most strata, he appreciates, have been disturbed or folded in varying degrees of complexity or compression. Lesley graphically explains how folded strata can almost exactly impress their underlying differences upon the form of the overlying surface. Ignoring complex variations, he divides folds into synclinal or anticlinal forms, and describes the features associated with each type. He realized that initially subsequent denudation of either a syncline or an anticline gave rise to many similar features; in each case when the formations are pierced by consequent streams the edges of the resistant beds stand out as sharp scarps while the softer rocks are broadened into large valleys. Yet there are fundamental differences which distinguish the two forms. First, in the case of anticlines, the scarps face inwards while those of a syncline face outwards: secondly, whereas an anticline loses its mountain form once erosion has begun its work and goes through a life cycle of incised valleys as each new bed is brought to the surface, the synclinal formation long stands out as an area of high land, and until a very late stage in dissection maintains its original broad valley forms. In more detail once the hard capping rock of an anticlinal dome has been pierced it is the anticlinal crest which is first removed by denudation and what was once a folded mountain becomes a valley. This downward cutting process continues until the next hard band is met with. When this is reached another anticlinal hump will stand out until the time when the processes of erosion are able to pierce its crest, whereupon another valley cycle begins. The syncline on the other hand stands up as a lipped tableland: it is the last area to be eroded and after erosion remains as a flat plain until the next hard stratum is reached. When this happens a new tableland is exposed and the process begins all over again (*Fig. 66b*).

In every shallow synclinal there is a high, narrow, flat-topped mountain, with precipices on both sides looking down upon the lowlands. Such is the structure of the Catskill, Towanda, Blossburg, etc., and other mountains in the north, and the Cumberland Mountains of Tennessee and Alabama. As the basins deepen, other higher sand-rocks come in above, and double the precipices and slopes; finally the whole is cleft in two, and the drainage, after traversing the parted mountain often for many miles, breaks out sideways into the plain. It is evident that it was due to the direction of the original currents setting along the centre of the geological basin before the cutting develops the present mountain.

The Sharp synclinal shows this more clearly, with this difference, however, that its longitudinal central cutting is never a ravine, but always a valley. The hard rocks pass off in diverging crests terraced and gashed, enclosing one another and giving place to mountain within mountain, in a series that has no limit but the number of hard beds in the system of formations. Their precise inclination with the horizon makes no essential difference in the action so long as the synclinal remains a simple one; but the moment this becomes compound then all manner of complications inaugurate themselves upon the surface and present a thousand puzzles to the skill of the geologist, as in the following set . . . , which more or less nearly represent the wrinkled compound synclinals of the anthracite coal region.

The Anticlinal structure is the reverse of the synclinal, and has its own definite system of forms equally subordinated to the general laws of denudation, and in many respects curious inverted parodies of those above. In fact, it may be seen in the next set . . . , the moment the huge back of the anticlinal mountain splits in two, we lose the anticlinal as a character of the mountain while it remains in the valley, until from the centre rises another mountain, again to be split lengthwise in its turn. In this case we represent the anticlinal as rising slowly, just as before we represented the synclinal as slowly sinking. The ridges on each side of a split anticlinal are called monoclinal, and are in fact the same as the ridges into which a parted synclinal mountain divides itself. Hence the forms are common to both. (*Lesley, 1856, pp. 128-9*)

Pursuing this reasoning Lesley went on to show that the line, height and evenness of the mountain crest could similarly be affected by variations in the dip of the strata. Normally he said a crest line is related to the uppermost hard band of exposed strata. If throughout the range the dip of the strata is regular then erosion will not radically alter the form of the mountain as it is worn down: it will lose mass but erode regularly maintaining its original shape. In practice, as he was careful to point out, such regularity of dip is rarely found. Therefore when the dip varies so will the form of the crest. And so in every mountain area the various small irregularities produced by differential erosion depend in large part upon the complexities or uniformities of the original structure (*Fig. 66c & d*).

The Crest line of a mountain is normally either a point or a horizontal line, according to whether it is a mountain of ejection under air, or a mountain of elevation and denudation under water. The type of the volcano is a cone: the type of any other mountain is a double sloping wall. Sedimentary mountains are the turned up edges of some hard sediment. The central hardest or

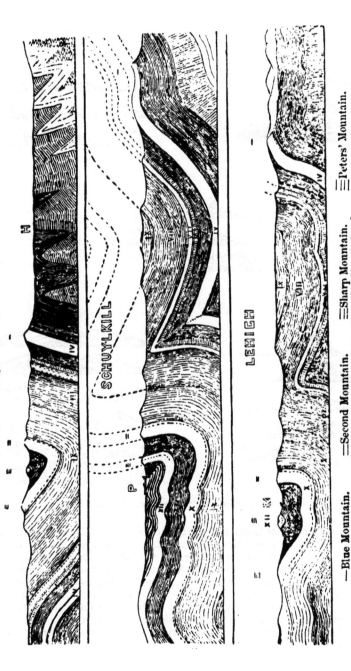

SUSQUEHANNA

SCHUYLKILL

LEHIGH

—Blue Mountain. ≡Second Mountain. ≡Sharp Mountain. ≡Peters' Mountain.
H. Harrisburg. P. Pottsville. M. Mauchchunk.

FIG. 66. Structural diagrams by J. P. Lesley (1856)

a. The folded Appalachians in the vicinity of the Susquehanna, Schuylkill and Lehigh rivers.

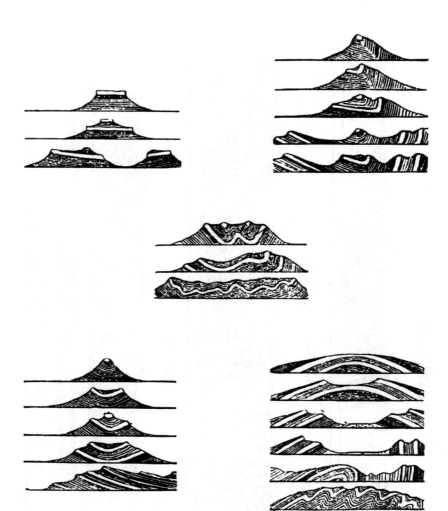

FIG. 66b. *Type examples of sections of eroded folds*

massivest layer of that sediment will of course form its crest. If there be subordinate layers, they will form terraces; and, at the gaps, ribs. If several central layers near together, be equally able to resist denudation the crest will move from one to the other along the line according to the cross-cutting. So long as the dip maintains itself steadily the mountain crest will be a straight and level line; when the dip changes, the crest line will fall off accordingly and suffer change in height and evenness. When the dip falls completely over or reverses itself, the crest line will double back upon itself, and do so as many times as the dip changes. At such a recurve, coves are formed; here valleys head up; here gaps usually break through the side walls; here also the

FIG. 66c. *Idealized diagram of the outcrop pattern associated with a series of plunging folds.*

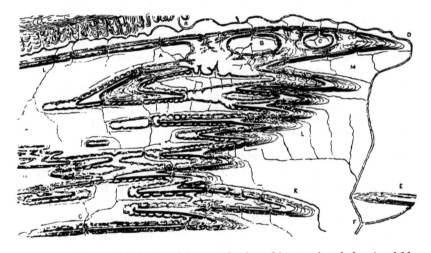

FIG. 66d. *Topographical map of an area dominated by a series of plunging folds*

mountain attains its maximum height. These doublings, if anticlinal, are in one direction; if synclinal, in the other. A series of them can only occur when a system of parallel anticlinal waves and synclinal basins pass under an outcrop . . . (*Lesley, 1856, p. 134*)

Lesley's analysis of folded topography was so effective that little has been added by later writers since he originally published his views. Yet despite his advanced thinking about structure he was as crude in his beliefs concerning erosion processes as many other contemporary geologists and a great deal cruder than some. From what he says it is clear that he did not subscribe to either the ideas of Lyell or to those of the fluvialists but had in mind some overwhelming deluge. At best his knowledge of the agents of erosion was scanty and the principles he had mastered were out of date and fraught with misconception.

> The arguments of force in proof that a cataclysmic deluge wrought out our topography, have been introduced into the body of illustrations on the fore-going pages. Conversely, those that are thought to prove that quiet, cyclical, ocean tidal or current action could not or actually did not work it out, have also been more or less clearly stated. The case of the Wind Gap has been cited as conclusive evidence that the present waters have had no part in the matter of denudation so far as our gaps and gorges are concerned. (*Lesley*, *1856, p. 167*)

Lesley was made Professor of Mining at the University of Pennsylvania in 1859 and three years later was again pressing the merits of the structural control of topography. In an article on the Appalachian region he attempts to demonstrate that the course of the New River across the Great Valley and into the Appalachians is determined solely by the junction of anticlinal structures on the north with faulted monoclines on the south.

> The New River heads in the Blue Range, crosses the Great Valley westward, breaks into (not out of) the Appalachians, striking the escarpment in its face, and flowing directly through and across it. . . . The cause of this phenomena is to be found in the change of structure at this line. Most of the valleys and mountains north of it . . . are unbroken anticlinals and synclinals. Most of the valleys and mountains to the south of it . . . are monoclinals, bounded by immense faults or downthrows. (*Lesley, 1862, p. 414*)

We should, perhaps, here point out that although Lesley quite properly drew attention to the fundamental influence of structure upon topography, he was misled in placing too much emphasis on this factor. Subaerial erosion will be guided initially by marked structural differences, but when the strata forming that structure have been stripped off by denudation and a new structure laid bare, the drainage may retain its old pattern though this bears no relation to the newly uncovered structure. Moreover if structure is to control, the differences must be very marked and near to the surface, otherwise the initial drainage will follow the main slope and not the structure, if the two happen to be in opposition to each other. Furthermore, in nature most landscapes occur from a combination of events and not because of any one factor; one influence may be predominant, but landscape evolution is too varied and too complex for it to be exclusive.

Lesley himself soon realised this and in 1866 he renounced his catastrophic views in his *Five Types of Earth Surface*. In 1874 he deservedly became the Director of the Second Geological Survey of Pennsylvania.

The catastrophic structuralists were by no means perturbed by Lesley's insistence on lithological or stratigraphical control nor by his defaulting from their more violent themes. It is perhaps surprising how little influence his intelligent observations had upon contemporary geologists.

Many stratigraphers continued to read more into the association of mountain building and marine submergence than actually existed. Because contorted strata had evidently experienced considerable pressure and because sedimentary strata were formed beneath the sea, they went on to make the further and unnecessary assumption that the topography was also the result of the combination of orogeny and marine erosion. Some, like Ramsay, Jukes, and Lesley, soon grew out of this false emphasis, but others, like Murchison and S. V. Wood, continued to make it the basis of all their theories.

Murchison seems to have had no more liking for lithological structuralism than he had for the fluvial theory. Despite all that had been said or written by Lesley on the one hand and by the fluvialists on the other, to Murchison a gorge still meant a volcanic eruption and rifting whereas an open valley meant a flood or marine devastation. Because a river took its course through both a gorge and a wide valley, he argued that the river had no connexion by birth with the valley but was running across features which had been prepared for it by another agent of erosion. He failed to appreciate that a river could simultaneously erode a gorge and a wide valley; in his mind the presence of one was inconsistent with the presence of the other, and he could not see that differences in rock resistance made this a natural result. Writing of the streams on the east flank of the Urals, he stated:

> ... when the geological traveller passes from the valley of the Serebrianka ... still more is he struck with wonderment at the unquestionable evidence, amidst intensely dislocated rocks, of the ruptures by which the deep narrow chasm has been formed in hard crystalline rocks, in which a lazy stream flows, which not descending from any altitude, has no excavating power whatever, and, like our own meandering Wye, has flowed on through clefts in limestone during the whole historic and prehistoric period, without deepening its bed. (*Murchison, 1863–4, p. 233*)

In 1867, as if to issue a challenge from the past against all that was new, Murchison emerges with a little-revised (fourth) edition of *Siluria* that is reminiscent of the geomorphic theories that were popular twenty or thirty years before. Hopkins' theory of violent eruptions and waves of translation is obviously the store-house from which he has gathered his main beliefs. No concession is made towards all the recent proofs of subaerial and glacial erosion.

> I may say that I never examined any extensive area without recognizing evidences of fracture, displacement, and occasionally inversion of the strata,

which no amount of gradual, continuous action could possibly explain. (*Murchison, 1867, p. 489*)

Let it not be supposed that we, who hold to the proofs of more powerful causation in ancient periods, do not fully admit that the former physical agencies were of the same nature as those which now prevail. We simply assert on the countless evidences of fracture, dislocation, metamorphism, and inversion of the strata, and also on that of vast and clean-swept denudations, that these agencies were from time to time infinitely more energetic than in existing nature, – in other words, that the metamorphisms and oscillations of the terrestrial crust, including the uprise of sea-bottoms, and the sweeping out of debris, were paroxysmal in comparison with the movements of our own era. (*Murchison, 1867, p. 490*)

Thinking along similar lines S. V. Wood decided that the alignment of structure in south-east England showed the influence of two tectonic revolutions. Before these earth-movements took place the region had been covered by thick deposits of Oolitic, Cretaceous and Tertiary strata.

. . . over the East of England, no irregularities of surface existed prior to the close of the drift epoch, other than those produced by erosion; and that . . . the entire surface covered by oolitic, cretaceous, and tertiary deposits remained, until the outburst of these convulsions, pretty nearly in the condition that their respective seas had left them . . . (*Wood, 1864, pp. 188–9*)

He thought that the valley system of south-eastern England could be divided into regular systems of concentric arcs, diverging from three main centres: (1) near Flamborough Head, (2) near the Isle of Wight and (3) near Canterbury. These arcs, he said, were the result of flexures produced by lateral pressures exerted from these several centres. The pressures themselves were caused by the marginal reaction to the earthquake disturbances occurring at these three centres beneath the 'Boulder Clay sea'. Wood held that the flexures determined the degree and direction of the denudation accomplished by the disturbed ocean water, the amount of denudation being proportionate to the angle of flexure. In his scheme the sequence of events was, firstly, uplift and marine erosion along the lines of flexures, with a virtual cessation of erosion when the areas had risen above sea level; secondly, in the case of those areas not affected by the first uplift, there was further uplift accompanied by rectilinear movements and fractures, the latter then being eroded at their margins by the sea.

This survey . . . led me to believe that the entire valley-system of the East of England originated in centres of arc-like or curvilinear disturbance, which immediately preceded the elevation of the bed of the sea from which was deposited the wide-spread deposit of Boulder-clay forming the latest of the Glacial beds of the South of England. My view is that, these disturbances having given an impress to the surface, the denudation accompanying and ensuing upon the elevation of the sea-bed, by wearing more deeply the impress thus imparted, made it more conspicuous in the parts where this denudation was most prolonged. I found also that the rectilinear upheaval along the Wealden line of disturbance appeared to have been a subsequent event to this curvilinear denudation, and that generally these rectilinear disturbances had occurred long after those which accompanied the first elevation of a part of the Glacial sea; so that, while a part of the valley-system was due solely to the action of the first of these agencies, the rest (and more especially the part of England lying south and south-west of London) had originated from the combined action of the two. (*Wood, 1867, p. 394*)

A more moderate view of the effect of structure is represented by the ideas of Scrope, the fluvialist. Scrope objected to Jukes' dogmatic assertion that internal causes have no direct effect on the external form of the ground. Jukes, as we shall notice later, underplayed the topographic influence of structural lines, especially when dealing with regions of superimposed drainage, where rivers run across the strike of the structure. It is interesting, and rather astonishing, to find such a pure fluvialist as Scrope rising to defend structuralism. Scrope considered that the occurrence of marine strata on Alpine crests and the origin of the Swiss 'Plain' as a large syncline between the Alps and Jura provided good reasons for supposing that structure dominated there as a factor in the evolution of the topography. What is more, he considered that structure controlled not only the grander mountain features of the earth but

. . . in respect even to some of the minor details, such as transverse valleys, that act as tributaries to these grander depressions of the surface, there seems good ground for believing many of them to owe their *origin*, and consequently the course of the superficial waters or ice-streams that have, since their emergence from the sea, widened and deepened them by erosion, to the transverse cracks and fissures which could not fail to accompany the violent elevation of more or less solid strata, even though effected by gradual throes. (*Scrope, 1866B, p. 242*)

The hint by Scrope that 'violent elevation' could be 'effected by gradual throes' is sufficient to point out the waning in the popularity of violent catastrophism, which forty years earlier had been almost the password for acceptance into the select circle of the Geologial Society. Yet a few die-hard catastrophic structuralists other than those discussed above continue to put in an occasional appearance (Brown, 1867, and Pattison, 1868).

By the late 1860's the leading catastrophic structuralists were either old or not in the leading rank of geologists. Thus, the Duke of Argyll (1868), who criticized Geikie's *Scenery of Scotland* (pp. 407–10), was not an eminent geologist although he had been for some years an active member of the Geological Society. He seems to have belonged to the old school of Sedgwick and Murchison, yet he does not object to Geikie's fluvialistic analysis except in so far as he considers its exclusive application to be exaggerated. He feels it would be more accurate to make some clearer mention of the influence of structure.

In fact, Geikie had not denied that valleys sometimes coincided with the faults and that sometimes the direction of valleys were actually governed by fractures. He was, however, more concerned with showing that, irrespective of the faults, weathering and river-erosion would form deep valleys. The fluvialists did not ignore the influence of structure; by now they felt that enough had been said of structure to make its value known and that their efforts could be spent more profitably in stressing the major part played by rain, rivers, and weathering.

THE STRUCTURAL DILEMMA

We have perhaps said enough of structuralism in mid-Victorian times to show its dual nature, catastrophic and stratigraphic. There is no doubt that by now structure is considered an indispensable factor or element in the evolution of landscapes. But whether by 'structure' is meant the folding and fracturing attendant on upheaval or merely lithological control over topography, or both, is quite another matter. Few would dare today to dissociate one from the other.

Even in 1960 our ignorance of the more violent aspects of structuralism, such as orogenesis, remains abyssal. Continents must be allowed to submerge and emerge; mountains must be elevated. Today we will even allow continents to float like corks (ISOSTASY) and to migrate horizontally.

In the late nineteenth century W. M. Davis emphasized the trilogy of landscape evolution as structure; process; stage; but to him structure was lithologic and he chose to ignore the orogenic aspect. On the other hand, Walther Penck in the 1920's preferred to stress the influence of upheaval. The dilemma is as real today as it was a century ago! The main difference lies in the fact that today all structuralists believe in the teachings of Hutton and in the powers of subaerial erosion. The gradual rise to general popularity of this omnipotent factor, and particularly of fluvialism, forms the topic of the next three chapters.

CHAPTER TWENTY

James Dana and the
Fluvialistic Survival in America

We have already shown how Lyell when in his prime swerved away from fluvialism and took with him all but a small minority of his contemporaries. The great man never fully returned to the fluvialistic camp but during the last thirty years of his lifetime the followers of Hutton gradually recovered much of the ground they had lost to the marine erosionists. By about 1866 the tide had definitely turned in favour of the fluvial theory. This survival and revival is especially associated with the work of James Dana, George Greenwood, Andrew Ramsay and J. B. Jukes but they had many helpers and derived much assistance also from two favourable circumstances.

First, there was a notable increase in the number and accuracy of the measurements made of the amount of material actually being carried off the land by the rivers. These calculations made it plain that rain and rivers together formed a very powerful agent of erosion.

Second, geological exploration and surveys in tropical countries began to show that in hot wet climates rain and rivers were by far the dominant, if not the only, factors in the fashioning of landscapes. Whereas in Britain the rainfall normally is gentle and evenly spread over the year and rarely gives rise to tremendous floods, in the tropics the rains are torrential and, over large areas either superabundant or concentrated in a few months of the year. In the stormy 'temperate' latitudes the seas are rough and wave-action assumes terrifying proportions whereas in the tropics the sea is often calm apart from the monotonous breaking of the coastal surf. Tropical rivers on the other hand frequently rise with a magnificent fury. During heavy rain and the wet season the streams are transformed with startling suddenness into raging torrents capable of astonishing destruction. Europeans faced with this evidence of the power of rivers and with what were obviously river-cut valleys had to revise drastically their preconceived ideas on fluvial erosion.

JAMES DANA: THE FIRST GREAT AMERICAN FLUVIALIST

The influence of tropical climate on landforms was first publicized to Europeans by James Dwight Dana, who ignored the general current of geologic opinion in Europe and America and came out strongly in favour of fluvialism. He seems to have been the first American student of landscape who followed a truly independent line of reasoning (Williams, 1895).

Born on the 12th of February 1813 at Utica, a town of about 1700 inhabitants, he went to the small local high school where the principal was Charles Bartlett, a very advanced educationist for the time. Among the staff was Asa Gray and the curriculum included Chemistry, Botany, Geology and Mineralogy. In 1830, Dana entered Yale University, attracted no doubt by the reputation of Benjamin Silliman. He obtained his B.A. within three years but as a scholar was not outstanding, probably because of his lack of training in Latin and classical Greek, then as much cherished in American as in British universities.

During his last year at college Dana secured an appointment as an instructor of midshipmen of the U.S. Navy. In August 1833 he joined the *U.S.N. Delaware* and toured the coast of West France and the Mediterranean. At this time crystallography seems to have been one of his main studies and after gleaning its main principles from Phillips's *Mineralogy* he worked out his own system of crystallographic symbols. He kept in touch with his old university by writing notes to his former teachers about what he had seen. After a visit to Mount Vesuvius in 1834 he wrote an article about it for *Silliman's Journal of Science*. When the tour was over he was appointed as Silliman's assistant in the chemical laboratory, but official duties did not occupy much of his time and by 1837 his first book, *System of Mineralogy and Crystallography*, had been published.

At this time the Americans, fired perhaps by the successes of the British exploration ship *Beagle*, planned to send a scientific squadron of six ships on a tour of the Pacific and Dana was invited to be one of the scientists of the expedition, his especial study to be geology, corals and crustacea. The squadron was some time in sailing and finally left on 18 August 1838, over a year after Dana's appointment. The tour which lasted nearly four years took him to Madeira, Brazil, Patagonia, the Magellan Strait, Chile, Taumotus, Tahiti, Samoa, Sydney, New Zealand, Tonga, Fiji, Hawaii, Vancouver, San Francisco, Manila, Singapore, the Cape of Good Hope and, finally, to New York. It proved the most formative period of his life and was the source of material for all his future writings. He brought to his work boundless enthusiasm. The world he was entering was entirely new in every way; it seemed to dazzle and elevate the level of his thoughts. No wonder he could write so glowingly of his impressions.

> These coral islands are truly fairy spots in the ocean. They rise but a few feet above the waters' surface, and are covered with a luxuriant tropical vegetation. On one of these, which was not inhabited, the birds were so tame that they permitted themselves to be taken from the bushes and trees, and flew about our heads so near us that we could almost take them with our hands. They did not know enough to fear. The whole island was almost a paradise. (*Gilman, 1899, p. 110*)

The scientific corps went everywhere with the squadron except when it visited Antarctica. The commander of the expedition though a keen and skilful officer,

FIG. 67. *James Dwight Dana*

had apparently little respect for the worth of his scientists and when the expedition was ready to set out for Antarctica they were left behind in New Zealand.

> The scientific corps were detached soon after as a worse than useless appendage to an expedition cruising among the ice; for we should find little or nothing in natural history in those frigid regions, and would only add to the other number of mouths that must be filled from the stock of provisions on board. We were satisfied of this ourselves and very gladly took advantage of the opportunity afforded to employ the season more profitably in Australia and New Zealand. (*Gilman, 1899, p. 114*)

At the end of the Pacific expedition Dana invested much of the money earned during the voyage in the family store at Utica. The decreasing interest of the government in the results of the expedition must have depressed him as must also the day-to-day interference of the Congressional Committee set up to supervise the writing of the *Report*. Instead of encouraging the preparation of a comprehensive and truly valuable document, the Committee seems to have confined its activities to niggling and unintelligent directives as to how the material was to be arranged. It soon became apparent that no real interest was felt and the few copies that finally emerged were mostly reserved for foreign dignitaries like the Emperor of China.

During this period of uncertainty about his future, Dana's fortunes changed once again. On 5 June 1844 he married Henrietta Silliman, daughter of his old professor. This marriage, as much as anything, ensured his return to Yale. Harvard had already offered him an appointment with the Lawrence Scientific School, which numbered Agassiz and Asa Gray among the members of its staff, but in 1850 when offered a post at Yale the choice was quickly made and he became their new Silliman Professor of Natural History and Geology. During his professorship he lived next door to his father-in-law, while the house beyond belonged to his brother-in-law.

Dana was slight in build and of moderate height, and physically supported himself by a mass of nervous energy, the supply of which periodically drained low, forcing him to rest. For all his love of science, he retained strong religious beliefs and, like Buckland, worked to reconcile the conflicting principles of the two as is evident in his publication *Science and the Bible*. He neither drank alcohol nor smoked and was eminently serious in his approach to most matters, particularly in his old age. After his return to New Haven he seldom left the area. It is possible that this simplicity of living and constant inwardness helped to emphasize the independence of approach always shown in his geological writing.

Dana did not begin his professorial duties until 18 February 1856 because of the need to prepare the government *Report*, which was published as three large folios, the second of which was confined to geology and contained conclusions of vital importance to the progress of geomorphology. What raised his writing far above the level of the ordinary geological treatise was his unequivocal emphasis on the part played by subaerial denudation. It was a brave statement for he stood almost alone. In Britain there was the sharp division between the marine erosionists of the Lyell group and those who favoured some more destructive agency such as earthquakes or waves of translation. On the Continent the current ideas were even more primitive and most scientists dallied with some modified notion of catastrophism. Dana had a young man's courage and integrity of opinion and in a country that generally accepted European thought he put forward views, which though not novel, were at least unpopular or discountenanced in Europe. His publication was a harbinger of America's growing scientific independence.

Dana's three special topics of investigation were volcanoes, corals and crustacea. The first provided most of the evidence for his insight into erosion processes. His tour of the Pacific Islands had led him to recognize four main causes which accounted for the form of the surface topography.

> The causes operating in the Pacific, which have contributed to valley-making, are the following:
> 1. Convulsions from internal forces, or volcanic action.
> 2. Degradation from the action of the sea.
> 3. Gradual wear from running water derived from the rains.
> 4. Gradual decomposition through the agency of the elements and growing vegetation. (*Dana, 1849A, pp. 380–90*)

He quotes many examples of fissures and craters which have been caused solely by the violent movement of underground volcanic forces giving rise in turn to fractures and subsidences in the earth's surface:

> The action of volcanic forces in the formation of valleys, is finely illustrated in the great rupture in the summit of Hale-a-Kala (Kauai). The valleys formed by the eruption are as extensive as any in the Hawaiian Group, being two thousand feet deep at their highest point, and one or two miles wide, they extend from the interior outwards towards the sea. Above they open into a common amphitheatre, the remains of the former crater, the walls of which are two thousand feet high.
> Other examples of volcanic action are to be seen in the pit craters of Mount Loa, among which Kilauea stands pre-eminent . . .
> The many fissures which are opened by the action of Kilauea, might be looked upon as valleys on a smaller scale, and the germs of more extensive ones. But with few exceptions, these fissures as soon as made are closed by the ejected lava, and the mountain is here no weaker than before. Those which remain open, may be the means of determining the direction of valleys afterward formed. (*Dana, 1849A, pp. 380–90*)

Of the potency of the sea he is quite unconvinced although it is obvious from his argument that he has both Lyell's and Hopkins' theories in mind. To the idea that the sea can slowly gnaw away the coast and etch out a rising landmass he counters with the assertion that wherever the sea was observed in the Pacific it could be seen to be removing the inequalities rather than emphasizing weaknesses. Even if a great rush of water is to be supposed it would be wildly unscientific to imagine that it could cut out the normal intricately winding and intersecting system of valleys.

> The waves tend rather to fill up the bays and remove by degradation the prominent capes, thus rendering the coast more even, and at the same time, accumulating beaches that protect it from wear. If this is the case on shores where there are deep bays, what should it be on submarine slopes successively becoming the shores, in which the surface is quite even compared with outline of the islands? Instead of making bays and channels, it could only give greater regularity to the line of coast.
> . . . The effects of the sea in making valleys have been much exaggerated as it is obvious from this appeal to existing operations, the appropriate test of truth in geology.

The action of a rush of waters in a few great waves over the land, such as might attend a convulsive elevation, though generally having a levelling effect, might it is true produce some excavations, yet it is obvious on a moment's consideration, that such waves could not make the deep valleys, miles in length, that intersect the rocks and mountains of our globe. . . . (*Dana, 1849A, pp. 380–90*)

He did not deny that the sea might at some time have washed over the land and worn much of the surface away. He merely denied that such a process could have produced the topographic features to be found in the Pacific Islands.

Although the sea can accomplish little along coasts towards excavating valleys, yet when the land is wholly submerged . . . the great oceanic currents sweeping over the surface and through channels between the islands, would wear away the rocks or earth beneath . . . the excavations formed would be very broad rounded valleys; . . . It is obvious that the valleys of the Pacific Islands have nothing in their features or positions attributable to such a cause. (*Dana, 1850A, p. 55*)

It is at this point that Dana displays the novel trend of his ideas. He states without qualification that the main work of valley erosion is done by falling rain collected into streams, which thus become powerful erosive agents. Two circumstances had apparently guided his conclusion. Where the lavas had recently spread across the surface he noted that the topography was relatively flat, but where they were older and the elements had had time to act the surface was deeply dissected. He noted a similar distinction between the rainy and the dry slopes of the mountains; on the rainy slopes the number and depth of the valleys was far more pronounced.

Running water of the land, and gradual decomposition. – Of the causes of valleys mentioned in the outset we are forced to rely for explanations principally upon running streams; and they are not only gouges of all dimensions, but of great power, and in constant action. There are several classes of facts which support us in this conclusion.

We observe that Mount Loa, whose sides are still flooded with lavas at intervals, has but one or two streamlets over all its slopes, and the surface has none of the deep valleys common to the other summits. Here volcanic action has had a smoothing effect, and by its continuation to this time, the waters have had scarcely a chance to make a beginning in denudation.

Mount Kea, which has been extinct for a long period, has a succession of valleys on its windward or rainy side, which are several hundred feet deep at the coast and gradually diminish upward, extending in general about half or two-thirds of the way to the summit. But to the westward it has dry declivities, which are comparatively even at base, with little running water. A direct connection is thus evinced between a windward exposure, and the existence of valleys; and we observe also that the time since volcanic action ceased is approximately or relatively indicated, for it has been long enough for a valley to have advanced only part way to the summit. Degradation from running water would of course commence at the foot of the mountain, where the waters are necessarily more abundant and more powerful in denuding action, in consequence of their gradual accumulation on their descent.

. . . The valleys of Mount Kea alone, extending some thousands of feet up its sides, sustain us in saying that time only is required for explaining the existence of any similar valleys in the Pacific. As in Tahiti, these valleys in general radiate from the centre, that is, take the direction of former slopes; they often commence under the central summits, and terminate at the sea level, instead of continuing beneath it . . . (*Dana, 1849A, pp. 380–90*)

Dana pushed his analysis further. He discovered that valleys differed in general appearance and that often each valley had the following succession of forms between its source and its mouth (*Dana, 1850A, p. 52*):

1. Narrow gorge, frequent cascades, valley sides 30–60 degrees. Source in a number of cascades in an amphitheatre.
2. Narrow gorge, almost vertical walls, flat uneven strip of land at the base over which the stream flows from side to side. Becomes type 1 headward.
3. Valley with extensive flat plain.

He set himself the task of accounting in detail for the several differences and was quick to realize that the velocity and transporting power of a river depended on the amount of water and the steepness of each slope. Where the amount of water was large or the slope steep the number of valleys and their depth would be at their maximum. If the slope were gentle then the maximum erosion would occur at its foot, where there was the greatest concentration of water. This regulatory factor he applied to the succession of forms of a river. As the headward slope of the river was often the steepest part, here the greatest erosion occurred and gorges, torrents and cascades were to be found. As distance downstream increased, the gentler slope was accompanied with a gradual broadening of the valley. Even here he recognized that erosion did not cease for it is often in this section of the river that the amount of water was at a maximum; it was the direction of the erosion that changed, instead of cutting downwards the river would laterally pare away its valley sides thereby accentuating the breadth. It was the presence or absence of these factors which he felt accounted for his three types of valley.

A brief review of the action and results of flowing waters will render the origin of these features intelligible.

(*a*) Suppose a mountain, sloping like one of the volcanic domes of the Pacific. The excavating power at work proceeds from the rains or condensed vapour, and depends upon the amount of water and rapidity of slope.

(*b*) The transporting force of flowing water increases as the sixth power of the velocity, – double the velocity giving sixty-four times the transporting power. The rate will be much greater than this on a descending slope, where the waters add their own gravity to the direct action of a progressive movement.

(*c*) Hence, if the slopes are steep, the water gathering into rills, excavates so rapidly that every growing streamlet ploughs out a gorge or furrow; and consequently the number of separate gorges is very large and their size comparatively small, though of great depth.

(d) But if the slopes are gradual, the rills flow into one another from a broad area, and enlarge a central trunk which with incessant additions from either side, descends towards the sea. The excavation above, for a while, is small; the greater abundance of water below, during the rainy seasons, causes the denudation to be greatest there, and in this part the gorge or valley most rapidly forms. In its progress, it enlarges from below upward, though also increasing above, while at the same time the many tributaries are making lateral branches.

(e) Towards the foot of the mountain, the excavating power gradually ceases when the stream has no longer in this part a rapid descent, – that is, whenever the slope is not above a few feet to the mile. The stream then consists of two parts, the torrents of the mountains and the slower waters below, and the latter is gradually lengthening at the expense of the former.

(f) After the lower waters have nearly ceased excavation, a new process commences in this part, that of widening the valley. The stream which here effects little change at low water, is flooded in certain seasons, and the abundant waters act laterally against the enclosing rocks. Gradually through this undermining and denuding operation, the narrow bed becomes a flat strip of land between lofty precipices, through which in the rainy seasons, the streamlet flows in a winding course. . . .

(g) The torrent part of the stream, as it goes on excavating is gradually becoming more and more steep. The rock-material operated upon, consists of layers of unequal hardness, varying but little from horizontality and dipping towards the sea, and this occasions the formations of cascades. Whenever a soft layer wears more rapidly than one above, it causes an abrupt fall in the stream . . .

(h) As the gorge increases in steepness, the excavations above deepen – the more rapid descent more than compensating, it may be, for any difference in the amount of water. Moreover, as the rains are generally most frequent at the very summits, the rills, in this part are kept in almost constant action through the year, while a few miles nearer the sea they are often dried up or absorbed among the cavernous rocks. The denudation is consequently at all times great about the higher parts of the valley, (especially after the slopes have become steep by previous degradation) and finally an abrupt precipice forms its head.

(i) The waters descending the ridges either side of the valley or gorge, are also removing these barriers between adjacent valleys, and are producing as a first effect, a thinning of the ridge at summit to a mere edge and as a second its partial or entire removal, so that the two valleys may become separated by a low wall, or terminate in a common head, – a wide amphitheatre enclosed by lofty mountain walls. (*Dana, 1849A, pp. 380–90*)

Dana's contribution to fluvial geomorphology was far greater than Ramsay's in this respect, and it was the more remarkable because Dana was a citizen of a country where geology was still in its infancy, where normally lip-service was paid to current European ideas. His tour of the Pacific had taken him to many of the areas visited by Darwin and had enabled him to compare his fluvial ideas with those of the marine erosion theory that Darwin had adopted from Lyell. In his article on New South Wales in 1850 he questioned Darwin's postulated marine origin of the valleys there, and recapitulated his own ideas of fluvial erosion (*Dana, 1850B*).

During the mid-nineteenth century Dana was the outstanding fluvialist. Unfortunately his early work was little known and debated on the European side of the Atlantic although his influence grew steadily at home and abroad in the next twenty years. Technically, his 1863 publication entitled *Manual of Geology* was the most important contribution since *The Principles*, which it resembles closely in methods of presentation. Dana had Lyell's capacity for detail and his treatment of minor landforms, such as river-terraces and flood-plains, was on a large scale. The result is a more mature and realistic picture. We may take for example river-terraces which Dana decided that, though caused mainly by elevation of the land (and this is a vital advance), could in particular cases be created largely by other agents. He realized that, more often than not, the level of the river at flood really determined the height and extent of the 'terrace' deposits.

The formation of the river-terraces . . . has been stated to be a consequence of the elevation of the land . . . Rivers in an open country have always both these two elements, a channel and a river-flat or flood-plain. The stream occupies the former during ordinary low water, but spreads over the latter during freshets. The sweeping violence of the flood determines the limits, other things being equal, and the flat surface of the flood-plain or river-flat.

If now the interior of a continent be raised, say 100 feet, while along the sea-coast it is little changed, the river will have an increased angle of slope, a quicker flow, and a greater power of erosion; and it will gradually wear down its channel, if there are no rocks to prevent it, until the old slope is again attained. The flood-plain will also sink at the same rate, although with more or less changed limits, owing to many causes of variation . . . It appears thus that each terrace was once part of a flood-plain of the river.

In the above explanation the terraces are supposed to correspond each to a separate period of elevation. This must be the case, and, when so, the same terrace would be traceable for great distances along the course of the larger rivers. But successive terraces may be formed in river-valleys, either (1) during a slowly-progressing elevation, or (2) in the course of the wear which may be in progress between periods of elevation; and it is often difficult to distinguish the accidental or intermediate plains from those that are distinct records of change of level . . . Some of the conditions producing them are the following: (1) changes in the river-channel to one side or the other of the river-valley, altering thereby the action of the flood-waters during freshets, and causing them to commence wear according to a new outline; (2) resistance to wear in a portion of the alluvium, owing to a degree of consolidation, or to some obstacle; (3) a permanent diminution in the waters of a stream, arising from changes about its sources, or in some other way.

It is important to observe also that the same terrace may differ in height ten to fifteen feet or more; because (1) the flood-plains of rivers (the original condition of the terrace-plains) often differ much in height in different parts; (2) the rains and streamlets often wear away the soft material of the terraces,

diminishing their height, and sometimes obliterating the plain altogether;
(3) the winds carry off the light soil of the surface, and in the course of
centuries may produce great results. (*Dana, 1863, pp. 555-6*)

Dana's knowledge of hydraulic principles is equally detailed and as he had
from the first believed that fluvial forces are the dominant agents in denudation, it is not surprising to find that this section of geomorphology receives
much attention. The following quotation will demonstrate the high quality
of his ideas on rivers.

> The river in this state, consists of its torrent portion, . . . and its river
> portion. . . . Along the former a transverse section of the valley is approxi-
> mately V-shaped, and along the latter nearly U-shaped, or else like a V
> flattened at bottom. The river-portion usually exhibits, even in its incipient
> stages, its two prominent elements, – a river-channel, occupied by waters in
> ordinary seasons, and the alluvial flat or flood-ground, which is most covered
> by the higher freshets. The two go together whenever the course of the

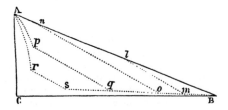

FIG. 68. *Dana's (1863) longitudinal stream profiles (dotted lines)
Development of profiles through time. Ap and Ar—Cascade-portions; lm to rs—
Torrent-portions; mB to sB—River-portions.*

> stream is not over and between rocks that do not admit of much lateral
> erosion and a widening thereby of the river-valley. . . .
> As the waters continue their work of erosion about the summits, where the
> mists and rains are most abundant and often almost perpetual through the
> year, the next step is the working down of a precipice under the summit or
> towards the top of the declivity. . . . The stream in this state has (1) a cascade-
> portion and (2) a torrent-portion; besides (3) its river-portion. The precipices
> thus formed are sometimes thousands of feet in height; and the waters often
> descend them in thready lines to unite below in the torrent. The mountain-
> top is chiselled out by these means into a narrow, crest-like ridge. . . .
> The next step in the progressing erosion is the wearing away of the ridge
> that intervenes between two adjoining valleys. This takes place about the
> higher portions nearest the mountain-crest where the descending waters are
> most abundant. Gradually the ridge thins to a crest, and finally becomes worn
> away for some distance, so that two valleys . . . have a common head. The
> erosion, continuing its action around the precipitous sides of the united head
> of the valleys, may widen it into a vast mountain amphitheatre. . . .
> The system is illustrated on a grand scale among the old volcanic islands of
> the Pacific, where the slope of the rocks at a small angle (5 to 10 degrees)
> from a centre has favoured a regular development. (*Dana, 1863, pp. 636–7*)

At the same time the forces which erode, he explains, also perform two other functions; they transport the eroded material away and, in doing so, separate, disintegrate and grind it down into smaller and smaller particles, as the distance from the source of origin increases.

> The amount of transportation going on over a continent is beyond calculation. Streams are everywhere at work, rivers with their large tributaries and their thousand little ones spreading among all the hills and to the summits of every mountain. And thus the whole surface of a continent is on the move towards the ocean. In the rainy seasons the streams increase immensely their force. Streamlets in the mountains that are almost dry in summer become destructive torrents during the rains.
>
> The process of transportation is also one of wear. The stones are reduced to sand and fine earth by the friction. The silt is nothing but the coarse material of the upper waters ground up. (*Dana, 1863, p. 643*)

Unlike his European contemporaries, Dana could already draw on the first results of the Western Explorations in the United States.

> Newberry attributes these profound gorges, and beyond doubt correctly, to erosion, each stream having made its own channel. The cliffs are so high that in general no undermining can set back the walls far enough to allow of alluvial plains along the bottom, even when the water is not too rapid; and when a channel is cut in granite, lateral wear is always small. (*Dana, 1863, pp. 638-9*)

Although a confirmed fluvialist, he did not deny the power of wave erosion and devoted a section to the erosive action of the sea and the formation of marine platforms.

> The wearing action of waves on a coast is mainly confined to a height between high and low tides. – Since a wave is a body of water rising above the general surface, and when thus elevated makes its plunge on the shore, it follows that the upper line of wearing action may be considerably above high-tide level.
>
> Again, the lower limit of erosion is the low-tide level for the waves have their least force at low-tide, and their greatest during the progressing flood; and when the waves are in full force, the rocks below are already protected by the waters up to a level above low-tide mark. There is, therefore, a level of greatest wear, which is a little above half-tide, and another of no wear which is just above low-tide.
>
> The feature of wave action, and the reality of a line of no wear above the level of low-tide, are well illustrated by facts on the coasts of Australia and New Zealand.
>
> In figure [69] (representing in profile a cliff on the coast of New South Wales near Port Jackson), the horizontal strata of the foot of the cliff extend out in a platform a hundred yards beyond the cliff. The tide rises on the platform and the waves unable to reach its rocks to tear them up, drive on to batter the lower part of the cliff. (*Dana, 1863, pp. 656-7*)

For Americans Dana set the pattern of geologic thought; he did for America what Lyell had done for Britain between 1830 and 1850. He showed them the content of geology, presented them with a method of approach and kept them abreast of the current theories. His *Manual of Geology* included every

FIG. 69. *Dana's (1863) illustrations of marine erosion platforms in New South Wales and New Zealand (the 'Old Hat')*

aspect of the subject, and for that reason it, of all the textbooks, was the one that set the fashion for those which were to follow. The work ran into several editions, the last appearing in 1895 after being completely rewritten. A scholar by tradition and character, Dana gave all his energies to ensuring the accuracy of his books. He was the first great American geologist; the first geomorphologically who 'sought materials of compensation in another hemisphere – (and) called the New World into existence to redress the balance of the Old'.

CHAPTER TWENTY-ONE

George Greenwood and the Fluvialistic Survival in Europe

> Discretion of speech is more than eloquence, and to
> speak agreeably with whom we deal is more than to
> speak good words or in good order.
>
> (Francis Bacon: *Essay on Discourse*)

In Europe, in contrast to the forthright views of Dana, the fluvial theory was poorly advocated at this time and its adherents were scattered and disorganized. A few such as De la Beche had no very fixed views on the problem of erosion but placed reliance equally on structure, unnatural causes and subaerial activity. Thus De la Beche in his report on the geology of the West Country quietly emphasized the major role played by rain and rivers in the removal and transport of the soil. He neither exaggerates nor compares his explanation with contradictory theories but merely sets down what he considered to be observable facts.

> In a hilly country, such as Cornwall, Devon and a large part of West Somerset, the descent of this detritus can be seen to be taking place in all directions, more particularly during heavy rains. According to the quantity of water, slope of ground, and other necessary conditions, will it be seen to travel. During every rain over the district there is a general movement of the particles composing a large portion of its surface from their previous places to lower levels. The amount moved may be comparatively insignificant during a year or a century, but the cubic contents of that which descends to lower ground, during the lapse of a great geological period, must be very considerable. (*De la Beche, 1839, p. 456*)

J. Prestwich was another British geologist with a leaning towards fluvial explanations. He seems to have been deeply impressed with the denuding and transporting capacity of a flood which he observed on 4 February 1852.

> ... if such are the remarkable effects of a temporary flood caused by a body of water comparatively so small, and along a valley where its power could not be maintained, we may form some conception of the enormous power which a more continuous flood, with more sustained action, could possess. (*Prestwich, 1852, p. 230*)

There were others such as R. Godwin Austen who thought that

> the agent which has produced the greater part of the existing valley – systems of the Chalk-area of the North of France has been merely meteoric. (*Godwin Austen, 1855, p. 118*)

But the fewness of these names serves to show that the European fluvialists were a small minority swamped by the marine erosionists. Herein lies the importance of George Greenwood.

COLONEL GEORGE GREENWOOD AND THE FLUVIALISTIC ASSAULT

Of all the figures that struggled for a proper recognition of Hutton's ideas, Greenwood was probably most faithful to their spirit and was certainly the most unsparing in his endeavours. Yet as a geological personality his work was little recognized and little liked. This disregard arose partly because Greenwood stood aloof from the main circle of British geologists and worked out his problems alone before presenting them in a somewhat dogmatic manner which probably deterred his contemporaries from making any conciliatory approaches. Indeed there is no doubt that Greenwood did nothing to spare the blushes of his opponents. As already mentioned (pp. 186–7), he was the first geologist to make a serious criticism of Lyell's ideas. Yet, despite the loneliness of his position as a writer dealing in concepts which received no immediate support, his arguments and the principle upon which they were based deserve keen attention. He preached the true gospel and if geologists did not hear him it was not because what he said was poorly expressed but that he suffered the reception that often befalls forthright tellers of inconvenient truths.

Greenwood had no training in geology. On completing his education at Eton, he started life as a soldier, being gazetted to the Second Life Guards in 1817. He seems, according to his son, to have made a memorable mark during his brief military career.

His wonderful feats of horsemanship, his brilliant courage and genial temperament, endeared him to the comrades by whom he was surrounded. He became Major and Lieutenant-Colonel in 1831, about which time it was the constant delight of William IV to send to Colonel Greenwood all the restive and insubordinate horses that the neighbourhood contained, with a view to their subjugation by his fine courage and finished horsemanship. Upon one occasion the Sailor King proposed Colonel Greenwood's health in St. George's Hall at Windsor, as being the most accomplished rider in the British army. . . . In 1837 he was made Colonel of the 2nd Life Guards, on the promotion of his predecessor, Colonel Edward Lygon, to be a Major-General. Colonel Lygon, who had command of the regiment at Waterloo, was always inquiring upon his return to barracks whether Colonel Greenwood, with whose dare-devil horsemanship he was well acquainted, had been killed as yet. But the questioner was never gratified by the report even of any accident to the fearless rider, who possessed to perfection the art of falling off, swimming rivers, and getting out of scrapes without sustaining damage. It was greatly regretted by Colonel Greenwood's many admirers that, acting upon the advice of Dr Chambers, who declared that he had a heart complaint, he left the army in 1840, and had therefore retired from active service before the Crimean War. But among the reforms which he introduced into the Household Cavalry it is still gratefully remembered that he reduced the weight of the helmet from 8 lb to 3 lb; Colonel Greenwood had experienced the intolerable burden of his headpiece when he sat upon his horse for several hours at the coronation of Queen Victoria in 1838. (*Greenwood, 1877, pp. ix–xi*)

When Greenwood left the army, he retired to his country property, Brookwood Park at Alresford in Hampshire where he divided his time between his favourite pursuits, hunting, aboriculture and geology. In all he reached a high standard of proficiency and on each he wrote vehemently as his *Hints on Horsemanship* and *The Tree-Lifter* will testify. He learnt his geology entirely by personal observation which may partly account for his lack of respect for academic opinions. The constant habit of command and dependence on his own judgment probably largely account for the passionate vigour and conviction with which he aired his views and accepted challenges. To the end of his life he never ceased to advocate his ideas and to defend them against the criticism of opponents. His son's picture of him probably provides the happiest outline of his proud, forceful and independent nature.

> Therefore the books that he consulted were the hills and vales, 'stones and the running brooks'. In the ploughed field, on the rain-washed road, in the sunken lane, on the chalk hill-side, on the sea-beach, he collected his materials. Map in hand and always on foot (for it was 'against the laws of the Medes and Persians,' he would say, that he should go otherwise!) he would scour the country for miles around carefully marking out the vallies and the 'water-partings' and noting the effects of 'subaerial denudation'. Did anyone allege special facts in opposition to a theory that he had started, he (at the age of seventy) would be off at a moment's notice to Norway or the Highlands, that such allegations might be put to the test. *Oculis subjecta fidelibus.* (*Greenwood, 1877, pp. xiii–xiv*)

The basic principle of his argument was that all denudation of the land could be explained by the simple combination of rain and rivers. The rain falling on the surface slopes carried down with it the loose soil. As the rain progressed towards the lower slopes it collected first into rills, then into streams and finally into broad rivers. The whole formed a unitary system which denuded the surface and carried off the debris for deposition on the flat sea-floor. These ideas, as he himself was not slow to repeat, were first published in 1853 and he had had no occasion to alter them since that date.

> The doctrines advocated in this book were first published by me in the second edition of the *Tree-Lifter*, Pt. III, 1853, before I had ever heard of Hutton. Then in the first edition of *Rain and Rivers*, 1857. (*Greenwood, 1876, p. vii*)

At first his views made no impression whatsoever on the trained and ordered phalanx of the marine erosionists, but as opinion swung gradually in his favour and his ideas were accepted one by one, more often than not in ignorance of the fact that he had published anything upon them, he repeatedly and rather arrogantly insisted on his prior claim to credit.

> In the same number of your *Magazine*, p. 568, Mr Hull says 'I adopt, though with some hesitation, the views of Professor Ramsay, Dr Foster and

Mr Topley regarding the subaerial denudation of the Weald'. If Mr Hull will do me the honour to read the chapter on the Weald in *Rain and Rivers*, I think that he will do me the justice to say that the above-named gentlemen have 'adopted' my principles, first published in 1853, and again in 1857. (*Greenwood, 1877, p. 160*)

This trait irritated his scientific contemporaries who might otherwise have accorded him proper respect. Much of his writing adopted a lordly tone of almost divine authority, patiently drawing the attention of lesser mortals to the correctness and prior publication of his revelations each time they were confirmed.

Page 449, *Report on Devon and Cornwall*, he . . . (Sir Henry de la Beche) . . . gives a disquisition headed 'Effects of Atmospheric Influences', and 'Transport of Detritus by Rivers', the doctrines of which are precisely those of *Rain and Rivers*. (*Greenwood, 1876, p. 2*)

The general tone of Greenwood's replies to his many critics may be judged from the following strictures on Murchison, who, for once, is given a dose of his own medicine in the correspondence column of a daily newspaper.

According to the *Morning Post*, of yesterday, Sir Roderick Murchison said, 'I do not apprehend that those who have examined the tract of Coalbrook Dale will contend that the deep gorge in which the Severn there flows has been eaten out by the agency of the river, the more so when the deep fissure is at once accounted for when we see the abrupt severance that has taken place between the rocks which occupy its opposite sides'; and 'when, indeed, we look at the lazy-flowing mud-collecting Avon, which at Bath passes along that line of valley, how clearly do we see that it never scooped out its channel; still more when we follow it to Bristol, and observe it passing through the deep gorge of mountain limestone at Clifton, everyone must be convinced that it never could have produced such an excavation. In fact, we know that from the earliest periods of history it has only accumulated mud, and has never worn away any portion of hard rock. From such data I conclude that we cannot apply to flat regions, in which water has no abrading power, the same influence which it exerts in mountainous countries.' Now the data are that Sir Roderick 'apprehends' and 'clearly sees' that a river cannot cut a gorge through limestone. But if Sir Roderick goes to the falls of Niagara he will 'apprehend' and 'clearly see' that a river can do this. Does Sir Roderick apprehend that the Niagara has not cut, and that it is not at this moment cutting the gorge below the fall? Does the 'abrupt severance' of the 'opposite sides' of the gorge prove it to be an igneous fissure, and is not the country through which the gorge is cut and the country above the gorge and below the gorge flat? Does not the river between Lake Erie and the falls flow (in Lyell's words) 'like an arm of Lake Erie,' over 'a flat-table-land'? Does not the same water which at the rapids and falls cuts the gorge through the solid rock become in Sir Roderick's words, 'a lazy-flowing mud-collecting river' between Queenstown (where it leaves the gorge) and Lake Ontario? How could it be otherwise? How could the river continue to fall over the hill after it has cut its channel through the hill and to the foot of the hill? When the further lowering of the bed of a valley is checked by the sea or a lake, or hard strata which cause a gorge or a shallow, denudation going on above produces

FIG. 70. *Colonel George Greenwood*
(*From a painting. By kind permission of Richard Cave, Esq.*)

a comparatively horizontal bed to the valley; this checks the flood-water descending from the comparatively steep upper valley, and the same flood-water which at first cut the valley down now deposits on it and fills the valley up. This is the origin of alluviums and deltas, on which gross ignorance prevails. The annual growth of (unembanked) alluviums and deltas in the flat part of valleys (which Sir Roderick adduces against aqueous denudation) is the very thing which proves that aqueous denudation is going on now. If not, how or from where does the deposit come? Can deposit take place without denudation? But these proofs of present denudation are not measures of it. As I have said, what is deposited on alluvial flats by the river is not a unit to a hundred thousand million of what flows out to sea from the present denudation of the atmosphere and rain. The foundation-stone of geology has yet to be laid. The first principle of geology is the law of stratification or deposit. But we have not perceived the cause of the first step towards this first principle. The first step to stratification or deposit is denudation. But geologists are ignorant of the cause of this first step, denudation, which (except the wear of coasts) is simply and solely from the atmosphere and 'rain and rivers'. These throw the entire surface of the earth into ridge and furrow; every furrow or valley, whether dry or having a river, ending only at the sea. Hence, stratification in the sea. People who are not geologists or physical geographers think that there must be a descent to the centre of these furrows or valleys. Our two great geologists and physical geographers, Lyell and Murchison, do not think this is necessary. They call 'the Wealden

Heights', on the anticlinal 'Forest Ridge', a valley, to the centre of which (Crowborough Beacon) there is an ascent of 800 feet. Is it not time for a reform here? (*Greenwood, 1877, pp. 92–94*)

It is this admixture of common sense, confidence and arrogance which makes Greenwood's work so irritating to his fellows and so intriguing to his followers.

'RAIN AND RIVERS; OR HUTTON AND PLAYFAIR AGAINST LYELL AND ALL COMERS'

As Greenwood launches into his theme he makes it perfectly clear that the key to the problem of denudation lies in the one factor, RAIN. Throughout his main book (1857) and in all his numerous counter-attacks on critics and opponents of it, he remains consistent to this principle.

> The alternation of the ravine and ridge, the hill and dale, the exquisite beauty of the earth's surface – of all that we see of the earth – the 'dread magnificence' of the Alpine chasm, and the rich loveliness of the alluvial vale, are ever-changing products resulting from . . . the daily 'wash of rain'. (*Greenwood, 1857, p. 9*)

Without rain, he reiterated, rivers would have no existence.

> Under this heading your number for December contains a paper by the Rev. O. Fisher. The opening sentence is: 'Upon the land-surface a certain amount of the fine material is being carried into the rivers, and by them deposited at the heads of the Broads, or where such do not exist, in the sea, this denudation by pluvial action is undoubtedly greater where the land is under the plough than it would be otherwise.' The wildest subaerialist will require nothing more than this. Grant this and time, and the entire land must be deposited beneath the sea. So far theoretically. Practically we know that it is so. Practically we know that the entire land has been under the sea, and has been covered with an enormous depth of stratified deposits. In fact, as I have headed a chapter in *Rain and Rivers*, 'It is only fire that keeps our heads above water'. Yet Mr Fisher, who admits the principle that rain ever has been and actually is now washing the entire land into the sea, begins a sentence . . . 'The windings of the valleys also appear to be on a larger scale than can be due to such rivers'. Why, the insignificant valleys which he mentions, nay, the largest valleys in earth, those of Amazon, Yang Tze, and Mississippi, might have been formed without any river at all, by atmospheric disintegration and the erosion of rain. That is, by the pluvial action mentioned by Mr Fisher himself. When these rivers are flooded by rain they are swollen to perhaps twenty times their usual volume; and these rain-floods would occur annually in their valleys whether the rivers existed there or not. This is, instead of constant rivers there would be periodical rivers in the valleys. I have said in *Rain and Rivers* that rivers are rain reappearing and returning to the sea. But Mr Fisher talks of rivers as if they were not rain; and if not, what are they? Evaporation condensed into rain is the *causa causarum*. Rain causes valleys. The largest rivers in the world are, by comparison, the niminy-piminy effects of this *causa causarum*, and are mere assistants in forming the wondrously magnificent valleys in which they flow

(for, perhaps, 4,000 miles), and which are the roads which carry the entire surface of the earth into the sea. This titanic traffic is brought to them entirely by rain. That is, owing to atmospheric disintegration everything on the surface of the earth which is not living is decaying. Hence, soil; and soil, which is rotted subsoil, is in perpetual formation over the entire surface of the earth, and is perpetually washed down the hill-side into the valley and along the valley into the sea.

    . . . In joining 'Rain and Rivers' together we must remember that rain is the cause, rivers the effect. In a chalk country like Norfolk there is not a single so-called river valley which does not begin with a dry valley or 'rain valley,' far above the highest springs of the river. Two opposite rain valleys constantly cut nearly through the dividing ridge. But as long as a water part-ing remains, and the waters run in opposite directions, we must consider them as two valleys. Mr Fisher talks of the valleys of the Waveney and Little Ouse, first as one valley, then as two valleys, then as two valleys inosculating. But the upper part of every so-called river valley on earth is always purely a 'rain valley or dry valley' *sine flumine vallis*, and in myriads of cases entire valleys are, especially in porous strata like chalk. And in nature, at the divid-ing ridge, each opposite dry valley or water-flow may be seen to stretch its fingers up each opposite water-slope to join hands across the intervening water-parting. Hence the low parts of a dividing ridge alternating with high parts, for which we have the modern northern terms, gap, saddle, col, etc. Hence the southern sierra or serra (saw), and the Latin *juga montium*, from the saw-like, or yoke-like ups and downs of dividing ridges. The very name of *jugum* (hill or yoke) originates here. But those opposite dry valleys, which run up to these low parts of the dividing ridge – these beginnings of valleys, are not caused by rivers. They are caused by the cause of rivers – rain. (*Greenwood, 1877, pp. 175–8*)

What is more, he points out that rain is not of inconsequential proportions as over a period of time the total quantities involved are equal to, if not in excess of, any amount of water required to accomplish the normal work of denudation.

    If only one year's rain could remain where it fell, the surface of the whole terraqueous globe would be covered a yard deep with fresh water. Is this so small a cause as to produce no effect? (*Greenwood, 1857, p. 29*)

With an agent of such power the necessity of searching for other causes vanishes.

    These cones, however, though the monument, are by no means the measure of this denudation, since they have themselves been melting away for 'an eternity of time', and are day by day vanishing *tenues in auras* at this instant. All this is going on now all over the world and all day long. These simple yet sublime truths are beneath the contempt of poor marvel-mongering human nature. We must have a cataclysm, or a glacial epoch, or a wave of transla-tion, or a diluvial period, or a boulder period, or a drift period, or a gravel period to account for every heap of rain-wash. (*Greenwood, 1876, pp. 17–18*)

He affirms, therefore, that it is not strictly accurate to think of rivers denuding the landscape. It is the combination of rivers and rain within one system that properly explains the process of denudation. As well as valley slopes there are also rain slopes which are really the beginnings of valleys or slopes that have still to form into valleys; both valley and rain slopes are integral features of a drainage system and they are intimately connected in such a way that all the material of denudation is smoothly carried away without any interruption in the process and disposed of at one lower outlet, the sea.

> A valley is, in my opinion, a water-slope or a system of water-slopes, converging to one outlet at the sea. (*Greenwood, 1866, p. 23*)
> Every valley, indeed, is a valley of denudation; which denudation has passed out of the valley through the single outlet or water-shed. For it is an essential characteristic of valleys, trunk or branch, that however infinite the tracery, and however countless the number of their ramifying upper inlets, each and every valley can have but one lower outlet. (*Greenwood, 1857, p. 35*)

This explanation of the unity of rain, rivers and landscape was essentially simple to follow but so vitally different from the various marine and catastrophic theories then popular, that Greenwood realized he would have to provide a reasoned and detailed justification for it. He did this, firstly, by exposing the weaknesses and inconsistencies of the alternative marine erosion theory, and secondly by applying his own theory to specific features.

Greenwood discredited Lyell's theory on three main counts:

1. The evidence of the shorelines showed that valleys did not coincide with lines of weakness.
2. Where the sea had encroached on the land the submerged surface did not show any similarity to the intricate form of the normal river basin but was represented by a flat platform.
3. The action of the sea did not encourage denudation but actually checked it, causing deposition to take place in the form of deltas and sandbars wherever the sea came in contact with freshwater outlets.

He went on to emphasize the significant opposition of the work of the sea and of rivers. Marine erosion could be carried forward to the point where the sea met the river and here even the weakest river was strong enough to prevent the further penetration of the sea. There was a strict line of demarcation between the two forces; the two systems or processes were quite distinct.

Greenwood had not advanced so far in his thoughts as to see a river as an instrument with powers of self-adjustment, but he was not really concerned with this aspect. His powers of observation were sufficiently acute to mark the changes that occurred when a river entered a wide expanse of water. The points which had chiefly affected his judgment are revealed in his more detailed discussion of particular minor features. Always he sees a procession of quantities of soil and of water going towards the sea, and as either varies in relation to the other so the changes will cause various differences in the topography. Thus, the rate of denudation is controlled by the steepness of the valley slopes, and whenever this flattens the amount of soil contained within the valley will accumulate despite the erosive processes.

And valleys always hold soil as rivers always hold water, though the soil and the water are both *in transitu*. Directly as each valley lengthens, its longitudinal slope becomes gentle. Directly as it widens, its lateral slope becomes gentle. And directly as the gentleness of the slope of each valley is the accumulation of soil. For as the rain takes the soil from the hill-top down the valley to the sea, in proportion of the flatness of the valley will deposit of soil exceed denudation. (*Greenwood, 1857, p. 7*)

This conception of a finely-weighted balance he applies equally to a river's alternation between denudation and deposition and to the formation of river-terraces. In the case of river-terraces he suggested that the benches were produced either by the intervention of the sea or by a bed of hard rock, which temporarily lessened the rate of denudation at one point, although in the region immediately upstream denudation still proceeded. As a result the river which was formerly able to transport all the denuded material finds itself unable to do so and begins to flood above the barrier. In time the barrier will be cut through and the change in slope will enable it to erode down into its own alluvium. The sequence is, of course, different where further erosion is checked by the sea because there can be no cutting of the barrier and erosion can only recommence if there is a fall in sea level.

Every barrier of hard rock which crosses the bed of a river or valley becomes a negative key to the depth of the river and the valley above the barrier. Thus deep shalt thou go and no deeper. But as denudation is ever at work, the bed of the river and of the soft valley above the barrier becomes horizontal at the same level as the barrier. The flood waters are checked at the barrier, overflow the horizontal plain, and form an alluvial patch. The barrier sinks and is widened from erosion. The alluvial patch is no longer over-flowed. The river is confined between walls of its own building. It sets to work to tear them down. It carries off the alluvial patch which it had deposited till it leaves only two terraces against the hill-sides. (*Greenwood, 1866, pp. 13–14*)

Parallel terraces, that is, one on each side of the valley, may be ancient shores; but the vast majority are the remains of alluviums where no lakes have ever been. These patches result from alternations of hard and soft strata, and as sure as there are alternations of hard and soft strata in the course of a river or valley, so sure will there be alternation of gorge and alluvial flat. For all alluvial flats are formed by the stoppage of the lowering of the valley. The valley above the stoppage is then worn down horizontal. The rain flood waters from the sides and inclined parts of the valley, checked at the flat plains, overflow and deposit alluvium. The sea stops the lowering of every valley, therefore the bed of every valley is flat and alluvial at the end next to the sea. (*Greenwood, 1876, pp. 18–19*)

But if the north of Africa were to rise ever so gradually, it would cause the Nile to fall into the sea, that is, directly as the rise of the land would be the fall of the river into the sea. Directly as its fall into the sea would be its power of deepening its channel. The river would deepen its channel, and a time would come when it could no longer overflow its banks. It would then in floods tear its banks down instead of building them up, and the banks would recede from the river in the form of two parallel terraces. But when the river had cut its banks down to the level at which it could again overflow them, it would again deposit on them, and would form a new alluvium at the new level of the river. If the rising of the land continued, this new alluvium, like the old one, would recede as two parallel terraces. And so, step by step would be formed, as long as the rising of the land continued, whether the rising was gradual or sudden. I have mentioned such terraces at Loch Ranza, in Arran, in the *Athenaeum*, July 22, 1865. (*Greenwood, 1877, pp. 202–4*)

The check which the sea placed on further denudation is used to explain the shape of the river valleys; rivers were narrow and deep at their sources and wide and flat at their mouths simply because sea level represented the maximum depth of fluvial erosion. For a river to erode there must be a slope and therefore erosion in the higher reaches would have to wait upon the progress of erosion in those lower. In brief, a river would erode backwards and the widest valleys would be found near its mouth. Anomalous stretches of wide valley would only be found in the middle reaches where there was a region of soft rock alternating with hard, the soft beds being eroded more rapidly.

> Rain, or a river, works at the whole length of a valley or river course at the same time. But the chief formation is backward, that is from the sea to the hill. For as the beds of rivers, if not horizontal, slope down to the sea, it is evident that the cutting down of the parts farthest from the sea must depend on the cutting down of the parts nearest the sea; that is, the parts farthest from the sea cannot be cut lower than the parts next the sea. So far the cutting of the parts farthest from the sea is dependent on the cutting of the parts next the sea. So that the chief longitudinal cutting of a valley or river-bed may be said to proceed from below upwards as regards the valley, or backward as regards the stream of the river; that is, it proceeds from the sea to the hill, not from the hill to the sea. (*Greenwood, 1857, pp. 173–4*)
>
> This determination of valleys to begin at the end, and to go backwards, is, as has been said, the cause why they are generally widest at the end next the sea. Because the longitudinal cutting being there first made, the sides will be there first exposed to disintegration and the lateral wash of rain. (*Greenwood, 1857, pp. 177–8*)

As has already been discussed, the apparent inconsistency between the wide longitudinal valleys and the narrow transverse valleys of the chalk regions had puzzled most early geologists. They could not understand why the river pattern drained in two different directions, when it was manifestly easier for the river to erode the clay vales than to cut through the resistant limestone and chalk ridges. Greenwood found no such difficulties when he came to examine these areas. He assumed the Weald to have been a dome or ridge and the Yorkshire–Lincolnshire uplands to have been an inclined plane, and imputed the present-day topographic features solely to the result of denudation by rain and rivers.

> The original form given to the 'Weald Valley' by fire, was one smooth hill, or hogs-back, running from the north of France westward, and covered with one single stratum or geological formation. The present form of the region is due to water. (*Greenwood, 1866, pp. 187–8*)

But the Weald is no valley at all. If it is any one thing, it is the reverse of a valley. It is a hill with, if not a quaquaversal dip, at least a quaquaversal waterflow; that is, its waterflow and denudation, like those of most other large hills, are carried to the sea by a great number of distinct trunk valleys. (*Greenwood, 1876, p. 51*)

At first, in the case both of the Weald and the Yorkshire and Lincolnshire Wolds, the rainwaters would drain off down the natural slope in the direction now followed by the transverse valleys. The first valleys (the consequents) would be drained in this direction. As these valleys cut down into the summit strata they would gradually reveal the underlying soft clay beds. On these, new tributaries would arise and, because of the lack of resistance, denudation would proceed at a rapid rate until the present wide longitudinal vales had been formed. Some of the old initial valleys would be beheaded by this differential erosion along the clay vales, but those that were firmly established would persist in their courses through gaps in areas that had now become escarpments and ridges.

Now these 'lateral valleys' (Longitudinal as regards the stretch of the strata) are simply the consequence of the erosion of rain and rivers in the soft strata above or behind the hard gorges, as I first said of the lateral valleys in the Weald clay above or behind the gorges of the North and South Downs. (*Greenwood, 1876, p. 22*)

But whether the Weald hill was originally capped with tertiary strata or not, it is allowed, by all that it was capped by the beds which I have enumerated from the chalk to the Hastings sand, and that these beds have vanished by denudation. Therefore, before this prodigious denudation took place, and while it was taking place, the ridge above the 'Forest Range', was incomparably the highest ground in the region. Naturally, then, the watershed, whether of the surface (rain and snow), or from under the surface (springs and rivers), would be from this highest point, the Forest Ridge, to the lowest point, the sea. And this waterflow is not only to the north and south . . . it is to the east and west also. From Tilgate Forest there is a quaquaversal watershed to the sea, by means of many different trunk valleys, formed and graduated by the wash of rain. And the courses of this watershed once made would be continued during the rise of the land, notwithstanding that, owing to a cause which will be explained, their starting point might have been denuded some feet lower, not than their beds, but than the heights through which their beds pass. That the main drainage of the Forest Ridge does pass through the North and South Downs, which are masses of uniform structure and hardness, is a fact which of itself proves that the watershed made its own courses. The sea could not have made them unless their original contents had been softer than their sides. (*Greenwood, 1857, p. 39*)

Greenwood strongly attacked the theories of Martin and Hopkins that the valleys of such regions follow and are based upon natural fractures. He could find no evidence of valley co-incidence with fractures nor did he feel that the widening of rivers towards their mouths fitted in well with Hopkins' 'mathematical' theory.

His examination of river-terraces and valleys had demonstrated to him that slope encouraged denudation while the inter-position of large masses of water or bands of hard rock caused it to cease. This involves the modern concept of baselevel and he clearly understood its operation. In his descriptions he includes a baselevel maintained by the sea and many local baselevels produced by variations in the hardness of the rock.

> ... the Ouse must cease to deepen its channel where it is level with the sea ...
> (*Greenwood, 1857, p. 52*)
> Every valley is always lengthening at the upper end or decreasing in height there; and all intermediate parts are perpetually in process of being planed down, or built up, to one uniform gradient from the head of the valley to its mouth, or to the end of the delta, if there be one. (*Greenwood, 1857, pp. 52–53*)
> Suppose a barrier of rock to run across any valley or river bed: when the bed of the valley or river on the upper side of the barrier has been worn down to a horizontal level with this barrier, it cannot go lower ... But as the barrier is cut through, the bed of the valley or river will be deepened backward ...
> (*Greenwood, 1857, p. 174*)

As we have already noticed, this consideration of hard strata acting as a baselevel profoundly affected Greenwood's views on the formation of scarp topography.

> A stream running through ridges, large or small, is the simple consequence of the differing hardness of the ground through which it runs. In all cases a stream cuts for itself a narrow channel, the depth of which is determined by the hardest part. For a stream cannot run down where its bed is soft, and up again where it is hard. But the wash of rain digs down where the ground is soft, and leaves hills or ridges where it is hard. And as a transverse stream cuts through a hard stratum, say the North or South Downs, the wash of rain is scooping out two longitudinal valleys behind it, that is, a valley behind each side of the gorge or ridge, (as in the Weald clay). (*Greenwood, 1857, p. 53*)
> The ridges then, instead of being considered as barriers to the river, have actually been formed by the river by the abstraction of intervening masses. (*Greenwood, 1866, pp. 58–59*)
> The great height of some of these ridges, through a proof, is a very inadequate measure of the enormous denudation which has gone on behind them, since the ridges themselves have suffered enormous denudation. The tops of all such ridges must originally have been lower than the hills from which the rivers rise. But whether all or any of them continue lower, and their comparative height one with the other, will depend simply on which is the hardest, that is, which is most quickly disintegrated and denuded; the hills from which the rivers rise might be denuded as low as the beds of the rivers in the gorges. (*Greenwood, 1866, pp. 59–60*)

Until Greenwood's work the emphasis placed on the weathering and the denuding action of rain had been much less. He insisted that rain not only forms rivers, but that it is in itself as important an agent of erosion as the river. Certain features like the sides of wide valleys and dry valleys he ascribes almost entirely to the work of rain-wash.

It is true that the direct action in waste and denudation of torrents and rivers is on lines only; and were it not for the atmospheric disintegration and the lateral wash of rain this their direct action would only cut ravines and channels to the sea, and the sides of all valleys, instead of sloping, would be cliffs; . . . But what widens this ravine into a broad valley with gently sloping sides? The lateral wash of rain into the longitudinal valley. And what forms the broad valley, even where there is no river at the bottom or within many miles? The longitudinal scooping power of the concentrated wash of rain, which in no respect differs from that of the torrent, except in its being a hundred-fold more powerful than the torrent. It is indeed intermittent: so is the real scooping force of the torrent, for torrents only really excavate when swollen by rain. (*Greenwood, 1857, pp. 102–3*)

But as sure as dry land stands betwixt high heaven and the sea, the waters of heaven will wash it into the sea. (*Greenwood, 1857, p. 104*)

This universal portage of soil by rain, the eternal effect of eternal causes, which in huge spaces of time results in such vast geological changes, may also be seen *oculis fidelibus* whenever a fence runs horizontally along the side of a hill. A natural terrace is then formed. For aqueous denudation goes on below the fence, and in chalk countries the ground becomes white; and not only does aqueous denudation cease above the fence, but aqueous deposit takes place, and the good soil which was on its way to the valley is arrested. (*Greenwood, 1857, pp. 120–1*)

It is much to his credit that he was the first geologist since Playfair to perceive the relative importance of rain.

So far Greenwood has been shown as an observer who had the knack of finding correct geological solutions and the unfortunate propensity of being able to vaunt his infallibility in the most uncompromising manner. Yet even he had his blind side and of all the fluvialists he was certainly the most extreme. His fanatical belief in his own ideas to the exclusion of others sometimes induced him to deny merit where it actually existed. He believed so strongly in his erosional theory that he would not contemplate any theory which involved a radical change in or interference with the existing physical processes. Anything which savoured of the abnormal, like an ice age or pluvial period, he attacked as antiquated or Wernerian mumbo-jumbo. Thus, as so often happens, what started out as a most praiseworthy appeal for a return to true Huttonian principles was marred by mis-applications of a principle that was never meant to be applied rigidly. But on this, as on all other matters, Greenwood, who died in November 1875, is quite capable of speaking for himself!

For instance, we are gravely assured in a review of Lyell's *Antiquity of Man*, that, because at the mouth of the Somme a freshwater deposit exists 30 feet below the level of the sea, the land must have been 30 feet higher than it is now when this deposit was formed. It is simply that the river which first formed the estuary, has since filled it up with the debris brought to it by the wash of rain in lengthening and breadthening the valley of the Somme. No change in the relative level of the sea and land has happened. In many places along our south coast estuaries are choked by travelling beaches. The fresh water soaks out through the beach, but the tide cannot soak in through the beach quickly enough to rise in the old estuary to near its height. Peat and marsh land instantly accumulate, and then firm ground by overflow of the flooded streams. So that trees grow far below high-water level. The erosion of the shore by the sea goes on. The shingle-bank is driven inland, and overwhelms the trees. The shingle is again driven inland, and the roots of the trees and the peat are by degrees found out at sea below high-water mark. Then comes a geologist and points to the so-called subterranean or submarine forest as a proof that the land is sinking, when, perhaps, neighbouring raised beaches show that it is rising. Such a submarine forest may be seen near Pevensey, another close to St Leonard's, another near Torquay, and a dozen more on our south coast. But they are all outside choked-up estuaries. Again, to form a drift-bed, or to move a boulder or erratic block, the late geologist must have a 'Cataclysm', as he calls a flood, or 'a wave of translation', or 'a great advancing wave from the north', or icebergs. True that 'nous avons change tout cela', and the present geologist substitutes for these that most monstrous assumption a 'glacial epoch'. Now the great majority of our drift-beds, including the so-called 'Northern drift' are simply old sea-shores or lake-shores, and erratic blocks are the result of the travelling of beaches, which has been totally overlooked and ignored by the geologist. There never was a 'drift period', or a 'boulder period', or a 'diluvial period', but there never was or will be a single day without drift and boulders being driven, and alluvium deposited. In fact, there never was any particular physical 'epoch', or 'period', or 'eara', in the geological history of this globe. The one grand, general, physical law in the formation and transformation of the surface of this globe, is the most peristent, most perpetual, and most minute graduality of change. This law ever was, is, and ever will be in force. (*Greenwood, 1877, pp. 33–34*)

CHAPTER TWENTY-TWO

Jukes and Ramsay and the Fluvialistic Revival

MAXIMUM PRECIPITATION (INCHES)

	Cherrapunji, Khasi Hills, India. (1906–40)	Greenwich, London. (1880–1950)
	Recorded in 24 hours	
Jan.	3·4	1·6
Feb.	3·6	0·8
Mar.	12·1	1·1
Apr.	18·2	0·7
May	32·0	1·0
June	36·4	1·6
July	33·0	1·4
Aug.	26·9	2·2
Sept.	24·8	1·3
Oct.	23·3	1·6
Nov.	13·1	1·4
Dec.	7·5	1·1

Recorded in one month

241 inches   7·65 inches

Recorded in one year

905 inches   35·54 inches

THE TROPICAL INFLUENCE

In an age when Britain's desire for trade was sending her ships and men over all the world it followed that as a matter of course scientific observers began to travel the same routes. In a surprisingly short time Survey departments were set up in the British colonies and geologists became essential members of exploring parties which examined the potentialities of the new regions. In this way the field of geological knowledge was widened rapidly. The new areas provided fresh features and also threw fresh light on old problems. Thus, the different balance that exists between fluvial and marine erosion in the tropics caused a healthy reconsideration of several erroneous assumptions on landscape evolution that had become engrained in the minds of European geologists. In the humid tropics the equivalent of the normal monthly rainfall

385

of a locality in the English lowlands could fall in one hour. Moreover, the warm heavy rain greatly assists weathering, and rocks, such as granite, which may form uplands in Britain may be degraded into lowlands in the wet tropics. The tremendous gorges, stupendous river-spates and great depth of rock-rotting impressed many early observers.

Oldham, a former colleague of Ramsay on the Geological Survey, carried out work in the Khasi Hills on the Bengal–Assam border. He believed in Ramsay's theory of marine planation but he could not help noticing the gorges, sometimes 3,000 ft. deep, which wind through the hills down to the plain. These, he asserts, could not have been produced by any action of the sea but must be attributed solely to river erosion. If this is so, he goes on, geologists must admit that rivers can over a long period of time perform prodigious feats of excavation.

> Another very peculiar feature in the Khasi Hills are the curiously deep and narrow gorges or valleys in which all the rivers, in the Southern portion of the hills, find their course to the plains. The level of the stream under Cherra Poonjee is some 3000 feet below that of the station . . .
>
> Now, although believing that marine denudation has exerted its powerful influence in modifying the features of these hills in former times and at different levels, as I have just stated, it is not possible to see how any littoral action, or any such ordinary marine action, could have produced those long, deep, and sinuous gorges here seen. On the contrary, these river gorges appear to me to have been excavated almost entirely by the force of the streams which have flowed and still continue to flow through them. And they appear to me to offer a magnificent instance of the almost incredible power of degradation and removal, which atmospheric force may exert under peculiar and favourable circumstances. (*Oldham, 1859, pp. 173–4*)

The rivers were more powerful than he had supposed.

> I took the opportunity of visiting one of the streams in these hills after a heavy and sudden fall of rain. The water had then risen only about thirteen feet above the level at which it stood a few days previously; the rush was tremendous – huge blocks of rock, measuring some feet across, were rolled along with an awful crashing, almost as easily as pebbles in ordinary streams. In one night, a block of granite, which I calculated to weigh upwards of 350 tons, was moved for more than 100 yards; while the torrent was actually turbid with pebbles of some inches in size, suspended almost like mud in a rushing stream. (*Oldham, 1859, p. 174*)

Henry Benedict Medlicott, who carried out surveys in the Himalayas, wrote along the same lines though he leans more strongly towards a purely fluvial interpretation. In his memoir, which was influenced by Jukes' 1857 textbook, he showed that the central chain of the Himalayas was older than those flanking it, because it had contributed the material for the formation of the flanking Pliocene chains. When referring to the post-Siwalik deposits, he agreed with Oldham that the gorges can only have been created by river-erosion.

It is only of late years that rivers have met with the attention they deserve as indicators of changes at the earth's surface . . . It has hitherto been the fashion to attribute the deep valleys, or rather gorges of the Himalayas, in a great measure to marine denudation, likening them to the deep fiords of the Norway coast . . . I feel assured that these valleys can all be most justly accounted for by river-action and atmospheric denudation generally, operating through the untold ages of the Sub-Himalayan epoch . . . (*Medlicott, 1860, pp. 157–8*)

He explains his belief by suggesting that too little importance has been allowed to river-erosion and cites the results of two observations. Firstly he describes rivers that have accurately adjusted themselves to slight variations in structure; and secondly, rivers that pass directly from a soft stratum on to a hard stratum when they could have avoided the resistant ridge altogether.

The Ravee in its bend round the termination of the Dhaoladhar gives a good instance of a river course adapting itself to the conditions of a rock structure. At innumerable places on every river and stream we may find instances of the deliberate contravention of this apparently necessary law of natural selection as applied to river courses, and which breaches of law may safely lead us to infer very remote conditions of the surface, very different from what is now apparent. As examples of this I may mention the case of the Blini, where its course turns out of the band of soft nummulitic strata to cut a narrow gorge across the strike of the hard Infra-Krol rocks, to fall again, after two more bends, into the course of the same valley of soft rocks. The Sutlej at Bubhor gives another instance of the same kind. It cuts a narrow defile across the Naina Devi ridge, which is composed of comparatively hard rocks, in which no sign of a crack or bend is traceable, whereas it might apparently, with much less trouble, have made its way round the point of the ridge, continuing throughout in the softer upper rocks. (*Medlicott, 1860, pp. 158–9*)

The second example of a river cutting a course, seemingly across the grain of relief, he explains in detail and, though he does not provide a complete description of the feature now known as *antecedent drainage*, it is obvious that his reasoning contained the germ of this idea. He will not accept the traditional explanation of a river following a line of fracture caused when the mountain chain was raised and contorted. Instead he suggests that the river always flowed in this direction, having its course determined at a time when circumstances were different from what they are now. As the land rose the river would maintain its course and, instead of the fracture deciding the direction of the river, the existence of the river might help to determine the position of the fracture.

At each of the great transverse river gorges there is a complete break in the continuity of the anticlinal flexure, (which forms the north side of the range) no doubt involving the transverse faulting. The stereotyped form of explanation for such coincidence is, that the pent up waters made a natural selection of these transverse fissures along which to carve out their course to the lower level. It seems to me to be open to discussion in this instance whether we should not thus be 'putting the cart before the horse', whether the rivers, for the existence of which in this position during the Sivalik period, we have such good evidence, may not have been the predetermining cause of these transverse fissures. (*Medlicott, 1860, p. 122*)

In the case of the Jumna there is nothing to interfere with the suggestion, that the irregularities in the actual state of disturbance in the region of the gorge may be, in great measure, owing to the unequal accumulation of deposits at the former river's mouth, and it may at least be asked if the river may not have had a more direct influence, if in the early stages of unheavement and contortion, the special erosion of the river course may not have had some influence in determining the position of these irregularities. (*Medlicott, 1860, pp. 126–7*)

W. Kingsmill in 1864 wrote an interesting article which illustrated the widespread effect of weathering on an exposed surface of rock on the southeast coast of China.

This granite, wherever it occurs, has been deeply disintegrated, sometimes to a depth of one or two hundred feet; . . . The original quartz veins of the granite, broken into small fragments by the forces which have operated on the surrounding rock, still traverse the disintegrated mass in all directions; . . . In the higher grounds the soft yielding matrix has generally been removed by denudation, leaving these pseudo-boulders perched all over the granite hills . . . (*Kingsmill, 1864, p. 2*)

From South Africa and Abyssinia the same ideas on fluvial and subaerial denudation begin to pour into Europe.

In South Africa, R. N. Rubidge concluded that the landscape had been formed subaerially and not by marine action.

The denudation undergone, then, is probably due to that series of causes comprised in the term subaerial, and into which the action of sun and wind, rain, ice, and the erosion by river and streams, enter as principals. (*Rubidge, 1866, p. 89*)

As evidence in support of erosion by subaerial forces he employs, first Scrope's original argument that incised meanders could not have been produced by faulting or by any means other than rivers (*Rubidge, 1866, p. 90*), and secondly, gives data of the amount of erosion actually being carried out, which provided a rough calculation of how long might have been required for the past denudation to have been accomplished if present rates were maintained throughout the periods involved.

Some years ago I took advantage of a three days' detention by a freshet of the Sunday's river to collect evidence with a view to the calculation of the approximate amount of denudation affected by it. The data . . . led me to a result of about ·8000 inch in a century over the area drained by the river. I have little doubt that the estimate is much too great, for it would require little more than 100,000,000 years to effect the amount of our denudation. (*Rubidge, 1866, p. 90*)

In the same way, William Blanford's account of the landscape of Abyssinia indulges in a minimum of theory and a lively recitation of observed facts.

Blanford argues, from what he has seen, that the fluvial theory alone is perfectly coincident with the scenic phenomena, and dismisses the action of the sea and of ice and of structure as a possible explanation of the topography.

The gorges of the Jitta and Bashilo . . . , impressed every one who saw them by their great depth and the excessive steepness of their sides, the breadth being singularly small in comparison . . . There cannot be a question but that these enormous hollows are simply channels cut by the streams which have run in them. The lapse of time necessary to have produced such an effect must have been very great.

If the action of such small streams as the Jitta and Bashilo has sufficed to sweep away the contents of ravines 3000 or 4000 feet in depth, what may not have been the effects of rivers like the Takkazzye and Mareb? How much of the Abyssinian highlands has been removed by these great torrents, and spread as an alluvial deposit over the basin of the Nile? (*Blanford, 1870, pp. 154–5*)

It is worthy of repetition that throughout this great denuded area, so far as it was possible to examine it, there is not a trace of marine denudation. The prevailing features of the country are deep ravines cut by rivers, and terraced hill-sides, moulded by the subaerial disintegration of the rocks of which they are composed. On all the slopes there are unequivocal marks of the unequal action of surface weathering on rocks of different chemical constitution. This is especially seen upon the traps.

How far the great scarp of the Abyssinian plateau towards the Red Sea is due to marine action is a very difficult question. If the sea ever exerted any great influence on the denudation of the plains at the base of the range, it must have been at a comparatively remote period; for so far as they were examined, the hills presented a very different aspect from that of a sea cliff. Still it is highly probable that, before the commencement of the volcanic outbursts which have left their traces along the whole southern portion of its shores, the sea extended farther west, and very probably it reached the foot of the hills in places. But tropical seas, and especially calm land-locked basins, like the Red Sea, are trifling agents of denudation when compared with such oceans as the Atlantic. Independently of the local causes which determine the size of the breakers, most tropical coasts are more or less protected by either coral reefs, where no sediment is brought down by rivers, or by the deposition of that sediment itself, which, owing to the more violent rainfall (a far more important item than the absolute amount, although the latter is also usually excessive in the tropics), and to the more rapid disintegration of rock, is far larger in quantity than in temperate regions. (*Blanford, 1870, pp. 155–6*)

I believe that the comparative importance of fresh-water and marine denudation is to this day misunderstood by many of the best geologists of Western Europe, and is only beginning to be appreciated by any of them, in consequence of the exceptional conditions there occurring. The whole of the circumstances attending the rainfall in tropical countries are far more favourable to denudation than in the temperate zone. . . . In the British Isles the average rainfall is about twenty-four inches distributed over the greater portion of the year. In India it averages over the whole country about fifty inches, by far the greater portion of which falls in three months. The showers are far heavier, and far more effective in sweeping soil, sand, and pebbles, from the surface of the country into the streams; and floods in the latter are of annual occurrence, instead of only happening at rare intervals. The effect of a river in full flood in sweeping detritus down into the sea compared with the usual denuding action, is as the comparison of the effect produced by the breakers of the ocean in a storm to those of an inland sea on an ordinary fair day. In flood, a river is liquid mud rather than water . . . It must be borne in mind that fresh-water denudation is distinctly antagonistic to marine; and where the former is large, the quantity of detritus carried down to the sea by rivers actually protects the rocks of the coast from destruction. . . . If over British India the effects of marine to those of fresh-water denudation in removing the rocks of the country be estimate at 1 to 100, I believe that the results of marine action will be greatly overstated.

Some years ago, before the question had assumed the present phase, before even Colonel Greenwood's pamphlet had appeared, I remember being struck by the absence of all signs of marine action and by the unmistakable evidence of immense fresh-water denudation in the Himalayas of Sikkim, where ravines from 6,000 to 15,000 feet in depth are evidently the excavations of the rivers running in them; so I am no new convert to a belief in the complete efficacy of rain and rivers to produce gigantic effects. But after seeing, both in India and Abyssinia, what the effects of these agents are in tropical countries, I do not feel surprised that their powers should be recognized with difficulty in regions where their effects are comparatively so dwarfed as in the British Isles, while the power of marine denudation is at its maximum from the enormous coast line exposed and the small amount of detritus furnished for its protection by rivers of small length, and in which floods are of exceptional occurrence. (*Blanford, 1870, footnote, pp. 156–8*)

Such factual accounts as these were of great assistance to the cause of the fluvial theory, though they were not conclusive nor were they accepted by all parties.

THE BRITISH INFLUENCE

The fluvialistic revival in Lyell's old age is deservedly connected with the work of Greenwood, Scrope, Jukes and Ramsay. Greenwood's sally on behalf of himself and Hutton and Playfair kept the fluvial standard flying in the face of the marine erosionists, but it was the work of Scrope, Jukes and Ramsay that gained a growing body of adherents. Greenwood was rather too militant to lead a successful reform; Scrope, on the other hand, was a more intelligent propagandist, while Jukes and Ramsay were actual converts to fluvialism.

GEORGE POULETT THOMSON SCROPE (1797–1876)

George Poulett Thomson was educated at Harrow and at Cambridge under Sedgwick. He extended his name on marrying the last representative of the Scropes, the old Earls of Wiltshire, in 1821, and, after his publications on France, he retired in the late 1820's to the family mansion at Castle Combe to devote himself to the estate, to public service and to the care of his wife who had become an invalid as the result of a carriage accident. During the period 1833–68 he was M.P. for Stroud and, although he seldom took part in debates in the House, he was a prolific writer of political tracts:

> He himself used to relate an amusing incident at his own expense. His great friend Lord Palmerston, on being greeted with the question 'Have you read my last pamphlet?' replied mischievously, 'Well, Scrope, I hope I have!' (*Judd, 1911, p. 40*)

However, Scrope kept in touch with the geological world mainly as a result of his correspondence with Lyell. His valuable reviews of *The Principles* and the first issue of his book on fluvial erosion in the Auvergne have already been discussed (pp. 125; 146; 169). Whereas this early appeal for the recognition of Hutton's and Playfair's ideas met with little response, the second edition, which was almost unaltered, made a more perceptible impact as during the intervening thirty years fluvialism had become less unpopular. The work (1858) drew attention once more to the manner in which lava-flows, by acting as a preservative cover for pre-existing topographies, allowed the geologist to distinguish valleys formed between each main outflow of lava. The present valleys had been cut since the last great eruptions.

> The amount of excavation which has taken place subsequently to the epoch of these eruptions can then have been only effected by the streams which still flow there; and as this quantity bears a very considerable proportion to the extent of the original excavation, there can be no reasonable grounds for hesitating to attribute the latter to the same agency which effected the former; it being only necessary to assign a longer duration to the process to account for the difference in magnitude of the result. (*Scrope, 1858, p. 206*)
> The leading idea which is present in all our researches, and which accompanies every fresh observation, the sound which to the ear of the student of Nature seems continually echoed from every part of her works, is –
>
> Time! – Time! – Time!
>
> (*Scrope, 1858, pp. 208–9*)

In later years Scrope occasionally replied to anti-fluvialistic writings.

JOSEPH BEETE JUKES (1811–69)

The geologic career of Joseph Beete Jukes followed a course like that of a comet; no sooner had he come into the gaze of the general public than he had burnt himself out and vanished. He joined the Survey in August 1846 but the true precocity of his ideas did not receive recognition until his work on the river valleys of South Ireland appeared in 1862, and by 1869 he was dead (Browne, 1871).

The eldest child and only son of a manufacturer, he was born near Birmingham on 10 October 1811. His father died when he was eight and his mother moved the family to Wolverhampton where she established and conducted a school for young ladies. At first Jukes attended a local Grammar School and then moved to King Edward's School, Birmingham, where he was not particularly outstanding and seemed to prefer athletics to study. Like

FIG. 71. *Joseph Beete Jukes*

many boys he had a passion for books on travel and it is said that he bought Cook's *Voyages* by saving up his pocket money. His mother, who does not appear to have been totally bereft of financial resources, sent her son to Cambridge with the idea that he would become a clergyman. Once again his youthful laziness of spirit prevented the successful culmination of this project and a good deal of his time at the University was absorbed in sporting pursuits. Yet, despite this apparent inattention, the course of his life's career was already fixed.

An aunt of his possessed a collection of Silurian fossils and while a boy he had shown a great interest in these. When therefore he went to Cambridge he attended the lectures of Sedgwick and was immediately so stirred by his teacher's enthusiasm and eloquence that he devoted all his working time to the study of geology. Throughout his life he retained his early admiration and respect for the old Cambridge Professor and even when he finally had to diverge from his ideas he did so with mingled regret. For some time he was unable to obtain a post and travelled up and down Britain giving lectures on geology at various Mechanics Institutions. During 1838 he investigated the geology of Derbyshire and the surrounding area and, later in the same year, seizing the chance to learn some practical surveying, worked without wages under a surveyor and his staff. With this freshly acquired technical skill he was able to accept the post of Geological Surveyor of Newfoundland.

For the next seven years his life is more like that of an explorer than of a geologist. At this period in history geologizing in the colonies meant journeys into the unknown. Whereas the frontiersman lived on the fringe of civilization, the geologist was expected to pursue his explorations some way beyond the pioneer fringe. In one year in Newfoundland Jukes faced shipwreck, frostbite and starvation. As a young man he made light of the privations of surveying in all weathers, which could involve anything from travelling in small schooners to living in the open in near-Arctic conditions, with little to sustain him except what might be shot with a gun or offered by the local inhabitants.

After he had returned to England he accepted an appointment as naturalist with the task of examining the country along the east coast of Australia, the Torres Strait and New Guinea. Here again his exploits have about them a savour of discovery. His skirmishes with the aborigines and his tales of finding regions and islands, where head-hunters were still happily employed in their murderous sport and animals were so unused to man that birds could be caught in the hand, compared accurately with many of Dana's descriptions. These marvellous and exotic scenes, however, seemed to increase Jukes' love for the English countryside.

> At all events, I shall have had quite enough of Australia. I hope you are having weather now at home somewhat resembling our climate here. If so, how beautiful England must be! No country equals it, even in this quality. The palm, the plantain, and the other far-famed trees of the tropics are certainly surpassingly lovely, especially by blue clear waters, in perpetual sunshine, with perhaps noble hills rising behind them; but their beauty is more like that of a picture to be looked at from a distance and at rest. A nearer view, when walking among them, most probably shows a dry, dusty, parched up soil, or a dank swamp; woods either scanty and devoid of shade, or matted and tangled with parasitical plants, so as to be impenetrable. No green lanes; no shady walks, above all no meadows, no lovely corners or smiling fields, such as England abounds in, and whose beauty is more easily felt than described. Still, even to relish English beauties you should have been abroad. (*Quoted by Browne, 1871, pp. 201-2*)

It is not surprising to find him, on his return, applying for a post in the Survey, and De la Beche reports to Ramsay as follows, on his acceptance:

> My Dear R—I have had a very satisfactory interview with Jukes, who appears a very fine fellow, and to love knowledge for its own sake. All is arranged; and as our corps is getting a thing desirable to join, I have found it desirable that he should join on the 1st October. The pressure for unfit persons is a matter to be avoided. I have explained his duties, and that his instructions would come from you as director for Great Britain. I expect it will be the day after

tomorrow before I leave London for Dol Jelly, via Shrewsbury and Dinas Mowddwy, I suppose.—Ever yours,

H. T. DE LA BECHE.

Jukes comes at 9s. od. per diem. He seems to care very little on what terms, so that he comes. Good this. (*Quoted by Browne, 1871, p. 272*)

In his early years with the Survey Jukes worked hard, as did all members of that department. His field-work alternated between Wales and the Midland coalfields. In September 1849 he was married and consequently forced to take an increasing interest in money. He received the tempting offer of the post of geological surveyor to the colony of New South Wales at £600 per annum but was reluctant to accept, partly because of the thought of leaving his wife so soon and partly because he had had his share of travelling. Instead, he induced the Survey to raise his salary from £200 to £300. His loyalty was further rewarded in 1851 by his promotion to the Directorship of the Irish Department of the Survey.

His sister asserts that this rise to a position of responsibility was accompanied by a weakening of his health and a lessening in the cheerfulness of his outlook. She shares the blame equally between the quantity and worry of his new duties and the troubled social background of the land in which he was obliged to carry them out.

It is this perpetual strife and atmosphere of contention in which one is obliged to live here that, fight against it as one will, laugh at it and bear up against it with good humour and courage in public as one may, yet, from the perpetual strain upon one's nerves and the gradual sap of one's heart and spirits, ultimately breaks one down!' (*Quoted by Browne, 1871, p. 458*)

Certainly the daily irritations of departmental supervision and ministerial interference began to press upon Jukes in a way that the rigours of exposure to heat and cold in Australia and Newfoundland had never done. Despite all, he continued to work hard and in 1853 and 1857 published two textbooks, *Popular Physical Geology* and *The Student's Manual of Geology*.

These two works are interesting because they show the early Jukes. In the space of five years his ideas were to undergo a radical change of emphasis.

When writing of the reasons for publishing the *Manual*, he explained that the works of Lyell, Murchison, Phillips and Portlock, though admirable, were too advanced for the beginner and his textbook was intended to bridge the gap. Apart from the text of the book itself, his choice of these particular great names in geology clearly indicated the tenor of his own reasoning. From the controlled catastrophes of Sedgwick he had made the natural progression to the marine erosion theory of Lyell. He still believed that erosion was primarily the work of the sea.

But when we feel ourselves entitled to take for granted that all cliffs at the foot of which the sea is now beating, have been produced by the erosive action of the waves, it only requires us to admit that the land may have stood formerly at lower levels, so as to allow the sea to flow over the lower parts of it, for us to see the probability that all inland cliffs, scars, precipices, valleys, and mountain passes, may have been produced in the same way. (*Jukes, 1862A, p. 101*)

The passes leading across the crests of great mountain chains could have been produced by no other cause than by the eroding action of the tides and currents as the mountains rose through the sea; what are now 'passes' being then 'sounds' or straits between islands. (*Jukes, 1862A, p. 101*)

Rivers form their own beds, but not their own valleys. Rivers are the result of their valleys, but they are their immediate results. The river could not be formed till after the valley, with all its tributary branches had been marked out. (*Jukes, 1862A, p. 105*)

From this conventional concurrence with popular opinion, Jukes suddenly emerged as a convinced fluvialist. Whether the conversion was caused by his change in interest from the complex stratigraphy of coalfields to the minute study of individual river-systems it is not possible to say. His early concentration on the size and distribution of adjacent strata would tend to make him reason, as it had done others, on a generous scale; the agencies of creation would seem to be of catastrophic proportions, because looking at a large syncline or a huge overfold the first thought would inevitably be of the tremendous thrusting power involved. On the other hand, the detailed study of a river system, with all its variations in valley-shape and discharge, would soon demonstrate that the river controls the shape of the landscape, rather than the landforms the river.

In his 1862 article on the river valleys of South Ireland Jukes has come a long way. The exclusive dependence on marine erosion is gone, and there is no more talk of cliff remnants and ancient estuaries. The sea is recognized only as a general levelling agent. He supported Ramsay's idea of plains of marine denudation in so far as he believed that the sea would level the land either beneath its surface or where it came within the action of its waves. But once the land was raised, the main work of erosion was carried out by rain-wash, streams and rivers.

1. Denudation is of two kinds, marine and atmospheric.
2. Marine denudation is effective only about the sea-level and along the margin of the land. It acts with a broad horizontal movement, tending to plane down the land to its own level. If the land be long stationary, it produces long vertical cliffs about its margin; if the land rises slowly and equably, it forms gentle slopes upon it.
3. Marine denudation cannot produce ravines or narrow winding valleys, except as gaps or passes upon the crests of ranges of hills, when the neighbouring summits were islands and the present gaps or passes were 'sounds' or 'straits' between them, traversed by strong tides and currents, and a narrow arm of the sea was thus made to assume a river-like action.
4. Atmospheric denudation acts vertically, either by weathering and disintegration of rock over the whole surface of land, or by the vertical cutting and grooving of ice in glaciers, and of running water in rills, brooks, and ravines. . . .

5. The present surface of the ground, where it differs from the original surface of deposition of the immediately subjacent rock, is in all cases the direct result of denudation, either atmospheric or marine, the internal forces of disturbance having only an indirect effect upon it, and having ceased to act long before the present surface was formed. (*Jukes, 1862B, pp. 391–2*)

The last part of the quotation clearly sets out what Jukes felt to be the correct relation between erosion and structure and it is advantageous to contrast his statement with the earlier structuralist ideas of Lesley. Jukes took a very broad view of erosion and in consequence he minimized the influence of structure.

The surfaces of our present lands are as much carved and sculptured surfaces as the medallion carved from the slab, or the statue carved from the block. (*Address to British Association, Cambridge 1862: quoted by Gregory, 1918*)

At the same time he did not deny that where differences in resistance existed, erosional processes would either be slowed down or accelerated, and therefore the structure of the surface must have an important bearing on the ultimate shape of the topography.

The simple and natural hypothesis is evidently that which supposes that, sometime after the formation of the highest beds of the coal measures, the rocks of the district became affected by movements acting from below, which gradually bent them into great longitudinal synclinal and anticlinal folds, and that the denuding agencies subsequently acted on these folded rocks, and eventually produced the present surface of the ground. It is possible that, as one result of the disturbing action, some of the upper beds were immediately raised into dry land, and began to suffer from both marine and atmospheric denudation. . . . It is, however, clear that the disturbing forces eventually ceased to operate, and the folding of the rocks became as complete as we now find it. The denuding agencies, however, are still in operation, and have never ceased to operate upon every portion of the country as long as it has been at or above the level of the sea.

The denudation will of course act upon rocks unequally, in accordance with the inequalities in their chemical composition or physical structure, and will of course produce a form of ground in accordance with these inequalities. (*Jukes, 1862B, pp. 392–3*)

Jukes was, of course, aware of the important fact that as much overlying strata had been removed, it is rare for the surface to be found in its original state: it was therefore fallacious to relate erosional systems to the present structural folds because those systems could have originated on a set of strata different in attitude from that upon which they were now acting. He was drawn towards this type of reasoning when he came to explain why transverse streams cut valleys straight across resistant anticlinal ridges (*Fig. 72*).

It would appear from the foregoing considerations, that the only possible way in which the transverse ravines on the present courses of the Rivers Bandon,

Lee and Blackwater could have been formed is by the erosive action of running water over dry land,—in other words, by river-action.

But we have already seen that no rivers could now commence to erode these ravines, supposing them not to exist and the adjacent ground to retain anything approaching to its present form and relative levels. We are, therefore, driven to the conclusion that the commencement of the erosion of these ravines took place upon a surface that had a different form and level from those which the present surface has. (*Jukes*, *1862B*, *p. 393*)

As Conybeare and Greenwood had done before him, Jukes raised the level of the land in his mind's eye until there was a slope that would both cross the resistant ridges and be able to carry away the water that fell upon them. Down this slope, he said, the predecessors of the present transverse valleys first took their courses. In time the continuing erosion would expose and lower the softer intermediate beds, and the whole erosional system would become more incised in proportion to the rate at which the transverse streams maintained their courses as steep gorges through the harder strata.

We are therefore led to look upon the valley of the Brinny as the primary valley, that was first formed, or at least first commenced, by the running of a stream from the dominant ridge on the north, southwards, towards the sea . . . over the surface considerably above any part of the present surface, formed a lateral N and S channel before any of the longitudinal E and W valleys and ridges became prominent, and thus marked out the winding southerly course it has ever since followed. As the longitudinal valleys were gradually formed by atmospheric denudation, the running waters of this brook always cut down across the intervening ridges faster than the general denuding agency lowered the longitudinal valleys. (*Jukes*, *1862B*, *p. 394*)

He said that because of marine planation, the original slope when it first emerged from the sea was of a uniform height. Later differential erosion of this surface was alone responsible for the form of the present detailed topography.

The surface left by the original marine denudation was considerably above the present one . . . when the original rivers first began to run over this high surface, (they) sought of course the lowest levels that then existed for their courses, and these lowest levels were nearly above the lines of their present channels . . . As the whole country wasted and sank under the wearing influence of the weather, their streams were strong enough always to cut their channels downwards, through whatever rocks became exposed, faster than the general degrading influence could lower the general surface of the country although these general influences lowered the limestone country to a greater extent than they were able to lower the country composed of other rocks which the river-channels traversed . . . Thus have been produced the present low limestone plains of the centre of Ireland, and the longitudinal limestone valleys of the south; and simultaneously with their production the ravines have been cut through the ridges of other rocks by which the drainage of these plains and valleys has always escaped to the sea. (*Jukes*, *1862B*, *pp. 399–400*)

His conclusion is a vigorous repetition of the point that rain and rivers alone,

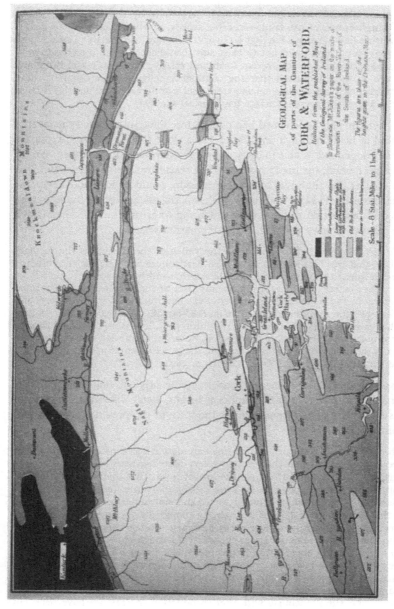

FIG. 72. *Jukes' (1862) map of part of southern Ireland, showing the relation between the drainage lines and the geology. The Brinny (p. 397) rises just south of Crookstown*

superimposed on a plain of marine denudation, combined to bring into existence the main features of this landscape:

> The rain and weather have disintegrated and lowered to some extent even the highest of the ridges left by the original marine denudation, and have impressed upon them the character of their own action, instead of that of the sea; but they have had still greater effect on the slopes, and the greatest of all on the channels of the streams that first commenced to drain the land. . . . When exercised on the quartzose ridges of the Old Red Sandstone, its greatest effect was confined within narrow limits, where the water ran rapidly, and thus produced glens or ravines. When the influence acted on the softer and more easily disintegrated argillaceous beds of the Carboniferous Slate, and still more when brought to bear on the soluble beds of the Carboniferous Limestone, it produced broader effects, and formed the larger longitudinal valleys. (*Jukes, 1862B, p. 397*)

Knowing the scepticism that will meet the publication of his views, he is almost apologetic in his enunciation of them.

> . . . I am fully aware that it will have rather a startling effect on some persons' minds, to be called on to believe that mere rain and other atmospheric influences can have washed away a thickness of some hundreds of feet of rock from off the surface of a whole country. (*Jukes, 1862B, p. 398*)

The possible application of his views to other areas did not escape him; he could see resemblances, for instance, in the Weald:

> I may also be allowed to ask whether it will not turn out to be a general law in all mountain-ranges in the world, that the lateral valleys are the first formed, running directly from the crests of the ranges down the steep slopes of the mountains, while the longitudinal valleys are of subsequent origin, gradually produced by atmospheric action on the softer and more easily eroded beds that strike along the chains. (*Jukes, 1862B, p. 400*)

This passage, important in itself, contained the first use of the word 'subsequent' in any way approaching its more modern geomorphological sense.

Four years later Jukes entered into the controversy aroused by the antifluvialistic writings of Mackintosh. He repeats his main canons of landscape evolution: that first the sea removed vast quantities of rock and produced sloping or virtually plane surfaces; secondly, on emergence, the master rivers flowed down the main slope of the plain of marine denudation, eventually cutting into the rocks, either in accordance with or in opposition to structural guidelines; thirdly, subaerial erosion, and particularly the developing stream-patterns proceeded to etch out the present relief.

> This conclusion, to which I found myself unconsciously and almost reluctantly brought, acted on me like a sudden revelation. It connected together and explained to me all that had been mysterious in the 'form of ground' in Wales and England . . . during my observations of the last thirty years, including many of the localities mentioned by Mr Mackintosh. (*Jukes, 1866A, p. 233*)

As an actual example of how subaerial erosion dissects a newly-elevated surface Jukes referred to Dana's observations of rain and river gullying on the windward flank of Pacific volcanic peaks.

Jukes (1866B) now became involved in an interesting argument with Scrope, a fellow-fluvialist, who considered, as already noticed on page 357, that more emphasis should be laid on the control which *structure* exercises upon topography. Jukes replied that this criticism seemed based on a misunderstanding of his ideas. Structure, he said, may have an effect on the landscape immediately after the land had been raised by subterranean movement but thenceforward erosion begins to change and to remove this initial form. Thus the surface rocks of the Alps are not those originally exposed to the atmosphere but have been laid bare as a result of massive subaerial erosion. Here we may notice that Jukes was right in affirming that erosion tended to eradicate the initial structural topography but wrong in minimizing the considerable extent to which erosion is guided by faults, alinement of major structural features and differences in rock hardness.

Scrope in reply admitted some misunderstanding, but still thought that Jukes had unnecessarily decried the importance of internal influences.

> All geologists, indeed, recognize as a natural law the tendency of all the external denuding forces to reduce the surface of the earth to one uniform level, so that but for the opposing agency of the subterranean forces no dry land could exist. It seems, therefore, like a paradox to deny to the latter power any 'direct' or considerable share in the external configuration of the earth. (*Scrope, 1866D, p. 380*)

Though a good answer to Jukes' views, it is still obvious that the two geologists were looking at the same problem from different angles. Jukes saw topography as a form which took shape from day to day, by the action of continuous processes, sometimes working in opposition to underlying structure; tectonic movements on the other hand, he regarded, as being of an intermittent nature and having little direct influence upon topography, except in so far as they raised the land above sea level. Scrope saw the forces of erosion and upheaval as being contemporary with and complementary to each other. Each geologist was correct in his own context, and in later years each idea found, and still finds, a wide circle of followers.

There is no doubt that Jukes' unexpected and complete conversion was very heartening to the fluvialists, who had for several decades been languishing in the shade. He himself realized, when replying in 1864 to a challenge by Dr Falconer, that fluvialistic views were unorthodox and unpopular.

> I am quite conscious of the weight of authority against the views which I advocate, and more especially that I am in this matter, and almost in this matter only, now compelled to depart somewhat from the Cambridge teaching of my dear old master Professor Sedgwick. If I can venture to dissent from him, my audacity is not likely to meet with any greater trial. But I am also

aware that my present views had never been without good authority on their side, and the number and weight of those who, like myself, are now adopting them seem to be greatly on the increase. (*Quoted by Browne, 1871, p. 543*)

I know that almost all the older fellows think us crazy—as they considered my address to the Geological section at Cambridge—but in the long run they will give in! (*Quoted by Browne, 1871, p. 554*)

Unhappily Jukes' active support was short-lived though it was exercised at a vital turning point of the science. Whilst staying at a small country inn he lost his footing on a badly-lit staircase and fell to the bottom, hitting his head on some limestone flags in the hall. He thought little of it at the time but a marked deterioration in his health could be traced from this date. He could no longer concentrate on his work and was in constant need of rest. For a man who had been an habitual early riser, often to be found working in his study before 6 a.m., this must have made life almost unbearable. Despite the care of his wife and the seeking of cures abroad, his condition gradually weakened. He died on 29 July 1869 and was buried at Selly Oak near Birmingham. For all too short a time he had occupied the front rank of geologists.

THE CONVERSION OF ANDREW RAMSAY

Ramsay, who by this time was confining the greater part of his activities to glaciology, seems to have undergone a change in outlook similar to that which had come to Jukes. After about 1860 there is no longer exclusive emphasis on the work of the sea; valley erosion is correctly related to the work of running water (Ramsay, 1863, pp. 103–10). Furthermore he grants pride of place to the work of rivers; they, he considers, have done the bulk of the denudation, both before and since the glacial epoch; the Alpine ice had deepened the initial features and exaggerated the existing landscape, rather than created fresh patterns of its own.

> No true geologist is likely to assert that these valleys have been mainly scooped out from end to end by ice, for the reason that, since the disappearance of the ice, running water, in the formation of gorges, &c., has comparatively effected so little. Given sufficient time, and, as old Ray long ago inferred, any amount of degradation may be produced by rain and running water. . . . For obvious reasons . . . it is evident that the period that has passed between the disappearance of the great glaciers and the present day is trifling compared with that which elapsed between the close of the Miocene and the commencement of the Glacial epoch of geologists; . . . It is not therefore to be expected that the later modifications of Alpine valleys produced by existing causes should be commensurate with the old, for time is wanting. (*Ramsey, 1862A, p. 378*)

Ramsay now appears also as an attacker of Murchison's misinformed criticisms of the fluvial theory. He selected statements by Murchison in which the latter said that all surface irregularities in the Alps were due to dislocations and denudation; that erosion followed fractures; and that many valleys were due entirely to fractures and subsidence. In answer Ramsay suggests that, on stratigraphic evidence alone, it is doubtful whether structural disruptions coincide with the majority of valleys. He quotes liberally from Hutton and Playfair in demonstrating that rivers over a period of time are powerful enough to carry out the quantity of erosion required to account for the present topographical forms. Ramsay went so far as to say that in his own mind the question had been decided:

> . . . enough has now been said to show that the theory of the formation of great systems of valleys by erosion in which water and ice are the main agents, is not a mere absurdity. . . . (*Ramsay, 1864, p. 297*)

Fortunately he lived to 1891 and, partly due to his own writings, saw the pendulum swing in favour of fluvialism.

OTHER FLUVIALISTS, NEW AND CONVERTED

Between 1860 and 1875 a rapid change of opinion on the work of rivers ensured that only the implacables and die-hards remained anti-fluvialist. The increasing popularity of fluvialism was aided by the increasing recognition of weathering agents as well as of the potency of rain-wash. G. Bischof between 1854 and 1859 wrote much on weathering processes and his three-volume work was translated as *Elements of Chemical and Physical Geology*. George Greenwood had also strongly championed this physical factor as part of his arguments for the power of rain, and in 1869 Kinahan gave a good account of how the composition of the soil progressively changes and develops as it is acted upon by the forces of weathering and how vegetation can act as a protection against natural denudation.

> The chemico-fluvial agents are undoubtedly the most universal performers in the great work of denudation. (*Kinahan, 1869B, p. 109*)

But the first detailed study of weathering by a British geologist was that published by David Ansted in 1871. Starting his description in the drier areas of the Mediterranean (Spain and Sicily), he explained how the prolonged heat in the absence of rain would crack the clay and sandstone deposits, thus making the task of the normal forces of erosion much easier during spells of heavy rain.

> In the loose tertiary marls that extend on the northern slopes of the Sierra Nevada the cracks, instead of being six or eight feet, are two or three hundred feet in depth, miles in length and several hundred yards wide. They are now valleys, but each valley originated in a fissure which has been gradually enlarged into a valley by the rain. (*Ansted, 1871, pp. 387–8*)
>
> The continued widening out of these ravines till they become valleys of considerable magnitude, is partly the effect of weathering on the vertical walls, partly the result of under-mining. (*Ansted, 1871, p. 388*)

The effects of heat, he continues, are not confined to the softer or more friable rocks but weathering can be seen to have acted along the joints in granite; and many of the joints in limestone, he believed, had a similar origin, though they may later have been enlarged by the action of rain-water. In short, because alternations between hot and cold, and wet and dry, are to be found all over the world, weathering is a factor of universal importance. He rightly cautions the geologist not to suppose that all cracks and joints in rock are the result of structural movements, but to give consideration first to the more likely possibility that weathering may have been the cause.

> Whenever and wherever the earth's surface is exposed above the water, there we may always discover in all climates the effect of disturbance. In such places we may always look for, and shall certainly find, some traces of the effect produced by sun and rain, heat and cold. It is not the earthquake or the volcano,—it is not the deluge or the storm waves, it is not the occasional but rare upheaval of a mountain chain, nor the depression of large areas by thousands of feet—that have produced the existing features of the earth. (*Ansted, 1871, p. 395*)

The study of weathering attracted much less attention than the growing interest in river-work. Rabid fluvialists, like Greenwood, were still rare and most of the new comers or converts to the creed believed also in other forms of erosion. We could quote from at least a score of writers in this category but must for brevity content ourselves with six authors.

Foster's and Topley's important work on the denudation of the Weald provided substantial support for the fluvial theory and a serious rebuttal of marine and structural theories. They rejected Hopkin's structural ideas mainly because neither the longitudinal nor the transverse valleys of the Weald coincided with strong lines of fracture. They objected to Lyell's marine dissection views for three principal reasons.

(1) The bases of the escarpments are not at a uniform level as they would have to be if they were sea cliffs, for the sea cannot tilt itself and approaches the coast at one height, the erosion consequently acting along a horizontal plane.

(2) There are no marine deposits at the foot of the escarpments.

(3) The present topography is of such an indented form that subsidence of the land would allow marine erosion to operate only very feebly.

Their whole argument was strongly in favour of erosion by subaerial forces. The presence of the high-level gravels along the valley of the Medway especially seems to have impressed them, for terraces in this position indicated that the river had flowed in the same direction when some 300 feet above its present level. The inference was that the Medway had created the greater part, if not all, of its own valley.

> And all of this denudation has been due to the action of rain and rivers; for we have shown that the Medway deepened its valley gradually – and not only are there no traces of marine erosion, but had the sea had access since the

gravel was deposited, surely it would have swept away such loose and incoherent deposits. If rain and rivers could do so much, if they could cut out
a valley 250 feet deep and seven miles broad, surely we may allow that by
giving them more time they could scoop out valleys 500 feet deep; in other
words, that, making every allowance for slight superficial inequalities produced by marine denudation, all existing inequalities in the basin of the
Medway, including the Greensand escarpment and the Chalk escarpment,
are entirely due to atmospheric denudation, that is to say, to the action of rain
and rivers. (*Foster and Topley, 1865, pp. 464–5*)

Their general explanation of the origin of the Weald closely paralleled that of
Greenwood with one important difference. Whereas Greenwood imputed
all denudation to the work of rain and rivers, Foster and Topley followed
Ramsay in suggesting that marine erosion had stripped off much of the
Tertiary and Upper Cretaceous strata, and when the land surface emerged it
was almost a plane or, at most, a gentle dome which sloped from the centre of
the Weald to the north and south.

The manner in which we consider the denudation of the Weald to have taken
place is as follows. After a large portion of the Tertiary and Upper Cretaceous
strata . . . had been removed by marine denudation, a comparatively plane
surface was formed, which gradually appeared above water. . . . The central
ridge determined the flow of the water that fell upon the area, streams began
to flow to the north and the south and in this manner the transverse valleys . . .
were first started. At the same time the longitudinal valleys along the strike
were formed, on account of the difference in hardness between the various
rocks. The moderately hard porous Chalk has suffered less than the soft
impervious Gault, and the hard porous Lower Greensand has been less
denuded than the soft impervious Weald Clay. As we are dealing with limestone beds, we must take into consideration the chemical action of rain
charged with carbonic acid . . . The mechanical atmospheric denudation,
however, exceeds the chemical denudation, and, in spite of the general
lowering of the Chalk and Kentish Rag, they still form escarpments. (*Foster
and Topley, 1865, p. 473*)

In the last few lines it is interesting to note that weathering, in this case
chemical weathering, is beginning to receive proper recognition.

During this same year, 1865, we meet with a newcomer named George
Maw who had been impressed by the unity of drainage systems. In terms
reminiscent of Playfair's he sees that the whole landscape is made up of a
series of river-basins.

What I wish particularly to notice is that the form of the whole land surface
with some trifling exceptions (as lake basins, which appear to admit of special
explanation) is merely a modification of the same principle of contour as the
true river valley, exhibiting a system of watersheds by which almost every
part of the land is connected with the sea by adjacent land on a graduated
series of lower levels than itself. . . . What can be more appropriate than the
apparent relation between the delicate graduation of levels from the river
mouths upwards to the watershed boundaries, and the exactly proportionate

concentration of water and consequent power of excavation downwards from the watershed lines to the river mouths? (*Maw, 1866A, p. 344*)

Maw doubts whether waves have eroded the land surfaces over any great area. The sea, he states, has two actions; it acts below its surface through currents, though here the dominant process is of deposition rather than of erosion; it also acts along the coast by the battering motion of the waves. If the waves had worn the surface down to a general plain as was supposed, he felt that there should be a greater topographic similarity along equivalent contours than seems to be the case. Instead there seems a marked division in character between the fluvial and marine features. Former sea cliffs and coastal features, he admitted, are to be found in inland positions but are anomalous when compared with fluvial features.

No one will dispute that many parts of the land represent cliffs and coasts eroded by the sea; but these appear to be altogether subordinate to the watershed system, and do not, as a rule, harmonize with its outlines. (*Maw, 1866A, p. 347*)

Maw did not confine his study solely to a refutation of Mackintosh's catastrophic views, and his descriptions throw light on some new features of river-valley systems which have hitherto not been appreciated. In his discussion of the origin of wide valleys he speaks of valley landslides, and suggests that a distinction should be drawn between the actual stream channel and the valley in which it is situated, for they are separate and subject to different forces. He also makes the point that valley sides steepen with the progressive distance downstream, which he considered to be due to the higher erosion rate resulting from the greater accumulation of runoff in the lower reaches of the stream. From these facts he was able to make the interesting observation that the valley sides which occurred in the lowest reaches of the river system would be the first to reach their limiting 'angle of repose'. His article also contained the first description of the exhumation of a fossil relief and was the first to name the feature of *superimposition* in its technical sense.

. . . let us, however, consider the case of an old land surface diversified with watershed ranges and river valleys buried beneath an overlying deposit. (*Maw, 1866B, p. 442*)
  Referring to Figures [73*a.*, *b.* and *c.*], representing a series of valleys superimposed in part unconformably over an ancient buried series, let us suppose the superincumbent deposit AA to be gradually denuded by subaerial agency, the main lines of waterflow will simply re-excavate and perhaps somewhat alter the old buried valleys over which they are conformable, but wherever the upper system of valleys CC runs transversely . . . to the buried system the separating ridges . . . will be cut through; denudation may proceed over ridge and valley, and gradually lower the entire surface to FF, after having removed the whole of the newer deposit, and will impress on the old deposit a contour partaking of that of both the old and new surfaces . . . (*Maw, 1866B, pp. 443–4*)

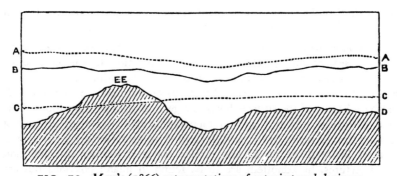

FIG. 73. *Maw's (1866) representations of superimposed drainage*

a. *Longitudinal section of the present superimposed valley system, transverse to the ancient valley system.*

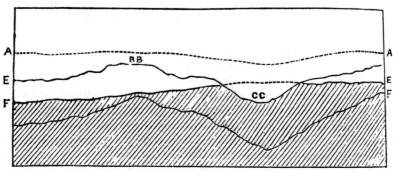

b. *Longitudinal section of the ancient valley system, transverse to the superimposed system.*

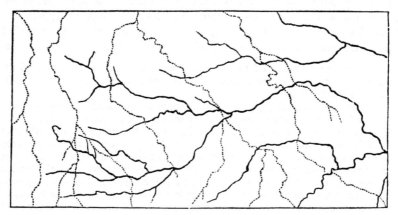

c. *Present drainage system (solid lines) developed on ancient buried channels (dotted lines).*

We will conclude our survey of the growth of fluvialism in the later years of Lyell by reference to four notable advocates of subaerial erosion.

Pride of place must go to Archibald Geikie whose *The Scenery of Scotland* (1865) not only reviewed the several erosive processes but also gave a scholarly exposition of the fluvial philosophy from the time of Hutton onwards.

Geikie, like Ramsay and Jukes, was one of the original members of the Ordnance Survey. Another Scot, most of his life's work was carried out in his native country. When Ramsay was appointed Director for England and Scotland, Geikie was made directly responsible for the work in Scotland. His lifelong association with the ideas of De la Beche, Ramsay and Jukes seems to have instilled him with a well-grounded belief in the fluvial theories. Probably the reactionary influence of Murchison, who did not become Director-General of the Survey until 1855, came too late in Geikie's life to have much effect on his ideas. When Ramsay and Jukes finally had to retire through ill-health, it was Geikie who took their place as the principal advocate of fluvialism.

His book, though nominally concerned with Scotland, is from beginning to end universal in context; its sole aim is to justify and establish the fluvial idea as the correct means of interpreting topography. His arguments and the reasons for his belief drew their inspiration mainly from the works of Hutton and Ramsay. His examples are based largely on acute and prolonged observations in the field as the following brief quotations on pot-holes and river-terraces (haughs) will show.

> In not a few places (in stream beds), too, we may notice cylindrical cavities, called pot-holes, in the bottom of each of which lie a few well-rounded and worn pebbles and boulders. These cavities are due to the circular movements of loose stones that have been caught in eddies, and have been kept whirling there till by their friction they have gradually worked their way downward into the solid rock. (*Geikie, 1865, p. 28*)
>
> These meadow-flats along the margin of streams bear in Scotland the name of 'haughs'. They have been formed out of the detritus brought down by the streams from the higher parts of the valleys. There are often traces of similar alluvial terraces at higher levels on either side of a valley. Three or four may sometimes be counted rising above each other. They mark each level which the river once occupied before cutting its way down through its own alluvium to its present channel. (*Geikie, 1865, p. 30*)

Geikie's explanation of the development of the physical landscape is little different from Hutton's except in emphasis; Hutton was concerned with informing the reader of the infinite succession of worlds, while Geikie follows Ramsay in stressing that a marine plain is the initial and ultimate event in the scheme of creation.

> The history of the superficial changes of a country might theoretically be summed up thus. The bed of the sea, which for the sake of illustration may be supposed to be a wide plain of marine denudation – the worn remnant of an

FIG. 74. *Archibald Geikie*
(*From a painting by R. G. Eves, A.R.A.*)

old land – is slowly raised above the waves. There is a point or line where the
elevation is greatest, and from which the ground slopes down to the sea-level.
Perhaps the elevatory force shows itself in the upheaval of one or more anticlinal folds. But whether in one great dome-shaped mass, or in a long ridge,
or in several ridges with several dividing hollows, the slopes are probably
gentle, and the elevation of the whole a quiet protracted process. No sooner
does the rock appear above the sea than it is attacked by the waves, and unless
the rate of elevation is more rapid than that of the marine waste, the rising
area can never get above the sea-level. But not only is it a prey to the breakers,
its surface begins to be carved out by the atmospheric agents. That surface is
not a mere dead level, so that when the rain falls upon it drainage necessarily

sets in from the highest parts down to the shores, the rain gathers into runnels, following the inequalities of the sea-worn slopes, and widening into brooks and rivers; or the moisture falls in the form of snow, and glaciers grind a path for themselves from the high grounds to the shore. Thus begins the scooping out of a system of valleys diverging from the higher parts of the rising land. These depressions are slowly dug deeper and wider, until at last the ancient elevated sea-bed is worn into a system of hills and mountains, valleys and glens. The land thus modelled may remain stationary for a vast interval, but in the end it descends again beneath the sea, is covered over with newer deposits, and its highest mountains perhaps buried beneath piles of their own ruin, worn from them by the sea, as they slowly sink under its waves. A subsequent elevation of this area into dry land exposes these later accumulations to a similar waste, and a new series of denudations is begun by the rains, streams, ice, and the sea, new valleys are excavated, and new hills are left standing out from them. By such a process of ceaseless change thus summarily stated, carried on during many successive geological periods, the present scenery of our country appears to have been produced. (*Geikie, 1865, pp. 87–89*)

The key which had unlocked the puzzle for Hutton had been deposition. He saw that the majority of rocks were sedimentary and that they must have been derived from the breaking down of earlier rocks; rain, rivers and sea were the normal agents of destruction and he concluded that as they acted now so they acted in former times. Geikie, following Ramsay, allowed his reasoning to be absorbed by another feature, the uniformity of summit levels, which when taken as an historical record of the existence of a former sea plain, supplied the basis of the marine planation theory. The plain was not necessarily flat in its entirety but might be undulating or with a few peaks or ranges, where the land had remained above the action of the waves. However the sea was regarded as a powerful agent of horizontal erosion, which over a period of time could flatten any surface however indented or angular. Even Hutton's theory, in so far as it was positively formulated on the point, seemed to suppose that the emerging continent would be flat and the valleys unmade. Essentially, according to Geikie, the sequence of events was a progression from plain to mountain and from mountain to plain, the subaerial elements working to restore the land to the sea while the sea worked over and levelled what it had regained. Wisely he adds:

Let me guard the reader against misconception arising from the use of this term plain. I do not at all mean to assert that the area of the Highlands was ever so levelled by the sea as to approach to anything like the flatness of a meadow. The marine denudation probably went on during many oscillations of level, and the general result would hence be the production of a great table-land, some parts rising gently to a height of many hundred feet above other portions, yet the whole wearing of that general sameness and uniformity of surface characteristic of a table-land where there are neither any conspicuous hills towering sharply above the average level, nor any valleys sinking abruptly below it. (*Geikie, 1865, footnote, pp. 106–7*)

At the end Geikie acknowledges his debt to Hutton and Jukes.

> Although I have long held the belief of Hutton, that our valleys are mainly
> the work of atmospheric waste, the history of their excavation was but dimly
> understood by me until the appearance of the admirable paper by my col-
> league, Mr J. B. Jukes, on the River-Valleys of the South of Ireland. (*Geikie,
> 1865, footnote, p. 141*)

There is no need to re-iterate that *The Scenery of Scotland* is a powerful commendation of fluvial notions and represents the sort of compromise that was
gradually being accepted by contemporary geologists.

From Scotland we must turn to America where Lesley, whom we have
already discussed as a leading protagonist of the dominant influence of
lithology and structure on topography, now recanted and admitted that he
was far more impressed by the powers of subaerial erosion. He illustrates
how these could effect so great a quantity of erosion that the influence of
other forces, such as the sea or ice-caps, was ephemeral and insignificant in
comparison.

> Behind these [the Appalachians], again, lies the Great Cumberland –
> Alleghany – Catskill Plateau, with its horizontal geology and its quaquaversal,
> arborescent drainage system, boldly contrasting with the Appalachian topo-
> graphy in front of it, and settling the questions of mode and agency in favour
> of slow aerial denudation. (*Lesley, 1866, p. 307*)
>
> It is a pity that we have none of the minute work yet published, which has
> been done here and there along the borders of the great coal area, for it
> would help greatly to explain the true nature of the erosive agency which has
> relieved an expanse of continent, amounting to one or two hundred thousand
> square miles . . . , from a superincumbent weight of Coal Measures at least
> two thousand and perhaps, if capped with Permian measures, five thousand
> feet in height. (*Lesley, 1866, p. 309*)
>
> . . . Erosion by wind and rain, sunshine and frost, slow chemical solution
> and spring and fall freshlets, has done the whole work. I have long taught that
> it could not have been accomplished under water level by oceanic currents,
> because the ocean is a maker and not a destroyer. But I must now abandon
> wholly the idea to which I have clung, with slowly relaxing grasp, so many
> years, that a complete erosion theory demanded some such forces as would
> have been supplied by the extra efficiency of an ocean translated across the
> upheaved surface through the air. At the same time, the above considerations
> make me all the less willing to admit the ice-cake theory of erosion as even
> approximately true. For if aerial erosion has been going on uninterruptedly
> ever since the uprise of the Coal Measure continent, how little of the whole
> effect can have remained over to be still produced at the time when the
> Glacial epoch set in and the supposed ice-cake began to take the work in
> hand. (*Lesley, 1866, p. 310*)
>
> The erosion of the wonderful Green River ravines, in the Colorado
> country, to a depth of five thousand feet, is not more evidently the product
> of ordinary meteorological causes than is the erosion of any given segment of
> the Appalachian or Alleghany district exhibited upon the map . . . (*Lesley,
> 1866, p. 311*)

Lesley also has something to add on chemical erosion of limestone areas and applies the suggestion that erosion is not confined to valleys but often begins in caverns, which when their roofs collapse become valleys.

I believe that this chemical element of erosive energy has been slighted even in discussion the most recent. I ascribe to it nine-tenths of the wastage of the Blue Grass area. I believe that the erosion went on, chiefly underground, along those narrow belts where the lime-rock formations approached the surface at whatever level the surface happened at the time to stand. The caverns grew; their roofs fell in; streams washed the debris continually away. As the surface thus kept falling piecemeal into the cavernous traps everywhere laid for it below, the general level of the area was slowly and insensibly let down to its present grade above the sea. While falling thus, the opposing outcrops of the upper rocks retired from one another, eastward and westward, and let the Blue Grass area between them widen slowly to its present size. And still the work goes bravely on. (*Lesley, 1866, pp. 311–12*)

Another geologist in the United States, H. A. Nicholson (1872), also has no reservations about the fluvial theory, and his examination of valleys in upstate New York led him on to the observation that where the hard strata forming the waterfalls dipped in the opposite direction to the flow of the stream, backcutting by the stream will ultimately destroy the break of slope.

Coming almost at the end of this period, it is perhaps fitting that Topley should have virtually the last word. He has not had to change his theory on the Weald though the turnings and twistings of argument have caused him to develop new points and place a stronger emphasis on the work of subaerial forces. Apart from changes in detail his *Geology of the Weald* (1875) is a triumphant repetition of what he had written with Foster ten years earlier.

He emphasizes that the Weald is not a unique feature but has parallels wherever the same conjunction of factors occurs.

If the North and South Downs were both continued across the Channel, the Wealden district of England would be continuous with the Lower Boulonnais, and there would be a closed area, surrounded everywhere by the Chalk escarpment, and sometimes by other lines of escarpment within that, these escarpments being interrupted only by the narrow river valleys.

It cannot be said that this is an improbable supposition. There are other anticlinal, or rather periclinal areas, similar in all respects to this ancient Weald, and in which the surrounding escarpments are continuous, breached only by the transverse valleys which drain the enclosed area. There are the Pays de Bray, which exactly resembles the Weald, exposing within its area similar Lower Cretaceous Beds, and bounded by a like escarpment. It differs from the Weald only in its smaller size, and in the fact that it is entire, not truncated by the coast line. Another well known example is the valley of Woolhope, surrounded by concentric escarpments which are breached by a transverse valley. The Upper Greensand inliers of Shalbourn and Kingsclere are other and smaller examples. (*Topley, 1875, p. 271*)

The method by which the topography is produced, he goes on, is for the land to be first smoothed into a plane by the action of the sea, and then to be etched into the alternation of valley and ridge by the action of subaerial forces. At this point he mentions Greenwood's work of 1857 and admits that his present analysis includes most of what Greenwood then said. He also quotes the second edition of Ramsay's textbook (1864) as representing the official Survey opinion.

Then, for a moment, Topley puts on one side general theory and concentrates on the English Weald which had been his special study throughout much of his life. He realized that the conjunction of broad valleys with steep scarp faces was the most perplexing feature, and it was on the explanation of this that he concentrated all his energies. He suggests that there were three main reasons which contributed towards the creation of the Wealden forms: (1) the alternation of hard and soft rocks; (2) the permeable nature of the Cretaceous rocks which form the escarpments; and (3) the relation between the rates of erosion of the scarp slope and that of the dip slope and also the angle of the dip itself both of which factors would decide whether the height of the scarp would increase or decrease with backward erosion.

In the case of the plain of marine denudation described above, we should have a comparatively plane surface sloping from a central line to the north and south. But the beds which crop to the surface at the centre or highest part of that plain, would be geologically the lowest; and as we pass to the north or south from that central line, higher beds would come on in succession; for the dip of the beds would exceed the slope of the plain. These beds would be alternately hard and soft, as we passed successively over the denuded edges of the Hastings Beds, Weald Clay, Lower Greensand, Gault, and Chalk.

Streams would flow down this slope, probably directed into their several channels by means hereafter described. . . . These streams would excavate their valleys and would carry off the materials which the more general subaerial agencies provided. It is to these agencies, and not directly to rivers or streams, that we must refer the escarpments.

Rain and frost acting upon the alternate hard and soft beds would soon exert an influence; and a hard bed, ending off in a thin edge along the plain, would rapidly assume a step-like form; which would be a small escarpment. This escarpment in wearing back would constantly get higher, because its base would be denuded lower.

But there is another point to consider, besides the relative hardness of the rocks. We find that escarpments are composed of porous rocks, whilst the plains at their feet are generally formed in impervious rocks. Therefore a great part of the rain which falls on the top, or dip slope, of the rock forming the escarpment soaks into it, and has little or no mechanical denuding effect. But a part of this water issues as springs at the base of the escarpment, where the impervious bed occurs, and thus tends to wear back the face.

The 'growth' of an escarpment depends principally upon the excess of the denudation of the face, over that of the dip slope; and it is evident that the requisite conditions are best fulfilled, when a hard and porous bed rests upon a soft and impervious bed. But it will not wholly depend upon this. We must also consider the relation of the slope of the marine plain, (roughly represented by the existing dip slope) to the true dip of the beds. If, for a certain distance, they are the same in amount, the escarpment will not increase in height

FIG. 75. (*above*) *The south-facing scarp face of the chalk of the North Downs, east of Trottiscliffe, Kent*

FIG. 76. (*below*) *The north-facing scarp face of the chalk of the South Downs at Fulking, Sussex*

(*Aerofilms and Aero Pictorial Ltd.*)

although it will still be worn back. It will rather tend to diminish in height, because of the slight denudation which the dip-slope may undergo. If, however, the dip of the beds exeeds the slope of the ground, the escarpment in receding, will necessarily increase in height. (*Topley, 1875, pp. 274-5*)

To show that he did not altogether disregard the influence of structure, Topley proceeds to admit that structure may help to account for some features, such as the main water-gaps.

It is certainly singular that the transverse valleys should correspond in position on opposite sides of the Weald, as they undoubtedly do; and in the Map of the Ancient Weald, . . . they are represented with a greater amount of agreement than has hitherto been allowed. It is generally supposed that these transverse valleys coincide with transverse anticlinal fractures, but I am not acquainted with any conclusive evidence in support of this. On the contrary, it can be shown that these transverse valleys, in some cases, coincide with transverse synclinals. (*Topley, 1875, pp. 276-7*)

Probably it is owing to these large transverse synclinals that escarpments are breached in so few places by transverse valleys. If it were not for some such cause as this, marking out the lines which the valleys must take, we should expect them to be very numerous. The innumerable streams which would first form on the old plain of marine denudation ought each to find its way down the greatest slope, and therefore the escarpment would be breached in a very large number of places. But the transverse synclinals have apparently determined the lines which these valleys must take, and hence they are few in number.

It was stated above that no marked disturbances are known along the Chalk escarpment. There are, however, some faults which are known to enter it, shifting the boundaries at its foot; and some of these seem to have a relation to the more important 'passes' or deep gaps in the escarpment. It is true that there are several passes, especially in Sussex, which are apparently independent of faults, and there are also faults intersecting the escarpment where there are no passes; but on the North of the Weald there are only three passes of consequence and these are evidently associated with faults. (*Topley, 1875, p. 278*)

There is a third kind of valley interrupting escarpments, through which the stream flows inwards, or towards the central country . . . against the general dip . . . I think that such valleys are likely to coincide with local transverse anticlinals, or – which is the truer way of expressing it – that a transverse anticlinal will probably give rise to such a valley. (*Topley, 1875, p. 279*)

Thus, Topley's position, like that of other fluvialists such as Jukes, was not to deny the influence of structure but to remedy the exaggerated emphasis placed upon it by some writers.

Topley proceeds to show that the main scarps were not old sea cliffs and that the evidence of the river-gravels afforded a strong argument for the acceptance of fluvial denudation.

To sum up the evidences of denudation offered by the river-gravels we find:
1st. That the gravels of the main trunk valley always contain pebbles of the rock formations drained by that river, and no others.

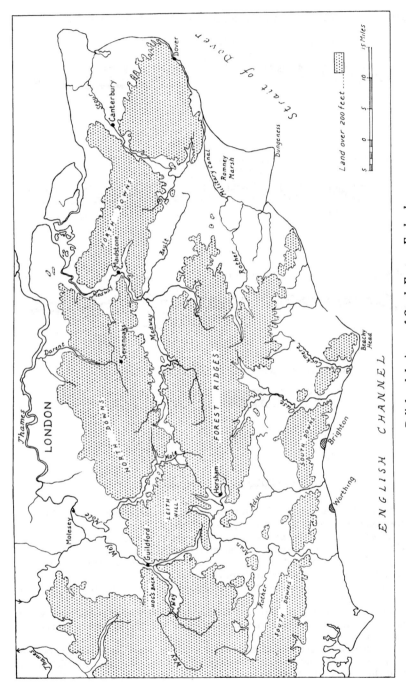

FIG. 77. *Relief and drainage of South-Eastern England*

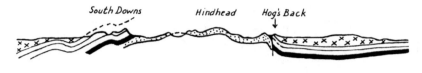

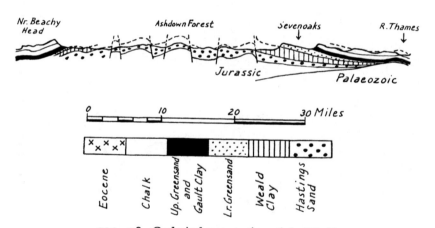

FIG. 78. *Geological cross-sections of the Weald*

2nd. That the gravels of each tributary stream likewise contain pebbles of the rocks drained by that stream, and no others.

From whence we infer that the gravels are the products of those several rivers or tributaries flowing in their present directions.

3rd. That undoubted river-gravel occurs at all heights, from the present level of the river up to 300 feet above that river.

Subaerial denudation has therefore excavated the valleys to a depth of at least 300 feet.

But a denudation of 300 feet in the lower part of the Medway valley represents more than half that denudation of the Weald which we refer to subaerial causes.

We cannot expect to find any direct evidence that the escarpments have been formed and worn back by subaerial agencies. We can only show, 1st, that there is no proof whatever of any other agencies having done the work;

2nd, that all other agencies with which we are acquainted are inadequate to the work done; 3rd, that the whole features of the district are such as can readily be explained by subaerial denudation. (*Topley, 1875, p. 300*)

Topley's book was important both as a detailed study of the Weald and as a powerful reassertion of the fluvial theory. The combination of fact, experience and theory made a stronger impression on his opponents than many of the wilder and more general criticisms of other fluvialists.

But here we must re-iterate that Topley even in 1875 was not prepared to go as far as Greenwood and impute all denudational features to the long-term action of rain and rivers. It is, he writes,

> . . . doubtful how far the reasoning of these writers (the utter fluvialists) can be held to be conclusive as it stands in their works. If the Weald, for instance, were upheaved with a dome of chalk extending entirely over the country, and upon this upheaved surface rain and rivers were to act, it is doubtful if the result would be the present system of longitudinal and transverse valleys. To obtain this it is necessary that the country should be planed across so that successive rocks should in turn crop up to the surface. (*Topley, 1875, p. 27*)

In fact, we begin to wonder if by the end of Lyell's long reign the knowledge of landscapes has not passed beyond the bounds of any single theory. Outside the wetter tropics, most landscapes seem to have been fashioned by two or three main forces, irrespective of structural influences, lithologic and tectonic. The picture is becoming increasingly complex. It would be little short of miraculous if in the future any geologist could devise a simple scheme that would dominate world thought in the way that Lyell's marine erosion theory had. But the history of geomorphology is full of minor miracles!

CHAPTER TWENTY-THREE

Quantitative and
Morphometric Advances

It is objected that my language is wanting in quantitative precision – that I use such terms as 'great', 'very great', 'small' and so forth without any statement of the units of comparison relatively to which these expressions are employed. No one reasoning on the combined influence of a multitude of physical causes could well avoid the almost continual use of such terms. Although my arguments are logical, few writers, I venture to say, have done more than myself to introduce definite quantitative exactness into the questions I have discussed.

(James Croll, *Climate and Cosmology*, 1889, p. 19)

The quantitative measurement of landscape processes made notable progress during the reign of uniformitarianism. About the mid-nineteenth century scientific methods in general were highly favourable to inquiries into the rates and nature of evolution of natural phenomena. Students of landscape had already pursued some scientific investigations (*see* Chapters 8 and 15) when Lyell published his *Principles* and he had seized upon them eagerly as providing fruitful support for uniformitarian suggestions. Because rivers are more amenable to scientific measurement and to physical laws than are most other landscape processes, much of the quantitative work on landscape-forming agencies tended to draw attention to the power of running water. Indeed it was the accumulation of factual observations that finally tipped the scale in favour of subaerial as opposed to submarine erosion.

We will attempt in the following pages to summarize the kind of measurements that had most influence on geomorphological thought.

RAINFALL AND RUNOFF

Towards the middle of the nineteenth century rainfall recordings began to give some idea of the great amounts of water falling on land-surfaces, particularly in the humid tropics. In Italy Castelli had recorded rainfall in 1639 and the first instrumental readings were made at Padua in 1725. In England Christopher Wren designed a rain gauge in 1661–2 and the first British observations of rainfall were begun in 1677 by R. Townley in Lancashire. By the beginning of the nineteenth century there were 12 stations in Europe and 5 in the U.S.A. accurately recording rainfall. Records for Madras, Bombay and Calcutta date from the 1820's (Blanford, 1889) and the Royal Observatory at Cape Town began recording rainfall in 1841 (Vorster, 1957). These tropical stations soon began to supply information on high-intensity rains, such as 17 inches in 12 hours (21 October 1846 at Madras) and 4·22 inches in 1 hour (3–4 a.m., 12 June 1847 at Bombay). By 1889 there were 451 stations recording daily rainfall in India, and 17 in South Africa by 1880. In the United States the Surgeon-General began regular weather observations in 1819 (Ward, 1925), and these were extended by Joseph Henry, the first Secretary of the Smithsonian Institution, in 1849. When the U.S. War Department assumed responsibility for the recordings in 1874 between 100 and 350 observers were at work. Climatic data for the United States were published by Forrey (1842), Blodget (1857) and Schott (1872).

The period 1850–65 saw great advances in meteorology associated with the researches of Fitz-Roy, Le Verrier, Buys Ballot and Ferrel; with the foundation of the British Meteorological Society in 1850; and with the facilitation of the correlation of data provided by the telegraph. Daily weather charts were published during the Great Exhibition of 1851, consolidating data from 27 British stations, although regular daily publication did not begin in Britain until 1872, nine years after a similar innovation in France. In 1859 G. J. Symons began to systematize the rainfall observations in Britain and by 1861 he was receiving reports from 168 stations of daily rainfall, which he published in the annual *Symons' British Rainfall*. By 1882 Symons was handling the reports of over 2000 observers. We need not elaborate the spread of the rain-gauge to most parts of the habitable earth, nor upon the fact that even the phlegmatic British were impressed by the devastation of the tropical deluge.

It was not long before runoff, too, began to be measured with some degree of accuracy. The first steps were to calculate the actual flow of the main rivers, as distinct from mere measurements of flood-heights which had been recorded for centuries in several areas. Between 1809 and 1821 Escher de la Linth computed the annual discharge of the upper Rhine near Basle. Defontaine measured the discharge of the Rhine and its tributaries from 1820 to 1833. Similarly from 1825 to 1836 Venturoli estimated the daily discharge of the Tiber at Rome and Baumgarten gauged that of the Garonne from 1837 to 1846. These computations were made from slope formulae aided by the measurements of velocity obtained from surface floats.

The second and more complicated stage was to compute the relation of stream discharge to rainfall. This was estimated at 20–25 per cent for the Buffalo River catchment in South Africa (Tripp, 1884–5); 60 per cent over the Vehar catchment in India (Conybeare, 1857–8); and 44 per cent (normal extended period), 54 per cent (during freshets) and approaching 100 per cent (falling on previously saturated ground) for the Upper Nepean, Cordeaux and

Cataract rivers in New South Wales (Coghlan, 1883–4). Formulae relating rainfall and runoff were given by O'Connell (1867–8);

$$y = M\sqrt{x}$$

and by Jackson (1875);

$$Q = n.100.x^{\frac{3}{4}}$$

where y = river discharge in cubic yards per second.

x = drainage basin area in square miles.

M = a local factor depending on basin slope, soil character, distribution of rainfall, etc. This was calculated at 13 for the Arkansas River, 57 for the Ohio River at Wheeling and up to 302 for the high-intensity runoff of some Indian rivers.

Q = flood discharge in cubic feet per second.

n = local factor depending on maximum daily rainfall.

AMOUNT OF MATERIAL CARRIED BY RIVERS

Having measured the flow or discharge of a river, it was easy by experiment or analysis to determine the amount of matter carried by it, as was done for example by Everest for the Ganges (pp. 281–2). These measurements were greatly extended in the second half of the nineteenth century. The matter carried in solution was carefully computed by Bischof, who with a German's love for exactness and accumulation of data, compiled tables showing the results of chemical analyses taken out of various river waters (Bischof, I, 1854, pp. 76–77; *see* also Jackson, 1875, Pt. 2, pp. 129–35). He did not offer any new method of analysis but followed the original mode of computation first introduced by Everest and Horner, of assessing the material removed in terms of pounds weight per month or per annum. If his approach lacked novelty it did at least provide important support and information for those geologists who were beginning to appreciate something of the prodigious dimensions of the work annually carried on by rivers.

> At first sight, the constituents which are carried into the sea in a state of solution, may appear inconsiderable; when, however, the large volume of water which is constantly flowing into the sea, is taken into consideration, this amount must be very great. Rivers like the Rhine, the Danube, the Rhone, and the Elbe, which contain of these constituents at least 1/8000, carry into the sea, in 8000 years, quantities as great as the weight of the water annually conveyed into it by them. But what are 8000 years, compared with geological periods, in which we must reckon millions of years? (*Bischof, I, 1854, pp. 85–86*)

The measurements of solid matter carried by rivers were added to by Login's work (1856–7) in the delta of the Irrawaddy. He showed that in March 1855 the proportion of suspended material was 1/5725 by weight, at a discharge of 75,000 cubic feet per second; yielding a mean water surface slope of 1·6 feet/mile and a mean surface velocity of 1·5 m.p.h. In flood time the discharge was measured at 750,000 c.f.s., the silt proportion 1/1700, the surface velocity 5 m.p.h. and the gradient 1·8 feet per mile. An average flow of 350,000 c.f.s. had a silt content of 1/3000, and a calculation on this basis gave 2000 million cubic feet of silt annually supplied to the delta mouth.

At the time there were also many attempts to measure the discharge and sediment content of the Mississippi. A. Brown and M. W. Dickeson (1848) took readings at Natchez, R. A. Marr at Memphis, C. G. Forshey in Louisiana and on their findings the influential *Ellet Report* was published in 1851. The amount of material carried to the sea was related by many writers to the advance of the delta. Lyell had always shown a keen interest in such studies as they provided fruitful support for uniformitarianism. In a work on the Mississippi delta in 1847, he described its minor features, including natural levees, and ascribed its continuing formation to a gradual and general subsidence of the area. In furtherance of this statement he referred to measurements made by Riddell at New Orleans. These had shown that the mean proportion of sediment brought down by the river was in the ratio of about 1/1245 in weight and 1/3000 in volume. From figures like these Lyell was able to estimate that it had taken 67,000 years to build the present delta (Lyell, 1847).

A year later, the American geologists Brown and Dickeson writing about the same river state that on the basis of their calculation for one particular year, assuming that the rate had been maintained throughout the period of the delta's formation, only 14,000 years or so would be required before the delta had attained its present size.

> Having ascertained the quantity of solid matter, annually brought down by the Mississippi River, to be 28,188,083,892 cubic feet, which would be equal to one square mile of the depth of 1056 feet in 381½ days, or one cubic mile in five years, eighty-one days, it therefore follows, that it would require a series of 14,203⅘ years for the river to effect the final formation of the present delta. (*Brown and Dickeson, 1848, p. 49*)

The advance of the Mississippi delta front, in feet per year, was variously estimated at 103 (Peck, 1850), 258 (Bayley, 1852), 365 (Montaigu, 1861), and 406 (Bache, 1852; Russell *et al.*, 1936). The material supplied to the lower delta had been estimated at 36 cubic miles per year (Bringier, 1821), 28,188,083,892 cubic feet per year (Brown and Dickeson, 1848), ⅛ of a cubic mile per year (St John, 1851) and 4 feet over 350 square miles per year (Carroll, 1866). The conventional methods of calculating the proportion of load moved to discharge yielded the following values: 1 part suspended load to 1158 parts water (Riddell, 1846), 1 part suspended load to 1808 parts water and 1 part bed load to 601 parts water (Forshey, 1878).

The study of river deposition and particularly of coastal deltas led to important findings on the nature of deposition and on the formation of meanders. The former is especially connected with an interesting experiment conducted by W. H. Sidell, yet another member of the Corps of Topographic Engineers. His tests with specially prepared solutions of sea and river water showed that the presence of sea water accelerated the precipitation of fresh water deposits. The article is no more than a description and it is obvious that its author did not fully understand all the implications embraced within the mechanism of the process:

> It is perhaps proper to mention in this place some experiments that were made to determine if the deposit of sediment were owing solely to the check of velocity of the current on meeting the outside waters. The conclusion was that the effect was not owing solely to this cause. Proper vessels had been provided for the experiments, and in these as many fit substances as were at hand were dissolved in a mixture with the water, each in a separate vessel. These substances were common salt, epsom salt, alum, sea-water, brine from the salt springs, and sulphuric acid. The river water alone took from ten to fourteen days to settle, while the solutions became perfectly limpid in from fourteen to eighteen hours, or from one-fifteenth to one-eighteenth part of the time. I know not to what cause to attribute the effect, unless it be action of these substances on the vegetable matter contained in the water, which aids in the suspension of the earthy matter . . .
>
> However, from these experiments we may conclude that the earthy matter is deposited more suddenly than would be the case if it depended on the check of velocity alone, that the bars will be formed just at the debouches, or where the salt-water is first met; and that the greater the quantity of water brought down, the sooner, on account of the sudden precipitation, will the bars be formed at the debouches. (*Sidell, 1861*)

The principle that Sidell had uncovered involves the phenomenon of *flocculation*. As Sidell had observed, most of the fine clay particles of sediment which the river carries in solution throughout its course, on coming into contact with sea-water (sodium chloride) coalesce into larger particles and thus settle more rapidly than they would otherwise do. Discoveries such as this, though they did not greatly assist the general development of geomorphology did at least illustrate the extensive ramifications and complexity of the subject.

Useful ideas on river deposition and meanders were postulated by J. Fergusson in an article on the Ganges. He discussed how the delta was formed and described how the minor deltaic features, like the natural levees, were produced.

> . . . It is owing to the existence of great sheets of still water in the low lands beyond the banks of the rivers. These, being still, have deposited their mud, if they ever had any in suspension; and being too massive to be set in motion by the rivers, they reduce the flowing streams to inaction the moment they leave their beds, and consequently force them to deposit their silt in their immediate proximity. (*Fergusson, 1863, pp. 324–5*)

More important perhaps than this explanation was his theory of the formation of meanders. He thought that meanders were created whenever the natural flow of the river was interfered with and he went on to draw the conclusion that there was a mathematical relationship between the angle of slope and the size of the meander curve.

> A river on the contrary is a body of water in unstable equilibrium, whose normal condition is that of motion down an inclined plane; and, if we could in like manner abstract all the natural conditions of inequality of surface or of soil, it would flow continuously in a straight line; but any obstruction or inequality whatever necessarily induces an oscillation, and, the action being continuous, the effects are cumulative, as those in a pendulum are discumulative; and the oscillation goes on increasing until it reaches a mean between the force of gravity tending to draw it on a straight line, and the force due to the obstruction tending to give it direction at right angles to the former.
>
> If this be so, it will immediately be perceived that the extent or radius of the curves will be directly proportional to the slope of the bed of the river . . . With a fall of 1 foot per mile, the radius of the curve is, as nearly as I can ascertain, double that of a river with a fall of 6 inches; and when the fall is about 3 inches per mile, the direct and tangential forces so nearly balance one another that the curves are practically semicircles. In the latter case the chord of the curves is practically four times the width of the river . . . Between a fall of 6 inches and 1 foot per mile, the oscillation is, apparently, once in about six times the width, above a foot it rises to one in ten or twelve . . . (*Fergusson, 1863, pp. 322–3*)

THE AMOUNT AND RATE OF SUBAERIAL DENUDATION

The amount or depth of rock removed by erosion in various localities had been estimated by many persons in the early nineteenth century. In later years it became commonplace for geologists such as Ramsay and Geikie to elaborate this theme. Thus Ramsay estimated that in Switzerland during Miocene times the Alps were lowered about 5500 feet by denudation (Ramsay, 1873–5). Geikie estimated that the River Doon in Scotland had removed 70 million cubic feet of rock in cutting a post-glacial gorge (Geikie, 1868). He also left his readers in no doubt of the quantitative work accomplished by rivers in general (pp. 426–7).

The rate of denudation was also discussed by Geikie but in this he was preceded by Tylor and Croll. Tylor (1853), using as his bases an investigation by C. Ellet (1851) on the Mississippi, departed from the normal method followed by earlier writers of calculating the volume removed, and instead worked out the depth by which the total drainage basin would be lowered annually by the combined operations of rain and rivers.

> The mere consideration of the number of cubic feet of detritus annually removed from any tract of land by its rivers, does not produce so striking an impression upon the mind as the statement of how much the mean surface level of the district in question would be reduced by such a removal. (*Tylor, 1853, pp. 260–1*)
>
> . . . the reduction of the mean level of the Ganges district is 1/1751 of a foot annually. The Mississippi . . . could occupy 9000 years at its present annual rate in reducing to the amount of one foot the mean surface-level of the district it drains. (*Tylor, 1853, p. 261*)

Assuming that there is three times as much sea as land, Tylor calculated that each time the land suffered a three-foot reduction the sea would only be raised one foot by the resulting deposition. From the figures available of alluvium and other sediment carried in the main rivers, he considered it possible that the general land surface could be lowered three feet in 40,000 years.

> The proportions of soluble salts in the water of the Thames is . . . 1 to 4117; while the proportion of alluvium suspended in the water of the Mississippi is as 1 to 3000. (*Tylor, 1853, p. 262*)
>
> . . . I now propose to consider that all these causes together might produce an elevation of the sea-level equal to 1 foot in 40,000 years, or 3 inches in 10,000 years. (*Tylor, 1853, p. 269*)

The statistical approach favoured by Tylor proved much to the liking of James Croll (1821–90), one of the most intriguing personalities among the early quantitative geologists. Croll was the son of a Perthshire crofter and had occupied a succession of menial jobs when in 1857, although quite self-taught, he started to publish on scientific subjects. Two years later he was appointed Curator of the Andersonian College and Museum in Glasgow at a minute salary. He made his name as the propounder of an astronomical theory of glacial periods and in 1867 became Secretary and Accountant to the Edinburgh Office of the Geological Survey. In 1881 ill-health forced him to retire and he was given a pension of only 30 shillings a week. He made three applications for a Civil List pension, one application being supported by 4 Dukes, 1 Marquis, 14 M.P.s; the President, 5 Vice-Presidents and 62 Fellows of the Royal Society; the Poet Laureate and 39 other notables. But he only received £100 from the Queen's Bounty and died in comparative poverty (Bailey, 1952).

Croll applied his mathematical reasoning to the work on the Mississippi:

> Time, as represented by geological phenomena, is deeply impressive; and when we attempt to express it in figures we are apt to be misled; for we can form but a very inadequate conception of immense duration represented in numbers. If a stream were to deepen its channel only one-tenth of an inch in a year, it would in 700,000 years cut a gorge nearly 6000 feet deep. It would deepen its channel nearly 600 feet if it were to scoop out only an inch in a century.
>
> The quantity of sediment discharged into the sea annually by the Mississippi river is 28,188,083,892 cubic feet. The area of drainage is 39,029,760,000,000 square feet. Consequently one foot is being removed off the face of the country every 1388 years and carried into the sea. If the rate of denudation be as great in this country as in America, then 500 feet must have been removed off the face of the country and carried by our rivers into the sea since the period of the boulder-clay, if we place that period 700,000 years back. (*Croll, 1867, p. 130*)

FIG. 79. *James Croll*

Perhaps the most famous and influential of the early papers on comparative quantitative erosion was that by Archibald Geikie, 'On denudation now in progress', 1868. In it he showed conclusively that marine erosion was relatively ineffectual compared with the work of subaerial agencies. His method was simple. He began with fluvial erosion and took, as the basis for his inquiry, the results obtained by previous workers of the amount of material carried either in suspension or rolled along the bottom of rivers. From these figures he was able to estimate roughly the length of time that it would take to erode the various continents to sea level.

> The extent to which a country suffers denudation at the present time is to be measured by the amount of mineral matter removed from its surface and carried into the sea. An attentive examination of this subject is calculated to throw some light on the vexed question of the origin of valleys and also on the value of geological time. Of the mineral substances received by the sea from the land, one portion, and by far the larger, is brought down by streams, and the other is washed off by the waves of the sea itself.
>
> 1. The material removed by streams is two-fold; one part being chemically dissolved in the water, the other mechanically suspended or pushed along by the onward motion of the streams. The former, though in large part derived from underground sources, is likewise partly obtained from the surface. In some rivers the substances held in solution amount to a considerable proportion. The Thames, for example, carries to the sea every year about 450,000 tons of salts invisibly dissolved in its waters. But the material in mechanical suspension is of chief value in the present inquiry. (*Geikie, 1868, p. 249*)
>
> But besides the materials held in suspension there must also be taken into account the quantity of sand and gravel pushed along the bottom. In the case of the Mississippi this was estimated by the United States Survey at 750,000,000 cubic feet. In our own rivers it is probably on the whole proportionately greater. Indeed the amount of coarse detritus carried down even by small streams is almost incredible . . .
>
> Comparing the measurements which have been made of the proportion of sediment in different streams we shall probably not assume too high an average if we take that of the carefully elaborated Survey of the Mississippi. This gives an annual loss over the area of drainage equal to 1/6000 of a foot. If then a country is lowered by 1/6000 of a foot in one year, should the existing causes continue to operate as now, it will be lowered 1 foot in 6000 years, 10 feet in 60,000 years, 100 feet in 600,000 years, and 1000 feet in 6,000,000 years. (*Geikie, 1868, pp. 250–1*)
>
> Under such a rate of denudation therefore Europe must disappear in little more than four million of years, North America in about four millions and a half, South America and Asia in less than seven millions. These results do not pretend to be more than approximative, but they are of value inasmuch as they tend to shew that geological phenomena, even those of denudation, which are often appealed to as attesting the enormous duration of geological periods, may have been accomplished in much shorter intervals than have been claimed for them. (*Geikie, 1868, p. 251*)

Geikie however was cautious and, in order to allow the maximum latitude for variations, he suggests that an adjustment to these figures should be made which would allow for the difference in river erosion which exists, between the areas of gorges and cataracts where the erosion rate is high, and the areas of the plains where the erosional rate is very much reduced.

> But it is obvious that the material so removed does not come equally from the whole area of drainage. Very little may be obtained from the plains and watersheds; a great deal from the declivities and valleys. (*Geikie, 1868, p. 251*)
> Let it be assumed that the waste is nine times greater in the one case than in the other (in all likelihood it is more); in other words, that while the plains and table-lands have been having one foot worn off their surface, the declivities and river-courses have lost nine feet. Let it be further assumed that one-tenth part of the surface of a country is occupied by its water-courses and glens, while the remaining nine-tenths are covered by the plains, wide valleys, or flat grounds. Now, according to the foregoing data, the mean annual quantity of detritus carried to the sea is equal to the yearly loss of 1/6000 of a foot from the general surface of the country. The valleys, therefore, are lowered by 1/1200 of a foot, and more open and flat land by 1/10800 of a foot. At this rate it will take 10,800 years before the level ground has had a foot pared off its surface, while in 1200 years the valleys will have sunk a foot deeper into the framework of the land. By the continuance of this state of things a valley 1000 feet deep may be excavated in 1,200,000 years. (*Geikie, 1868, pp. 251–2*)

If subaerial denudation can do all this, it is unnecessary and muddling to suggest further causes, he concludes.

> It seems an inevitable conclusion that those geologists who point to deep valleys, gorges, lakes, and ravines, as parts of the primeval architecture, referable to the upheavals of early geological time, ignore the influence of one whole department of natural forces. For it is evident that if denudation in past time has gone on with anything like the rapidity with which it marches now, the original irregularities of surface produced by such ancient subterranean movements must long ago have been utterly effaced. That the influence of these underground disturbances has often controlled the direction in which the denuding forces have worked, or are now working, is obvious enough, but it is equally clear that under the regime of rain, frosts, ice and rivers, there must have been valley-systems wherever a mass of land rose out of the sea, irrespective altogether of faults and earthquakes. No one who has ever studied rocks in the field is likely to overlook the existence of faults and other traces of underground movement. But he meets everywhere with proofs of the removal of vast masses of rock at the surface, which no amount of such movements will explain. At this present rate of excavation the 'gentle rain from heaven', and its concomitant powers of waste, will carve out deep and wide valleys in periods, which by most geologists, will be counted short indeed. And if an agency now in operation can do this, it seems as unnecessary as it is unphilosophical to resort to conjectural cataclysms and dislocations for which there is no evidence, save the phenomena which they are invented to explain. (*Geikie, 1868, p. 252*)

Having set out all his points in favour of subaerial denudation, Geikie turns to examine the alternative force, the sea. He adopts the same method as before; a known rate of erosion is used to calculate how long it would take the sea to erode away the continent of Europe. His conclusion is damning to the cause of the marine erosionists.

II. The detritus wasted from the land is carried away not only by streams, but in part also by the waves and currents of the sea. Yet if we consider the abrasion due directly to marine action, we shall be led to perceive that its extent is comparatively small. In what is called marine denudation, the part played by the sea is rather that of removing what has been loosened and decomposed by atmospheric agents than that of eroding the land by its own proper action. Indeed, when a broad view of the whole subject is taken, the amount of denudation which can be traced to the direct effects of the sea alone is seen to be altogether insignificant. (*Geikie, 1868, p. 253*)

Let us suppose that the sea eats way a continent at the rate of ten feet in a century – an estimate which probably attributes to the waves a very much higher rate of erosion than can on the average be claimed for them, – then a slice of about a mile in breadth will require about 52,800 years for its demolition, ten miles will be eaten away in 528,000 years, one hundred miles in 5,280,000 years. But we have already seen that on a moderate computation such a continent as Europe will, at the present rate of subaerial waste, be worn away in about 4,000,000 years. Hence, before the sea could pare off more than a mere marginal strip of land between 70 and 80 miles in breadth, the whole land would be washed into the ocean by atmospheric denudation. (*Geikie, 1868, p. 254*)

THE NATURE OF WATER-MOTION AND TRANSPORT BY WATER

It was not long before the attention of hydrologists began to be concentrated less and less on the discharge of rivers and on the amount of material carried by them. Additional factors such as the size and variety of debris transported, the slope of the river bed, the river's powers of self-adjustment and a host of other hydrological aspects were gradually investigated. We have for example already mentioned the notable *Principes d'Hydraulique* of L. G. Du Buat in 1779 (3rd edn, 1816) and the detailed study of A. Surell, *Étude sur les torrents des Hautes-Alpes* published in 1841. To the earliest Italian and early French and Swiss hydrological texts were now added several remarkable studies by American and British investigators. These were strongly influenced by principles already drawn up by Continental authors.

In 1851, N. Beardmore's *Manual of Hydrology* contained statements about the behaviour of rivers, the geomorphic significance of which did not become apparent until a relatively modern period in the history of geomorphology. Like some of the earlier workers, whose findings have already been set out, he observed that the speed of a stream was greater at its surface (s) than near its bed (b) and in its middle section than near its banks. He stated a formula which served to express the former relationship; $b = (\sqrt{s} - 1)^2$, and also showed that the mean velocity of a stream can be taken to be eight-tenths of its maximum velocity. Again, like earlier observers, he was beginning to realize that the changing shape of a river was partly a matter of self-regulation. This was important, for up till now most geologists only credited rivers with an ability to erode to a limited extent and to carry away a small amount of the material detached by this erosion. Except in Surell's work, there was no hint as yet that a river was capable of controlling, or at least having some say in, its own development; where changes had taken, or were taking place at some part of the river's course, the cause was invariably sought in some earth movement or structural control. Yet the hydrologist Beardmore treated a river as if it were a human organism, with all the concomitant powers of being able to compensate itself for losses and to adjust the remainder of its system in response to changes affecting any one part. Though Beardmore was right, for geomorphologists his statement was revolutionary and passed largely unnoticed.

> If any river pass at a greater rate than the banks will bear, it is a beautiful law of nature, and most certain in its effects, that a greater sectional area is cut out; thus the hydraulic mean depth being increased, the surface slope becomes flatter, and the general velocity and scouring action is reduced. (*Beardmore, 1851, p. 6*)

Following Brahms, he noted the mathematical relationship between velocity gradient and the cross-sectional area. For the main principles involved in this relationship he quotes the following points from Du Buat's *Principes d'Hydraulique*:

1. In water there is no increase of friction as pressure increases.
2. The sole cause of stream movement is the surface inclination of the water.
3. When the mean velocity is constant, the accelerating force is in equilibrium with the resistance of the wetted border (i.e. wetted perimeter).
4. The filling of surface irregularities with water means that the wetted border does not vary in resistance with its own roughness.
5. Thus the resistance of the wetted border is proportional to the square of the velocity; but the resistance decreases as the velocity increases.
6. Perimeter resistance is transmitted to the whole water cross-section. Thus, at equal inclinations, the velocity will be proportional to the square root of the area to the perimeter (i.e. to the square root of the wetted perimeter).
7. Least resistance, and therefore greatest velocities, occurs at the greatest distance from the perimeter.

If Beardmore's study had stopped there it would not have been more important than those of earlier writers. But he went further and compiled tables showing the weight, volume and specific gravity of materials which water could move at certain velocities. He prepared tables which related discharge, velocity and slope in streams, and also gave a formula expressing velocity in terms of channel inclination. This led on naturally to the conclusion that stream discharge was dependent primarily upon two factors, the amount of rainfall, and the type of underlying bed-rock. It was becoming apparent that topography is not the fortuitous product of unconnected events, but was adjusted to provide the most economical surface runoff. Beardmore in fact supplied data on the amount of discharge to be expected from actual drainage basins and made calculations of the runoff per unit area. In search for confirmation of his ideas, he examined the discharge of several river systems, and especially of the river Po, for which he described the material being transported, the deposits lining the river channel, and changes in course. Also he demonstrated that the slope of the river increased during times of flood. Other techniques he employed involved detailed drawings of longitudinal profiles of the Po and its tributaries and cross-sections of the Nile during periods chosen to bring out the maximum differences in the amount of discharge.

In the same year that Beardmore produced his *Manual*, T. J. Taylor published *An Inquiry into the Operation of Running Streams and Tidal Waters* which threw much light on the laws by which a river maintains or changes its course. He started by accepting that a river is able to regulate itself and that each separate feature or topographic form to be found along the river's course can be accounted for as a stage in this process of self-regulation. Although he appears to have derived some of his conclusions from the works of Italian river engineers from the sixteenth century onwards, he exemplifies the state of hydrological knowledge in the middle of the nineteenth century;

> In short, new terms of power as well as of resistance, continually occur for investigation, according to the variable nature of the circumstances alluded to; but rivers have thus the distinction of being not merely obedient to laws, but of being governed, to a great extent, by those of their own creation. (*Taylor, 1851, p. 3*)
>
> ... if the currents of undivided rivers be permitted to waste their energies on shoals and other obstructions of the bed, the result is, *pro tanto*, similar to that of withdrawing a portion of water. Neither does Guglielmini omit to notice this very circumstance, for he says; 'If we restrict the bed of a river by Art, we cause it to deepen its bed; while if the bed be too wide, or divided into several branches, its bottom will be raised in proportion. (*Taylor, 1851, p. 4*)

Taylor felt constrained to discuss at length the reasons that had persuaded him to accept the above and he traces first the known facts. Rivers, he says, must transport the materials which are to be found on their beds because otherwise they would become choked up, and this is not the case in nature. The rate at which the materials are moved must be slow because they will necessarily travel at the speed of the largest particles, which shelter or impede the smaller ones:

> But the question arises, in what manner does gravel disappear from the beds of rivers? It is assumed to be worn away by attrition: a position adopted

by the writer; with this qualification, however, that the time required to wear down the gravel is longer than has been usually assigned . . . stones shifted from a particular point are probably laid down again at the next expansion of the current, or even before they reach that expansion. Neither does the smaller gravel travel fast, for it is sheltered by the heavier stones, and the friction during floods must be regarded as an average due to all the fragments on the bed, since the smaller and larger are so mixed together that they must move together; the whole bed must be considered as endued with one movement and the motion of the whole bed down stream is very slow: . . . (*Taylor, 1851, pp. 7–8*)

Viewing the river from source to mouth he says it can be divided into sections, each with its particular group of characteristic features.

The torrent trenches its mountain almost perpendicularly: the Torrent River receiving the torrent has already flowed some distance among the hills, and has flattened its bed considerably, but still possesses a great deal of fall: further on, we arrive at a stage of the river where the angular fragments of gravel have disappeared, and those found in the bed are rounded and much reduced in size: this is the river with the proper gravelly bed, whose fall is a good deal less than that of the torrent river. The fourth and last stage is that of the sandy bed, whose fall is very moderate, and in the case of first class rivers hardly appreciable.

In the manner described there is established a succession of inclined planes, which, constantly diminishing in their rate of fall constitute the beds of rivers in their progress towards the sea: the entire extension of bed has been, therefore, compared to an asymptotic line, which, though always approaching another line, yet never becomes exactly coincident with the latter: the resemblance, however, like the most of our attempts to measure nature by rule and scale, fails in being quite exact, since the line is disturbed by various local actions, and, if accurately adjusted to all the variations occasioned by tributaries, changes of course and of geological formation, would present an irregular and serrated outline. (*Taylor, 1851, p. 13*)

It is on the basis of these facts that Taylor arrives at his first conclusions. He perceived that not only would the size of the materials to be transported have an effect on the steepness of slope but so also would the discharge of water to be carried off.

The truth, however, of the general proposition, that the coarser the materials of a river bed the greater is its rate of fall, is a manifest consequence of the relation between cause and effect. A certain declivity is established, because the current cannot, without the help of that declivity, press forward the materials that encumber its bed. The heavier the materials therefore the greater the fall, and contra, the lighter the materials the less the fall. (*Taylor, 1851, pp. 13–14*)

The motion of rivers towards the outfall is in fact, governed more by the slope of the surface than that of the bed: thus we perceive the justness of the rule laid down by the Italian engineers: – 'The greater the volume of water the less is the slope of the bed;' but also, 'The greater the volume of water, the greater is the slope of the surface'. (*Taylor, 1851, p. 17*)

The next conclusion follows logically from the above propositions. As the size of material affected the angle of slope, so the erosion of the valleys would be intimately related to the locations of the source from which the material was removed, and therefore the channel erosion could be expected to be mathematically proportionate to the rate at which the hills were denuded and receded. In other words, as the hills receded the distance the gravel would have to travel would lengthen and the particles would become smaller; this in turn would mean a lessening in the gradient and a diminution in the erosion. At the same time, as the hills were denuded the amount of gravel available to the rivers would decrease and this would serve to augment the rate at which erosion lessened. In this way Taylor began to make it clear that the portions of a river system did not operate independently of each other, or, for that matter, of the surrounding landscape. A river developed as a unit, with a fine balance between its parts.

As an example of this balance between velocity, discharge and slope he discussed the conditions when two rivers unite. Often he says when a tributary joins a main river a steeper slope will occur just above the junction. He also noted, as had the Italians and Lyell, that where two rivers united their channels increased mainly in depth (1851, p. 38). Taylor then passed on to a new theme, in which he asserts that the true interpretation of a stream is not to be sought from the average conditions, but from the times of flood. Such conditions, he believed, provided the key to all the features that were to be observed at intermediate periods when the river was flowing at a lower level or discharge.

> On comparing these various conditions, it becomes evident that the flood or high water state is the really governing state of rivers. We cannot, indeed, appreciate the effects produced by rivers from their average condition, and still less from their lowest state; neither can we comprehend the established relation between them and their beds, except by referring it to the flood state. If, for example, the bed of a gravelly river be examined when the water is low, there will be observed a succession of streams and of pools: in the streams the water moves with considerable velocity; in the pools it is almost still. Why then are not the pools filled up? How are we to reconcile the principle of great velocity and an encumbered bed, with that of a slow velocity and the contrary? On further examination we find that the bed, wet and dry together, of the stream, is much wider than that of the pool; and the consequence is, that when the waters are in a state of flood, the conditions of the velocity are exactly reversed: where the stream is, the current has then a wide bed, and where the pool is situated, a much narrower one: in other words, the section of the pool is then the smallest, and the flood current, therefore, moves through it with greater velocity than it does over a stream bed; the mutual relations being thus entirely changed on a comparison of the flood with the ordinary state. (*Taylor, 1851, p. 19*)

To show the credibility of his argument he goes to some length to point out how different are conditions during a flood. The difference is so vast as to make consideration of the average conditions misleading.

By viewing the subject in this light, we can easily understand how a few large floods in a twelvemonth may constitute themselves the law-givers, as it were of a river; and we may also appreciate the fallacy of those calculations so often made, which profess to determine the energy of rivers from the average quantity of their waters: . . . (*Taylor, 1851, p. 24*)

It will be recollected, that an inch of rain is equal to a fall of almost exactly 100 tons of water on every acre; wherefore brooks, regarded as insignificant and which collect the water of only a few square miles of basin, may yet, during heavy rains, discharge very considerable volumes of water, and are then possessed of corresponding, and indeed, surprising power. (*Taylor, 1851, p. 25*)

We may take an example of this kind from the Tyne itself. It is not unusual for this river, during a land flood, to discharge 36 millions of tons of water in 24 hours, being equivalent to a net quantity of half an inch of rain over the entire extent of its basin. The highest sources are about 1200 feet above the sea level; but the mean elevation of the basin may be taken at 500 feet. Now 36 million tons in 24 hours are 25,000 tons per minute; a horse-power being 33,000 lbs., equal to 14·7 tons, we have . . . 850,340 horses' power. (*Taylor, 1851, p. 26*)

Taylor also noted that where the downward erosion was checked, lateral erosion might commence or shoals and islands might form.

It would appear that when Nature cannot bring the bed and current into a state of equilibrium by windings, she does so by means of island and shoals. (*Taylor, 1851, p. 75*)

From this point he is drawn into a discussion of the characteristics of meanders and he combats forcibly the looseness of phrase of some geologists who spoke as if water had elastic properties.

Water, being entirely non-elastic, does not rebound from a surface which is struck by it: but on the contrary, establishes itself against that surface with whatever energy its current possesses. If the river edge be corrodable, a curved bank is thus formed, the resultant of two forces, one of which endeavours to act upon the bank and the other opposes that action, and in this way a new direction is given to the current: but the application, to the movement of water, of such phrases as 'rebound', 'resilience', etc. implying a degree of elasticity, is manifestly incorrect.

When a river acts on one side of its bed only, a sand bank grows on the opposite shore, and is attached to that shore: (*Taylor, 1851, pp. 83–85*)

He then stresses, as did Beardmore, the principle that, other circumstances being equal, the discharge of a river will depend upon the acreage of its drainage basin. This in turn leads him to consider the question of runoff and the relative distribution of water between the various portions of the basin. He appreciated that as the source region involved a smaller unit area so the volume of water which was carried away from there would be less than that carried by the lower reaches. In the same way he also appreciated that heavy falls of rain or snow would have more serious effects in the higher reaches than in the broader and lower reaches.

HSL—P

It is obvious from the foregoing considerations, that the quantity of water in a river constantly diminishes in approaching its source; for the nearer the source the less is the remaining surface of supply; and *e contra*, the further from the source, the greater the surface and the corresponding volume. Further, the volume is not a direct proportion to the length of course, but a question of area, that is, of length multiplied by breadth; since we have a right to assume that the width of basin is proportional to its length. (*Taylor*, *1851, pp. 31–32*)

The nearer the source the more rapid is the rise of a flood, because the waters, on account of the shortness of the run, and the declivities of the country, collect very rapidly: but for the same reason they pass away rapidly, whence the nearer the source the shorter is the duration of floods. (*Taylor*, *1851, p. 32*)

The other points Taylor makes are less novel and less important. They are based upon the various factors of resistance which the flow of a river must overcome. Both friction and the size of the particles are examined in order to discover what proportion of this resistance they cause.

The few premises to be attended to are as follows:

1. The resistance encountered by a current, is as the surface exposed to its action, that is, as the wetted surface multiplied by the length of channel.
2. The resistance must be assumed to be divided amongst all the particles of fluid contained in a cross section, and as the number of particles so contained is proportional to the area of the section, the effective resistance is, therefore, so much the less as the area is larger, and vice versa . . .
3. The force with which an object is struck with a fluid is as the volume of fluid multiplied by its velocity; . . . This is what is meant when it is stated that the impulse of a fluid is as the square of the velocity: and since friction must be regarded in the same light as that of a force . . ., on the ground that action and reaction are equal, so it follows that the friction of a fluid is as the square of its velocity. $\dfrac{slv^2}{a}$ is therefore a universal expression for the resistance of fluids, where

$$s = \text{the wetted surface,}$$
$$l = \text{the length of channel,}$$
$$v = \text{the velocity,}$$
$$a = \text{the area of the cross-section.}$$

If we put h for a motive power or water pressure and change the terms, then v^2 is proportional to $\dfrac{ha}{ls}$, . . . (*Taylor, 1851, pp. 39–40*)

He is led to conclude from his investigation that only a small proportion of hydraulic energy is utilized in transporting the stream-bed material, for by the sixth-power law a small increase in velocity will effect the movement of much larger particles of material.

Now we shall see immediately that the velocities due to the transporting of even the large gravel of river beds are by no means great, relatively with the theoretical force of the river. The balance of that force is therefore absorbed by other elements of resistance. (*Taylor, 1851, p. 45*)

Taylor turns to an examination of friction, and, following Du Buat, sets out a table showing the velocities of the current read at different heights above the river bed.

> One of the effects of friction is to cause a less velocity at the sides and bottoms of rivers than at the surface and in the middle . . . I have arranged thirty-one experiments of Du Buat, from 10·91 inches to 48 inches per second of surface velocity, with the following mean results:
>
> 1. Velocity at surface . . . . 27·81 inches
> 2. Velocity at the bottom . . . . 18·07 „
> 3. Actual mean velocity . . . . 21·54 „
> 4. Mean velocity calculated at 4/5ths of the velocity of the surface . . . . 22·24 „
> 5. Mean velocity, considered as the arithmetical mean between the surface and bottom velocities . . . . . 22·92 „
>
> (*Taylor, 1851, pp. 49–50*)

It is quite clear that Taylor believed most of the energy of the stream was consumed by the internal friction of the water particles.

It is certain that the investigations and writings of engineers such as Beardmore and Taylor would, if read widely by geologists, have effectively established the erosive ability of streams. It seems, however, that the geomorphological significance of many of these more scientific contributions were not appreciated for many decades.

The amount and depth of hydraulic engineering research increased steadily after the mid-nineteenth century. In Britain, J. D. Forbes (1856–57), for instance, conducted a series of experiments with half-inch deep water flowing over a bed of various-sized material, showing:

1. That a water velocity of 125 feet/minute would not pick up moist clay, but that a flow of 15 feet/minute would disturb clay mixed with water and allowed to settle.
2. That fresh water sand sank in still water at a rate of 10 feet/minute, and required a velocity of 40 feet/minute over the bed to disturb it.
3. Sea sand settled at 11·707 feet/minute and was disturbed by a velocity of 66·22 feet/minute.
4. Pea-size, rounded pebbles settled at 60 feet/minute and were disturbed by a velocity of 120 feet/minute.

On the Continent, Cunit (1855) described the *courbe de régularisation* of the longitudinal stream-profile which he considered asymptotic to the mouth and closely related to the bed material to be moved.

Dausse (1857, p. 759) employed the term *pente d'équilibre* and followed Surell in postulating that a river tends to adjust the slope of its bed throughout its whole course (Woodford, 1951).

Darcy, basing his findings partly on the experiments of Hagen (1839) and Poiseuille (1846), who determined that the relation of loss of head to velocity of flow is different in small and large tubes (Tolman, 1937), published equations for the velocity of fluid flow in various soils. Darcy's Law of Permeability of granular material having a prismatic form (1856) was as follows:

$$Q = K.F.s$$

where Q = discharge.

s = hydraulic gradient (i.e. loss of head divided by the length of flow).

F = cross-sectional area of the prism.

K = the coefficient of permeability (i.e. the velocity of percolation under a hydraulic gradient of unity).

The equation for percolation or fluid-flow through water-bearing sands was given as

$$V = \frac{kh}{l} = k \sin \theta^4$$

where V = velocity of flow.

h = difference of head in the column.

l = length of the column.

θ = angle of slope of the pressure surface.

k = constant of permeability, depending on the properties of the material.

A few years later, Darcy, with Bazin, published the results of hydraulic experiments involving a straight wooden flume (Darcy and Bazin, 1865), and Bazin developed a formula expressing mean flow velocity in a channel:

$$V_m = V_x - 14\sqrt{RS}$$

where V_m = mean velocity in the channel cross-section.

V_x = maximum velocity in the channel cross-section.

R = hydraulic radius.

S = channel slope.

These authors also appreciated the need to give Coulomb's expression for channel resistance (p. 90), different values according to the roughness of the wetted perimeter.

In the United States the progress of hydrological studies now became phenomenal. Whereas the Italians and Swiss had presented the world with studies mainly of small rapid streams, the Americans produced a detailed work on a tremendous river, including a vast flood-plain. In 1850 the United States Congress authorized a study of the hydrology of the Mississippi river to assist flood control, and Capt. A. A. Humphreys was put in charge, to be joined in 1854 by Lieut. H. L. Abbot. They produced in 1861 a definitive work on hydrology, wide in its scope and of immense geomorphic significance – a significance which Archibald Geikie was not slow to exploit.

This monumental work began with a complete summary of previous

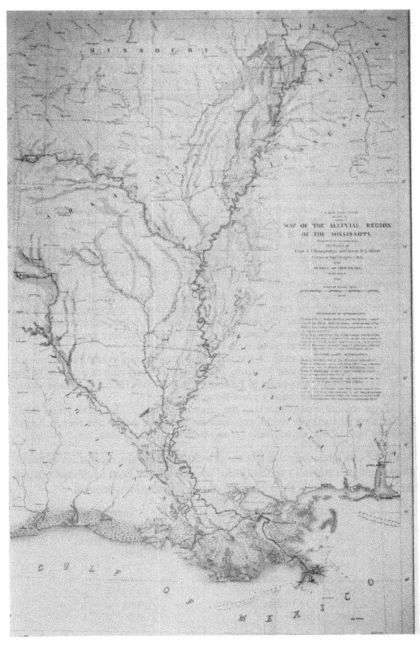

FIG. 80. *Map of the alluvial valley of the Mississippi by Humphreys and Abbot*
(*1861*)

hydrological work; went on to examine the basic principles of hydraulics and to add some new ones. It showed, for example, how the depth and width of a river varies with its discharge, and also that the curves of both surface and vertical velocities are parabolic. From the point of view of contemporary geomorphology, however, their most important contribution was to demonstrate the amount of work which the Mississippi river was performing:

> *Conclusions respecting proportion of sedimentary matter.* A comparison of these different results leads to the belief that no material error will result from assuming that the sediment of the Mississippi is to the water, by weight, nearly as 1 to 1500 and, by bulk, nearly as 1 to 2900; providing long periods of time be considered.
>
> *Annual amount transported to the gulf.* If this be so, and if the mean annual discharge of the Mississippi be correctly assumed at 19,500,000,000,000 cubic feet, it follows that 812,500,000,000 pounds of sedimentary matter, constituting one square mile of deposit 241 feet in depth, are yearly transported in a state of suspension into the gulf.
>
> *Observations upon materials rolling along the bottom of the river.* Besides the amount held in suspension, the Mississippi pushes along into the gulf large quantities of earthy matter.
>
> . . . A keg similar to that used in collecting water below the surface was sunk to the bottom of the river. The current immediately overturned it, and the valves opening allowed the water to pass freely through. After remaining a few minutes it was drawn suddenly up, and was invariably found to contain material such as gravel, sand, and earthy matter . . .
>
> No exact measurement of the amount of the annual contributions to the gulf from this source can be made, but from the yearly rate of progress of the bars into the gulf, it appears to be about 750,000,000 cubic feet, which would cover a square mile about 27 feet deep.
>
> *Total annual contributions of the river to the gulf.* The total yearly contributions of the river to the gulf amount then to a prism 268 feet in height with a base of one square mile (*Humphreys and Abbot, 1861, pp. 148–50*)

A few years later there was a spate of writings about the mechanics of fluids. These included Russell's article (1869) which used Du Buat's findings relating the velocity of river-bottom currents to the size of grains moved.

Material	*Competent velocity*	
Soft clay	0·25 feet/sec.	
Fine sand	0·50 ,,	,,
Coarse sand and pea-size gravel	0·70 ,,	,,
Gravel, the size of French beans	1·00 ,,	,,
Gravel, 1 inch diameter	2·25 ,,	,,
Pebbles, 1½ inches diameter	3·33 ,,	,,
Heavy shingle	4·00 ,,	,,

From it, Russell emphasized the direct connexion between a river's velocity and the composition of its bottom sediments. A more novel relationship was contained in his formula for the movement of sediment by a stream, which he

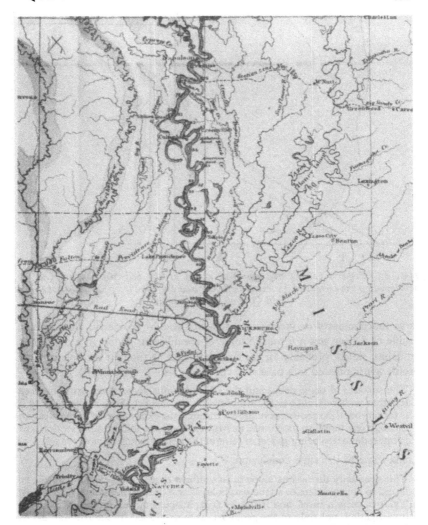

FIG. 81. *Detail of Mississippi river meanders between Napoleon and Natchez, from the map by Humphreys and Abbot*

directly related to the amount of rainfall. $S_1 = \frac{r_1}{r} . S$; where $S =$ present movement of sediment; $S_1 =$ future movement of sediment; $r =$ present rainfall; and $r_1 =$ future rainfall.

The result of this quantitative research by chemists, engineers, and physicists in the Old and New Worlds was a continued interest in processes as well

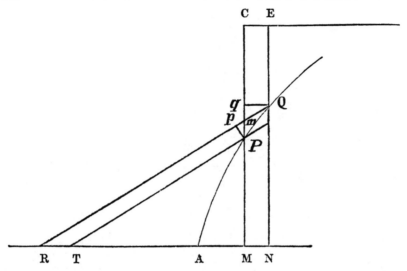

FIG. 82. *Fisher's (1866) mathematical model of the bed-rock surface (APQ) developed by the retreat of a vertical cliff (CP and EQ) skirted by a talus slope (PT and QR) of constant angle*

as in morphometry or forms; in mechanics and in geometry. The average profile of a river was a favourite topic of speculation. Tylor, in a discussion on deltas, stated, probably for the first time, that he thought the normal river profile approximated to a parabolic curve (*see* also Tylor, 1869B).

> The surfaces of these deltas and the alluvial plains above them are compared together with reference to their inclination and height above sea level; and it is found that a parabolic curve drawn through the extremities of each river and through one point on its course, nearly represents its longitudinal section, the greatest deviation being 30 feet in some of the largest deltas. (*Tylor, 1869A, p. 7*)

Login came to the same conclusion.

> This gradually decreasing slope is so uniform that . . . I believe it would be found to approximate to a true parabola, as was first demonstrated by Mr A. Tylor in certain cases, and if this be true, it opens out an important field for investigation, not only to the geologist, as it would aid in determining dates, but to the hydraulic engineer, as it would throw much light on the vexed question of scour and deposit. Should this law be a correct one, it at once shows how the slope is dependent on the load carried by flowing water . . . (*Login, 1872, p. 193*)

About the same time Dausse made a most significant contribution by extending the idea of equilibrium stream slopes to all landform slopes (*Kesseli, 1941, p. 568*):

> This formula: The gradient of the stream and the form of each of its sections are the result of an equilibrium between the force of the current and the resist-

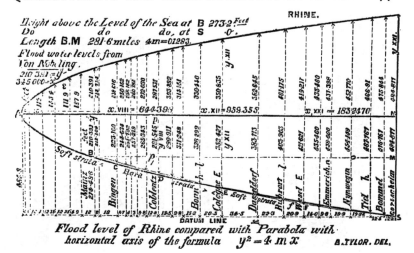

FIG. 83. *Tylor's (1875) calculated parabola fitted to the flood level of the Rhine*

ance of the materials of the river beds, expresses not only a hydraulic axiom. The forms of the solids which suffer attack from atmospheric agents also follow from it; they are also the result of an equilibrium between the forces which endeavour to carry off the particles of these solids and the forces which hold them back. (*Dausse, 1872, p. 321*)

We must, however, point out that in England Fisher (1866B) had already made an attempt to develop a mathematical model with which to infer changes with the passage of time. He assumed that a vertical cliff (height $= a$) would retreat parallel to itself by the removal from its face of material which would accumulate at the cliff base at a constant angle of repose (α). He thereby deduced that the form of the sub-talus rock surface would be a semi-parabola ($y^2 = 2a . \tan \alpha . x$) with its vertex at the base of the original cliff profile. This attempt by a geologist to reduce geologic forms and processes to mathematical equations was revolutionary and far ahead of its time. Like many new ideas it was too novel and remained for the time being quite neglected.

The matter was pursued further by Tylor, who had long thought that many landscape problems might be solved by mathematical means in conjunction

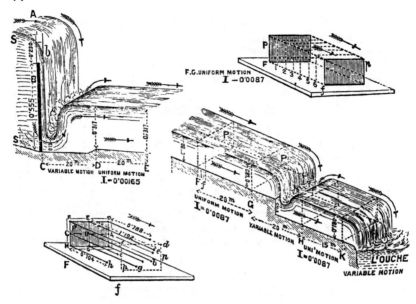

FIG. 84. *Tylor's diagrams illustrating the motion of running water*

with observations. He advanced beyond the bare announcement of a parabolic curve for the average river profile and in 1875 attempted to explain the
factors which produced such a form. In an important geomorphological contribution he confirmed that the velocity of flow is directly proportional to the slope
of the river channel and the quantity of water flowing. In the actual case of
the river Rhine he related its longitudinal profile at flood level to a parabola
of the form: $y^2 = 4mx$. What is more he found also that he could apply
geometry to other geomorphic features. For instance, he took Lyell's example
of a tributary joining the main river and came to the same conclusion (viz.
that the slope of the stream bed changes below the point of junction), with the
addition that he was able to pronounce a calculable mathematical relationship.
By comparing the hydraulic geometry of the two channel cross-sections,
where the two cross-sectional areas are A and a, the respective velocities V
and v, the discharges Q and q, and the slopes I and i, he was able to advance
from the simple initial equation $Q = AV$ to the more complicated relation-
ship: $\dfrac{I}{i} = \dfrac{1}{2}\left(\dfrac{A}{a} + \dfrac{q}{Q}\right)$ which showed by how much the slope changed. He
noted that the slope similarly underwent a change with a variation in the dis-
charge. Just how advanced his ideas were can be gauged from the fact, that in
this paper he did not limit his attempts at quantification to river processes,
but sought to identify binomial curves among the erosive shapes of hills.

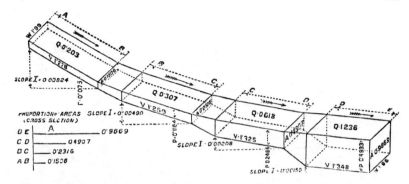

FIG. 85. *Tylor's diagrams illustrating the hydraulic geometry of channel cross-sections*
Above: *Velocity kept more or less constant along a reach, but discharge (Q) varied.*
Below: *Main channel branching into three distributaries.*

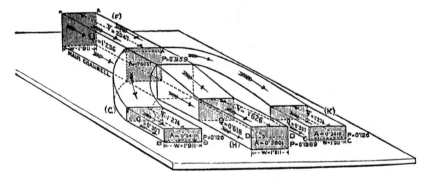

Nor should we let pass unnoticed Tylor's diagrammatic skill, the remarkably high quality of which may be judged from the few examples chosen from his numerous drawings (Tylor, 1875).

Several other authors with a strictly mathematical or scientific outlook contributed research or suggestions on river-flow and river-forms about this time. But in Europe the two most important not yet mentioned are H. Sternberg in Germany and J. Thomson in Britain (*see* pp. 445 & 607). Sternberg in 1875 showed that there was a relationship between the curve of the river profile, the velocity of the river at any point and the size of the material it could carry. His work on river geometry continued along the lines already suggested by Surell, Cunit, Taylor, Dausse, Russell, Fergusson, Tylor, and Login. It stands out because he was able to synthesize their ideas. Sternberg followed earlier writers in assuming that in order to transport material the velocity of the river must match the size of the particles to be carried, but he

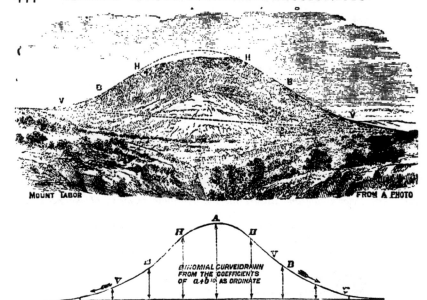

FIG. 86. *Binomial curve fitted by Tylor to the outline of Mount Tabor, Galilee*

refined this assumption by adding the fact that as the particles moved down-stream they would decrease in size by attrition and at the same time the maximum velocity required to move them would also decrease. This he in turn related to the forms of actual river profiles.

Stream valleys and stream beds have their forms determined almost entirely by the work of running water . . . Flowing water has the capacity under favourable circumstances, to carry solid bodies (debris particles) with it, if the frictional resistance, which opposes their movement, is overcome by the impact of the water. With debris particles of similar surface forms, the resistance increases as the cube, the impact of water only as the square, of their diameters. It follows that for a given water velocity all the debris of various sizes on the stream bed must fall into two groups, of which one moves with the current, the other is left behind. The pebbles or other particles left behind raise the stream bed until, through increase of velocity or of gradient, velocities are attained which will move the particles along. It is therefore clear that at any point on a stream bed the size of the debris particles corresponds to a certain velocity, as a rule the maximum velocity, of the water flowing over it. But as the debris moves along, wear and comminution take place according to definite laws dependent on the physical properties of the rock. In general, therefore, the maximum velocity of the water necessary for the progress of the bed load becomes less the farther the load has travelled from its place of production, i.e. from rocky mountains. At a given point on a stream course the quantity of water discharged remains essentially unchanged, and extraordinarily long periods of time must have elapsed after the initiation of valley formation, so that there have developed the appropriate maximum water velocities, the sorting of the load according to particle size, the progress of the load along the bed, etc., in regularly continued or periodically recurring fashion, and thus all the characteristics of the stream are in equilibrium and mutually determined. That such equilibria prevail in actual streams according to general laws is shown in striking fashion by their longitudinal profiles: as long as their beds lie in alluvium, the profiles are lines concave upward, which become even steeper toward the source regions. The lines are broken at places where a stream bed is for any reason less mobile than the load which is being carried over it, whether the obstruction be bedrock or hardened and cemented debris masses. Below such irregularities – breaks in slope and stream masses – the proper curve of the long profile is again continued. (*Sternberg, 1875; translated by A. O. Woodford in Mather and Mason, 1939, pp. 477–8*)

Sternberg realized that his law provided a means whereby river changes could be accurately calculated.

Knowledge of the laws prevailing (in debris transport) is of general value. It will be especially useful to the hydraulic engineer . . . to be able to predict the changes of long profile which must occur as the result of stream control works which alter the properties of the stream and especially the width of the water surface. The lowering of the water surface and the deepening of the stream bed, almost invariably observed in the course of a few years after the completion of control works, undoubtedly belong here. (*Sternberg, 1875, translated by A. O. Woodford in Mather and Mason, 1939, pp. 477–8*)

Working on the profile of the Rhine between Basle and Mannheim, he discovered that it could be represented by an exponential curve, which he associated with a geometrical decrease in grain size of the materials carried as they were worn smaller with distance travelled by the various processes of attrition. In addition he recognized the importance of the stream cross-section, and calculated the optimum form which would be necessary if a stream were to be so regulated as to prevent the formation of bars at all stages by the keeping constant of the length of the hydraulic radius. The exponential curve that he proposed for the middle Rhine was:

$$y_1 = y \, \frac{l - e^{\frac{3}{4\lambda n x_1}}}{l - e^{\frac{3}{4\lambda n x}}}$$

where x_1, y_1, x and y are the horizontal and vertical co-ordinates, respectively.
$\lambda = $ (a small fraction) is a function of pebble size and resistance to attrition.
$n = $ either 2 or 4 (Woodford, 1951).

Sternberg's work is important because it set forth his famous 'law of attrition', advanced the mathematical study of longitudinal profiles, related bed slope to bed particle size and was a particularly clear statement of ideas on the graded condition. His synthesis shows plainly enough that the motion of rivers was no longer regarded as being either simple or mysterious; that all the features of a river course had a natural explanation which, if wished, could be translated into algebraic and geometrical terms.

Many students of landscape may not share this liking for a scientific approach to geomorphology and it is certain that the general public still prefers the artistic or more descriptive side. Yet all geomorphologists would do well to remember that by the 1870's a large storehouse of quantitative and geometric data already existed. Thereafter any propounder of theories on landscape evolution would be wise to consult and to accept the findings of physicists and hydraulic engineers.

CHAPTER TWENTY-FOUR

Chronological, Climatic and Comparative Aspects

The study of landscape evolution involves more than a detailed knowledge of present processes, though that alone would suffice to keep all geomorphologists busy for the next hundred years. The time-scale or actual history of the existing landscape is also of vital importance. By 1850 it was already beginning to be realized that in some regions this might involve several changes in process and that the changes might be structural and climatic. The utilization of present denudation rates to form a time-scale was soon found to be full of difficulties (more so than the depositional time-scale) and much of the denudational chronology inevitably became comparative. The relative position or height of landscape features assumed great significance once it was recognized that topography arose mainly from the incision of subaerial forces operating with reference to changing sea levels. There was, however, by now a chronology of the earth in the time of man and a chronology of the earth in pre-human times.

ARCHAEOLOGICAL INVESTIGATIONS AND THE AGE OF GRAVEL TERRACES

Before 1840 the interpretation of prehistory was hampered by the catastrophist ideas of the geologists, which were confined to Mosaic chronology (Daniel, 1950). Buckland's excavations in Goat's Hole Cave, Gower and at Burrington Coombe, Somerset, had yielded remains which he considered 'post-diluvian', and Father MacEnery's work in Kent's Cavern, Torquay, led him to the conclusion by 1829 that man and extinct animals had existed together prior to 4004 B.C. Yet the only really important work on the antiquity of man before 1840 was that leading to the recognition of the Stone, Bronze and Iron Ages by the Scandinavians, Thomsen, Worsaae and Nilsson. After 1840 geological advances on uniformitarian lines assisted the development of archaeology, and it is interesting that by the 1860's archaeological information was bringing home to geologists the recent operation of normal fluvial processes and providing them with a means of establishing chronologies for valley-cutting by rivers.

William Pengelly, a Devon schoolmaster, found flint tools beneath a sheet of stalagmite containing fossil animal remains at Windmill Hill cave near Brixham in 1858–9, and his work had a profound effect upon that of Lyell and Prestwich. Pengelly himself was strongly uniformitarian in outlook and, in a geological work published in 1864, held that the valleys of east Devon were excavated by streams and that the Triassic outliers had been isolated by denudation (*Pengelly, 1864, pp. 47 and 50*).

The most significant results leading to a better appreciation of fluvial action using archaeological methods, however, came from the study of the Somme valley near Abbeville. In 1838 Boucher de Perthes published the results of his excavations showing that artifacts were associated with extinct animal remains down to a depth of 40 feet in the gravel deposits (*Daniel, 1959*).

Prestwich, with the antiquary John Evans, visited de Perthes' site in 1859 and confirmed his work on the antiquity of man as did Lyell who saw the site in the same year. Prestwich (1862–63A, 1862–63B, 1864 and 1865) decided that the gravel deposits of the Somme were of local origin, that the upper terraces were the coarser and the older, and that they were normal features in the gradual downcutting of the river-valley. The existence of these gravels at high levels was due, he thought, to a severer climate at the time of their formation; the cold conditions of the Pleistocene era had produced a much greater runoff of surface water.

> The conditions, taken as a whole, are compatible only with the action of rivers flowing in the direction of the present rivers, and in operation before the existing valleys were excavated through the higher plains, of power and volume far greater than the present rivers, and dependent upon climatological causes distinct from those now prevailing in these latitudes. (*Prestwich, 1865, p. 286*)

As we have already noticed (p. 338), Lyell commented favourably on this work in the eleventh edition of his *Principles*, and, after virtually quoting from it, continued

> Lastly we arrive at the still older monuments of the Palaeolithic period properly so called, which consist chiefly of unpolished stone implements buried in ancient river-gravels and in the mud and stalagmite of caves. Both the gravel and the caves are now so situated in their relation to the present drainage and geography of the countries where they occur as to imply a great lapse of intervening time during which the erosive power of rivers has been active in deepening the valleys. The implements of this age in Western Europe are chiefly composed of chalk-flint – more rarely of chert from the greensand. Besides being unpolished they differ in shape from those of the Neolithic age. They are associated with remains of the mammoth, the woolly-haired rhinoceros, the hippopotamus, the musk-ox, and many other quadrupeds of extinct and living species. No pottery has been found strictly referable to this era, and there is an entire absence of metallic weapons, as in the later Bronze period of coins. (*Lyell, 1872, Vol. II, p. 566*)

Among several geologists who challenged Prestwich's main theme was A. Tylor who considered high- and low-level gravels were of one formation, closely connected in age and dating from just before historical times. He assumed the valley had been cut before any gravel was deposited and that there had been a 'pluvial period' during which the heavy rainfall had caused the rivers to rise to the level of the highest gravel terraces (Tylor, 1866 and 1868). A few years later Prestwich was beginning to accept the occurrence of a pluvial period but had no need to change his views on the age of terraces. What is more, Col. Greenwood (1867) rushed to his aid and wrote the following stringent comments on Tylor's articles:

> . . . Mr Tylor, while he considerately spares us a 'gravel period', creates a brand new period of his own – a pluvial period. With this implement (notwithstanding that 'a valley of Somme had assumed its present form prior to the deposition of any of the gravel or "loess" now to be seen there'), he floods the valley 'eighty feet above the present level of the Somme'. These prodigious bodies of water do not in the least erode the soft chalk sides, or the bed of the valley, but, on the contrary, they deposit the gravel terraces at their high-water mark. Flints, therefore, in the pluvial period, must have been lighter than water, and must have floated on the surface to their present position. In periods other than the pluvial one drift is driven along the beds of rivers and valleys. And these terraces of the Somme have been the beds of the river or valley, as I have had the honour to state in the *Geological Magazine* for May 1867. (*Greenwood, 1877, pp. 166–7*)

A work by C. E. de Rance (1874–6) shows further advance in the knowledge of terrace deposits, and of glacial and post-glacial conditions. It showed also that the factors that might have to be considered were numerous. From the elementary argument as to whether or not they had been formed by the sea or by rivers, the study of terraces had rapidly evolved into a highly specialized topic, which as knowledge of glaciology became more accurate began to fit in more easily with the remaining pattern of geomorphological knowledge.

> It has been shown that in the case of the river Ribble a wide and extensive valley has been cut out entirely since the deposition of the Upper Boulder Clay, that terraces fringe the sides, and that peat beds occur near the bottom of the lowest alluvium, which are continuous with extensive peat beds under-lying the sea level, and yet resting on post-Glacial marine beds, proving that the land was lower after the Glacial Period and the formation of the terraces than it is at present, and that afterwards it became higher, when neolithic man emigrated from more southern districts over continuous land. (*Rance, 1874, p. 252*)
> . . . there being the strongest reason to believe that the whole of these terraces, like those of the northern valleys, were formed since the deposition of the Glacial deposits . . . (*Rance, 1874–6, p. 252*)

DENUDATION CHRONOLOGY

Whereas the recent or post-Pliocene chronology was largely bound up with the study of gravel terraces and of prehistoric remains, the older chronology was concerned mainly with the detection of ancient rock-platforms, the development of drainage systems and the changes in erosional processes.

Hildreth and Ramsay had noted summit accordances and the latter, by associating each main summit-level with an ancient sea level, had laid the foundations of studies in denudational chronology. Even before the death of

Lyell, the idea that the landscape had emerged rapidly ready-made from beneath the sea was no longer as popular as Ramsay's theory that in the first place the land was raised slowly above the sea as a plain of marine denudation, and that upon this flat or gently-undulating surface subaerial forces proceeded to etch out the present topography. A. Lebour in 1869 follows this pattern exactly in an article on Brittany, concluding that:

1. With the exception of the central mountains of elevation, the north and south sides of western Brittany are made up of two plains of marine denudation.
2. These plains were last exposed to the waves in Upper Miocene times.
3. That the valleys are due to rivers and other subaerial agents. The dimensions of the valleys are inversely related to the time during which they were submerged and directly to the time during which they have been exposed to subaerial agents.

This is merely an intelligent application of Ramsay's principles to a particular area.

Early studies in the development of drainage systems will always be associated with Jukes and Ramsay. During the later years of his life Ramsay turned his attention to the elucidation of the problems of river chronology. In the case of a number of individual rivers he attempted to piece together the various phases in the history of each, working back from the evidence available, in order to determine how far external causes had influenced the river's development. Geikie in his memoir of Ramsay, writes at length of this research.

During the last ten years of his official life the physical history of river-valleys exercised a peculiar fascination on Sir Andrew's mind. The subject had for many years engaged his attention, but not until the appearance in 1862 of his friend Jukes's remarkable memoir on the river-valleys of the south of Ireland did he realize how the problem might be satisfactorily attacked. He was led to the conclusion that the denudation of the Weald had been effected by the subaerial waste, and that the cause of the flow of the rivers, from the central low tract through the encircling rim of chalk downs, was to be sought in the ancient topography of the region, when the streams descended from the central, still unremoved dome or ridge of chalk. Extending this process of reasoning, he afterwards discussed the main causes whereby the rivers of England had been led to flow in the courses they now follow. There was undoubtedly a good deal of speculation in this discussion, but his treatment of the subject was full of suggestiveness, and pointed out the direction in which, with perhaps a larger array of facts, the question might eventually be solved.

Subsequently he attacked the history of individual rivers, working it out in more detail along the same lines as he had already followed. In this manner he traced the successive stages which, in his opinion, had led to the excavation of the present valley of the Rhine, showing that in Miocene time the flow of the drainage between the Black Forest and the Vosges had been from the north to south, or towards the great hollow lying to the north of the Alps, that subsequent disturbance and elevation of the Alpine chain tilted the ground in such a manner that the drainage was reversed, and the streams from the tract of the Alps were collected into a river which found its way northward, and gradually excavated the valley and gorge in which the present Rhine still flows. Though it cannot be demonstrated that such have been the successive stages in the history of the course of this river, the available evidence makes Ramsay's explanation highly probable. (*Geikie, 1895, pp. 359–60*)

An article in 1872 showed the typical results of these researches. It dealt with the geologic causes that determined the ages and directions of the main rivers of England and Wales, and described how there had been a great Cretaceous overlap from east to west on top of the tilted New Red Sandstone and younger rocks, and how this lay against the Carboniferous and older uplands of Wales; how Alpine movements later gave a west and north-west tilt to the whole area, forming a gentle slope towards the Welsh uplands; how the Severn was formed between these two elements with the Avon forming later as a tributary; how a new disturbance again tilted the whole area this time eastwards, the runoff collecting and forming into the Thames which flowed east over the Cretaceous cover; and finally how the chalk escarpment was worn back eastward, and the edges of the lower tilted strata became etched out into valleys and further escarpments. The Humber, he concluded, was formed in the same way and at the same time as the Thames. As he says himself the process of river evolution has not finished.

> The escarpments both of Chalk and Oolite are still slowly changing and receding eastwards, and as that of the Oolite recedes, the area of drainage will diminish and the Thames decrease in volume. (*Ramsay, 1872, p. 153*)

There is no need for us to emphasize the very 'modern' nature of Ramsay's attempts (1872, 1873–5, 1874 and 1876) at denudational chronology.

The recognition of frequent changes in the dominant erosional processes of certain areas grew up at the same time as the recognition of alternating structural influences. The complexity of the erosional history of countries in so-called temperate latitudes was carefully expressed by many authors. James Geikie, an uncommitted member of the younger school of geologists, in an article on post-glacial denudation in Scotland (1868), showed clearly how glacial studies were quite indivisible from studies of general erosion. When describing erosion, he coupled together as agents of equal rank, weathering, streams and the sea, but he also hastened to draw proper attention to the influence of structure or rock-type. His most significant conclusion was that the topography of Scotland had been produced before the Ice Age and that glacial action had only modified the original outline; in post-glacial time subaerial erosion had caused considerable inroads in deposits of boulder clay but had made relatively little impression on the vast body of solid rock.

Another writer who expressed clearly the multi-genetic nature of old landscapes was J. C. Ward (1870) who in his explanation of the denudation of the Lake District considered that the vast amount of erosion performed could

FIG. 87. *James Geikie*
(*Drawn and etched by William Hole, 1884. National Galleries of Scotland*)

only have been accomplished with the aid of subaerial forces. He stresses the length of time involved and the power of rivers in flood and even refers to past changes in climate which have accelerated pluvial processes. He concluded that the Lake District landscape was a product of various forces: marine planation in the Carboniferous Period, glacial erosion during the Permian and Pleistocene epochs and subaerial erosion at all other times.

MID-VICTORIAN IDEAS ON CHANGES OF CLIMATE AND SEA LEVEL

The general, or almost general, acceptance of the occurrence of Ice Ages and of changes in sea level dates back to about 1870. Soon a few geologists were correlating oscillations in sea level with fluctuations in climate and a start had been made on a complex inter-relationship which still baffles geologists.

Croll, writing in 1868, made interesting observations on the rate of land erosion, the result of which forced him to change the date he had assigned to the occurrence of the glacial period. Quoting Humphreys and Abbot, he assumed that

> ... the amount removed is equal to one foot of rock in about 6,000 years. The average height of the North American continent above the sea-level, ... is 748 feet; consequently, at the present rate of denudation, the whole area of drainage will be brought down to the sea-level in less than 4,500,000 years if no elevation of the land takes place. (*Croll, 1868, p. 379*)

Yet if these figures were correct, then he was forced to admit that his original estimate of the end of the glacial period occurring 700,000 years ago could not possibly be sustained.

> It would therefore follow that the general features of the country must now be totally different from what they were at the close of the glacial epoch, a conclusion which we know from geological evidence is incorrect. (*Croll, 1868, pp. 366–7*)

If the effects of the glacial period were not really understood, the evidence for the existence of such a period was still marked, indeed to such an extent that it was obvious that subaerial erosion had had little time since in which to act upon the surface and to obliterate the rough glacial markings. Croll, therefore, changes his date to a mere 80,000 years ago. Throughout his account he stresses the rapidity with which streams erode, and gives it as his opinion that in the past geologists have allowed too much time to geological periods. His work not only strongly supported the fluvial theory but also was among the first to attempt to determine a precise geological chronology. Hitherto various dates and lengths had been assigned to the Ice Age and most geologists did not consider time at all. Croll, in another article (1869), showed a keen appreciation of the relationship between glacial and fluvial erosion when he described two river-channels buried under drift. These, he suggested, were originally excavated by rivers during pre- and inter-glacial times.

This aspect of climatic change was touched upon by Tylor and others. Tylor, as we have already mentioned, believed in a 'pluvial period' when enormous rivers filled their valleys. In 1869, he pursued this idea in relation to the building of deltas. He had found that the deeper deltaic deposits were coarser and therefore concluded that they were laid down by rivers running more rapidly in a pluvial epoch. As he saw it, during the glacial period the sea level fell about 600 feet and left as evidence the subaerial features, river channels and littoral conditions found in the sea down to about the same depth as the coarser deltaic deposits. The lowering of the sea level was in his scheme coincident with the Pluvial Period, which raised the levels of the rivers to a height about eighty feet in excess of normal limits. This idea was important because it represented a major contribution to Agassiz's original glacial theory.

Tylor amplified these conclusions in a further article on the phenomena of the glacial period (1872). In this, he drew support from recent works on sea-level fluctuations and described the deltas of the Po, Mississippi, Ganges and Volga which demonstrated this fluctuation. He repeated his reasons for belief in a pluvial period, which he said he had derived from his study of the Somme.

> No one can see the great valley of the Somme . . . without being convinced that in the Quaternary Period these wide and deep valleys, excavated out of solid chalk, were filled by large rivers (*Tylor, 1872, p. 395*)

The inter-connexion between changes of sea level and river incision was cleverly outlined by T. M. Reade in a study of the deposits and organic remains in Lancashire and Cheshire (1871–2). He discussed the oscillation of land level during the post-glacial period, and how evidence for this could be found in the peat and marine deposits. He also referred to the amount of river-bed excavation which took place during this period of changes in sea level. In Reade's article there was a simultaneous employment of botany, palaeontology, stratigraphy, glaciology and fluvial erosion. None of these subjects was new but as knowledge increased their relationship was becoming far more intimate.

COMPARATIVE LANDFORMS AND RELATIVE STAGES

The faint beginnings of two other aspects of landscape study can be discerned in the late 1860's. The first was the idea of comparing the nature of similar landforms throughout the world; the second, was to express the relation between a landform and its possible age or its stage of development.

The early classification of surface features and the comparison of their morphology is especially connected with Oscar Peschel (1867 and 1870).

Peschel's aim throughout was to discover a satisfactory causal classification of the earth's relief features. He believed each form was part of or a stage in a progression of forms and that if one were able to classify the forms one would be able to understand the sequence of the progression. He was inspired with this view by the successful morphological studies of Cuvier in anatomy and of Bopp in philology, whose method of comparing the similarities of particular objects he introduced into physical geography under the name of *vergleichende Erdkunde*, a term coined by Karl Ritter for use in anthropo-geography. Thus Peschel's scheme was a natural outcome of the contemporary scientific background.

His main idea is well expressed in the following quotation

> Let us keep in mind our conjecture that landforms are not produced by chance, but that, on the contrary, even the slightest segmentation of outline and elevation, and all striving of the surface sideways or upwards has a hidden meaning which we ought to ascertain. To do so we must only look for the similarities in nature as they are presented by the cartographer. After having determined a larger number of such similarities, we usually come to under-

stand their mode of origin by studying their distribution. (*Peschel, 1870, pp. 4–5*)

Whenever Peschel has to deal with a landform problem, he begins by comparing as many as possible of the phenomena in question, and then proceeds to inquire after their mode of origin. Thus when discussing coastal morphology he writes:

> Here, the first problem of comparative geography is to find such coastal stretches as were evidently raised or submerged; the second is to look for some general or, at least, frequent signs of this process, so that a quick eye might detect the responsible powers even on the strength of the visible forms alone. Thereby a map becomes an historical picture showing by the mere outlines of all the fixed expanses the spectacle of that silent struggle between the once victorious, once defeated natural forces. (*Peschel, 1870, pp. 90–91*)

However, Peschel's method of comparative morphology was, as we have already noticed, not confined to the explanation of the origin of landforms; it was also a means to distinguish between young, intermediate and old-age stages in their formation. Whenever such a division seemed impossible from a comparison of the forms of the surface alone, then other features of the surface, particularly animal and plant life, were also considered. Thus old islands in the open sea will show a greater wealth of species than young islands.

> . . . if there are old islands of Tertiary origin, there will be also plants on them of this very period; for whatever plants were floated or flown to these islands, they will not only have found a new home, but also have been protected from the adversities . . . which on the continents came to destroy their co-species one after the other. (*Peschel, 1870, p. 50*)

In some of his studies Peschel was wrong, as for example in his belief that submarine relief was relatively flat in all parts of the ocean, but these errors were common to his time. On the whole his work was a successful attempt to employ contemporary methods of science in the field of physical geography. As Tatham rightly observed, the principle involved fitted in perfectly with the line of thought that Davis so successfully explored.

> Peschel's genetic classification of relief types provided the concept from which the formulation of geomorphology could and did proceed. It started the development that culminated in the work of W. M. Davis and the recognition of the Cycle of Erosion. (*Tatham, 1951, p. 62*)

Yet one ought not to go as far as Tatham; for this certainly was not the most fundamental work that influenced Davis and we doubt whether it influenced him at all.

The idea of expressing a relationship between a landform and its age or stage of development was also conceived in a more localized way by Ludwig Rutimeyer (1825–95), the Professor of Zoology at Basle. In a book dealing with valley and lake formation (*Ueber Thal- und Seebildung, 1869*) he attempted to show that the largest Alpine valleys had been produced by stream erosion acting over long periods of geologic time. This was of course the normal belief of the fluvial analysts. But Rutimeyer did not finish there. He suggested that, though the erosion began with the first uplift of the mountains, it varied in intensity both in time and throughout the river's course, the area of greatest erosion working backwards from the foot of the mountains towards the higher levels; consequently, he said, the different sections of a river's course might be marked by distinct types of erosional forms, such as water-falls, lakes, meanders, or flood plains (Rutimeyer, 1869; Von Zittel, 1901). This, as the modern geologist knows, was in line with Davis's theory of progressive headward 'grading'. Hutton and Greenwood had said that subaerial forces produced a sequence of erosion by wearing down the mountain till it became a plain but they had not carried their observation further. They had not discussed the possibility that each different topographic form could be an intermediate stage in such a total denudation and that a succession of stages could be ascertained, which was what, in effect, Rutimeyer said.

1846–75 IN RETROSPECT

The period just discussed coincided with the later years of Lyell but we hope we have given sufficient proof that it was not merely an age of Lyellian dictatorship. It was, in reality, a period of mass discussion. Geomorphological ideas were still in a muddled state, particularly those connected with glaciation, but many of the older fallacies and simple theories had gone almost beyond recall. The scope of landscape study had advanced to the state where complexity was becoming the norm. In modern jargon, landscape studies were rapidly assuming a polycyclic and multi-genetic basis. Weathering, soil formation, stream mechanics, and glacial features were all beginning to be considered as major topics. The fluvial theory now formed an accepted part of geomorphology. In addition new themes such as the classification of landforms and landscape geometry were appearing as promising fields of future research. The evolution of landscape now attracted many more scientists than artists; the Leonardos and Ruskins had given way to hydrologists and chemists; mining experts were less important than hydraulic engineers. Where modern geomorphology has advanced, it has not been in contradiction of this period but by the expansion and justification of its ideas. There were many lines of possible advance but, as in the past, the future trends will be much influenced by the expanding geographical horizon – the impact of the surveying of the tremendous topography of the Western United States has yet to be recorded.

References: Part Three

ANON. (1869) Review of D. Mackintosh's 'The Scenery of England and Wales: its character and origin' Geol. Mag., Vol. 6, pp. 465–70.

— (1872) Review of J. Phillips 'Geology of Oxford, 1871' Geol. Mag., Vol. 9, pp. 80–85.

— (1875) Review of 'Valleys and their relation to fissures, fractures, and faults; by G. H. Kinahan', Geol. Mag., N.S., 2 Dec., Vol. 2, pp. 131–2.

ANSTED, D. T. (1871) 'On some phenomena of the weathering of rocks, illustrating the nature and extent of sub-aerial denudation' Trans. Camb. Phil. Soc., Vol. 11, Pt. 2, pp. 387–95.

ARGYLL, DUKE OF (1868) 'On the physical geography of Argyllshire in connexion with its geological structure' Quart. Jour. Geol. Soc., Vol. 24, pp. 255–73.

AUSTEN, R. A. C. (1850) 'On the valley of the English Channel' Quart. Jour. Geol. Soc., Vol. 6, pp. 69–97.

— (1851) 'On the superficial accumulations of the coasts of the English Channel, and the changes they indicate' Quart. Jour. Geol. Soc., Vol. 7, pp. 118–36.

AVELINE, W. TALBOT (1866) 'The Longmynd and its valleys' Geol. Mag., Vol. 3, p. 279.

BACHE, A. D. (1851) 'Annual Report Supt. of the Coast Survey', 32 Cong., 1st Sess., H. Doc. 26, pp. 74–78.

BAILEY, E. (1952) 'Geological Survey of Great Britain' (Murby, London), 278 pp.

BAKEWELL, R. (1847) 'Observations on the Whirlpool, and on the Rapids before the Falls of Niagara' Amer. Jour. Sci., 2nd Ser., Vol. 4, pp. 25–36.

— (1857) 'Observations on the Falls of Niagara, with reference to the changes which have taken place, and are now in progress' Amer. Jour. Sci., 2nd Ser., Vol. 23, pp. 85–95.

BAULIG, H. (1926) 'La notion de profil d'équilibre; histoire et critique' Congrès Internat. Géog. (1925), C.R., Vol. 3, pp. 51–63.

BAYLEY, G. W. R. (1852) Review of Charles Ellet's 'Report on the overflow of the delta of the Mississippi' De Bow's Review, Vol. 13, pp. 166–76.

BEARDMORE, N. (1851) 'Manual of Hydrology' (London). (Notes from edition of 1862.)

BEVAN, G. P. (1860) 'On the South Wales coal-field' Geologist, Vol. 3, pp. 90–99.

BISCHOF, G. (1854–9) 'Elements of Chemical and Physical Geology' (Trans. Cavendish Soc., London), 3 vols.

BLANFORD, H. F. (1889) 'A Practical Guide to the Climate and Weather of India, Ceylon and Burmah' (Macmillan, London), 369 pp.

BLANFORD, W. T. (1870) 'Observations on the Geology and Zoology of Abyssinia' (Macmillan, London), 487 pp.

BLODGET, L. (1857) 'Climatology of the United States and of the Temperate Latitudes of the North American Continent' (Philadelphia), 536 pp.

BRINGIER, L. (1821) 'On the region of the Mississippi' *Amer. Jour. Sci.*, Vol. 3, pp. 15–46.

BROWN, A. and DICKESON, M. W. (1848) 'The sediment of the Mississippi river' *Proc. Amer. Assn. Adv. Sci.*, Vol. 1, pp. 42–54.

BROWN, T. C. (1867) 'Denudation of valleys' *Geol. Mag.*, Vol. 4, pp. 139–140.

BROWNE, C. A. (1871) 'J. B. Jukes: Letters and Extracts from Occasional Writings' (London), 596 pp.

CAMPBELL, J. F. (1865) 'Frost and Fire' (Edinburgh), 2 vols.

CAMPBELL-IRONS, J. (1896) 'Autobiographical Sketch of James Croll' (London), 553 pp.

CARROLL, W. J. (1866) 'New theory in that branch of natural science known as river hydraulics' *Natchez Democrat*, Feb. and March.

CATE, A. (1959) 'J. Peter Lesley, a biographical sketch' *Geotimes*, Vol. 4, No. 4, pp. 18–19 and 45.

CHAMBERS, R. (1848) 'Ancient sea margins, as memorials of changes in the relative level of sea and land' (Edinburgh), 338 pp.

— (1848–9) 'Geological notes on the valleys of the Rhine and Rhone' *Proc. Roy. Soc. Edin.*, Vol. 2, pp. 189–90.

— (1849–50) 'Personal observations on terraces, and other proofs of changes in the relative level of sea and land in Scandinavia' *Proc. Roy. Soc. Edin.*, Vol. 2, pp. 247–8.

COGHLAN, T. A. (1883–4) 'Discharge of streams in relation to rainfall, New South Wales' *Min. and Proc. Inst. Civil Engineers*, Vol. 75, pp. 176–94.

CONYBEARE, H. (1857–8) 'Description of the works, recently executed, for the water supply of Bombay, in the East Indies' *Min. and Proc. Inst. Civil Engineers*, Vol. 17, pp. 555–68.

CROLL, J. (1867) 'On the excentricity of the earth's orbit and its physical relations to the Glacial Epoch' *Phil. Mag.*, 4th Ser., Vol. 33, pp. 119–31.

— (1868) 'On geological time, and the possible date of the Glacial and the Upper Miocene Period' *Phil. Mag.*, 4th Ser., Vol. 35, pp. 365–84; Vol. 36, pp. 141–54 and 362–86.

— (1869) 'On two river channels buried under drift' *Trans. Edin. Geol. Soc.*, Vol. 1, pp. 330–45.

CUNIT, C. (1855) 'Études Sur Les Cours D'Eau à Fond Mobile' (Grenoble).

DANA, J. D. (1849A) In 'United States Exploring Expedition, during the years 1838, 1839, 1840, 1842 and 1843, under the command of Charles Wilkes', Vol. 10, pp. 380–90 (Philadelphia).

— (1849B) Review of Chambers' 'Ancient sea margins . . .' *Amer. Jour. Sci.*, 2nd. Ser., Vol. 7, pp. 1–14 and Vol. 8, pp. 86–9.

— (1850A) 'On denudation in the Pacific' *Amer. Jour. Sci.*, 2nd Ser., Vol. 9, pp. 48–62.

— (1850B) 'On the degradation of the rocks of New South Wales and formation of valleys' *Amer. Jour. Sci.*, 2nd Ser., Vol. 9, pp. 289–94.

— (1856) 'On American geological history' *Amer. Jour. Sci.*, 2nd Ser., Vol. 22, pp. 305–34.

— (1863) 'Manual of Geology', 1st Edn (Philadelphia), 798 pp.

DANIEL, G. E. (1950) 'A Hundred Years of Archaeology' (Duckworth, London), 344 pp.

— (1959) 'The idea of man's antiquity' *Sci. Amer.*, Vol. 210, No. 5, pp. 167–76.

DARCY, H. (1856) 'Les Fontanes Publiques de la Ville de Dijon' (Paris).

DARCY, H. and BAZIN, H. (1865) 'Recherches Hydrauliques' (Paris).

DARWIN, C. (1848) 'On the transportal of erratic boulders from a lower to a higher level' *Quart. Jour. Geol. Soc.*, Vol. 4, pp. 315–23.

DAUBENY, C. (1866) 'On the origin of valleys' *Geol. Mag.*, Vol. 3, pp. 278–9.

DAUSSE, M. (1857) 'Note sur un principe important et nouveau d'Hydrologie' *Acad. Sci., C.R.*, Vol. 44, pp. 756–66.

— (1872) 'Études relatives aux inondations et à l'endiguement des rivières' *Inst. de France; mémoires présentés par divers savants a l'Académie des Sciences*, Vol. 20, pp. 287–507.

DAVIES, G. L. (1962) 'Joseph Beete Jukes and the rivers of Southern Ireland' *Irish Geography*, Vol. 4, pp. 221–33.

DAVIS, W. M. (1896) 'Plains of marine and sub-aerial denudation' *Bull. Geol. Soc. Amer.*, Vol. 7, pp. 377–98.

— (1915) 'Biographical memoir of Peter Lesley, 1819–1903' *Nat. Acad. of Sci. Washington, Biog. Memoirs*, Vol. 8, pp. 153–240.

DE LA BECHE, H. T. (1839) 'Report on the Geology of Cornwall, Devon, and West Somerset' (H.M.S.O. London), 648 pp.

ELLET, C. (1851) 'On the alluvial valley of the Mississippi', Rept. to the Bureau of Topographic Engineers, War Dept. Washington.

EYTON, MISS (1867) 'On the glacio-marine denudation of certain districts' *Geol. Mag.*, Vol. 4, pp. 545–9.

FERGUSSON, J. (1863) 'On recent changes in the delta of the Ganges' *Quart. Jour. Geol. Soc.*, Vol. 19, pp. 321–54.

FISHER, O. (1861) 'On the denudation of soft strata' *Quart. Jour. Geol. Soc.*, Vol. 17, pp. 1–4.

— (1866A) 'On the probable glacial origin of certain phenomena of denudation' *Geol. Mag.*, Vol. 3, pp. 483–7.

— (1866B) 'On the disintegration of a chalk cliff' *Geol. Mag.*, Vol. 3, pp. 354–6.

— (1868A) 'Denudation, and its agents' *Geol. Mag.*, Vol. 5, pp. 34–6.

— (1868B) 'On the denudations of Norfolk' *Geol. Mag.*, Vol. 5, pp. 544–58.

— (1869) 'On denudation, and the crags' *Geol. Mag.*, Vol. 6, pp. 141–3.

— (1871) 'On phenomena connected with denudation observed in the so-called Coprolite pits near Haslingfield, Cambridgeshire' *Geol. Mag.*, Vol. 8, pp. 65–71.

FORBES, J. D. (1856–7) 'On the abrading power of water at different velocities' *Proc. Roy. Soc. Edin.*, Vol. 3, pp. 474–5.

FORREY, S. (1842) 'The Climate of the United States and its Endemic Influences'.

FORSHEY, C. G. (1878) 'The Physics of the Gulf of Mexico and its Chief Affluent, the Mississippi River' (Salem), 42 pp.

FOSTER, C. LE NEVE and TOPLEY, W. (1865) 'On the superficial deposits of the valley of the Medway, with remarks on the denudation of the Weald' *Quart. Jour. Geol. Soc.*, Vol. 21, pp. 443–74.

FRISI, P. (1764) 'Nature of Torrents' (trans. by J. Garstin, 1861).

GEIKIE, A. (1865) 'The Scenery of Scotland' 1st Edn (Macmillan, London), 360 pp.

— (1868) 'On denudation now in progress' *Geol. Mag.*, Vol. 5, pp. 249–54. (See also *Trans. Geol. Soc. Glasgow*, Vol. 3, 1868, p. 164.)

— (1873) 'Earth sculpture and the Huttonian School of Geology' *Trans. Edin. Geol. Soc.*, Vol. 2, 247–67.

— (1895) 'Memoir of Sir Andrew Crombie Ramsay' (Macmillan, London), 397 pp.

GEIKIE, J. (1868) 'On denudation in Scotland since glacial times' *Geol. Mag.*, Vol. 5, pp. 19–25.

GILMAN, D. C. (1899) 'Life of J. D. Dana' (Harper's, New York).

GODWIN AUSTEN, R. (1855) 'On land surfaces beneath the drift gravel' *Quart. Jour. Geol. Soc.*, Vol. 11, pp. 112–19.

GREEN, A. H. (1868) 'Sea-cliffs and escarpments' *Geol. Mag.*, Vol. 5, pp. 40–42.

GREENWOOD, G. (1844) 'The Tree Lifter . . .' (2nd Edn, 1853).

— (1857) 'Rain and Rivers', 1st Edn (London), (2nd Edn, 1866, 237 pp., 3rd Edn, 1876, 247 pp).

— (1867) 'Valley terraces' *Geol. Mag.*, Vol. 4, pp. 205–10.

— (1868) 'Denudation of the Weald' *Geol Mag.*, Vol. 5, pp. 37–39.

— (1869) 'Denudations of Norfolk' *Geol. Mag.*, Vol. 6, pp. 45–47.

— (1871) 'Terraces of Norway', *Geol. Mag.*, Vol. 8, pp. 190–1, 430–2, 475–6 and 574.

— (1875) 'Denudation of the Weald' *Geol. Mag.*, N.S., Dec. 2, Vol. 2, pp. 282–3.

— (1877) 'River Terraces' (London), 247 pp.

GREGORY, H. E. (1918) 'Steps of progress in the interpretation of land forms' *Amer. Jour. Sci.*, 4th Ser., Vol. 46, pp. 104–32.

GUNN, W. (1876) 'Subaerial erosion versus glacial erosion' *Geol. Mag.*, N.S., Dec. 2, Vol. 3, pp. 97–105.

HAGEN, G. (1839) 'Ueber die Bewegung des Wassers in engen cylindrischen Rohen' *Poggendorff Annalen*, Vol. 46, pp. 423–42.

HASWELL, J. (1864) 'On the denudation of Arthur's Seat' *Geologist*, Vol. 7, pp. 93–99.

HIMMELFARB, G. (1959) 'Darwin and the Darwinian Revolution' (Chatto and Windus, London), 422 pp.

HITCHCOCK, E. (1849) 'On the river terraces of the Connecticut valley, and on the erosions of the earth's surface' *Proc. Amer. Ass. Adv. Sci.*, pp. 148–56.

HOPKINS, W. (1848) 'On the elevation and denudation of the district of the lakes of Cumberland and Westmorland' *Quart. Jour. Geol. Soc.*, Vol. 4, pp. 70–98.

— (1849) 'On the transport of erratic blocks' *Trans. Camb. Phil. Soc.*, Vol. 8, Pt. 2, pp. 220–40.

— (1852) 'Annual Address' *Quart. Jour. Geol. Soc.*, Vol. 8, pp. xxi–lxxx.

HORNER, L. (1861) 'Anniversary Address' *Quart. Jour. Geol. Soc.*, Vol. 17, pp. xxxi–lxxii.

HUBBARD, O. P. (1850) 'The condition of trap dikes in New Hampshire, an evidence and measure of erosion' *Amer. Jour. Sci.*, 2nd Ser., Vol. 9, pp. 158–71.

HULL, E. (1855) 'On the physical geography and Pleistocene phenomena of the Cotteswold Hills' *Quart. Jour. Geol. Soc.*, Vol. 11, pp. 477–94.

— (1866A) 'Denudation of valleys' *Geol. Mag.*, Vol. 3, pp. 474–7.

— (1866B) 'Modern views of denudation' *Pop. Sci. Monthly*, Vol. 5, pp. 453–61.

HUMPHREYS, A. A. and ABBOT, H. L. (1861) 'Report on the Physics and Hydraulics of the Mississippi River' Corps of Topographic Engineers, Prof. Paper No. 4 (Govt. Printing Office, Washington), 691 pp.

JACKSON, L. D'A. (1875) 'Hydraulic Manual' 3rd Edn (Allen, London), 2 parts.

JAMIESON, T. F. (1860) 'On the drift and rolled gravel of the North of Scotland' *Quart. Jour. Geol. Soc.*, Vol. 16, pp. 347–71.

— (1863) 'On the parallel roads of Glen Roy' *Quart. Jour. Geol. Soc.*, Vol. 19, pp. 235–59.

JENKINS, H. M. (1870) 'On the surface geology, denudation, and "form of the ground" of Belgium' *Geol. Mag.*, Vol. 7, pp. 199–203.

JONES, D. (1871) 'Denudation of Coalbrook-dale coal-field' *Geol. Mag.*, Vol. 8, pp. 201–8.

JONES, R. (1871) 'On the geology of the Kingsclere valley' *Geol. Mag.*, Vol. 8, pp. 511–15.

JONES, T. R. (1859) 'Note on some granite-tors' *Geologist*, Vol. 2, pp. 301–12.

JUDD, J. W. (1911) 'The Coming of Evolution' (Cambridge), 171 pp.

JUKES, J. B. (1862A) 'The Students' Manual of Geology' 2nd Edn (Black, Edinburgh), 764 pp. (1st Edn, 1857).

— (1862B) 'On the mode of formation of some of the river-valleys in the South of Ireland' *Quart. Jour. Geol. Soc.*, Vol. 18, pp. 378–403.

— (1866A) 'Atmospheric vs. marine denudation' *Geol. Mag.*, Vol. 3, pp. 232–5.

— (1866B) 'Reply to Mr G. Poulett Scrope's article' *Geol. Mag.*, Vol. 3, pp. 331–3.

— (1867) 'On the gorge of the Avon; at Clifton' *Geol. Mag.*, Vol. 4, pp. 444–7.

— MISS. (1871) 'Letters and extracts from the addresses and occasional writings of J. B. Jukes' (London).

KESSELI, J. E. (1941) 'The concept of the graded river' *Jour. Geol.*, Vol. 49, pp. 561–88.

KINAHAN, G. H. (1866) 'Ancient sea margins in the counties Clare and Galway' *Geol. Mag.*, Vol. 3, pp. 337–43.

— (1869A) 'Notes on the growth of soil' *Geol. Mag.*, Vol. 6, pp. 263–8 and 348–51.

— (1869B) 'Suggestions about denudation' *Geol. Mag.*, Vol. 6, pp. 109–15.

— (1870) 'Notes on the features of Devon, Cornwall, and Galway' *Geol. Mag.*, Vol. 7, pp. 310–14.

— (1875) 'Valleys and their relation to fissures, fractures, and faults' (Trubner, London), 240 pp.

KINGSMILL, W. (1864) 'Notes on the geology of the east coast of China' *Jour. Geol. Soc. Dublin*, Vol. 10, pp. 1–6.

LAGANNE, A. (1868) 'Note sur les érosions des calcaires denudés de la vallée de la Vézère et ses affluents' *Geol. Mag.*, Vol. 5, p. 371.

LEBOUR, G. A. (1869) 'On the denudation of western Brittany' *Geol. Mag.*, Vol. 6, pp. 442–6.

LESLEY, J. P. (1856) 'Manual of Coal and its Topography' (Philadelphia), 224 pp. (See especially pp. 121–87.)

— (1862) 'Observations on the Appalachian region of southern Virginia' *Amer. Jour. Sci.*, 2nd Ser., Vol. 34, pp. 413–15.

— (1863) 'A remarkable coal mine or asphalt vein in the coal measures of Wood Co., W. Va.' *Trans. Amer. Phil. Soc.*, Vol. 9, pp. 183–97.

— (1866) 'Notes on a map intended to illustrate Five Types of Earth-surface in the United States' *Trans. Amer. Phil. Soc.*, N.S., Vol. 13, pp. 305–12.

LOGIN, T. (1856–57) 'On the delta of the Irrawaddy' *Proc. Roy. Soc. Edin.*, Vol. 3, pp. 471–4.

— (1872) 'On the most recent geological changes of the rivers and plains of northern India' *Quart. Jour. Geol. Soc.*, Vol. 28, pp. 186–200.

LUCY, W. C. (1868) 'Denudation now in progress' *Geol. Mag.*, Vol. 5, pp. 343–4.

LYELL, C. (1847) 'On the delta and alluvial deposits of the Mississippi' *Amer. Jour. Sci.*, 2nd Ser., Vol. 3, pp. 34–39.

— (1850) 'On craters of denudation, with observations on the structure and growth of volcanic cones' *Quart. Jour. Geol. Soc.*, Vol. 6, pp. 207–34.

— (1872) 'Principles of Geology' 11th Edn (Murray, London), 2 vols.

MACKIE, S. J. (1860) 'Geological localities—No. 1 Folkestone' *Geologist*, Vol. 3, pp. 201–7.

MACKINTOSH, D. (1865A) 'Marine denudation illustrated by the Brimham rocks' *Geol. Mag.*, Vol. 2, pp. 154–8.
— (1865B) 'A tourist's notes on the surface-geology of the Lake-district' *Geol. Mag.*, Vol. 2, pp. 299–306.
— (1866A) 'The sea against rain and frost; or the origin of escarpments' *Geol. Mag.*, Vol. 3, pp. 63–70.
— (1866B) 'The sea against rivers; or the origin of valleys' *Geol. Mag.*, Vol. 3, pp. 155–60.
— (1866C) 'Origin of valleys' *Geol. Mag.*, Vol. 3, p. 235.
— (1866D) 'Denudation.—Reply to Mr G. Poulett Scrope and Mr J. B. Jukes' *Geol. Mag.*, Vol. 3, pp. 280–2.
— (1866E) 'The denudation controversy' *Geol. Mag.*, Vol. 5, p. 334.
— (1866F) 'The sea against the plough.—Reply to Mr G. Poulett Scrope' *Geol. Mag.*, Vol. 3, pp. 381–2.
— (1866G) 'Results of observations on the cliffs, gorges and valleys of Wales' *Geol. Mag.*, Vol. 3, pp. 387–98.
— (1867) 'Railway geology. No. 1—From Exeter to Newton-Bushnell and Mortonhampstead' *Geol. Mag.*, Vol. 4, pp. 390–401.
— (1869) 'The Scenery of England and Wales, its character and origin' (London), 292 pp.
— (1870) 'Terraces on inland slopes; Reply to G. Poulett Scrope, Esq.' *Geol. Mag.*, Vol. 7, pp. 25–26.
MARTIN, P. J. (1840–1) 'On the relative connection of the eastern and western chalk denudations' *Quart. Jour. Geol. Soc.*, Vol. 3, pp. 349–51.
— (1851–4–6–7) 'On the anticlinal line of the London and Hampshire Basins' *Phil. Mag.*, 4th Ser., Vol. 2, pp. 41–51, 126–34, 189–98, 278–88, 366–75 and 471–7; Vol. 7, pp. 166–71; Vol. 12, pp. 447–52; Vol. 13, pp. 33–39 and 109–16.
MATHER, K. F. and MASON, S. L. (1939) 'A Source Book in Geology' (McGraw-Hill, N.Y.), 702 pp.
MAW, G. (1866A) 'On watersheds' *Geol. Mag.*, Vol. 3, pp. 344–8.
— (1866B) 'Notes on the comparative structure of surfaces produced by sub-aerial and marine denudation' *Geol. Mag.*, Vol. 3, pp. 439–51.
MEDLICOTT, H. B. (1860) 'On the geological structure and relations of the southern portion of the Himalayan range between the rivers Ganges and Ravee' *Mem. Geol. Surv. India*, Vol. 3, Art. 4, 206 pp.
MERRILL, G. P. (1924) 'The first One Hundred years of American Geology' (Yale), 773 pp.
MILLER, J. F. (1849) 'On the meteorology of the Lake District of Cumberland and Westmorland' *Phil. Trans. Roy. Soc. London*, Pt. 2.
MONTAIGU, R. (1861) 'Project of a ship canal between the Mississippi River and the Gulf of Mexico' (Lamarre, New Orleans).
MURCHISON, R. I. (1850) 'On the distribution of the superficial detritus of the Alps, as compared with that of Northern Europe' *Quart. Jour. Geol. Soc.*, Vol. 6, pp. 65–69.
— (1851) 'On the distribution of the flint drift of the south-east of England' *Quart. Jour. Geol. Soc.*, Vol. 7, pp. 349–98.
— (1863–4) 'Address to the Royal Geographical Society' *Proc. Roy. Geog. Soc.*, Vol. 8, pp. 170–254.
— (1867) 'Siluria' 4th Edn (John Murray, London), 566 pp.
NICHOLSON, H. A. (1872) 'Notes on some valleys of erosion' *Geol. Mag.*, Vol. 9, pp. 318–21.

O'CONNELL, P. P. L. (1867–8) 'On the relation of the fresh-water floods of rivers to the areas and physical features of their basins' *Min. and Proc. Inst. Civil Engineers*, Vol. 27, pp. 204–14.

OLDHAM, T. (1859) 'On the geological structure of a portion of the Khasi Hills, Bengal' *Mem. Geol. Surv. India*, Vol. 1, pp. 99–210.

PARKER, J. (1874–6) 'On the relationship between the Somme River and the Somme Valley' *Proc. Geol. Assn.*, Vol. 4, pp. 286–307.

PATTISON, S. R. (1868) 'Formation of valleys; A description of Heudshope' *Geol. Mag.*, Vol. 5, pp. 161–3.

PECK, H. J. (1850) 'The levee system of Louisiana' *De Bow's Review*, N.S., Vol. 2, pp. 101–5.

PENGELLY, H. (1897) 'A Memoir of William Pengelly of Torquay, F.R.S., Geologist, with a selection from his correspondence' (Murray, London).

PENGELLY, W. (1864) 'The denudation of rocks in Devonshire' *Rept. and Trans. Devonshire Assn.*, Vol. 1, Pt 3, pp. 42–59.

PESCHEL, O. (1867) 'Das Wesen und die Aufgaben der Vergleichende Erdkunde'.

— (1870) 'Neue Probleme der Vergleichende Erdkunde als Versuch einer Morphologie der Erdoberfläche' (Duncker & Humblot, Leipzig).

PHILLIPS, J. (1853) 'The rivers, Mountains, and Sea-Coast of Yorkshire' (London), 291 pp.

— (1855) 'Manual of Geology' 1st Edn (Griffith, London), 669 pp.

— (1871) 'Geology of Oxford and the Valley of the Thames' (Oxford), 524 pp.

POISEUILLE, J. (1846) 'Recherches éxperimentales sur le mouvement des liquides dans les tubes de très petite diamètre', *Mém. Savants Étrange*, Vol. 9.

PRESTWICH, J. (1852) 'On some effects of the Holmfirth flood' *Quart. Jour. Geol. Soc.*, Vol. 8, pp. 225–30.

— (1862–3A) 'Theoretical considerations on the conditions under which the drift deposits containing the remains of extinct mammalia and flint implements were accumulated; and on their geological age' *Proc. Roy. Soc.*, Vol. 12, pp. 38–52.

— (1862–3B) 'On the loess of the valleys of the south of England and of the Somme and the Seine' *Proc. Roy. Soc.*, Vol. 12, pp. 170–3.

— (1863–4) 'On some further evidence bearing on the excavation of the valley of the Somme by river action' *Proc. Roy. Soc.*, Vol. 13, pp. 135–7.

— (1865) 'Theoretical considerations on the conditions under which the (drift) deposits containing the remains of extinct mammalia and flint implements were accumulated, and on their geological age' *Phil. Trans. Roy. Soc.*, Vol. 154, pp. 247–310.

RAMSAY, A. C. (1846) 'The denudation of South Wales' *Mem. Geol. Surv. Great Britain*, Vol. 1 (H.M.S.O., London), pp. 297–335.

— (1847) 'On the origin of the existing physical outline of a portion of Cardiganshire' *Trans. Brit. Ass. Adv. Sci.*, pp. 66–67.

— (1852) 'On the superficial accumulations and surface markings of North Wales' *Quart. Jour. Geol. Soc.*, Vol. 8, pp. 371–6.

— (1855) 'On the occurrence of angular, subangular, polished, and striated fragments and boulders in the Permian breccia of Shropshire, Worcestershire, &c.' *Quart. Jour. Geol. Soc.*, Vol. 11, pp. 185–205.

— (1862A) 'The excavation of the valleys of the Alps' *Phil. Mag.*, 4th Ser., Vol. 24, pp. 377–80.

— (1862B) 'On the glacial origin of certain lakes in Switzerland' *Quart. Jour. Geol. Soc.*, Vol. 18, pp. 185–204.

RAMSAY, A. C. (1863) 'The physical Geology and Geography of Great Britain' 1st Edn (Stanford, London), 145 pp.
— (1864) 'On the erosion of valleys and lakes' *Phil. Mag.*, 4th Ser., Vol. 28, pp. 293–311.
— (1872) 'On the river-courses of England and Wales' *Quart. Jour. Geol. Soc.*, Vol. 28, pp. 148–60.
— (1873–5) 'The Pre-Miocene Alps, and their subsequent waste and degradation' *Proc. Roy. Inst.*, Vol. 7, pp. 455–7.
— (1874) 'The physical history of the valley of the Rhine' *Quart. Jour. Geol. Soc.*, Vol. 30, pp. 81–95.
— (1876) 'On the physical history of the Dee, Wales' *Quart. Jour. Geol. Soc.*, Vol. 32, pp. 219–29.
RANCE, C. E. DE (1874–6) 'On the relative age of some valleys in the north and south of England, and of the various and Post-Glacial deposits occurring in them' *Proc. Geol. Ass.*, Vol. 4, pp. 221–53.
READE, T. M. (1871–2) 'On the geology and physics of the post-glacial period, as shown in the deposits and organic remains in Lancashire and Cheshire' *Proc. Liverpool Geol. Soc.*, Vol. 2, pp. 36–88.
— (1872–3) 'The buried valley of the Mersey' *Proc. Liverpool Geol. Soc.*, Vol. 2, pp. 42–65.
— (1873–4) 'Tidal action as a geological cause' *Proc. Liverpool Geol. Soc.*, Vol. 2, pp. 50–72.
RIDDELL, J. L. (1846) 'Deposits of the Mississippi and changes in its mouth' *De Bow's Review*, Vol. 2, pp. 433–48.
ROGERS, H. D. (1849) 'On the origin of the drift' *Proc. Amer. Ass. Adv. Sci.*, pp. 239–55.
— (1858) 'Final Report on the Geology of Pennyslvania' (Lippencott, Philadelphia), 1631 pp.
RUBIDGE, R. N. (1866) 'On the denudation of South Africa' *Geol. Mag.*, Vol. 3, pp. 88–91.
RUDLER, F. W. (1887–8) 'Fifty years' progress in British geology' *Proc. Geol. Ass.*, Vol. 10, pp. 234–72.
RUSKIN, J. (1865) 'Notes on the shape and structure of some parts of the Alps, with reference to denudation' *Geol. Mag.*, Vol. 2, pp. 49–54 and 193–6.
RUSSELL, R. (1869) 'On the flow of rivers and the measure of river sediments' *Geol. Mag.*, Vol. 6, pp. 268–71.
RUSSELL, R. J. et al. (1936) 'Lower Mississippi River Delta', *Dept. of Conser., Louisiana Geological Survey*, Geol. Bull. No. 8, 454 pp.
RUTIMEYER, L. (1869) 'Ueber Thal- und Seebildung' (Basle).
SCHOTT, C. A. (1872) 'Tables and results of the precipitation in rain and snow, in the United States', Smithson. Contrib. to Knowledge, No. 222.
SCROPE, G. P. (1858) 'The Geology and Extinct Volcanoes of Central France', 2nd Edn (Murray, London), 258 pp.
— (1866A) 'On the origin of valleys' *Geol. Mag.*, Vol. 3, pp. 193–9.
— (1866B) 'On the origin of hills and valleys' *Geol. Mag.*, Vol. 3, pp. 241–3.
— (1866C) 'The terraces of the Chalk Downs' *Geol. Mag.*, Vol. 3, pp. 293–6.
— (1866D) 'Atmospheric forces' *Geol. Mag.*, Vol. 3, pp. 379–80.
— (1869) 'On the pretended "raised sea-beaches" of the inland slopes in England and Wales' *Geol. Mag.*, Vol. 6, pp. 535–42.
SHARPE, D. (1856) 'On the last elevation of the Alps, with notices of the heights at which the sea has left traces of its action on their sides' *Quart. Jour. Geol. Soc.*, Vol. 12, pp. 102–23.

SIDDELL, W. H. (1861) 'Report on the Physics and Hydraulics of the Mississippi River' (Appendix A); Corps of Topographic Engineers, Prof. Paper No. 4 (Gov. Printing Office, Washington).

SINGER, C. et al. (1957) 'A History of Technology', Vol. 3 (Oxford), 766 pp.

SORBY, H. C. (1869) 'Note on the excavation of the valleys in Derbyshire' Geol. Mag., Vol. 6, pp. 347–8.

STEPHENSON, J. J. (1903) 'Memoir of J. Peter Lesley' Bull. Geol. Soc. Amer., Vol. 15, pp. 532–41.

STERNBERG, H. (1875) 'Untersuchungen über Längen- und Querprofil geschiebeführender Flüsse' Zeit. fur Bauwesen, Vol. 25, pp. 483–506.

ST. JOHN, S. (1851) 'Elements of Geology' (New York).

STODDART, D. R. (1960) 'Colonel George Greenwood; the father of modern subaerialism' Scot. Geog. Mag., Vol. 76, pp. 108–10.

TATHAM, G. (1951) 'Geography in the nineteenth century', Chapter 2 of Geography in the Twentieth Century (Ed. by G. Taylor), (Methuen, London), 630 pp.

TAYLOR, T. J. (1851) 'An Inquiry into the Operation of Running Streams and Tidal Waters' (London), 119 pp.

THOMSON, J. (1876–7) 'On the origin of windings of rivers in alluvial plains, with remarks on the flow of water round bends in pipes' Proc. Roy. Soc., Vol. 25, pp. 5–8.

TOLMAN, C. F. (1937) 'Ground Water' (McGraw-Hill, N.Y.), 593 pp.

TOPLEY, W. (1866) 'Notes on the physical geography of East Yorkshire' Geol. Mag., Vol. 3, pp. 435–9.

— (1867) 'The origin of escarpments' Geol. Mag., Vol. 4, pp. 184–6.

— (1875) 'Geology of the Weald' (H.M.S.O., London), 503 pp.

TRIPP, W. B. (1884–5) 'The river Buffalo, total flow . . . compared with rainfall' Min. and Proc. Inst. Civil Engineers, Vol. 81, pp. 241–51.

TYLOR, A. (1853) 'On the changes of sea-level effected by existing physical causes during stated periods of time' Phil. Mag., 4th Ser., Vol. 5, pp. 258–81.

— (1866) 'Remarks on the interval of time which has passed between the formation of the upper and lower valley-gravels of part of England and France' Quart. Jour. Geol. Soc., Vol. 22, pp. 463–8.

— (1868) 'On the Amiens gravel' Quart. Jour. Geol. Soc., Vol. 24, pp. 103–25.

— (1869A) 'On the formation of deltas; and on the evidence and cause of great changes in the sea-level during the Glacial Period' Quart. Jour. Geol. Soc., Vol. 25, pp. 7–11.

— (1869B) 'On Quaternary gravels' Quart. Jour. Geol. Soc., Vol. 25, pp. 57–100.

— (1872) 'On the formation of deltas and on the evidence and cause of great changes in the sea-level during the Glacial Period' Geol. Mag., Vol. 9, pp. 392–9 and 485–501.

— (1875) 'Action of denuding agencies' Geol. Mag., N.S., Dec. 2, Vol. 2, pp. 433–73.

VON ZITTEL, K. (1901) 'History of Geology and Palaeontology' (London), 562 pp.

VORSTER, J. H. (1957) 'Trends in long range rainfall records in South Africa' South Af. Geog. Jour., Vol. 39, pp. 61–66.

WARD, J. C. (1870) 'On the denudation of the Lake District' Geol. Mag., Vol. 7, pp. 14–17.

WARD, R. DE C. (1925) 'The Climates of the United States' (Ginn, Boston), 518 pp.

WESTON, C. H. (1850) 'On the diluvia and valleys in the vicinity of Bath' *Quart. Jour. Geol. Soc.*, Vol. 6, pp. 449–51.

WHEWELL, W. (1847) 'On the wave of translation in connection with the Northern Drift' *Quart. Jour. Geol. Soc.*, Vol. 3, pp. 227–32.

WHITAKER, W. (1867A) 'On subaerial denudation, and on cliffs and escarpments of the Chalk and Lower Tertiary beds' *Geol. Mag.*, Vol. 4, pp. 447–454 and 483–93.

— (1867B) 'On subaerial denudation, and on cliffs and escarpments of the Tertiary strata' *Quart. Jour. Geol. Soc.*, Vol. 23, pp. 265–6.

— (1868) 'Subaerial denudation' *Geol. Mag.*, Vol. 5, pp. 46–47.

WILLIAMS, H. S. (1895) 'James Dwight Dana and his work as a geologist' *Jour. Geol.*, Vol. 3, pp. 601–21.

WITCHELL, E. (1868) 'On the denudation of the Cotteswolds' *Geol. Mag.*, Vol. 5, pp. 280–1.

WOOD, S. V. (1864) 'On the formation of river- and other valleys of the east of England' *Phil. Mag.*, 4th Ser., Vol. 27, pp. 180–90.

— (1866A) 'On the structure of the Thames valley and of its contained deposits' *Geol. Mag.*, Vol. 3, pp. 57–63 and 99–107.

— (1866B) 'On the structure of the valleys of the Blackwater and Crouch' *Geol. Mag.*, Vol. 3, pp. 348–54 and 398–406.

— (1867) 'On the structure of the postglacial deposits of the South-east of England' *Quart. Jour. Geol. Soc.*, Vol. 23, pp. 394–417.

— (1868) 'Reply to Mr W. Boyd Dawkins on the Thames Valley deposits &c; and to Mr A. H. Green, on the Ouse Valley at Buckingham' *Geol. Mag.*, Vol. 5, pp. 42–45.

— (1871) 'On the evidence afforded by the detrital beds without and within the north-eastern part of the valley of the Weald as to the mode and date of the denudation of that valley' *Quart. Jour. Geol. Soc.*, Vol. 27, pp. 3–27.

WOODFORD, A. O. (1951) 'Stream gradients and Monterey Sea Valley' *Bull. Geol. Soc. Amer.*, Vol. 62, pp. 799–852.

WYNNE, A. B. (1867) 'On denudation with reference to the configuration of the ground' *Geol. Mag.*, Vol. 4, pp. 3–11.

PART FOUR

The Western Explorations

FIG. 88. *The Grand Canyon of the Colorado – 6,200 feet deep*
(Frontispiece from Powell, 1875)

CHAPTER TWENTY-FIVE

The Western Pioneer Geologist: John Wesley Powell

INTRODUCTION

The period of geomorphic investigation about to be discussed brought with it a freshness of approach and opinion largely because it involved the exploitation of a novel environment. Some of its discoveries provided no more than confirmation of theories evolved in Europe, but it would be misleading to allow the assessment to end there. The explorers of this new environment, for they were explorers, viewed the evidence of nature with minds that were largely free from foreign preconceptions. What they saw around them was a world different from anything that they had ever seen or imagined. Even if foreign ideas had been applicable, the geologic evidence of the western United States provided an explanation of its own evolution that was at once so simple and clear that no other interpretation was needed. All who visited the area were impelled to the same conclusion. Historical events largely decided the moment when this striking geological evidence would be discovered. The period 1790 to 1890 was the great era of westward expansion in the United States but this unique and fascinating story lies outside our main theme. We are concerned only with the first scientific investigators of these wild, inland territories. Yet thereby we plunge at once into the epic of the West and, as perhaps might have been expected, meet straightaway a new breed of geologists; not the studious, often-theorizing specimens roaming the hills of western Europe but strong, independent men filled with a fire and spirit acquired in the hardy trials of an expanding pioneer society.

The area into which American geologists penetrated included the plateaus, the mountain cordilleras and the deserts of the West. They went usually at the behest of the government to survey routes for the establishment of new forts against the Indians or under the patronage of government financiers with an eye for minerals and railway routes. That the West had so much to offer to academic geologists and students of landscape was a fortunate coincidence.

JOHN WESLEY POWELL

Of all the geologists who explored this region John Wesley Powell comes first to mind, not because he was in any way superior to several others, nor because his theories were more detailed and advanced, but because his upbringing and his feat of navigating the Colorado Canyon are truly representative of the principles, ambitions and standards of his age. By starting with him we enter straightway into the full atmosphere of the nineteenth-century American West.

Powell's origins were not exceptional (Darrah, 1951). His father, a fervent Methodist, was a mixture of Welsh and English ancestry who originally came from Shrewsbury and emigrated to America in 1830. A strong and active man, he allowed his proselytizing activities to take precedence over his family responsibilities and his trade of tailoring. John Wesley, the second son, was born on 24 March 1834. By the time he was of school age the family had left New York State and moved west to Ohio, where his father bought a building lot for 200 dollars. The school which Powell attended was not graded in classes and one teacher taught pupils of all ages, usually in return for clothing and food supplied by the children's parents.

When Powell was nine, George Crookham, a wealthy neighbour, undertook his education. George was a typical frontiersman, entirely self-taught and of unorthodox educational views. He seems to have encouraged Powell to explore and reason, but otherwise allowed him a free rein. George himself was interested in archaeology and, no doubt in a simple way, geology, as he was a friend of William Mather, state geologist of Ohio. Most important of all he had a good library of books in English and Latin, including scientific works. By the age of ten Powell had read Gibbon's *The Decline and Fall of the Roman Empire* and Hume's *History of England*.

The Powells spent nearly eight years at Jackson, Ohio, and each year the number of children increased by one. The family was now prosperous, as despite the calls of his missionary work, the father made a combination of farming and tailoring profitable. However, he repeatedly felt the call of the frontier which already seemed to have moved so far from him. With the thought of more souls to save and new virgin lands to be ploughed, he sold his property at Jackson and again uprooted his household. This time he bought a large farm in Wisconsin and left John Wesley, now twelve, to work and manage it.

The young boy had to cut a clearing in the primeval forest; to burn or dig out the tree stumps before ploughing the soil; to protect the few stock and his family from the threat of wild animals and Indians; to take the grain to market; and to bargain and buy provisions for the household. Though an obedient and capable son, Powell had little heart for farming. In 1850, as soon as his younger brother was old enough to look after the farm, he left home and went to school at Jamesville. He attended classes during the day and paid for his lodgings by tending the cattle and doing household chores at night and in the early morning. He found the instruction easy in spite of his deficiency in mathematics, grammar and the classical tongues.

Unfortunately this advancement of his education did not last long as his father moved to Illinois, and young Powell had to go there to help prepare the new site. Powell's father was not hostile to his son's wish for learning, but wanted to direct his obvious talents towards evangelism. He offered to send his son to Oberlin College if he would become a Minister. This young Powell refused and the two men quarrelled. Shortly after he managed to gain a post as a teacher to a small frontier community, at a salary of 14 dollars a month. He lived frugally, saving what money he could to pay for his future tuition at college. Meanwhile his father had moved to Wheaton and induced him to return home by telling him of the new college about to be founded in the town. However Powell, now twenty, lost interest on hearing that there would not be a science faculty and went off teaching again, this time for 24 dollars a month. He improved his knowledge of classics by taking evening lessons and eventually, in 1855, enrolled in the scientific department of Illinois College, where he learnt algebra, English grammar, geometry and elocution. Here he became friendly with Johnathan B. Turner, an educationist and horticulturist, who later was influential in turning Powell's thoughts towards scientific topics. After a short while, Powell left the college and returned to teaching at a salary of 60 dollars a month.

However, it seems that he soon tired of this and in 1856 he went by rowing boat down the Mississippi, from the falls of St Anthony to the river's mouth, collecting molluscs. In 1857 he made a similar trip from Pittsburgh to the mouth of the Ohio River and in the following year travelled down the Illinois and up the Des Moines River. These trips used up most of the money he had saved while teaching and he returned to Wheaton College for a short time. In 1858 he entered Oberlin College, taking classics and botany as his two subjects. However, here the militant Christianity of the college bored him and he left, but was soon able to obtain a teaching post at Hennapin at a salary of 100 dollars a month. His friend Turner had meantime founded a local natural history society and, through his influence, Powell was appointed secretary, a position he retained for some years. The society collected plants, snakes, minerals and fossils and this work did not interfere with his teaching, which was mostly confined to the winter months. In 1860 he became the Principal of the public school at Hennapin.

He still led a semi-nomadic life though he was already courting his cousin Emma Dean and if he sensed the approach of domesticity, it was not betrayed in his habits. Instead he was more concerned with the approaching Civil War, which he felt was inevitable. He began to study mapping, engineering (particularly bridge-building) and military tactics and in May 1861 volunteered as a private in the Federal Army. His intelligence immediately marked him out for promotion; within a few days he was a sergeant major – and within five weeks a second lieutenant. On 28 November he married Emma, who henceforth travelled with him whenever possible. Powell was keen to have his mettle tested in action but, unfortunately almost his first taste of conflict made him a casualty. At the battle of Shiloh, on 6 April 1862, his artillery company was cut off and in the ensuing engagement his right arm, raised as a signal to fire, was hit by a half-spent minié-ball. It is characteristic of Powell that some years later he became friendly with the Confederate officer Colonel Hooker, who had lost his left arm at the same battle. Subsequently whenever either of them bought a pair of gloves, he sent the unused glove to the other (*Davis, 1915, pp. 13–14*). Powell's arm was amputated and he suffered a great deal of pain from the stump throughout the rest of his life. During convalescence he acted as a recruiting officer and while he was doing this he obtained permission for his wife to be with him. The loss of his arm had not lessened his desire for military glory and by April 1863 he was back on active service with General Grant's army in the advance south. During the siege of Vicksburg he was able to combine war with his hobby, for he is said to have collected molluscs while engaged in forays and operations around the beleaguered fortress. His newly severed arm was giving him considerable pain and he was forced to have another operation on the stump. In September 1864 his talent and enthusiasm saw his promotion to Major and in his last big action he commanded with distinction an artillery force at the battle of Nashville. The war was by now virtually decided and in January 1865 his wife managed to persuade him to resign his commission.

When he had regained health Powell was fortunate enough to be offered a choice of jobs and accepted the Professorship of Geology at the Illinois Wesleyan University, Bloomington, at a salary of 1000 dollars a year. However, as well as being a lover of science he was still ambitious and was anxious to reach a status of independence that would allow him to pursue his scientific explorations in any way he chose. He worked hard and his first term of lecturing so raised the number of students that his salary was increased by 500 dollars a year. Powell with complete self-confidence taught with equal aplomb and enthusiasm Botany, Cellular Histology, Anatomy and Physiology, Zoology, Natural Philosophy, Logic of Natural Science, Geology and Mineralogy. Yet rising academic fame did not satisfy his real desire, which was to explore and test current theories in the field. He had set his heart on exploring the mountains of the West. To this end he gave lectures at Normal University in his free hours and at the same time campaigned for the establishment of a museum to be subsidized by the government. His hard upbringing had perhaps made him a powerful executor of schemes which were likely to further his personal aims. Henceforth he appears as a skilful leader of men and an equally skilful manipulator of congressional committees. After personally pleading before the House of Representatives, he succeeded in obtaining the establishment of the museum with himself as its curator. He immediately planned a trip to the Rockies. The college made a grant of 500 dollars, the government gave aid in the form of rations and railway passes, other scientific institutions, like the Smithsonian, provided grants and instruments in return for a promise of measurements and duplicate specimens.

FIG. 89. *Major Powell and a Pai Ute in southern Utah, 1872*
(Photograph by J. K. Hillers, U.S. Geological Survey)

Powell's first expedition was in the Pike's Peak region in 1867. The party, made up of wealthy students, fellow-professors, a few experienced hunters and his wife, climbed the Peak and thoroughly explored the locality as well as the headwaters of the Grand River. On his return he organized a second expedition, during which, with the aid of grants of 600 dollars from Normal University and of military supplies for 25 men from the government, he explored Long's Peak, the Uinta Mountains and the source of the Green River. This provided good experience for his approaching navigation of the Colorado, a project which long seems to have been in his mind.

On his third expedition, in 1869, Powell committed his party to the passage of the Grand Canyon. Little or nothing definite was known about the river and Powell was relying heavily on his geomorphic deductions. Its lower reaches had been explored by an army survey under Lieut. Ives with Newberry as geologist, and Powell himself had travelled through the regions around the sources of the Green and the Grand. The reports of Hayden and Newberry provided the only scientific information about the area and neither geologist had actually travelled along the course Powell intended to take. The dryness and inaccessibility had deterred settlement by even the Indians. Wild tales existed, like that of James White who in 1867 claimed he had been swept through the Grand Canyon on a raft. As no one knew what the Canyon was like, his tale, although doubted, could not be disproved. Legend and fancy wove the unknown quantities into an object of supernatural qualities which was treated with awe and mystical reverence. Powell explains this in his report.

> These cañon gorges, obstructing cliffs and desert wastes, have prevented the traveller from penetrating the country, so that, until the Colorado River Exploring Expedition was organized, it was almost unknown. Yet enough had been seen to foment rumor, and many wonderful stories have been told in the hunter's cabin and prospector's camp. Stories were related of parties entering the gorge in boats, and being carried down with fearful velocity into whirlpools, where all were overwhelmed in the abyss of waters; others, of underground passages for the great river, into which boats had passed never to be seen again. It was currently believed that the river was lost under the rocks for several hundred miles. There were other accounts of great falls, whose roaring music could be heard on the distant mountain-summits. There were many stories current of parties wandering on the brink of the cañon, vainly endeavouring to reach the waters below, and perishing with thirst at last in sight of the river which was roaring its mockery into dying ears. The Indians, too, have woven the mysteries of the cañons into the myths of their religion . . .
> More than once have I been warned by the Indians not to enter this cañon. They considered it disobedience to the gods and contempt for their authority, and believed that it would surely bring upon me their wrath. (*Powell, 1875, pp. 6–7*)

With so little true knowledge to depend upon, Powell's confidence in his own assumptions seems amazing. After his success it was easy to minimize his achievement and give the credit for his discovery to earlier pioneers, but it was not possible to take away from him his claim to have been the first to make the passage.

> . . . there have been some who appear to be inclined to withhold from Major Powell the full credit which is his for solving the great problem of the Southwest, and who, therefore, make much of the flimsy story of White, and even assume on faint evidence that others fathomed the mystery even before White. There is, in my opinion no ground for such assumptions. Several trappers, like Pattie and Carson, had gained a considerable knowledge of the general course and character of the river as early as 1830, but to Major Powell and his two parties undoubtedly belongs the high honour of being the first to explore and explain the truth about it and its extraordinary canyon environment. (*Dellenbaugh, 1902, Preface p. vii*)

The party started as ten men spread among four boats, specially built to Powell's specifications so as to be able to withstand the hazard and force of

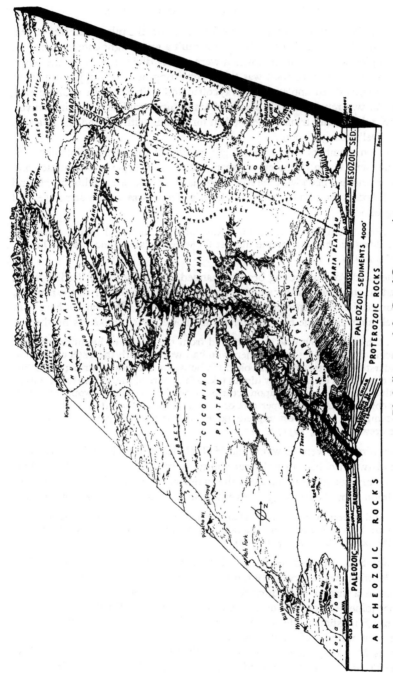

FIG. 90. *Block diagram of the Grand Canyon region*
(Drawn by Erwin Raisz)

the river. Made of oak, three were 21 feet, and one, the pilot boat, was 16 feet long. Into the bow and stern of each was built a water-tight compartment for the supplies and instruments, which were evenly divided between the boats. Ten months' ration of food was taken consisting of army supplies of flour, sugar, beans, dried apples, salt pork, coffee and a little tea. In addition each man had two revolvers with plenty of ammunition and two hunting knives. Calico, tobacco and trinkets were taken for trading with the Indians (Darrah, 1951; Dellenbaugh, 1909).

The ten members were a mixed group. Powell shared the pilot boat with Sumner and Dunn. Sumner, 29 years old, was an experienced hunter and trapper who had been with the Major on his two previous expeditions. He and Powell both kept a daily record of the trip. Dunn was of the same age and background;

> Dressed habitually in dirty buckskin, with raven hair hanging down his shoulders, he had a supreme disgust for water, scissors, and razor. (*Darrah*, *1951, p. 114*)

Bradley and Powell's young brother manned the second boat. Bradley was a skilled boatman, who after failing to find adventure as a soldier on the frontier, welcomed the trials anticipated in Powell's expedition. He is said to have held himself aloof from the rest. Walter Henry Powell was, unlike his brother;

> . . . a moody fellow who still suffered from the effects of his imprisonment during the war. Most of the men who had worked with the 1868 expedition understood Walter and overlooked his behavior. He was of strong build but his disposition did not allow him to take a fair share of the work. He was gifted with a fine bass voice and upon occasion, when he was willing, would entertain his companions with a repertoire of popular ballads, but often he was sullen, ill-tempered, and at best unpredictable. He was the least popular member of the crew. (*Darrah, 1951, p. 115*)

In the third boat were Hawkins and Hall. Hawkins, a fugitive from justice in the mid-twenties, had no technical ability so was made the cook. Andy Hall, a Scot, only twenty years old, was the junior of the party. Too young to join the Union Army he had sought adventure in the wars with the Plains Indians.

> . . . and generally raised hell wherever he found it. . . . With judgment far beyond his years and with a humor that never failed him, Andy was the life of the party. (*Darrah, 1951, p. 114*)

The last boat was manned by the Howland brothers and Goodman. The elder Howland was a fervent Methodist, who before joining the expedition had worked as a printer and editor for many western newspapers. His younger brother joined the expedition at the elder's suggestion. Frank Goodman was the only Englishman. He was allowed to join the party at his own request, apparently in hopes of an exciting trip.

On 24 May 1869 the party left Green River City, which had been reached by the Trans-Continental railroad the previous year. The start of their journey was fortified by the cheers of the few local inhabitants and the dinner provided the night before by the local Chinese cook. The first few days of travelling presented few dangers. Powell always led in the pilot boat and they made easy progress downstream interrupted only by occasional landings to take readings and to study the geology. The men were in good heart and they sang as they manœuvred their small craft. Sometimes Powell would entertain them by reading from the Bible or from one of the classics. Though born and fostered in a harsh atmosphere, he had a keen sense for the beautiful and his report is constantly touched with passages that are sharp and moving in their power of description.

> During the afternoon, we run down to a point where the river sweeps the foot of an overhanging cliff, and here we camp for the night. The sun is yet two hours high, so I climb the cliffs and walk back among the strangely carved rocks of the Green River bad-lands. These are sandstones and shales, gray and buff, red and brown, blue and black strata in many alternations, lying nearly horizontal, and almost without soil and vegetation. They are very friable, and the rain and streams have carved them into quaint shapes. Barren desolation is stretched before me; and yet there is a beauty in the scene. The fantastic carving, imitating architectural forms, and suggesting rude but weird statuary, with the bright and varied colors of the rocks, conspire to make a scene such as the dweller in verdure-clad hills can scarcely appreciate.
>
> Standing on a high point, I can look off in every direction over a vast landscape, with salient rocks and cliffs glittering in the evening sun. Dark shadows are settling in the valleys and gulches, and the heights are made higher and the depths deeper by the glamor and witchery of light and shade.
>
> Away to the south, the Uinta Mountains stretch in a long line; high peaks thrust into the sky, and snow fields glittering like lakes of molten silver; and pine forests in somber green; and rosy clouds playing around the borders of huge black masses; and heights and clouds, and mountains and snow fields, and forests and rock lands are blended into one grand view. (*Powell, 1875, pp. 9-10*)

The attraction of the landscape tended to conceal the dangers and unpleasant-nesses associated with it. Towards the end of the trip, when the threat of starvation excluded all other thoughts, such poetic ecstasies are markedly absent.

On 30 May the party reached the beginning of Flaming Gorge and Lodore Canyon where the Green River cuts through the Uinta Mountains. Soon afterwards they are shooting their first rapid and gaining some idea of the elements which they must learn to master.

> Entering Flaming Gorge, we quickly run through it on a swift current and emerge into a little park. Half a mile below, the river wheels sharply to the left and we turned into another cañon cut into the mountain. We enter the narrow passage. On either side, the walls rapidly increase in altitude. On the left are overhanging ledges and cliffs five hundred – a thousand – fifteen hundred feet high.
>
> On the right, the rocks are broken and ragged, and the water fills the channel

FIG. 91. *Lodore Canyon of the Green River*
(From Powell, 1875)

from cliff to cliff. Now the river turns abruptly around a point to the right, and the waters plunge swiftly down among great rocks; and here we have our first experience with cañon rapids. I stand upon the deck of my boat to seek a way among the wave beaten rocks. All untried as we are with such waters, the moments are filled with intense anxiety. Soon our boats reach the swift current; a stroke or two, now on this side, now on that, and we thread the narrow passage with exhilarating velocity, mounting the high waves, whose foaming crests dash over us, and plunging into the troughs, until we reach the quiet water below; and then comes a feeling of great relief. Our first rapid is run. Another mile, and we come into the valley again. (*Powell, 1875, p. 14*)

As the next passage shows, Powell was an exceptional person. To pass from a mood of brooding over his responsibilities to the elucidation of a trivial phenomenon shows powers of scientific detachment of more than ordinary calibre.

As the twilight deepens, the rocks grow dark and somber; the threatening roar of the water is loud and constant, and I lie awake with thoughts of the morrow and the cañons to come, interrupted now and then by characteristics of the scenery that attract my attention. And here I make a discovery. On looking at the mountain directly in front the steepness of the slope is greatly exaggerated, while the distance to its summit and its true altitude are correspondingly diminished. I have heretofore found that to properly judge of the slope of a mountain side, you must see it in profile. In coming down the river this afternoon, I observed the slope of a particular part of the wall, and made an estimate of its altitude. While at supper, I noticed the same cliff from a position facing it, and it seemed steeper, but not half as high. Now lying on my side and looking at it, the true proportions appear. This seems a wonder and I rise up to take a view of it standing. It is the same cliff as at supper time. Lying down again it is the cliff as seen in profile with a long slope and distant summit. Musing on this, I forget ' the morrow and the cañons to come'. I find a way to estimate the altitude and slope of an inclination as I can judge of distance along the horizon. The reason is simple. A reference to the stereoscope will suggest it. The distance between the eyes forms a base-line for optical triangulation. (*Powell, 1875, pp. 15–16*)

A grim relic of the river's constant threat was seen on June 2nd.

This morning we make a trail among the rocks, transport the cargoes to a point below the falls, let the remaining boats over, and are ready to start before noon.

On a high rock by which the trail passes we find the inscription: 'Ashley 18–5'. The third figure is obscure – some of the party reading it 1835, some 1855.

James Baker, an old time mountaineer, once told me about a party of men starting down the river, and Ashley was named as one. The story runs that the boat was swamped, and some of the party drowned in one of the cañons below. The word 'Ashley' is a warning to us, and we resolve on great caution.

Ashley Falls is the name we give to the cataract.

The river is very narrow; the right wall vertical for two or three hundred feet, the left towering to a great height with a vast pile of broken rocks lying between the foot of the cliff and the water. Some of the rocks broken down from the ledge above have tumbled into the channel and caused this fall. One great cubical block, thirty or forty feet high, stands in the middle of the stream, and the waters parting to either side, plunge down about twelve feet, and are broken again by the smaller rocks into a rapid below. Immediately below the falls the water occupies the entire channel, there being no talus at the foot of the cliffs. (*Powell, 1875, pp. 17–18*)

The party were not discountenanced by Ashley's failure and Powell expresses their mood of continuing confidence with another effusion of poetic description.

This morning we spread our rations, clothes &c., on the ground to dry, and several of the party go out for a hunt. It takes a walk of five or six miles up to a pine grove park, its grassy carpet bedecked with crimson, velvet flowers set in groups on the stems of pear shaped cactus plants; patches of painted cups are seen here and there, with yellow blossoms protruding through scarlet bracts; little blue-eyed flowers are peeping through the grass; and the air is filled with fragrance from the white blossoms of a Spiraea. A mountain brook runs through the midst, ponded below by beaver dams. It is a quiet place for retirement from the raging waters of the cañon. (*Powell, 1875, p. 18*)

Sometimes one is inclined to think that Powell was compounded of an inflexible zeal in the furtherance of science and often forgot the suffering of his companions who were not similarly compensated. But it seems from his writings that this was not so.

After a good drink, we walk out to the brink of the cañon, and look down to the water below. I can do this now, but it has taken several years of mountain climbing to cool my nerves, so that I can sit, with my feet over the edge, and calmly look down a precipice 2000 feet. And yet I cannot look on and see another do the same. I must either bid him come away, or turn my head. (*Powell, 1875, pp. 20-21*)

On 8 June the party suffered their first setback in Lodore Canyon.

During the afternoon, we come to a place where it is necessary to make a portage. The little boat is landed, and the others are signalled to come up.

When these rapids or broken falls occur, usually the channel is suddenly narrowed by rocks which have been tumbled from the cliffs or have been washed in by lateral streams. Immediately above the narrow, rocky channel, on one or both sides, there is often a bay of quiet water, in which we can land with ease. Sometimes the water descends with a smooth, unruffled surface, from the broad, quiet spread above, into the narrow angry channel below, by a semicircular sag. Great care must be taken not to pass over the brink into this deceptive pit, but above it we can row with safety. I walk along the bank to examine the ground, leaving one of my men with a flag to guide the other boats to the landing-place. I soon see one of the boats make shore all right and feel no more concern; but a minute after, I hear a shout, and looking around, see one of the boats shooting down the centre of the sag. It is the *No-Name*, with Captain Howland, his brother, and Goodman. I feel that its going over is inevitable, and run to save the third boat. A minute more, and she turns the point and heads for the shore. Then I turn down stream again, and scramble along to look for the boat that has gone over. The first fall is not great, only ten or twelve feet, and we often run such; but below, the river tumbles down again for forty or fifty feet, in a channel filled with dangerous rocks that break the waves into whirlpools and beat them into foam. I pass around a great crag just in time to see the boat strike a rock, and, rebounding from the shock, careen and fill the open compartment with water. Two of the men lose their oars; she swings around, and is carried down at a rapid rate, broadside on, for a few yards, and strikes amidships on another rock with great force, is broken quite in two, and the men are thrown into the river; the larger part of the boat floating buoyantly, they soon seize it, and down the river they drift, past the rocks for a few hundred yards to a second rapid, filled with huge boulders, where the boat strikes again, and is dashed to pieces and the men and fragments are soon carried beyond my sight. Running along, I turn a bend and see a man's head above the water, washed about in a whirlpool below a great rock.

It is Frank Goodman, clinging to it with a grip upon which life depends. Coming opposite, I see Howland trying to go to his aid from an island on which he has been washed. Soon, he comes near enough to reach Frank with a pole, which he extends toward him. The latter lets go the rock, grasps the pole, and is pulled ashore. Seneca Howland is washed farther down the island, and is caught by some rocks, and, though somewhat bruised, manages to get ashore in safety. This seems a long time, as I tell it, but it is quickly done.

And now the three men are on an island, with a swift dangerous river on either side, and a fall below. The *Emma Dean* is soon brought down, and Sumner, starting above as far as possible, pushes out. Right skilfully he plies the oars, and a few strokes set him on the island at the proper point. Then they all pull the boat up stream, as far as they are able, until they stand in water up to their necks. One sits on a rock, and holds the boat until the others are ready to pull, then gives the boat a push, clings to it with his hands, and climbs in as they pull for mainland, which they reach in safety. We are as glad to shake hands with them as though they had been on a voyage around the world, and wrecked on a distant coast. (*Powell, 1875, pp. 23–25*)

This disaster was most serious, even though the men were saved, for, besides the loss of the craft itself, all the barometers by some miscalculation were on the *No-Name*. They were able to make camp on the shore and survey the situation. 'No sleep comes to me in all those dark hours,' writes Powell. To meet with such a reverse at so early a stage was very discouraging but Powell had counted on disaster, and, as he was never given to repining, as soon as breakfast was eaten the next morning he cast about for a way to rescue the barometers which were in a part of the wreck that had lodged among some rocks a half mile below. Sumner and Dunn volunteered to try to reach the place with the small boat, and they succeeded. When they returned, a loud cheer went up from those on shore, and Powell was much impressed with this exhibition of deep interest in the safety of the scientific instruments, but he soon discovered that the cheer was in celebration of the rescue of a three-gallon keg of whiskey that had been smuggled along without his knowledge and happened to be on the ill-fated *No-Name*. (*Dellenbaugh, 1909, pp. 199–200*)

Though unfortunate, the loss of one boat did not prevent the party continuing, although they soon learnt that misfortune could come in a more unexpected form.

Sometimes the danger was of a novel and unexpected character as on June 16th, when the dry willows around camp caught fire. Powell had started for a climb of investigation and looking down on the camp he perceived a sudden tremendous activity without being able for some moments to discover the cause. So rapidly did the fire spread that there was no escape by the boats. Some had their clothing burned and their hair singed, while Bradley even had his ears scorched. The cook in his haste stumbled with his arms full of culinary utensils and the load disappeared beneath the waters, ever on the alert to swallow up man, boat or beast. Just below the camp was a rapid, and, casting off, they were forced to run this without stopping to examine it. No harm was done to the boats, and they landed at the first opportunity. When the fire had burned out they went back along the rocks to pick up what had been left behind and was unconsumed. (*Dellenbaugh, 1909, p. 202*)

Such losses, although not causing personal injuries, were to have an increasing effect as the supplies of food ran low. Each time a boat overturned more food was spoilt or lost and perhaps a blanket or another sleeping-bag swept away in the current. Danger is often only half-size when met on a full stomach and in dry clothing but when encountered after weeks of living on a diminishing diet of rancid bacon and beans may seriously weaken the will. On 18 June they reached the junction with the Yampa River, where they spent a few days repairing the boats, taking geological observations and hunting for fresh meat to relieve the monotony of their daily rations. The party were still in good spirits.

> June 21. – We float around the long rock, and enter another cañon. The walls are high and vertical; the cañon is narrow; and the river fills the whole space below, so that there is no landing-place at the foot of the cliff. The Green is greatly increased by the Yampa, and we now have a much larger river. All this volume of water, confined, as it is, in a narrow channel, and rushing with great velocity, is set eddying and spinning in whirlpools by projecting rocks and short curves, and the waters waltz their way through the cañon, making their own rippling, rushing, roaring music. The cañon is much narrower than any we have seen. With difficulty we manage our boats. They spin about from side to side, and we know not where we are going, and find it impossible to keep them headed down the stream. At first this causes us great alarm, but we soon find there is but little danger, and that there is a general movement of progression down the river, to which this whirling is but an adjunct; and it is the merry mood of the river to dance through this deep, dark gorge; and right gaily do we join in the sport. (*Powell, 1875, pp. 35–36*)

On 28 June they reached the junction with the Uinta River, where they again broke the journey for a few days while some of them made observations and others visited the nearby Indian Agency. They hoped by trading or with money to replenish their supplies but to their disappointment they found the agency comprised only a few degenerate Indians who had barely enough to live on themselves. Frank Goodman, the Englishman, left the party at the agency and returned to the duller life from which he had come. He had had enough excitement.

On 6 July the party set off again. The section of the river upon which they were to enter was wholly unknown. On 18 June Powell himself had a lucky escape while climbing.

> The adventures had occasionally a smack of grim humour about them. On one occasion Powell and one of his men began the ascent of the canyon wall just below the mouth of the Yampa, a tributary of the Green River above its junction with the Uinta. After ascending five or six hundred feet they came to a sheer precipice, up which they managed to climb till the top was nearly reached. 'Here,' says Powell, 'by making a spring I gain a foothold in a little crevice, and grasp an angle of the rock overhead. I find I can get up no further,

and cannot step back, for I dare not let go with my hand, and cannot reach foot-hold below without. I call to Bradley for help. He finds a way by which he can get to the top of the rock over my head, but cannot reach me. Then he looks around for some stick or limb of a tree, but finds none. Then he suggests that he had better help me with the barometer case; but I fear I cannot hold on to it. The moment is critical. Standing on my toes, my muscles begin to tremble. It is sixty or eighty feet to the foot of the precipice. If I lose my hold I shall fall to the bottom, and then perhaps roll over the bench, and tumble still further down the cliff. At this instant it occurs to Bradley to take off his drawers, which he does, and swings them down to me. I hug close to the rock, and let go with my hand, seize the dangling legs, and, with his assistance, I am enabled to gain the top. (*Cadell, 1887, p. 450; quoting Powell, 1875, pp. 33–34*)

Powell never allowed his one-armedness to be a handicap. On 11 July he received his own first baptism by water.

We soon approach another rapid. Standing on deck, I think it can be run, and on we go. Coming nearer, I see that at the foot it has a short turn to the left, where the waters pile up against the cliff. Here we try to land, but quickly discover that, being in swift water, above the fall, we cannot reach shore, crippled, as we are by the loss of two oars; so the bow of the boat is turned down stream. We shoot by a big rock; a reflex wave rolls over our little boat and fills her. I see the place is dangerous, and quickly signal to the other boats to land where they can. This is scarcely completed when another wave rolls our boat over, and I am thrown some distance into the water. I soon find that swimming is very easy, and I cannot sink. It is only necessary to ply strokes sufficient to keep my head out of the water, though now and then, when a breaker rolls over me, I close my mouth, and am carried through it. The boat is drifting ahead of me twenty or thirty feet, and, when the great waves are passed, I overtake it, and find Sumner and Dunn clinging to her. As soon as we reach quiet water, we all swim to one side and turn her over. In doing this, Dunn loses his hold and goes under; when he comes up, he is caught by Sumner and pulled to the boat. In the mean time we have drifted down stream some distance, and see another rapid below. How bad it may be we cannot tell, so we swim toward shore, pulling our boat with us, with all the vigor possible, but are carried down much faster than distance toward shore is gained. At last we reach a huge pile of drift-wood. Our rolls of blankets, two guns and a barometer were in the open compartment of the boat, and, when it went over, these were thrown out. The guns and barometer are lost, but I succeeded in catching one of the rolls of blankets, as it drifted by, when we were swimming to shore; the other two are lost, and sometimes hereafter we may sleep cold. (*Powell, 1875, pp. 48–49*)

The passage was becoming much harder. On 12 July twenty portages were necessary in order to pass round ten fierce rapids. Portages were more exhausting than dangerous as each meant that the boats had to be unpacked and the items carried separately. With the increase in the number of rapids the canyon walls were rising in height.

We continue our journey. In many places the walls, which rise from the water's edge, are over-hanging on either side. The stream is still quiet, and we glide along, through a strange, weird, grand region. The landscape every-where, away from the river, is of rock – cliffs of rock; tables of rock; plateaus of rock; terraces of rock; crags of rock – ten thousand strangely carved forms. Rocks everywhere and no vegetation; no soil; no sand. In long, gentle curves, the river winds about these rocks.

When speaking of these rocks, we must not conceive of piles of boulders, or heaps of fragments, but a whole land of naked rock, with giant forms carved on it; cathedral shaped buttes, towering hundreds or thousands of feet; cliffs that cannot be scaled, and cañon walls that shrink the river into in-significance, with vast, hollow domes, and tall pinnacles, and shafts set on the verge overhead, and all highly colored – buff, grey, red, brown, and chocolate; never lichened; never moss-covered, but bare, and often polished. We pass a place, where two bends of the river come together, an inter-vening rock having been worn away, and a new channel formed across. The old channel ran in a great circle around to the right, by what was once a circular peninsula; then an island; then the water left the old channel entirely, and passed through the cut, and the old bed of the river is dry. So the great circular rock stands by itself, with precipitous walls all about it, and we find but one place where it can be scaled. Looking from its summit, a long stretch of river is seen, sweeping close to the overhanging cliffs on the right, but having a little meadow between it and the wall on the left. The curve is very gentle and regular. We name this Bonita Bend. (*Powell, 1875, pp. 54–55*)

Though conditions were worsening Powell reports on 15 July that the party were still in good spirits. Even so the hard going was making them realize that they were being tested to the limit and that any greater danger might be too much for them.

Our way, after dinner, is through a gorge, grand beyond description. The walls are nearly vertical; the river broad and swift, but free from rocks and falls. From the edge of the water to the brink of the cliffs it is one thousand six hundred to one thousand eight hundred feet. At this great depth, the river rolls in solemn majesty. The cliffs are reflected from the more quiet river, and we seem to be in the depths of the earth, and yet can look down into waters that reflect a bottomless abyss. We arrive, early in the afternoon, at the head of more rapids and falls, but, wearied with past work, we determine to rest, so go into camp, and the afternoon and evening are spent by the men in discussing the probabilities of successfully navigating the river below. The barometric records are examined, to see what descent we have made since we left the mouth of the Grand, and what descent since we left the Pacific Railroad, and what fall there yet must be to the river, ere we reach the end of the great cañons. The conclusion to which the men arrive seems to be about this: that there are great descents yet to be made, but, if they are distributed in rapids and short falls, as they have been heretofore, we will be able to over-come them. But, may be, we shall come to a fall in these cañons which we cannot pass, where the walls rise from the water's edge, so that we cannot land, and where the water is so swift that we cannot return. Such places have been found, except that the falls were not so great but that we could run them with safety. How will it be in the future! So they speculate over the serious probabilities in jesting mood, and I hear Sumner remark, 'My idea is, we had better go slow, and learn to paddle.' (*Powell, 1875, p. 62*)

On 17 July they had reached the junction of the Green and Grand Rivers where they broke the journey for a few days. During this break they over-hauled the rations. The musty flour was sieved through mosquito netting and when the lumps had been thrown away there was little more than two month's supply left. Considering they had completed only half the journey this meant a considerable diminishment in rations. By 19 July there were the first signs of general concern about food. The continuous diet of mouldy bread and spoiled bacon was beginning to reveal longings for the more tasty fare of ordinary life. However, the party left the junction on 21 July having dined well on beaver-tail soup. By this time several newspapers, including the *Chicago Inter-Ocean*, had fastened on to a wild rumour spread by an imposter who claimed to be the sole survivor of the expedition and they printed obitu-ary notices for the whole party. Though nothing had been heard from them, Mrs Powell had sufficient confidence in her husband to disbelieve this tale and her faith was rewarded when she received a letter from him written and posted at the Uinta agency.

Though they had a long way to go before they reached the Grand Canyon the party were now entering upon the most testing section of their journey. As they laboriously progressed from Cataract Canyon through Glen Canyon and on to the Grand Canyon itself the cataracts became more numerous and more violent and the canyon walls grew steeper until occasionally they seemed almost to shut out the sky. On 21 July so many portages were necessary that the party only progressed two or three miles. In the future this was often to be the average rate of advance. On 22 July the boats had to be caulked again and new oars made out of driftwood to replace those broken against rocks or swept away in the periodic capsizes. There was no question of master-ing the Colorado; the party had to go how and where it took them. Only by adapting themselves to its violence and withdrawing themselves from its worst dangers did they manage to survive.

> July 24th. – We examine the rapids below. Large rocks have fallen from the walls – great, angular blocks, which have rolled down the talus, and are strewn along the channel. We are compelled to make three portages in suc-cession, the distance being less than three fourths of a mile, with a fall of seventy five feet. Among these rocks, in chutes, whirlpools, and great waves, with rushing breakers and foam, the water finds its way, still tumbling down. We stop for the night, only three fourths of a mile below the last camp. A very hard day's work has been done, and at evening I sit on a rock by the edge of the river, to look at the water, and listen to its roar. Hours ago, deep shadows had settled into the cañon as the sun passed behind the cliffs. Now, doubtless, the sun has gone down, for we can see no glint of light on the crags above. Darkness is coming on. The waves are rolling, with crests of foam so white they seem almost to give a light of their own. Near by, a chute of water strikes the foot of a great block of limestone, fifty feet high, and the waters pile up against it, and roll back. Where there are sunken rocks, the water heaps up in mounds, or even in cones. At a point where rocks come very near

the surface, the water forms a chute above, strikes, and is shot up ten or fifteen feet, and piles back in gentle curves, as in a fountain; and on the river tumbles and rolls. (*Powell, 1875, pp. 62–63*)

But, with a leader made of Powell's determination, the party could not be deterred:

About ten o'clock Powell, Bradley, Howland, Hall, and myself start up a side cañon to the east. We soon come to pools of water; then to a brook, which is lost in the sands below; and, passing up the brook, we find the cañon narrows, the walls close in, are often overhanging, and at last we find ourselves in a vast amphitheater, with a pool of deep, clear, cold water on the bottom. At first, our way seems cut off; but we soon discover a little shelf, along which we climb, and, passing beyond the pool, walk a hundred yards or more, turn to the right, and find ourselves in another dome shaped amphitheater. There is a winding cleft at the top, reaching out to the country above, nearly two thousand feet overhead. The rounded, basin shaped bottom is filled with water to the foot of the walls. There is no shelf by which we can pass around the foot. If we swim across, we meet with a face of rock hundreds of feet high, over which a little rill glides, and it will be impossible to climb. So we can go no farther up this cañon. Then we turn back, and examine the walls on either side carefully, to discover, if possible, some way of climbing out. In this search, every man takes his own course, and we are scattered. I almost abandon the idea of getting out, and am engaged in searching for fossils, when I discover, on the north, a broken place, up which it may be possible for me to climb. The way, for a distance, is up a slide of rocks; then up an irregular amphitheater, on points that form steps and give handhold, and then I reach a little shelf, along which I walk, and discover a vertical fissure, parallel to the face of the wall, and reaching to a higher shelf. This fissure is narrow, and I try to climb up to the bench, which is about forty feet overhead. I have a barometer on my back, which rather impedes my climbing. The walls of the fissure are of smooth limestone, offering neither foot nor hand hold. So I support myself by pressing my back against one wall and my knees against the other, and, in this way, lift my body, in a shuffling manner, a few inches at a time, until I have, perhaps, made twenty five feet of the distance, when the crevice widens a little, and I cannot press my knees against the rocks in front with sufficient power to give me support in lifting my body, and I try to go back. This I cannot do without falling. So I struggle along sidewise, farther into the crevice, where it narrows. But by this time my muscles are exhausted, and I cannot climb longer; so I move still a little farther into the crevice, where it is so narrow and wedging that I can lie in it, and there I rest. Five or ten minutes of this relief, and up once more I go, and reach the bench above. On this I can walk for a quarter of a mile, till I come to a place where the wall is again broken down, so that I can climb up still farther, and in an hour I reach the summit. I hang up my barometer, so give it a few minutes time to settle, and occupy myself in collecting resin from the Piñon pines, which are found in great abundance. One of the principal objects in making this climb was to get this resin, for the purpose of smearing our boats; but I have with me no means of carrying it down. The day is very hot, and my coat was left in camp, so I have no linings to tear out. Then it occurs to me to cut off the sleeve of my shirt, to tie up at one end, and in this little sack I collect about a gallon of pitch. After taking observations for altitude, I wander back on the rock, for an hour or two, when suddenly I notice that a storm is coming from the south. I seek a shelter in the rocks; but when the storm bursts, it comes down as a flood from the heavens, not with gentle drops at first, slowly increasing in quantity, but as if suddenly poured out. I am thoroughly drenched, and almost washed away. It lasts not more than half an hour, when the clouds sweep by to the north and I have sunshine again.

In the mean time, I have discovered a better way of getting down, and I start for camp, making the greatest haste possible. On reaching the bottom of the side cañon, I find a thousand streams rolling down the cliffs on every side, carrying with them red sand; and these all unite in the cañon below, in one great stream of red mud.

Travelling as fast as I can run, I soon reach the foot of the stream, for the rain did not reach the lower end of the cañon, and the water is running down a dry bed of sand; and, although it comes in waves, several feet high and fifteen to twenty feet in width, the sands soak it up, and it is lost. But wave follows wave, and rolls along, and is swallowed up; and still the floods come on from above. I find that I can travel faster than the stream; so I hasten to camp, and tell the men there is a river coming down the cañon. We carry our camp equipage hastily from the bank, to where we think it will be above the water. Then we stand by, and see the river roll on to join the Colorado. Great quantities of gypsum are found at the bottom of the gorge; so we name it Gypsum Cañon. (*Powell, 1875, pp. 63–65*)

The canyon walls now varied in height from two thousand to three thousand feet. Food was becoming an increasing problem but at this stage the party had a stroke of luck.

Late in the afternoon, we pass to the left, around a sharp point, which is somewhat broken down near the foot, and discover a flock of mountain sheep on the rocks, more than a hundred feet above us. We quickly land in a cove, out of sight, and away go all the hunters with their guns, for the sheep have not discovered us. Soon, we hear firing, and those of us who have remained in the boats climb up to see what success the hunters have had. One sheep has been killed, and two of the men are still pursuing them. In a few minutes, we hear firing again, and the next moment down come the flock, clattering over the rocks, within twenty yards of us. One of the hunters seizes his gun, and brings a second sheep down, and the next minute the remainder of the flock is lost behind the rocks. . . .

We lash our prizes to the deck of one of the boats, and go on for a short distance; but fresh meat is too tempting for us, and we stop early to have a feast. And a feast it is! Two fine, young sheep. We care not for bread, or beans, or dried apples tonight; coffee and mutton is all we ask. (*Powell, 1875, p. 66*)

This was the last successful catch the party made, and except a few unsatisfying squashes filched from deserted Indian gardens, was the last substantial supplement to their food supply. Amidst the grimness of the daily scene they were still able to glimpse views of exceptional beauty.

Riding down a short distance, a beautiful view is presented. The river turns sharply to be east, and seems inclosed by a wall, set with a million brilliant gems. What can it mean? Every eye is engaged, every one wonders. On coming nearer, we find fountains bursting from the rock, high overhead, and the spray in the sunshine forms the gems which bedeck the wall. The rocks below the fountain are covered with mosses, and ferns, and many beautiful flowering plants. We name it Vasey's Paradise, in honour of the botanist who travelled with us last year. (*Powell, 1875, p. 76*)

The canyon walls became lower at the entrance of the Frémont River but rose again as the junction was left behind. The party tried to climb out of Glen Canyon at its junction with the San Juan River but failed. By 4 August they had reached the junction with the Paria River and entered Marble Canyon. Below began the granite section whose presence the party learnt to fear because it meant difficult rapids, unscaleable precipices and black forbidding walls. Food was getting still scarcer; the narrowness of the canyon made fuel almost impossible to find; and all the boats were leaking badly. On 7 August to take the men's minds off the monotony of privation Powell organized a group to make readings of an eclipse. Unfortunately they were foiled by rain and losing their way on the return had 'to weather out' a very stormy night. On 13 August they reached the head of the Grand Canyon.

August 13. – We are now ready to start on our way down the Great Unknown. Our boats, tied to a common stake, are chafing each other, as they are tossed by the fretful river. They ride high and buoyant, for their loads are lighter than we could desire. We have but a month's rations remaining. The flour has been resifted through the mosquito net sieve; the spoiled bacon has been dried, and the worst of it boiled; the few pounds of dried apples have been spread in the sun, and reshrunken to their normal bulk; the sugar has all melted, and gone on its way down the river; but we have a large sack of coffee. The lightening of the boats has this advantage: they will ride the waves better, and we shall have but little to carry when we make a portage.

We are three quarters of a mile in the depths of the earth, and the great river shrinks into insignificance, as it dashes its angry waves against the walls and cliffs, that rise to the world above; they are but puny ripples, and we but pigmies, running up and down the sands, or lost among the boulders. (*Powell, 1875, p. 80*)

August 14. – At daybreak we walk down the bank of the river, on a little sandy beach, to take a view of a new feature in the cañon. Heretofore, hard rocks have given us bad river; soft rocks, smooth water; and a series of rocks harder than any we have experienced sets in. The river enters the granite!

We can see but a little way into the granite gorge, but it looks threatening.

After breakfast we enter on the waves. At the very introduction, it inspires awe. The cañon is narrower than we have ever before seen it; the water is swifter; there are but few broken rocks in the channel; but the walls are set, on either side, with pinnacles and crags; and sharp, angular buttresses, bristling with wind and wave polished spires, extend far out into the river. (*Powell, 1875, p. 81*)

Sometimes the walls were so steep that portages were impracticable.

About eleven o'clock we hear a great roar ahead, and approach it very cautiously. The sound grows louder and louder as we run, and as last we find

FIG. 92. *Noon-day rest in Marble Canyon*
(From Powell, 1875)

ourselves in a long, broken fall, with ledges and pinnacles of rock obstructing the river. There is a descent of, perhaps, seventy five or eighty feet in a third of a mile, and the rushing waters break into great waves on the rocks, and lash themselves into a mad, white foam. We can land just above, but there is no foot-hold on either side by which we can make a portage. It is nearly a thousand feet to the top of the granite, so it will be impossible to carry our boats around, though we can climb to the summit up a side gulch, and, passing along a mile or two, can descend to the river. This we find on examination; but such a portage would be impracticable for us, and we must run the rapid, or abandon the river. There is no hesitation. We step into our boats, push off and away we go, first on smooth but swift water, then we strike a glassy wave, and ride to its top, down again into the trough, up again on a higher wave, and down and up on waves higher and still higher, until we strike one just as it curls back, and a breaker rolls over our little boat. Still, on we speed, shooting past projecting rocks, till the little boat is caught in a whirlpool, and spun around several times. At last we pull out again into the stream, and now the other boats have passed us. The open compartment of the *Emma Dean* is filled with water, and every breaker rolls over us. Hurled back from a rock, now on this side, now on that, we are carried into an eddy, in which we struggle for a few minutes, and are then out again, the breakers still rolling over us. Our boat is unmanageable, but she cannot sink, and we drift down, another hundred yards, through breakers; how, we scarcely know. We find the other boats have turned into an eddy at the foot of the fall, and are waiting to catch us as we come, for the men have seen that our boat is swamped. They push out as we come near, and pull us in against the wall. We bail our boat and on we go again. (*Powell, 1875, pp. 82–83*)

A thousand feet of this is up through granite crags, then steep slopes and perpendicular cliffs rise, one above another, to the summit. The gorge is black and narrow below, red and grey and flaring above, with crags and angular projections on the walls, which, cut in many places by side cañons, seem to be a vast wilderness of rocks. Down in these grand, gloomy depths we glide, ever listening, for the mad waters keep up their roar; ever watching, ever peering ahead, for the narrow cañon is winding, and the river is closed in so that we can see but a few hundred yards, and what there may be below we know not; but we listen for falls and watch for rocks, or stop now and then, in the bay of a recess, to admire the gigantic scenery. And ever, as we go, there is some new pinnacle or tower, some crag or peak, some distant view of the upper plateau, some strange shaped rock, or some deep narrow side cañon. (*Powell, 1875, p. 83*)

The labour was continual and always hazardous.

August 15. – This morning we find we can let down for three or four hundred yards, and it is managed in this way: We pass along the wall, by climbing from projecting point to point, sometimes near the water's edge, at other places fifty or sixty feet above, and hold the boat with a line, while two men remain aboard, and prevent her from being dashed against the rocks, and keep the line from getting caught on the wall. In two hours we have brought them all down, as far as it is possible, in this way. A few yards below, the river strikes with great violence against a projecting rock, and our boats are pulled up in a little bay above. We must now manage to pull out of this, and clear the point below. The little boat is held by the bow obliquely up the stream. We jump in, and pull out only a few strokes, and sweep clear of the dangerous rock. The other boats follow in the same manner, and the rapid is passed.

It is not easy to describe the labor of such navigation. We must prevent the waves from dashing the boats against the cliffs. Sometimes, where the river is swift, we must put a bight of rope about a rock, to prevent her being snatched from us by a wave; but where the plunge is too great, or the chute too swift, we must let her leap, and catch her below, or the under-tow will drag her under the falling water, and she sinks. Where we wish to run her out a little way from the shore, through a channel between rocks, we first throw in little sticks of drift wood, and watch their course, to see where we must steer, so that she will pass the channel in safety. And so we hold, and let go, and pull, and lift, and ward, among rocks, around rocks, and over rocks. (*Powell, 1875, pp. 84–85*)

Powell finds relief in describing the fanciful contortions of nature.

Clouds are playing in the cañon today. Sometimes they roll down in great masses, filling the gorge with gloom; sometimes they hang above, from wall to wall, and cover the cañon with a roof of impending storm; and we can peer long distances up and down this cañon corridor, with its cloud roof overhead, its walls of black granite, and its river bright with the sheen of broken waters. Then, a gust of wind sweeps down a side gulch, and, making a rift in the clouds, reveals the blue heavens, and a stream of sunlight pours in. Then, the clouds drift away into the distance, and hang around crags, and peaks, and pinnacles, and towers, and walls, and cover them with a mantle, that lifts from time to time, and sets them all in sharp relief. Then, baby clouds creep out of side cañons, glide around points, and creep back again, into more distant gorges. Then, clouds, set in strata, across the cañon, with intervening vista views, to cliffs and rocks beyond. The clouds are children of the heavens, and when they play among the rocks, they lift them to the region above. (*Powell, 1875, pp. 85–86*)

By 17 August the smallness of the food supply was beginning to worry Powell and made him consider abandoning any further advance.

August 17. – Our rations are still spoiling; the bacon is so badly injured that we are compelled to throw it away. By an accident this morning, the saleratus is lost overboard. We have now only musty flour sufficient for ten days, a few dried apples, but plenty of coffee. We must make all haste possible. . . .

We make ten miles and a half, and camp among the rocks, on the right. We have had rain, from time to time, all day, and have been thoroughly drenched and chilled; but between showers the sun shines with great power, and the mercury in our thermometers stands at 115°, so that we have rapid changes from great extremes, which are very disagreeable. It is especially cold in the rain tonight. The little canvas we have is rotten and useless; the rubber ponchos, with which we started from Green River City, have all been lost; more than half the party is without hats, and not one of us has an entire suit of clothes, and we have not a blanket apiece. So we gather drift wood, and build a fire; but after supper the rain, coming down in torrents, extinguishes it, and we sit up all night, on the rocks, shivering, and are more exhausted by the night's discomfort than by the day's toils. (*Powell, 1875, pp. 88–89*)

FIG. 93. *The inner gorge of the Grand Canyon, looking east from Toroweap*
(From Powell, 1875)

From now on the trip became a contest against time and a weakening bodily spirit.

August 18. – The day is employed in making portages, and we advance but two miles on our journey. Still it rains.

While the men are at work making portages, I climb up the granite to its summit, and go away back over the rust coloured sandstones and greenish yellows shales, to the foot of the marble wall. I climb so high that the men and boats are lost in the black depths below, and the dashing river is a rippling brook; and still there is more cañon above than below. All about me are interesting geological records. The book is open, and I can read as I run. All about me are grand views, for the clouds are playing again in the gorges. But somehow I think of the nine day's rations, and the bad river, and the lesson of the rocks, and the glory of the scene is but half seen. (*Powell, 1875, p. 89*)

Since we left the Colorado Chiquito, we have seen no evidences that the tribe of Indians inhabiting the plateaus on either side ever come down to the river; but about eleven o'clock to day we discover an Indian garden, at the foot of the wall on the right, just where a little stream, with a narrow flood plain, comes down through a side cañon. Along the valley, the Indians have planted corn, using the water which burst out in springs at the foot of the cliff, for irrigation. The corn is looking quite well, but is not sufficiently advanced to give us roasting ears; but there are some nice, green squashes. We carry ten or a dozen of these on board our boats, and hurriedly leave, not willing to be caught in the robbery, yet excusing ourselves by pleading our great want. We run down a short distance, to where we feel certain no Indians can follow; and what a kettle of squash sauce we make! True, we have no salt with which to season it, but it makes a fine addition to our un-leavened bread and coffee. Never was fruit so sweet as these stolen squashes. (*Powell, 1875, pp. 95–96*)

Some of the others in the party did not share Powell's optimism or his scientific devotion. Bradley wrote in his diary;

August 2nd . . . doomed to be here another day, perhaps more than that for Major has been taking observations ever since we came here and seems no nearer done now than when he began. He ought to get the latitude and longitude of every mouth of a river not known before and we are willing to face starvation if necessary to do it but further than that he should not ask us to wait, and he must go on soon or the consequences will be different from what he anticipates. If we could get game or fish we should be all right, but we have not caught a single mess of fish since we left the junction. (*Darrah, 1951, pp. 135–6*)

As the privations continued a note of resentment entered into the feelings of the men, and by 11 August Bradley's tone is more bitter:

If Major does not do something soon I fear the consequences, but he is contented and seems to think that biscuit made of sour and musty flour and a few dried apples is ample to sustain a laboring man. If he can only study geology he will be happy without food or shelter, but the rest of us are not afflicted with it to an alarming extent. (*Darrah, 1951, p. 137*)

Even temporary improvements in the grim conditions did not raise the spirits of the party.

> The trip was resumed on August 17th with almost continuous rapids and two gruelling portages. In one let-down the *Maid* struck a rock, loosening her head block. The boats had deteriorated too much to withstand much more battering. After the damage was repaired the men examined several old Pueblo ruins and collected fragments of pottery.
>
> The rations were now barely sufficient to sustain life, the diet being reduced to heavy bread and coffee – reason enough for sinking morale. But the treacherous river, demanding hard work with disappointing forward progress to show for it, helped not one bit – only nineteen miles passed in three days. Even with a fair run of eight miles on the 20th, and better than twenty next day, the dashing wildness of the river and the gnawing hunger reminded them that such progress was not enough. The canyon walls stood more than four thousand feet high and the labyrinth of lateral canyons, some almost as large as the main canyon, cast a shadow of dread over the minds of the miserable men on the river below.
>
> On the 23rd for the first time, some of the party showed signs of losing heart. Below the camp spread a rapid which Bradley regarded as easy and safe. The Major was noncommital, but most of the men preferred to make a portage rather than to risk a run.
>
> Next morning the majority did not feel any more willing to attempt it, so the day began with a long haul. Twenty-two miles were accomplished on the 24th and thirty-five miles on the 25th. According to the Mormon estimate the distance to Grand Wash from the Little Colorado was seventy to eighty miles. Yet they had already run more than 120 miles; how much more remained they had no idea. That day the last sack of flour was opened, with somber ceremony, at a cheerless supper. (*Darrah, 1951, pp. 139–40*)

On 20 August they had passed out of the granite and on to slate. This allowed better going and, in spite of continuous rapids, the boats could run as much as 35 miles a day. On 24 August they were still making good progress and wasting none of the daylight but the lack of food remained a major concern. On 27 August granite was encountered again and Bradley described his own reactions.

> The water dashes against the left bank and then is thrown furiously back against the right. The billows are huge and I fear our boats could not ride them if we could keep them off the rocks. The spectacle is appalling to us. We have only subsistence for about five days and have been trying half a day to get around this one rapid while there are three others in sight below. What they are we cannot tell, only that they are huge ones. If we could let our boats down to that point and then have a foothold all the rest of the way but we have tried all the P.M. without success. Shall keep trying tomorrow and I hope by going farther back in the mountains and then coming down opposite we may succeed. Think Major has now gone to try it. There is discontent in camp tonight and I fear some of the party will take to the mountains but hope not. This is decidedly the darkest day of the trip but I don't despair yet. I shall be one to try to run it rather than take to the mountains. (*Darrah, 1951, p. 140*)

At this stage the elder Howland, his brother and Dunn announced that they were not prepared to go any farther by water and suggested that the party should climb out of the canyon and make for one of the nearby Mormon settlements where they could be certain of obtaining fresh food. This was the moment of crisis for Powell. He knew that the success and failure of the expedition were almost evenly balanced. His report clearly shows the thoughts that were passing through his mind and the conclusions that ultimately dominated his decision.

After supper Captain Howland asks to have a talk with me. We walk up the little creek a short distance, and I soon find that his object is to remonstrate against my determination to proceed. He thinks that we had better abandon the river here. Talking with him, I learn that his brother, William Dunn, and himself have determined to go no farther in the boats. So we return to camp. Nothing is said to the other men.

For the last two days, our course has not been plotted. I sit down and do this now, for the purpose of finding where we are by dead reckoning. It is a clear night, and I take out the sextant to make observation for latitude, and find that the astronomic determination agrees very nearly with that of the plot – quite as closely as might be expected, from a meridian observation on a planet. In a direct line we must be about forty five miles from the mouth of the Rio Virgen. If we can reach that point, we know that there are settlements up that river about twenty miles. This forty-five miles, in a direct line, will probably be eighty or ninety in the meandering line of the river. But then we know that there is comparatively open country for many miles above the mouth of the Virgen, which is our point of destination.

As soon as I determine all this, I spread my plot on the sand, and wake Howland, who is sleeping down by the river, and show him where I suppose we are, and where several Mormon settlements are situated.

We have another short talk about the morrow and he lies down again; but for me there is no sleep. All night long, I pace up and down a little path, on a few yards of sand beach, along by the river. Is it wise to go on? I go to the boats again, to look at our rations. I feel satisfied that we can get over the danger immediately before us; what there may be below I know not. From our outlook yesterday, on the cliffs, the cañon seemed to make another great bend to the south, and this, from our experience, heretofore, means more and higher granite walls. I am not sure that we can climb out of the cañon here, and, when at the top of the wall, I know enough of the country to be certain that it is a desert of rock and sand, between this and the nearest Mormon town, which, on the most direct line, must be seventy five miles away. True, the late rains have been favorable to us, should we go out, for the probabilities are that we shall find water still standing in holes, and, at one time, I almost conclude to leave the river. But for years I have been contemplating this trip. To leave the exploration unfinished, to say that there is a part of the cañon which I cannot explore, having already almost accomplished it, is more than I am willing to acknowledge, and I determine to go on.

I wake my brother, and tell him of Howland's determination, and he promises to stay with me; then I call up Hawkins, the cook, and he makes a like promise; then Sumner, and Bradley, and Hall, and they all agree to go on.

August 28. – At last daylight comes, and we have breakfast, without a word

being said about the future. The meal is as solemn as a funeral. After breakfast, I ask the three men if they still think it best to leave us. The elder
Howland thinks it is, and Dunn agrees with him. The younger Howland
tries to persuade them to go on with the party, failing in which, he decides to
go with his brother.

Then we cross the river. The small boat is very much disabled, and un-
seaworthy. With the loss of hands, consequent on the departure of the three
men, we shall not be able to run all of the boats, so I decide to leave my *Emma
Dean*.

Two rifles and a shot gun are given to the men who are going out. I ask
them to help themselves to the rations, and take what they think to be a fair
share. This they refuse to do, saying they have no fear but that they can get
something to eat; but Billy, the cook, has a pan of biscuits prepared for
dinner, and these he leaves on a rock.

Before starting, we can take our barometers, fossils, the minerals, and some
ammunition from the boat, and leave them on the rocks. We are going over
this place as light as possible. The three men help us lift our boats over a rock
twenty five or thirty feet high, and let them down again over the first fall,
and now we are ready to start. The last thing before leaving, I write a letter
to my wife, and give it to Howland. Sumner gives him his watch, directing
that it be sent to his sister, should he not be heard from again. The records
of the expedition have been kept in duplicate. One set of these is given to
Howland, and now we are ready. For the last time, they entreat us not to go
on, and tell us that it is madness to set out in this place; that we can never
get safely through it; and, further, that the river turns again to the south into
the granite, and a few miles of such rapids and falls will exhaust our entire
stock of rations, and then it will be too late to climb out. Some tears are shed;
it is rather a solemn parting; each party thinks the other is taking the
dangerous course.

My old boat left, I go on board the *Maid of the Cañon*. The three men
climb a crag, that overhangs the river, to watch us off. The *Maid of the Cañon*
pushes out. We glide rapidly along the foot of the wall, just grazing one great
rock, then pull out a little into the chute of the second fall, and plunge over it.
The open compartment is filled when we strike the first wave below, but we
cut through it, and then the men pull with all their power toward the left
wall, and swing clear of the dangerous rock below all right. We are scarcely a
minute in running it, and find that, although it looked bad from above, we
have passed many places that were worse.

The other boat follows without more difficulty. We land at the first
practicable point below and fire our guns, as a signal to the men above that
we have come over in safety. Here we remain a couple of hours, hoping that
they will take the smaller boat and follow us. We are behind a curve in the
cañon, and cannot see up to where we left them, and so we wait, until their
coming seems hopeless, and push on. (*Powell, 1875, pp. 98–100*)

This was the one sad episode of the expedition, for, though the three deserters
climbed out of the canyon safely and reached an Indian settlement, they were
killed by the Indians who mistook them for miners who had lately assaulted
their womenfolk. If they had not left when they did there is no doubt that
they would have lived to share the party's success. There were further anxious
moments before this success was achieved.

And now we have a succession of rapids and falls until noon, all of which we run in safety. Just after dinner we come to another bad place. A little stream comes in from the left and below there is a fall, and still below another fall. Above, the river tumbles down, over and among the rocks, in whirlpools and great waves, and the waters are lashed into mad, white foam. We run along the left above this, and soon see that we cannot get down on this side, but it seems possible to let down on the other. We pull up stream again, for two or three hundred yards, and cross. Now there is a bed of basalt on this northern side of the cañon, with a bold escarpment, that seems to be a hundred feet high. We can climb it, and walk along its summit to a point where we are just at the head of the fall. Here the basalt is broken down again, so it seems to us, and I direct the men to take a line to the top of the cliff, and let the boats down along the wall. One man remains in the boat, to keep her clear of the rocks, and prevent her line from being caught on the projecting angles. I climb the cliff, and pass along to a point just over the fall, and descend by broken rocks, and find that the break of the fall is above the break of the wall, so that we cannot land; and that still below the river is very bad, and that there is no possibility of a portage. Without waiting further to examine and determine what shall be done, I hasten back to the top of the cliff, to stop the boats from coming down. When I arrive, I find the men have let one of them down to the head of the fall. She is in swift water, and they are not able to pull her back; nor are they able to go on with the line, as it is not long enough to reach the higher part of the cliff, which is just before them; so they take a bight around a crag. I send two men back for the other line. The boat is in very swift water, and Bradley is standing in the open compartment, holding out his oar to prevent her from striking against the foot of the cliffs. Now she shoots out into the stream, and up as far as the line will permit, and then, wheeling, drives headlong against the rock, then out and back again, now straining on the line, now striking against the rock. As soon as the second line is brought, we pass it down to him; but his attention is all taken up with his own situation, and he does not see that we are passing the line to him. I stand on a projecting rock, waving my hat to gain his attention, for my voice is drowned by the roaring of the falls. Just at this moment, I see him take his knife from its sheath, and step forward to cut the line. He has evidently decided that it is better to go over with the boat as it is, than to wait for her to be broken to pieces. As he leans over, the boat sheers again into the stream, the stem-post breaks away, and she is loose. With perfect composure Bradley seizes the great scull oar, places it in the stern rowlock, and pulls with all his power (and he is an athlete) to turn the bow of the boat down stream, for he wishes to go bow down, rather than to drift broadside on. One, two strokes he makes, and a third just as she goes over, and the boat is fairly turned, and she goes down almost beyond our sight, though we are more than a hundred feet above the river. Then she comes up again, on a great wave, and down and up, then around behind some great rocks, and is lost in the mad, white foam below. We stand frozen with fear, for we see no boat. Bradley is gone, so it seems. But now, away below, we see something coming out of the waves. It is evidently a boat. A moment more, and we see Bradley standing on deck, swinging his hat to show that he is all right. But he is in a whirlpool. We have the stempost of his boat attached to the line. How badly she may be disabled we know not. I direct Sumner and Powell to pass along the cliff, and see if they can reach him from below. Rhodes, Hall, and myself run to the other boat, jump aboard, push out, and away we go over the falls. A wave rolls over us, and our boat is unmanageable. Another great wave strikes us, the boat rolls over, and tumbles and tosses, I know not how. All I know is that Bradley is picking us up. We soon have all right again, and row to the cliff, and wait until Sumner and Powell can come. After a difficult climb they reach us. We run two or three miles farther, and turn again to the northwest, continuing until night, when we have run out of the granite once more. (*Powell, 1875, pp. 100–2*)

Then at 12 o'clock on 29 August they emerged from the Grand Canyon at the Virgen River and success was theirs. On 30 August they met Mr Asa and two of his sons who had been sent out from the neighbouring Mormon settlement to search for the wreckage of the supposedly ill-fated expedition. By 31 August they had sent back word to the settlement and the party were dining on fresh supplies. The relief was great and the feeling of success exhilarating.

> Now the danger is over; now the toil has ceased; now the gloom has disappeared; now the firmament is bounded only by the horizon; and what a vast expanse of constellations can be seen!
> The river rolls by us in silent majesty; the quiet of the camp is sweet; our joy is almost ecstacy. We sit still long after midnight, talking of the Grand Cañon, talking of home, but chiefly talking of the three men who left us. Are they wandering in those depths, unable to find a way out? are they searching over the desert lands above for water? or are they nearing the settlements? (*Powell, 1875, p. 103*)

Powell was justified, but the margin had been very fine. When they met Asa at the junction with the Virgen River there were left 10 lb. of flour, 15 lb. of dried apples and 70 lb. of coffee. They had descended five thousand feet in the most hazardous conditions. It was a fine achievement of endurance and made at a cost that was almost too great.

Such was the spirit of the American pioneers, and the type of environment they had to combat. The plateau region consists almost entirely of many plateaus built of layer upon layer of horizontal strata sometimes extending without a break for hundreds of miles. Upon its surface vegetation was sparse and the rocks were exposed to the eye in their natural colours and geologic formations. Here and there lava had intruded providing a preservative cover for the underlying older formations. Below this plateau scene of structural grandeur, the deep canyons pursued their tortuous courses and revealed the world's highest vertical geological sections.

Though the region was peopled by Indians they were mainly Utes who were warlike but so primitive that they could not match the white men. The other local tribe was the more or less peaceful Navajos who lived in the 'Four Corners Area' at the junction of the states of Colorado, Utah, New Mexico and Arizona. Even before they were expelled from the territory the Indians did not form a real obstacle to entry as they were relatively few. The dryness of the soil and the difficulty of passage across the dissected terrain deterred them from settling as much as it did the white man.

FIG. 94. *Exit of the Colorado River from the Grand Canyon*
(Photograph by W. T. Lee, U.S. Geological Survey)

Thus many factors were auspicious to the disclosure of an astounding wealth of geomorphic information once white men penetrated the area. Consequently, the Western Explorations, and particularly the study of the erosional features of the Colorado plateaus, introduced a completely new dimension into geomorphology. Up to this time the most spectacular relief features had been studied in regions of intense crustal deformation, like the Alps and North Wales, in regions of well-marked fracturing and large-scale faulting, or of ancient or modern igneous activity, like the Lake District and the Massif Central. In all these mountain landscapes, moreover, there was undisputable evidence of what Ramsay and others recognized as extensive glacial excavation. Nowhere before in these latitudes had it been possible to dissociate so obviously fluvial erosional features from structural, tectonic and glacial influences. But the Colorado canyons are cut deep into great thicknesses of horizontal beds which have been uplifted without great complications due to fracturing, folding or igneous activity. Their obviously fluvial pattern and form could be observed to bear little relationship to any gross structural influences, except to the hypothetical original slope of the surface. The paucity of soil cover and of vegetation enabled geologists to get at a glance extremely accurate ideas of the geological structures, especially as many of the individual thick formations are most distinctively coloured and extend over hundreds of square miles without any marked lithological change. Moreover, because of the arid climate and the relative rapidity of stream incision, the inner canyon walls have not been worn back excessively, and so they tightly enclose the rivers, mirror their curves and provide clear proof that the canyons were cut by the streams themselves.

Here in tremendous magnification were the answers to the doubts that had been troubling European geologists for a century. After the Western Explorations there could be no more question of doubting the ability of rivers to erode, no more confusing talk of structural control and marine planation. Here was river erosion at its grandest, creating topographic forms on a gigantic scale. Historically the explorations mark the end of a geological epoch, in other circumstances they could well have marked the beginning of one:

> Had the cradle of Geology been in the Far West, the science would doubtless have grown stronger and reached a riper manhood than it has yet attained. The history of the rocks is there so simple that a child could read, and rival theories with their accompanying *odium geologicum* are out of the question. But great discoveries are only received when the spirit of the age is ready to accept them, . . . It is only lately that we have reached the stage when the wonders of the West can be rightly appreciated. (*Cadell, 1887, p. 456*)

CHAPTER TWENTY-SIX

The Earliest Western Geological Explorations

PRE-CIVIL WAR

Although John Wesley Powell is archetypal of Western pioneer geologists, he was by no means the first. The expeditions to the West in the early nineteenth century were mainly reconnaissances to discover the geographical location of resources and to test the possibilities of trade. The reports brought back encouraged the government to send out its own survey teams, the first being organized in 1834. But these parties lacked the services of skilled geologists. Consequently, from our point of view, the first notable journeys were those of Newberry and Hayden.

John Strong Newberry (1822–92) was born at Windsor, Connecticut, and in 1848 became a Doctor of Medicine at Cleveland where, after a spell in Paris, he opened a practice. While studying at the medical college he had been interested in Coal Measure plants and he showed an unfailing devotion to palaeontology all his life (Stevenson, 1893).

In 1855 he was chosen as surgeon and naturalist for the first of the truly scientific expeditions into the West. This exploration, led by Lieut. Williamson, traversed northern California and Oregon with a view to surveying possible routes for future railways. Newberry's presence enabled the party to discern much of the significance of the topographic features seen.

In 1857 he joined Lieut. Joseph Christmas Ives in an expedition to the lower Colorado River, which, it was hoped, would prove a water route up which supplies could be carried to the new forts in Utah and Arizona. The party sailed from the mouth of the river at midnight on 31 December 1857 in the iron steamboat *Explorer*. Just after leaving Fort Yuma, on 11 January 1858, the *Explorer* ran aground on the first of many sandbanks while still in view of the jeering garrison and a series of similar groundings and rapids caused them to forsake the river at the head of Black Canyon (in the lower part of the present Lake Mead) on 14 March. Ives, Newberry and the German topographer Von Egloffstein then proceeded overland and descended into the Grand Canyon from the south side at Diamond Creek. Ives, probably the first white man to set foot on the floor of the Grand Canyon, subsequently described his impressions:

> . . . the increasing magnitude of the colossal piles that blocked the end of the vista, and the corresponding depth and gloom of the gaping chasms into

FIG. 95. *John Strong Newberry*

which we were plunging, imparted an unearthly character to a way that might
have resembled the portals of the infernal regions. Harsh screams issuing
from aerial recesses in the canyon sides, and apparitions of goblin-like
figures perched in the rifts and hollows of the impending cliffs, gave an odd
reality to this impression. (*Ives, 1861, p. 100*)

After spending two perilous days in following an Indian trail to the canyon
rim, the party subsequently visited Cataract Canyon before finally returning
to Santa Fé. The report of the journey was published in 1861, and in Part 3
Newberry stressed the monopoly of river erosion:

I may say here, however, that, like the great cañons of the Colorado, the
broad system of valleys bounded by high and perpendicular walls belong to

FIG. 96. *The first illustration of the Grand Canyon, drawn in a 'Gothic' manner by the German topographer F. W. von Egloffstein in 1857 from the floor of the canyon near Diamond Creek. Von Egloffstein was a member of the Ives Expedition (Yale University Library)*

a vast system of erosion, and are wholly due to the action of water. Probably nowhere in the world has the action of this agent produced results so surprising, both as regards their magnitude and their peculiar character. It is not at all strange that a cause, which has given to what was once an immense plain, underlaid by thousands of feet of sedimentary rocks, conformable throughout, a topographical character more complicated than that of any mountain chain; which has made much of it absolutely impassable to man, or any animal but the winged bird, should be regarded as something out of the common course of nature. Hence the first and most plausible explanation of the striking surface features of this region will be to refer them to that embodiment of resistless power – the sword that cuts so many geological knots – volcanic force. The Grand Canyon of the Colorado would be considered a vast fissure or rent in the earth's crust, and the abrupt truncation of the steps of the table-lands as marking lines of displacement. This theory though plausible, and so entirely adequate to explain all the striking phenomena, lacks the single requisite to acceptance, and that is truth.

Aside from the slight local disturbance of the sedimentary rocks about the San Francisco mountain, from the spur of the Rocky Mountains, near Fort Defiance, to those of the Cerbat and Aztec mountains on the west, the strata of the table-lands are as entirely unbroken as when first deposited. Having this question constantly in mind, and examining with all possible care the structure of the great cañons which we entered, I everywhere found evidence of the exclusive action of water in their formation. The opposite sides of the deepest chasm showed perfect correspondence of stratification, conforming to the general dip, and nowhere displacement; and this bottom rock, so often dry and bare, was perhaps deeply eroded, but continuous from side to side, a portion of the yet undivided series lying below. (*Newberry, 1862, p. 398*)

. . . the erosion of rocks is always subaerial, or at least never takes place more than forty feet below the ocean surface, . . . (*Newberry, 1862, p. 399*)

However, Newberry believed that the mountain ranges had impounded the river in a series of basins which, on overflowing, had led to the cutting of the canyons . . . 'through which its turbid waters now flow with rapid and almost unobstructed current from source to mouth'. A further factor explaining the origin of the local geology he saw in a change of climate:

Everything indicates that the table-lands were formerly much better watered than they now are. (*Newberry, 1861, p. 47*)

In 1859 Newberry went on his third and last expedition. The party, under Capt. John N. Macomb, went from Santa Fé to explore a new military route into Utah, penetrating to within 6 miles of the junction of the Green and Grand Rivers. The subsequent report has become famous for its graphic descriptions of the scenery but geologically it added nothing to what Newberry had said in 1861.

The plain which stretches westward from the Sierra de la Plata . . . is part of an immense plateau which once stretched continuously far beyond the course of the Colorado. How it has been divided by the cañons of the draining streams, and how, by erosion, its plateau character has been so modified as to be locally lost, will fully appear in the progress of our geological narra-

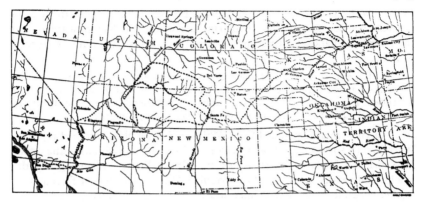

FIG. 97. *Map of routes traversed by Newberry, Ives and Macomb*

tive; yet no one who observes the orderly and unbroken arrangement of its underlying rocks, the perfect correspondence of its sections on opposite sides of the profound cañons which cut it, would hesitate to assent to the assertion that an unbroken table-land once stretched from the base of the Sierra de la Plata all the way across to the mountain chains west of the Colorado, and that from this plateau, grain by grain, the sedimentary materials which once filled the broad and deep valleys of the Colorado and San Juan have been removed by the currents of these streams. The mind is awestruck in the contemplation of the magnitude of the element of time which enters into the analysis of the process by which these stupendous monuments of erosion have been produced; but if the numerical faculty is baffled in the effort to count the years or ages which must have been consumed in the erosion of the valleys and cañons to which I have referred, even the imagination itself is lost when called upon to estimate the cycles on cycles during which the much grander features of the high table-lands were wrought from a plateau which once over-spread most of the area of the Colorado Basin, burying the present Sage-plain 2000 feet beneath its upper surface. (*Newberry, 1876, p. 84*)

In fact, the publication of this report was interrupted by the Civil War and it was not printed until 1876, when its content had lost much of its force through the simultaneous publication of similar and better accounts on the same area. After the Civil War Newberry was appointed Professor of Geology in the Columbia School of Mines and he drops out of the history of landscape studies in the West as he devoted himself entirely to stratigraphy and palaeontology.

The most notable rival of Newberry as a Western geologist in the pre-Civil War period was Ferdinand Vandeveer Hayden (1829–1887). Like Newberry, he was a doctor and had in fact studied surgery under him. Hayden had worked in the Dakota badlands and the upper Missouri in 1853–4, but his first real opportunity came when he joined the expedition led by Lieut. Gouverneur Kemble Warren which left Fort Pierre, Dakota on 3 June 1856 and went up the Missouri to above the junction with the Yellowstone in order to assist with the Northern Pacific Railroad preliminary survey. In the following year Hayden and Warren, exploring the Platte and Niobrara rivers and the Dakota badlands, ventured into the southern Black Hills but were turned back by:

> . . . a large force of the Dakotas who made such earnest remonstrances and threats against our proceeding into their country that I did not think it prudent for us, as a scientific expedition, to venture further in that direction. (*Report by Lieut. Warren, quoted in Goetzmann, 1959, p. 414*)

In 1859 Hayden accompanied Capt. William F. Raynolds, who had replaced Warren, on a survey of natural resources and Indian dispositions in the upper Missouri and Yellowstone areas. At the Upper Platte Indian Agency the whole command, except the Commanding Officer who was a fervent teetotaller, became hopelessly drunk, and the subsequent court martial forced them to spend the winter there.

The results of some of this work were expressed in reports published in 1861 and 1862. Hayden's explanation of the mountain landscape went deeper than Newberry's, and to explain the presence of the canyons he invoked the principle of antecedence.

> Another illustration of the gradual and long continued rise of the country may be found in the immense chasms or cañons which have been formed by the streams along the mountain sides. We can only account for them on the supposition that as the anticlinal crest was slowly emerging from the sea, the myriad sources of our great rivers were seeking their natural channel, and that these branches or tributaries began this erosive action long before the great thoroughfares, the valleys of the Mississippi and Missouri, were marked out. The erosion would go on as the mountains continued slowly rising at an almost imperceptible rate, and in process of time the stupendous channels which everywhere meet us along the immediate sides of the mountains would be formed. (*Hayden, 1862, p. 312*)

He thought that the vertical uplift began at the close of the Cretaceous period and continued well into the Tertiary. The limits of the region affected by uplift were correspondingly marked by sharp flexures. He noticed also the amount of horizontal erosion that had taken place simultaneously with the vertical incision of the rivers.

> . . . the White River Tertiary beds, . . . extend southward along the Laramie Mountains . . . they also extend up the North Platte to the Box Elder Creek, and even beyond are small outliers, showing that much has been removed by erosion. (*Hayden, 1861, p. 241*)

There are really no other prominent Western geologists at this time. Gouverneur Kemble Warren, who has already been mentioned in connexion with Hayden, gave in his report an interesting but vague description of certain features of the Niobrara River:

> The slope of all this part of the plains . . . has therefore its line of greatest descent in a northeast direction, and north of the Niobrara; this is the direc-

tion in which a majority of the rivers flow till they join with the Missouri or Yellowstone. To the south of the Niobrara the greatest slope of the plains is to the southeast, . . . and this is the direction pursued there by nearly all the rivers of the plains. Thus the Niobrara would seem, as it were, to run along a swell or ridge on the surface. The average slope of the plains from the Missouri to the mountains nowhere make an angle with the horizon greater than one-half degree.

A remarkable feature with regard to this change of slope which occurs in the neighbourhood of the course of the Niobrara is the shortness of its tributaries, the surface drainage seeming to be away from and not towards its banks. A result of this is the absence of the amphitheatre-like valley which rivers generally have, and which enable us to look down at the stream often many miles distant. Through the greater portion of the middle half of its course you have scarcely any indication of it as you approach, till within close proximity, and then you look down from the steep bluffs, and catch, at the distance of two hundred to five hundred yards, only here and there a glimpse of the river below, so much is it hidden by the precipitous bluffs which at the bends stand at the water-edge. So strongly was I impressed with the fact that the surface drainage could never have been directed along its course so as to have worn out this channel, that I think a portion of it must have originated in a fissure in the rocks which the waters have since enlarged and made more uniform in size, and which the soft nature of the rock would render easy of accomplishment. (*Warren, 1859, pp. 379–80*)

POST-CIVIL WAR

The Civil War caused the cessation of geologic expeditions to the West and the drafting of their members into one or other of the armies. Lieut. Ives survived but died shortly afterwards, Hayden served throughout as a surgeon, Warren became a General and Powell a major.

Yet during the war the pioneers still spread out through the West and at the close of hostilities abandoned projects were quickly resumed. Though the war had been long and the losses severe on both sides it was only in the south that the social organism was disrupted and the economy brought to a standstill. Furthermore, now that rivalry between north and south had been removed, the whole of the West lay open for exploitation. Soon Americans turned to the West as a possible means of replenishing their depleted coffers.

The labour of surveying the vast, unknown areas was undertaken by several independent parties. Four main surveys were organized under the leadership respectively of King, Hayden, Powell and Wheeler. All were directly or indirectly under the auspices of the government.

Clarence King was in 1867 appointed to conduct a survey of mineral wealth along the route to be traversed by the Pacific Railroad. His party surveyed a strip about 100 miles wide along the fortieth parallel between the Colorado Front Range and the Sierra Nevada. King was an efficient geologist but of rather an unassuming nature.

Hayden was the reverse. Though a gifted geologist, he had a fondness for publicity which in later years made him hurry work which would have profited from a more sustained concentration. At the end of the war Hayden accepted the post of Professor of Mineralogy and Geology at the University of Pennsylvania. In 1867 he also became a government geologist and during the next eleven years undertook surveys in Colorado, Wyoming, Montana and Idaho. Annual reports described the results of these efforts. One of his most gifted colleagues was A. R. Marvine who directed the Middle Park section of the Colorado survey.

Lieut. G. N. Wheeler as the Army's geologist was ordered in 1869 to make a military and topographic reconnaissance of south-west Nevada and western Utah (Merrill, 1924; Smith, 1924). On 16 September 1871 he took a party of 34, including 14 Indians, in flat-bottomed rowing boats from Camp Mohave up the Colorado to Diamond Creek in the Grand Canyon, which was reached on 19 October. The expedition was not a success because the craft chosen were ill-designed for the 208 rapids which were encountered, and the party returned overland. G. K. Gilbert was a member of this expedition and worked with Lieut. Wheeler for three years in all (Davis, 1922). During these three field seasons Gilbert covered the following areas:

21st April, 1871–3rd January, 1872; The Great Basin of Nevada, the Arizona Plateaus as far east as Mt. San Francisco, and down the Gila River to Yuma.
June, 1872–12th December 1872; South-western Utah and north-western Arizona, north of the Grand Canyon.
July, 1873–December 1873; Western New Mexico and Eastern Arizona.

Eventually he found military jurisdiction irksome, particularly the slow publication of his written reports, and he joined Powell's party. Marvine was also for a while a member of Wheeler's team.

Powell was the only survey leader not directly under government control. He received a government grant and had to send a share of specimens to the Smithsonian Institution, but strictly his party was under the control of his college and virtually under his own independent direction. After two journeys down the Colorado, he lived with the Indians from 1873 onwards, studying their habits and language, and acting as a special commissioner for their establishment in reservations.

By now Powell's staff included such talented geologists as Gilbert, Dutton and C. A. White. In 1875, with their help, he published a report on the Colorado and followed this in 1877 with an account of the arid regions. The latter appeared at a time when the public were keenly interested in the future of the West. Whereas in 1865 there were only 3,600,000 settlers west of the Mississippi, by 1877 the population had risen to 13,000,000. Already the cattlemen had absorbed large areas of land and were contending with the miners and homesteaders. By fraudulent appropriation and other more honest means large companies were swallowing up the remaining land and the all-important

FIG. 98. *The Wheeler Expedition encamped near Belmont, Nevada*
(U.S. Geological Survey)

FIG. 99. *The Hayden Survey Party encamped at Red Buttes, Natrona County,*
Wyoming, in August 1870
(Photograph by W. H. Jackson, U.S. Geological Survey)

water supplies. Powell believed that surveying should be carried out with an
eye to the land's future utilization and for this purpose the land should not be
merely explored but examined in every respect. As each survey adopted
different methods, Powell began to campaign for a consolidation of all independent parties, which included the other three groups already mentioned
as well as the Land Parcelling Survey under the Department of the Interior
and the Coast and Geodetic Survey under the Department of the Treasury.
Powell pointed out how wasteful it was that he and King should publish their
results on maps of a scale of four miles to the inch with contours while others
used crude hachuring and scales of eight miles to the inch. He also complained
that other parties received a larger share of government appropriations which
often went to finance expeditions of little scientific value and generally unrelated to work already done. Hayden also campaigned in a similar vein and
each party's particular merits were debated, often with bitter invective and
personal recrimination.

Hon. J. A. Garfield.   Jan. 20. 1877.
My Dear Sir;
    I know you are terribly busy and very much engrossed in the settlement of
the exciting Presidential question; but I venture to ask you to turn from that
for a moment and listen to a few words in behalf of science.

FIG. 100. *The Hayden Field Party at La Veta Pass, southern Colorado, in August 1877. Hayden is seated on the right, Sir Joseph Hooker on the left with Asa Gray on the ground next to him* (*U.S. Geological Survey*)

I learn that the matter of appropriations for Western explorations and surveys is soon to come before Congress, and it is reported that there will be some retrenchment in the expenditure for this purpose.

Now while sharing the conviction which I know you entertain that the condition of the country requires the most rigid economy on the part of both people and government, and fully approving the spirit that prompts such retrenchment, it seems to me important that retrenchment if resolved upon, may be wise and discriminating and may not be so applied as to cripple good men, and useful enterprise, and leave the less worthy to flourish.

You know there are two civilian parties now engaged in Western Explorations, those of F. Hayden and Major Powell; and as I am intimately acquainted with these men, and the work of each, I take the liberty of reporting to you my judgment of their respective merits.

Briefly, for I know brevity will please you best, Hayden has come to be so much of a fraud that he has lost the sympathy and respect of the scientific men of the country and it may well be questioned whether he and his enterprizes should be generously assisted as they have been. In former times he was an energetic and successful explorer, and although his individual work had little scientific value, he has been the means of causing much good work to be done by others. Of late years, however, he has come to be simply the political manager of his expeditions, has spent most of his time in Washington where he has in some way accumulated a handsome property. His influence in Congress has been maintained by the hearts of the lobbyist, such as giving employment to the relatives of those by whose influence he was assisted, making large expenditure for photographs, which were distributed with an eye to political effect. For years a son and brother-in-law of Sen. Logan were connected with his parties and he thereby gained an efficient friend in the Senate. Mr Holman was propitiated by the same means, and I could give you other similar cases. At the same time his work has deteriorated. By the death of Marvine, and the withdrawal of Gardner; whom you know, he lost his best men and has not one who can do in geology and topography what they did. The most important contribution lately made to science through his agency are papers on Paleontology prepared by experts from materials obtained through his collectors. These, however interesting to scientists, seem hardly to belong to the category of necessities for which the expenditures of government should be chiefly made in these hard times.

Major Powell, however, belongs to quite a different class. He and his assistant, Mr G. K. Gilbert, are men of first rate ability and are inspired by true scientific enthusiasm. Their work has been mainly done in the field by themselves and is chiefly geologic and topographic. In quality it ranks with the best done anywhere. It is also cheaply and honestly done. Major Powell, instead of being enriched by his connections with the government, has spent quite a large supply of his fortune. In these circumstances it seems to me that if entrenchment must come, it should fall on Hayden rather than Powell and my own judgment is that it would be wise to increase rather than diminish Major Powell's appropriation.

I need hardly say to you that I have no relations with either of the gentlemen whose names I have mentioned which could modify my opinion of their merit, unless indeed in F. Hayden's favor. He was formerly a student of mine, and I have been his sincere and efficient friend through many years. Still I now feel compelled to speak of him as I do.

With Major Powell my relations are only those of scientific friends and the good words I say of him and Mr Gilbert are truthful expression of my estimate of their merits.

If your judgment in this matter should coincide with mine, you will perhaps say a kind word for these gentlemen as you may have opportunity.

Yours very truly,

J. S. NEWBERRY.

(Darrah, 1951, pp. 240–1)

Powell, Hayden and Wheeler were questioned by a Congressional committee but nothing was done until 1879 when Congress formed a Geological Survey and Land Office under the Department of the Interior. King was appointed the first director and Powell the director of the newly-created Bureau of Ethnology. After one year King resigned his post and was succeeded by Powell.

Thus the task of surveying the United States was not consolidated into one coherent system until 1879 and American geological ideas had received a great stimulus long before that event. The three men principally responsible for these early advances were Powell, Gilbert and Dutton. It was Powell's survey group that provided most of the theories upon which future American geomorphology was based;

(The Powell Survey) was the least pretentious of the four operating during this period – it covered less area, expended less public money, and published

FIG. 101. *The floor of the Grand Canyon near Paria Creek, looking west*
(Photograph by W. Bell, U.S. Geological Survey, in 1872)

much less – but its contribution to American geology is not to be measured
by miles or pages but by ideas. Its physical environment favored this
survey, and in the work of Powell, Dutton and Gilbert can be seen the
beginnings of physiography on the heroic scale exemplified in the Grand
Canyon and the High Plateaus. The first use of terms like 'base level of

erosion', 'consequent and antecedent drainage', and 'laccolith' marked the introduction of new ideas in the interpretation of land sculpture and geologic structure. The daring boat trip of Powell was no less brilliant than his simple explanation of the Grand Canyon itself. (*Smith, 1924, p. 205*)

They did this from their own observations and apparently without noticeable assistance from previous work done in Europe.

. . . neither Powell nor Gilbert had had any considerable measure of geological training before they entered the Western field: hence they were little trammeled by the conventions of geological science as it was then developed, and were consequently freer than they might otherwise have been to entertain new points of view, and to develop new lines of thought appropriate to their new fields of work; and as this freedom was coupled with exceptional powers of observation and of generalization it resulted in notable progress. (*Davis, 1925, p. 182*)

CHAPTER TWENTY-SEVEN

The Scientific Work of J. W. Powell

Powell's early life has already been discussed. From unpromising beginnings he progressed to the Directorship of the new Geological Survey. Yet from that moment his theoretical inspiration seems to have suffered, largely, no doubt, because his duties as Director involved him in endless administrative problems and in many political canvassing campaigns. As these increased, his personal contributions to geological theory gradually lessened. Consequently, most of Powell's landscape theories are to be found in his reports on the Colorado River and on the Uinta Mountains published in 1875 and 1876 respectively.

EXPLORATION OF THE COLORADO RIVER OF THE WEST (1875)

The report on the Colorado was based on two expeditions down the river; the first in 1869, and the second in 1871–2. The second was as difficult and probably as important as the first but lacked its novel and heroic interest. They differed too in personal aspects for the 1871–2 expedition began in style with Powell sitting in a chair roped to the leading boat and reading Tennyson's *Lady of the Lake* and finished in an atmosphere of discontent and recrimination. When the party reached the mouth of the Uinta River Powell learnt that his wife was ill and he left the expedition in camp while he went to see her. Even after he returned he left the party on several occasions either to see his wife or for other reasons. In January 1872 he travelled to Washington leaving Thompson in charge and did not rejoin the party till August. Thompson was neither an imposing leader nor a firm disciplinarian and his control of the party weakened as Powell's absences became more prolonged. The trip was abandoned on 8 September 1872 before the expedition had travelled as far down stream as on the first occasion. The main reason for the slowness of progress was that wide surveys were carried out upon the surrounding country whereas during the earlier expedition such work was restricted to the immediate gorge.

The vital geomorphological content of the report on the Colorado concerned three aspects: *the principle of base level*, the significance of which has since been greatly magnified because of its importance to later geological theories; *the nature and relative potency of the processes of erosion*; and, lastly, *the generic classification of landforms*.

We must, however, begin with Powell's exquisite description of the region.

The Colorado River is formed by the junction of the Grand and the Green. The Grand River has its source in the Rocky Mountains, five or six miles

FIG. 102. *The start of Powell's Second Colorado River Expedition, Green River Station, Wyoming, in May 1871. Major Powell is standing in the middle boat, the 'Emma Dean'*
(Photograph by J. K. Hillers, U.S. Geological Survey)

FIG. 103. *Marble Canyon, with Powell's boat in the foreground*
(Photograph by J. K. Hillers, U.S. Geological Survey)

west of Long's Peak, in latitude 40° 17' and longitude 105° 43' approximately. A group of little alpine lakes, that receive their waters directly from perpetual snow-banks, discharge into a common reservoir, known as Grand Lake, a beautiful sheet of water. Its quiet surface reflects towering cliffs and crags of granite on its eastern shore; and stately pines and firs stand on its western margin.

The Green River heads near Frémont's Peak, in the Wind River Mountains, in latitude 43° 15' and longitude 109° 45' approximately. This river, like the last, has its sources in alpine lakes, fed by everlasting snows. Thousands of these little lakes, with deep, cold, emerald waters, are embosomed among the crags of the Rocky Mountains. These streams, born in the cold, gloomy solitudes of the upper mountain-region, have a strange, eventful history as they pass down through gorges, tumbling in cascades and cataracts, until they reach the hot, arid plains of the Lower Colorado, where the waters that were so clear above empty as turbid floods into the Gulf of California.

The mouth of the Colorado is in latitude 31° 53' and longitude 115°.

The Green River is larger than the Grand, and is the upper continuation of the Colorado. Including this river, the whole length of the stream is about two thousand miles. The region of country drained by the Colorado and its tributaries is about eight hundred miles in length, and varies from three hundred to five hundred in width, containing about three hundred thousand square miles, an area larger than all the New England and Middle States with Maryland and Virginia added, or as large as Minnesota, Wisconsin, Iowa, Illinois and Missouri.

There are two distinct portions of the basin of the Colorado. The lower third is but little above the level of the sea, though here and there ranges of mountains rise to an altitude of from two to six thousand feet. This part of the valley is bounded on the north by a line of cliffs, which present a bold, often vertical step, hundreds or thousands of feet to the table-lands above.

The upper two-thirds of the basin rises from four to eight thousand feet above the level of the sea. This high region, on the east, north, and west, is set with ranges of snow-clad mountains, attaining an altitude above the sea varying from eight to fourteen thousand feet. All winter long, on its mountain-crested rim, snow falls, filling the gorges, half burying the forests, and covering the crags and peaks with a mantle woven by the winds from the waves of the sea – a mantle of snow. When the summer-sun comes, this snow melts, and tumbles down the mountain-sides in millions of cascades. Ten million cascade brooks unite to form ten thousand torrent creeks; ten thousand torrent creeks unite to form a hundred rivers beset with cataracts; a hundred roaring rivers unite to form the Colorado, which rolls, a mad, turbid stream, into the Gulf of California.

Consider the action of one of these streams: its source in the mountains, where the snows fall; its course through the arid plains. Now, if at the river's flood storms were falling on the plains, its channel would be cut but little faster than the adjacent country would be washed, and the general level would thus be preserved; but, under the conditions here mentioned, the river deepens its bed, as there is much through corrasion and but little lateral degradation.

So all the streams cut deeper and still deeper until their banks are towering cliffs of solid rock. These deep, narrow gorges are called cañons.

For more than a thousand miles along its course, the Colorado has cut for itself such a cañon; but at some few points, where lateral streams join it, the

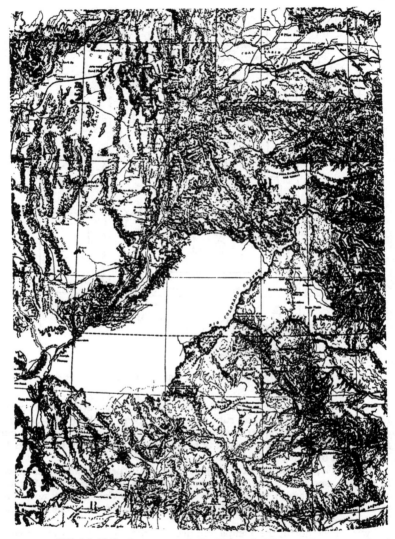

FIG. 104. *Official U.S. Army map of the Colorado River region, 1868, showing the large stretch of unknown territory through which Powell's first expedition passed* (*Yale University Press*)

cañon is broken, and narrow, transverse valleys divide it properly into a series of cañons.

The Virgen, Kanab, Paria, Escalante, Dirty Devil, San Rafael, Price, and Uinta on the west, the Grand, Yampa, San Juan and Colorado Chiquito on the east, have also cut for themselves such narrow, winding gorges, or deep

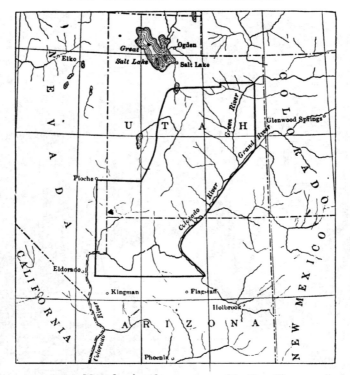

FIG. 105. *Map showing the area covered by Powell's surveys*

cañons. Every river entering these has cut another cañon; every lateral creek has cut a cañon; every brook runs in a cañon; every rill born of a shower, and born again of a shower, and living only during these showers, has cut for itself a cañon; so that the whole upper portion of the basin of the Colorado is traversed by a labyrinth of these deep gorges.

Owing to a great variety of geological conditions, these cañons differ much in general aspect. The Rio Virgen, between Long Valley and the Mormon town of Schunesburgh, runs through Pa-ru-nu-weap Cañon, often not more than twenty or thirty feet in width, and from six hundred to one thousand five hundred feet deep.

Away to the north, the Yampa empties into the Green by a cañon that I essayed to cross in the fall of 1868, and was baffled from day to day until the fourth had nearly passed before I could find my way down to the river. But thirty miles above its mouth, this cañon ends, and a narrow valley, with a flood plain, is found. Still farther up the stream, the river comes down through another cañon, and beyond that a narrow valley is found, and its upper course is now through a cañon and now a valley.

All these cañons are alike changeable in their topographic characteristics.

The longest cañon through which the Colorado runs is that between the mouth of the Colorado Chiquito and the Grand Wash, a distance of two hundred and seventeen and a half miles. But this is separated from another above, sixty five and a half miles in length, only by the narrow cañon-valley of the Colorado Chiquito.

All the scenic features of this cañon land are on a giant scale, strange and weird. The streams run at depths almost inaccessible; lashing the rocks which beset their channels, rolling in rapids, and plunging in falls, and making a wild music which but adds to the gloom of the solitude.

The little valleys nestling along the streams are diversified by bordering willows, clumps of box-elder, and small groves of cottonwood.

Low *mesas*, dry and treeless, stretch back from the brink of the cañon, often showing smooth surfaces of naked, solid rock. In some places, the country rock being composed of marls, the surface is a bed of loose, disintegrated material, and you walk through it as in a bed of ashes. Often these marls are richly colored and variegated. In other places, the country rock is a loose sandstone, the disintegration of which has left broad stretches of drifting sand, white, golden and vermilion.

Where this sandstone is a conglomerate, a paving of pebbles has been left, a mosaic of many colors, polished by the drifting sands and glistening in the sunlight.

After the cañons, the most remarkable features of the country are the long lines of cliffs. These are bold escarpments, often hundreds or thousands of feet in altitude, great geographic steps, scores or hundreds of miles in length, presenting steep faces of rock, often quite vertical.

Having climbed one of these steps, you may descend by a gentle, sometimes imperceptible, slope to the foot of another. They will thus present a series of terraces, the steps of which are well-defined escarpments of rock. The lateral extension of such a line of cliffs is usually very irregular; sharp salients are projected on the plains below, and deep recesses are cut into the terraces above.

Intermittent streams coming down the cliffs have cut many cañons or cañon valleys, by which the traveler may pass from the plain below to the terrace above. By these gigantic stairways, you may ascend to high plateaus, covered with forests of pine and fir.

The region is further diversified by short ranges of eruptive mountains. A vast system of fissures – huge cracks in the rocks to the depths below – extends across the country. From these crevices, floods of lava have poured, covering *mesas* and table lands with sheets of black basalt. The expiring energies of these volcanic agencies have piled up huge cinder-cones, that stand along the fissures, red, brown, and black, naked of vegetation, and conspicuous landmarks, set, as they are, in contrast to the bright, variegated rocks of sedimentary origin. (*Powell, 1875, pp. 3–6*)

Every ingredient necessary to write a comprehensive study of inland geomorphology was present in the area. Powell came very near to achieving this completeness.

The key, as he saw it, existed not in any special structural formation but in a readjustment of the river system to an alteration in the level of the land. He realized that if the surface of the land rose slowly, the antecedent rivers would be able to maintain their original courses by cutting down through the rising mountain folds or plateaus.

FIG. 106. *Physiographic diagram of the Colorado Plateaus*
(Drawn by Erwin Raisz)

To a person studying the physical geography of this country, without a knowledge of its geology, it would seem very strange that the river should cut through the mountains, when, apparently, it might have passed around them to the east, through valleys, for there are such along the north side of the Uintas, extending to the east, where the mountains are degraded to hills, and, passing around these, there are other valleys, extending to the Green, on the south side of the range. Then, why did the river run through the mountains?

The first explanation suggested is that it followed a previously formed fissure through the range; but very little examination will show that this explanation is unsatisfactory. The proof is abundant that the river cut its own channel; that the cañons are gorges of corrasion. Again, the question returns to us, why did not the stream turn around this great obstruction, rather than pass through it? The answer is that the river had the right of way; in other words, it was running ere the mountains were formed; not before the rocks of which the mountains are composed, were deposited, but before the formations were folded, so as to make a mountain range.

The contracting or shriveling of the earth causes the rocks near the surface to wrinkle or fold, and such a fold was started athwart the course of the river. Had it been suddenly formed, it would have been an obstruction sufficient to turn the water in a new course to the east, beyond the extension of the wrinkle; but the emergence of the fold above the general surface of the country was little or no faster than the progress of the corrasion of the channel. We may say, then, that the river did not cut its way down through the mountains, from a height of many thousand feet above its present site, but, having an elevation differing but little, perhaps, from what it now was, as the fold was lifted, it cleared away the obstruction by cutting a cañon, and the walls were thus elevated on either side. The river preserved its level, but mountains were lifted up; as the saw revolves on a fixed pivot, while the log through which it cuts is moved along. The river was the saw which cut the mountains in two. (*Powell, 1875, pp. 152-3*)

It will thus be seen that the upheaval was not marked by a great convulsion, for the lifting of the rocks was so slow that the rains removed the sandstones almost as fast as they came up. The mountains were not thrust up as peaks, but a great block was slowly lifted, and from this the mountains were carved by the clouds – patient artists, who take what time may be necessary for their work. (*Powell, 1875, p. 154*)

This uplift, the existing drainage and the dry conditions were the reasons for the form of the landscape. Fluvial erosion is therefore the main agent operative along lines which are often in defiance of the structural trends.

Though the entire region has been folded and faulted on a grand scale, these displacements have never determined the course of the streams. The cañons are seen to cut across them, either directly or obliquely, here and there, and in a few instances, I have observed cañons to follow the course of faults for a short distance. They have also been observed to run back and forth across a fault; but such instances are surprisingly rare. In all the cañons where the streams are not so large as to cover the bottom, the continuity of the strata below has been apparent; and in the cañons traversed by the larger streams, the beds on either side have been found at the same altitude; and if

FIG. 107. *Parunuweap Canyon*
(*From Powell, 1875*)

it is supposed that these water-ways were determined by fissures, then such fissures were made without displacement, and did not extend to the depths now reached by the streams. If it is possible to conceive of such fissures, they must have been quite narrow; in fact, the whole supposition is evidently absurd. All the facts concerning the relation of the water-ways of this region to the mountains, hills, cañons, and cliffs, lead to the inevitable conclusion that the system of drainage was determined antecedent to the faulting, and folding, and erosion, which are observed, and antecedent, also, to the formation of the eruptive beds and cones. (*Powell, 1875, p. 198*)

The arid conditions, Powell explains, produce their own special group of geomorphic forms. The lack of moisture prevents the growth of thick vegetation and most of the region is bare rock. For a similar reason the rate of weathering is slow and fluvial agencies are easily able to remove the weathered material. Such conditions happen to be ideal for permitting an accurate assessment of stratigraphical relationships, as wherever beds outcrop there is little or no vegetation to obscure the geologic pattern. The extent of what remains and what has been removed can be clearly seen.

All the mountain forms of this region are due to erosion; all the cañons, channels of living rivers and intermittent streams, were carved by the running waters, and they represent an amount of corrasion difficult to comprehend.

But the carving of the cañons and mountains is insignificant, when compared with the denudation of the whole area, as evidenced in the cliffs of erosion. Beds hundreds of feet in thickness and hundreds of thousands of square miles in extent, beds of granite and beds of schist, beds of marble and beds of sandstone, crumbling shales and adamantine lavas have slowly yielded to the silent and unseen powers of the air, and crumbled into dust and been washed away by the rains and carried into the sea by the rivers.

The story we have told is a history of the war of the elements to beat back the march of the lands from ocean depths.

And yet the conditions necessary to great erosion in the Valley of the Colorado are not found to exceed those of many other regions. In fact, the aridity of the climate is such that this may be considered a region of lesser, rather than greater, erosion. We may suppose that, had this country been favored with an amount of rain-fall similar to that of the Appalachian country, and many other districts on the surface of the earth, that the base level of erosion of the entire area would have been the level of the sea; and, under such circumstances, though the erosion would have been much greater than we now find, the evidences of erosion would have been more or less obliterated. As it is, we are able to study erosion in this country, and find evidences of its progress and its great magnitude, from the very fact that the conditions of erosion have been imperfect.

It is proper to remark here that erosion does not increase in ratio to the increase of the precipitation of moisture, *caeteris paribus*, as might be supposed; for, with the increase of rains there will be an increase of vegetation, which serves as a protection to the rocks, and distributes erosion more evenly, and it may be that a great increase of rains in this region would only produce a different series of topographic outlines, without greatly increasing the general degradation of the Valley of the Colorado.

FIG. 108. *Bird's-eye view of the terrace canyons*
(*From Powell, 1875*)

To a more thorough discussion of this subject I hope to return at some future time.

From the considerations heretofore presented, it is not thought necessary to refer the exhibition of erosion shown in the cañons and cliffs to a more vigorous action of aqueous dynamics than now exists, for, as I have stated, a greater precipitation of moisture would have resulted in a very different class of topographic features. Instead of cañons, we should have had water-gaps and ravines; instead of valleys with cliff like walls, we should have had valleys bounded by hills and slopes; and if the conclusions to which we have arrived are true, the arid conditions now existing must have extended back for a period of time of sufficient length to produce the present cañons and cliffs. But there are facts which seem to warrant the conclusion that this condition has existed for a much longer period than that necessary for the production of the present features; that is, the characteristics of the present topography have existed for a long time. There are evidences that the lines of cliffs themselves have been carried back for great distances as cliffs by undermining, which is a process carried on only in an arid region. (*Powell, 1875, pp. 208–10*)

When a great fold emerges from the sea, or rises above its base level of erosion, the axis appears above the water (or base level) first, and is immediately attacked by the rains, and its sands are borne off to form new deposits. It has before been explained that the emergence of the fold is but little faster than the degradation of its surface, but, as it comes up, the wearing away is extended still farther out on the flanks, and the same beds are attacked in the new land which have already been carried away nearer the centre of the fold. In this way the action of erosion is continued on the same bed from the up-turned axis toward the down-turned axis, and it may and does often happen that any particular bed may be entirely carried away, with many underlying rocks, near the former line, before it is attacked near the latter. (*Powell, 1875, pp. 171–2*)

The commonest feature is the canyon, and Powell describes how canyons originate and how the conjunction or retreat of canyon walls leaves residual buttes:

> In some places the waters run down the face of the escarpment, and cut narrow cañons, or gorges, back for a greater or less distance into the cliffs, until what would, otherwise, be nearly a straight wall, is cut into a very irregular line, with salients and deep re-entering angles.
>
> These cañons which cut into the walls also have their lateral cañons and gorges, and sometimes it occurs that a lateral cañon from each of two adjacent main cañons will coalesce at their heads, and gradually cut off the salient cliff from the ever retreating line. In this way buttes are formed. The sides of these buttresses have the same structural characteristics as the cliffs from which they have been cut. (*Powell, 1875, p. 173*)
>
> Climb the cliff at the end of Labyrinth Cañon, and look over the plain below, and you see vast numbers of buttes scattered about over scores of miles, and every butte so regular and beautiful that you can hardly cast aside the belief that they are works of Titanic art. It seems as if a thousand battles had been fought on the plains below, and on every field the giant heroes had built a monument, compared with which the pillar on Bunker Hill is but a mile stone. But no human hand has placed a block in all those wonderful structures. The rain drops of unreckoned ages have cut them all from the solid rock. (*Powell, 1875, p. 174*)
>
> Away from the river, on either side, there are broad stretches of naked sandstone, carved by the rains into gentle billows or mounds. As the rains gather into streams, the little valleys, or grooves, between the mounds become

gulches, and where the smaller streams gather into larger the gulches become cañons, often having vertical or even overhanging walls.

When, in the progress of corrasion, these streams have cut through harder beds, and reach softer, the channels are seen to widen. The manner in which

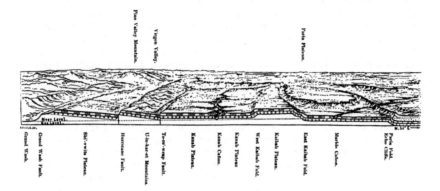

FIG. 109. *West to east section and block diagram across the plateaus north of the Grand Canyon, showing the faults and monoclines* (*From Powell, 1875*)

this widening occurs is curious. The streams are everywhere tortuous, and, as the power of the water is constantly exerted in corrasion, the streams are not only made deeper, but the curves are increased by methods well known to those who have studied the origin and change of river channels; so the walls are often undermined on the outer side of curves, and here overhanging cliffs are found. (*Powell, 1875, p. 179*)

Powell recognized that this progressive denudation had a vertical limit.

There is a limit to the effect of these conditions, for it should be observed no valley can be eroded below the level of the principal stream, which carries away the products of its surface degradation; . . . (*Powell, 1875, p. 163*)

This limit or control he called 'base level'. He recognized a permanent base level in the sea and temporary base levels in the beds of principal streams and outcrops of very hard rock. Powell viewed 'base level' more as a theoretical concept than as a physical feature because it is rarely that a river can be found which has stopped eroding vertically. He saw that the rate of erosion was dependent upon the discharge of the stream and the inclination of the slope, both in turn being dependent upon the amount of moisture falling on the surface of the region and the height of that surface above the base level. Examples of past realizations of complete vertical erosion he found in unconformities or fossil erosion surfaces.

> In Illustration [110] we have a section of the rocks of the Grand Cañon. A,A represents the granite; a,a dikes and eruptive beds; B,B these nonconformable rocks. It will be seen that the beds incline to the right. The horizontal beds above, C,C are rocks of Carboniferous Age, with underlying conformable beds. The distance along the wall marked by the line x,y, is the only part of its height represented by these rocks, but the beds are inclined, and their thickness must be measured by determining the thickness of each bed. This is done by measuring the several beds along lines normal to the planes of stratification; and, in this manner, we find them to be 10,000 feet in thickness.
>
> Doubtless, at some time before the Carboniferous rocks, C,C were formed, the beds B,B extended off to the left, but between the periods of deposition of the two series, B,B and C,C there was a period of erosion. The beds, themselves, are records of the invasion of the sea; the line of separation, the record of a long time when the region was dry land. The events in the history of this intervening time, the period of dry land, one might suppose were all lost. (*Powell, 1875, p. 212*)
>
> In a district of country, the different portions of which lie at different altitudes above the sea, the higher the region the greater the amount of rainfall, and hence the eroding agency increases in some well observed, but not accurately defined, ratio, from the low to the high lands. The power of running water, in corrading channels and transporting the products of erosion, increases with the velocity of the stream in geometric ratio, and hence the degradation of the rocks increases with the inclination of the slopes. Thus altitude and inclination both are important elements in the problem.
>
> Let me state this in another way. We may consider the level of the sea to be a grand base level, below which the dry lands cannot be eroded; but we may also have, for local and temporary purposes, other base levels of erosion, which are the levels of the beds of the principal streams which carry away the products of erosion. (I take some liberty in using the term level in this connection, as the action of a running stream in wearing its channel ceases, for all practical purposes, before its bed has quite reached the level of the lower end of the stream. What I have called the base level would, in fact, be an imaginary surface, inclining slightly in all its parts toward the lower end of the principal stream draining the area through which the level is supposed to extend, or having the inclination of its parts varied in direction as determined by tributary streams.) Where such a stream crosses a series of rocks in its course, some of which are hard, and others soft, the harder beds form a series of temporary dams, above which the corrasion of the channel through the softer beds is checked, and thus we may have a series of base levels of erosion, below which the rocks on either side of the river, though exceedingly friable, cannot be degraded. In these districts of country, the first work of rains and rivers is to cut channels, and divide the country into hills, and perhaps, mountains, by many meandering grooves or water-courses, and when these have reached their local base levels, under the existing conditions, the hills are washed down, but not carried entirely away.
>
> With this explanation I may combine the statements concerning elevation

FIG. 110. *Powell's (1875) section of the wall of the Grand Canyon showing the unconformities which he associated with past erosion to base level*

and inclination into this single expression, that the more elevated any district of country is, above its base level of denudation, the more rapidly it is degraded by rains and rivers. (*Powell, 1875, pp. 203–4*)

Though rivers that had attained a permanent base level cannot be found in the Colorado region, temporary base levels can be seen and these are distinguished by the presence of incipient flood plains.

> No valley is found along the course of the Colorado, from the Grand Wash toward the sources of the river, until we reach the head of Labyrinth Cañon. For this entire distance the base level of erosion is below the general surface level of the country adjacent to the river, but at Gunnison's Valley, we have a local base level of erosion which has resulted in the production of low plains and hills for a number of miles back from the stream. North of the Cañon of Desolation and south of the Uinta Mountains, another local base level of erosion is found, so near to the general surface of the country that we find a district of valleys and low hills stretching back from Green River, up the Uinta to the west, and White River to the east, for many miles. North of the Uinta Mountains a third local base level of erosion is seen, but its influence on the topographic features is confined to a small area of two or three hundred square miles. Going up the chief lateral streams of the Colorado, we find one or more of these local base levels of erosion, where the streams course through valleys.
>
> Where these local base levels of erosion exist, forming valley and hill regions, the streams no longer cut their channels deeper, and the water of the streams, running at a low angle, course slowly along and are not able to carry away the products of surface wash, and these are deposited along the flood-plains, in part, and in the valleys, among hills, and on the gentler slopes. (*Powell, 1875, pp. 206–7*)

As Powell himself says 'base level' is not so much an actual feature as an ideal; it is a perfect adjustment of the principal stream and its tributaries along an average slope where from the mouth to the heads of the tributaries erosion has ceased in each part of the system. In this, his concept differs from the theoretical base level idea of Davis, and hints at the peneplain.

Powell's study of the Colorado prompted him to describe three types of rivers by a genetic classification. To each type he gave a technical name which is still in use today:

> I have endeavored above to explain the relation of the valleys of the Uinta Mountains to the stratigraphy, or structural geology, of the region, and, further, to state the conclusion reached, that the drainage was established antecedent to the corrugation or displacement of the beds by faulting and folding. I propose to call such valleys, including the orders and varieties before mentioned, *antecedent valleys*.
>
> In other parts of the mountain region of the west, valleys are found having directions dependent on corrugation. I propose to call these *consequent valleys*. Such valleys have been observed only in limited areas, and have not been thoroughly studied, and I omit further discussion of them.
>
> In the great metamorphic belt extending through the Territory of Colorado, comprising the Rocky Mountain chain of this Territory, the structural

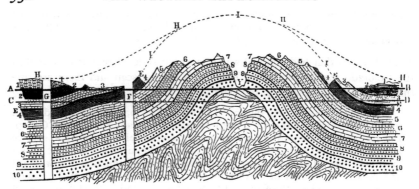

FIG. 111. *Cross-section through the Uinta Mountains from north (A) to south (B)*
(From Powell, 1875)

geology is exceedingly complex, while the drainage is comparatively simple, and only to a limited extent does it seem to be governed by geological structure. The conclusions to which I arrived were that the present drainage was established in rocks now carried away from the higher regions, but still seen to be turned up against the flanks of most of the ranges. (*Powell, 1875, p. 163*)

There can be no doubt that the present courses of the streams were determined by conditions not found in the rocks through which the channels are now carved, but that the beds in which the streams had their origin when the district last appeared above the level of the sea, have been swept away. I propose to call such *superimposed valleys*. Thus the valleys under consideration, if classified on the basis of their relation to the rocks in which they originated, would be called *consequent valleys*, but if classified on the

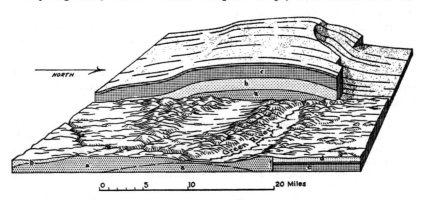

FIG. 112. *Block diagram of the Uinta Mountains uplift. The rear part gives an impression of the character of the uplift without erosion (Pre-Cambrian – a; Palae-ozoic – b; Mesozoic – c; Tertiary – d); the front part shows the present landscape, including Lodore Canyon*
(After Powell, 1875)

basis of their relation to the rocks in which they are now found, would be called *superimposed valleys*. (*Powell, 1875, pp. 165-6*)

He essayed other classifications but these are less interesting because they are purely generic.

It will thus be seen that the relation of the direction of the streams to the dip of the rocks is very complex, and, for convenience of description, I have elsewhere classified these valleys, on the basis of these relations, in the following manner:

Order first. Transverse valleys, having a direction at right angles to the strike.

Order second. Longitudinal valleys, having a direction the same as the strike.

Of the first order, three varieties are noticed:

a, *diaclinal*, those which pass through a fold. (Fig. 113a)

b, *cataclinal*, valleys that run in the direction of the dip. (Fig. 113b)

c, *anaclinal*, valleys that run against the dip of the beds. (Fig. 113c)

Of the second order, we have, also, three varieties:

A, *anticlinal* valleys, which follow anticlinal axes. (Fig. 114a)

B, *synclinal* valleys, which follow synclinal axes. (Fig. 114b)

C, *monoclinal* valleys, which run in the direction of the strike between the axes of the fold – one side of the valley formed of the summits of the beds, the other composed of the cut edges of the formation. (Fig. 114c)

Many of the valleys are thus simple in their relations to the folds; but, as we may have two systems of displacements, a valley may belong to one class, in relation to one fold, and to another in its relation to a second. Such we designate as complex valleys.

Again, a valley may belong to one class in one part of its course and to another elsewhere in its course. Such we designate as compound valleys. It will be further noticed that valleys may have many branches, but in relegating a valley to its class, we consider only the stem of the valley proper, and not its branches. (*Powell, 1875, p. 160*)

Having in view the forms which are produced by erosion, it will be convenient to classify the methods of erosion as follows: First, corrasion by running streams; and, second, erosion by rains; the first producing channels along well defined lines, the second producing the general surface features of the landscape.

Of the first class we have two varieties:

A. The corrasion of water-gaps.

B. The corrasion of cañons.

Of the second class we have three varieties:

A. Cliff erosion, where the beds are slightly inclined, and are of heterogeneous structure, some soft and others hard; and for the production of the best marked forms, the climate should be arid. Here the progress of erosion is chiefly by undermining.

B. Hog-back erosion, where the beds have a greater inclination, but are still of heterogeneous structure. Here the progress of erosion is by undermining and surface washing, and the typical forms would require an arid climate.

C. Hill and mountain erosion, where the beds may lie in any shape, and be composed of any material not included in the other classes, and the progress

FIG. 113. *Powell's (1875) illustrations of transverse valleys,
running at right angles to the strike:*
a. *A diaclinal valley – cutting through a fold.*

b. *A cataclinal valley – with the river flowing down-dip.*

c. *An anaclinal valley – with the river flowing up-dip.*

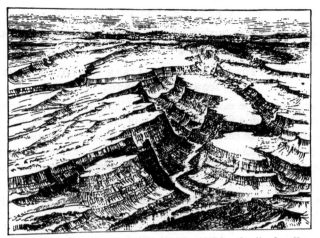

FIG. 114. *Powell's (1875) illustrations of longitudinal valleys, running parallel to the strike:*

a. *An anticlinal valley – following an anticlinal axis.*

b. *A synclinal valley – following a synclinal axis.*

c. *A monoclinal valley – following the strike along the steeply dipping beds of a monocline.*

of erosion is chiefly by surface washing. The typical forms are found in a moist climate. (*Powell, 1875, p. 205*)

REPORT ON THE GEOLOGY OF THE EASTERN PORTION OF THE UINTA MOUNTAINS (1876)

Powell's account of the eastern part of the Uinta Mountains is written in a similar style but with a fuller explanation of erosion processes and less emphasis on the theory of base level. He begins with a vivid historical summary of the region's structural origin and emphasizes the quantity of erosion that has taken place, so focussing attention on the ephemeral nature of mountain ranges.

In the latter part of Mesozoic time the greater part of the Basin Province was dry land. The Plateau Province was an open but shallow sea. In the Park Province a chain of islands extended to the south. The Cenozoic time was inaugurated by a series of movements, which, continued to the present time, have produced the topographic features now observed. This part of the crust of the earth, and I mean by the term 'crust' simply that portion of the earth which we are able to study by actual observation in truncated folds and eroded faults – this portion of the crust, then, was gradually broken and contorted. The Plateau and Park Provinces were cut off from the sea, and great bodies of fresh water accumulated in the basins, while to the east in the region of the Great Plains, in earlier Tertiary times at least, there was an open sea. Slowly through Cenozoic times the outlines of these lakes were changed, doubtless in two ways: first, by the gradual displacement of the rock beds in upheaval and subsidence here and there; and, second, by the gradual desiccation due to the filling up of the basins by sedimentation and the erosion of their barriers; and the total result of this was to steadily diminish the lacustrine area. But the movements in the displacement extended over the Basin Province, for that region was then a comparatively low plain, constituting a general base level of erosion to which that region had been denuded in Mesozoic and early Tertiary time when it was an area of dry land; for I think that from the known facts we may reasonably infer that the Basin Ranges, though composed of Paleozoic and Eozoic rocks, are, as mountains, of very late upheaval. For some purposes, and in broad generalization, erosion furnishes a valuable measure of geological times. A mountain, as a mountain, is comparatively ephemeral. The evidence of this is found on every hand as we study the Rocky Mountain region. There can be no conclusion reached from reasoning on geological data more certain than that the Uinta upheaval began at the close of Mesozoic time, and has continued intermittently near to the present, and during that time this upheaval has suffered a degradation in areas of maximum erosion of no less than 30,000 feet; and there is evidence also which leads to the conclusion that the conditions for great erosion were not persistently maintained during this time. (*Powell, 1876, pp. 32–33*)

The area of degradation which I have often for convenience called the region of uplift, embraces a little more than 2800 square miles. From this about 8300 cubic miles of rock have been carried away by rains and rivers – a mean degradation of about three cubic miles to the square mile. But a part of the region embraced on that plate will be discussed hereafter under the head of Yampa Plateau, and in what I said concerning the Uinta Mountains above, this region has not been considered. Taking the Uinta Mountain region proper, then, we have an area of about 2000 square miles from which about 7100 cubic miles of rock have been taken, giving a mean degradation of $3\frac{1}{2}$ cubic miles to every square mile of surface. But this has not been taken from all points equally; a greater amount has been carried from the axial region than from the districts along the flanks; and in the axial region a greater amount has been taken from the eastern than from the western end. Here where the displacement lines are carried highest, the surface lines are lowest, so that the degradation is more, not only by the amount of greater uplift but also by an additional amount in the deeper excavation of the valley. The region of highest uplift is the region of lowest degradation, and here more than 25,000 feet of beds have been removed. (*Powell, 1876, p. 181*)

Powell now discusses the various means whereby rocks are changed from a solid to a clastic state:

The principal forces of disintegration are gravity, heat, crystallization and chemical reaction. Gravity disintegrates the rocks directly where they break from cliffs and ledges partly by their own weight, and are further disintegrated by the fall, and indirectly through the agency of water in abrasion. Heat disintegrates the rocks by change of temperature, and probably by expansion of water permeating the rocks. Crystallogenic force also acts through the agency of water, for where the water which has permeated the rocks is frozen, the expansion due to this crystallization breaks them asunder. In chemical reaction the rocks are broken up through the agency of water which acts directly in dissolving them, or indirectly in promoting other chemical reactions. (*Powell, 1876, p. 183*)

He then proceeds to describe in some detail the transport of this loosened material. First, he stresses the inter-relationship of stream velocity, the specific gravity of the particles in suspension and their total mass. A variation in the quantity of each will give rise to a different geomorphic form. Thus if the particles are very small, stream velocity is relatively unimportant and the amount of the total mass of particles removed depends on the discharge. On the other hand, if the particles are large, velocity becomes the vital factor. Today the explanation seems obvious but at the time it helped to explain how a river could be of high velocity in its upper reaches, removing annually tons of coarse material, and perhaps have a lower mean velocity in its lower reaches and yet not become choked by its own material.

We come now to the principal method of transportation; that is, that by flotation. Here the rock power is used in transportation, and the water is the vehicle. If the matter to be floated is of the same or less specific gravity than the water, a condition seldom obtaining, flotation is perfect, and the water power is not used either in transportation or to promote flotation. But when the floating matter is of greater specific gravity than the water, then the water power is used in promoting flotation. With a given amount of water and sufficient supply of load, the extent to which the water power will be utilized in promoting flotation will depend on two conditions: First, the power of the water, which is measured by fall into mass, or which may be expressed as velocity or again as declivity; second, it will depend on the relation which exists between the floating surface, or surface presented downward, and the mass of each particle of the load. In other words, it will depend on the specific gravity and comminution of the load. If the specific gravity of the load is but little greater than water, the velocity of the water becomes a very small factor, and the amount which can be transported will be chiefly limited by the containing capacity of the water, but this is a condition not actually found in nature. The difference between the specific gravity of water and load is great, and variable within such small limits that the variability may be neglected; but the relation between the floating surface and the mass of each particle of load may be determined by another condition than that of specific gravity, i.e. size; for the ratio between the floating mass of a body, or the surface presented downward, and the mass of the body, increases with the diminution of a body. Then if the body is smaller, the ratio between the floating surface and mass is larger; and hence comminution promotes flotation. If this comminution is great, approaching complete or molecular comminution, the velocity of the water again becomes a small factor, and the amount of transportation chiefly depends on the containing capacity of the water; but as comminution is less and the size of the particles larger, some force must intervene to promote flotation, and this is derived from the water power; and to sustain the same amount of flotation more of this force must be utilized as the size of the particles increase. The amount of this force from which must be drawn the supply in supporting flotation depends, *caeteris paribus*, on the velocity or declivity, but practically the whole force is rarely utilized. (*Powell, 1876, pp. 185–6*)

Powell also realized that a steep slope assisted the transport of material in two ways: firstly it increased the stream's velocity and secondly the greater velocity caused increased turbulence, or powerful eddies, which helped to lift and move larger particles of debris which the main mass of water had failed to shift.

And in obedience to well known laws of friction this heterogeneity of flow is greatly increased by intensifying the flow, or in other words, increasing the velocity of the water; and velocity is due to declivity. And that heterogeneity of channel, which by producing heterogeneity of flow utilizes the water power in lifting the load, is also due to declivity; and hence it remains, first, that the water power, *caeteris paribus*, is a function of declivity; and, second, the utilization of the water power, *caeteris paribus*, is a function of declivity; so that with a given amount of water and sufficient supply of load, the rate of transportation through the agency of flotation depends on declivity; and the amount of transportation through driving also depends on declivity. Therefore the rate of transportation of all mechanically comminuted matter is determined by declivity. (*Powell, 1876, pp. 186–7*)

In fact, Powell believed that the angle of slope was probably the most important factor in the promotion of transportation; with its increase went an automatic increase in erosion and the rate of removal.

The lesser or greater rapidity of erosion depends chiefly on three conditions: first, elevation above the base level of erosion; second, the induration of the rocks; and, third, the amount of rain fall. (*Powell, 1876, p. 34*)

Rain which, under gravity, supplied the motive power for transport did not cause an automatic increase in the same ratio as did increase of slope. The reason for this discrepancy was that although an increase in rainfall produced a greater flow of water, it also promoted the growth of vegetation which was a definite hindrance to erosion and transport of material.

> We have already seen in the former analysis of transportation that with a given quantity of water, transportation will depend on declivity, but in the transportation belonging to this method of degradation the quantity of water is a factor of transportation only to a limited extent, for increased rainfall promotes the growth of vegetation which serves as a protection to the soil. Nor is this protection inconsiderable, for it preserves the rocks from the beating of the storms, and prevents the waters from gathering rapidly into rills and brooks, and strains the water of its earthy sediments. I have many times witnessed the action of a storm in an arid region where the disintegrated rocks were unprotected by forests, shrubbery, or turf, and as often have I been impressed with the wonderful power of the infrequent storm to gather up and carry away the land, as compared with the frequent storm in the prairie or forest of a land more richly clad. The same contrast may be observed in a region brought under the dominion of man by cultivation where the surface of a plowed field is swept away by a storm, and the furrows are the channels for floods of mud, while the meadow receives the rain with out-stretched arms of verdure, which bear it gently to the earth, where it is gathered into quiet rills, which feed a stream made turbid it is true, but pure when compared with the stream of mud flowing from the field from which the plowman was driven by the storm.
>
> Erosion, then, or surface degradation is not greatly promoted by increased rainfall, and it may be that its effect is rather to retard the process; but the difference between greater or lesser rainfall is plainly manifest in the topo-graphic features produced – little rainfall giving angular reliefs; much, rounded reliefs.
>
> Neglecting such a hypothetic condition as no rainfall, we have in nature to consider only greater or lesser rainfall. With greater rainfall we have a greater power, but a lesser utilization of the power; with lesser rainfall we have lesser power, but greater utilization; and in these varying conditions, just where maximum degradation is found I am not able to state. Hence, in the process of degradation which I have called erosion, we have simply to consider declivity, with exceptions so minute that they may be neglected. (*Powell, 1876, pp. 188–9*)

When a stream transports material it is not usually restricted to this one function and it will continue to corrade its bed and banks. By this combination of functions, streams continually lower the level of the surface across which they flow and the lowering proceeds until base level is reached.

> All the processes of erosion and sapping serve to obliterate the evidences of corrasion, and the latter appears more plainly as its progress exceeds that of the other methods; but still the evidences of corrasion rarely disappear until the land is buried by the sea; for wherever an area of land is above its base level of degradation, there corrasion will be manifest by deepening its channel;

and wherever the dry land has been brought down near to its base level, there corrasion is manifest by widening its channel. (*Powell, 1876, p. 191*)

Finally, from a comparison of erosion processes acting on a newly-elevated surface and on one which has almost reached base level, Powell again demonstrates the geological impermanence of mountains and his concept of the fluvial erosion surface.

> There are conditions of degradation in the extremes of declivity worthy of mention. In high degrees of declivity transportation in a horizontal direction is limited, the cliffs soon tumble down, and degradation by this process ceases. In a very low degree of declivity approaching horizontality the power of transporting material is also very small. The degradation of the last few inches of a broad area of land above the level of the sea would require a longer time than all the thousands of feet which might have been above it, so far as this degradation depends on mechanical processes – that is, driving or flotation; but here the disintegration by solution and the transportation of the material by the agency of fluidity come in to assist the slow processes of mechanical degradation, and finally perform the chief part of the task. We may now conclude that the higher the mountain, the more rapid its degradation; that high mountains cannot live much longer than low mountains, and that mountains cannot remain long as mountains; they are ephemeral topographic forms. Geologically all existing mountains are recent; the ancient mountains are gone. But existing mountains may be old or young as compared with other existing mountains. We may speak of the age of mountains, referring to the age of the rocks of which they are composed but this will have no reference to the age of the mountain form. We may speak of the age of a mountain with respect of the inception of the upheaval which exposed the rocks to that degradation which has produced the mountain form; and this epoch will not be very long ago, geologically, for the rate of upheaval must be greater than the rate of degradation, else mountain forms will not be produced. We may speak of the age of mountains, referring to the completion of the upheaval by which the mountain forms were produced through degradation; the time which has elapsed since the epoch to which we then refer must be short indeed; but if, in speaking of the age of mountains, we refer to the time when those topographic forms were produced, they are all newly born. (*Powell, 1876, p. 196*)

We should, perhaps, here point out that there were *two* sides to Powell's base level concept. On the one side were the erosive or levelling processes; on the other, the agent of uplift. Without uplift, whether sudden, intermittent or gradual, erosion would not exist; uplift destroys the level of equilibrium which erosion works to restore. The nature of this uplift became most important in future geomorphic studies and in the case of Davis formed the foundation of his whole theory.

The remaining part of Powell's Uinta report is of little significance as it consists mainly of another generic classification of mountain forms.

I. – APPALACHIAN STRUCTURE.
(Not found in the three provinces)

II. – SIMPLE ANTICLINAL STRUCTURE.

Primary topographic form. Plateau with rounded vertical outline.
Concomitant forms: 1. Monoclinal Ridges on the Flanks.
2. Monoclinal Ridges only.
3. Inclined Plateaus.

III. – UINTA STRUCTURE.

Primary topographic form. Plateau with rounded summit and abrupt shoulders on the flank.
Concomitant forms: 1. Subsidiary Plateaus.
2. Projecting Ridges.
3. Axial Peaks.
4. Flanking Peaks.
5. Interrupted Monoclinal Ridges.

IV. – KAIBAB STRUCTURE.

Primary topographic form. Plateau with angular outlines.
Concomitant forms: 1. Cliffs of Displacement.
2. Slopes of Displacement.
3. Interrupted Monoclinal Ridges on the Flanks.
4. Monoclinal Ridges with Plateau Carried Away.
5. Projecting Ridges.
6. Cliffs of Erosion.
7. Buttes.
8. Cameo Mountains.

V. – BASIN RANGE STRUCTURE.

Primary topographic form. Monoclinal Ridges of displacement.
Concomitant forms: 1. Monoclinal Ridges on the Back.

VI. – ZONES OF DIVERSE DISPLACEMENT.

Topographic form. Irregular hills.

(Powell, 1876, pp. 21–22)

POWELL'S LATER GEOMORPHIC ESSAYS

Though he worked indefatigably for the good of the survey, Powell's own ideas made little further advance on those in his two main reports. His account of physiographic features published in 1896 presented the same approach to erosion processes. He also persisted with his generic/genetic classification of landforms;

A. PLAINS
1. Sea plains – areas reduced to a level plain with respect to the base level of the sea.
2. Lake plains – areas reduced with reference to the level of a lake.
3. Stream plains – plains of lateral stream corrasion.
4. Flood plains.

B. PLATEAUS
1. Diastrophic plateaus – due to uplift en bloc.
2. Volcanic plateaus – built up by lava flows.

C. MOUNTAINS

    1. Volcanic mountains.
        a. Volcanoes.
        b. Laccolithic mountains.
        c. Table mountains – protected from erosion by lava-flow cappings.
        d. Imbricated mountains – complex mountains.
    2. Diastrophic mountains – folded mountains, metamorphic mountains, and mountains with metamorphic cores.

D. VALLEYS

    1. Volcanic valleys – those between volcanic cones.
    2. Diastrophic valleys – downfaulted valleys.
    3. Valleys of gradation – cut by stream-erosion.

E. HILLS.
F. CLIFFS.
G. SPECIAL FORMS – buttes, monuments, dike walls, volcanic necks, etc.
H. STREAM CHANNELS AND CATARACTS.
I. FOUNTAINS – springs, geysers, wells, etc.
J. CAVERNS.
K. LAKES.
L. MARSHES.

(Powell, 1896)

The weakness of such a classification exists in its failure to touch the imagination. Though the name of the landform may indicate its origin, each form is separate and has no stated connexion with the others. In short it is little more than a catalogue or index and its content is therefore difficult to assimilate. Such an elaborate description was almost inevitably doomed to disuse once the simple genetic classification of Davis had apparently connected all landforms into one integrated system.

ASSESSMENTS OF POWELL'S SIGNIFICANCE

Powell's contributions to landscape studies have been admirably summed up by his colleague Gilbert.

> He published the stratigraphy, structure, and part of the areal geology of the Colorado plateau and the Uinta Mountains. In connection with the field studies in these districts he developed a new classification of mountains, by structure and genesis; a structural classification of dislocations; a classification of valleys; and a genetic classification of drainage systems. His classification of drainage recognized three modes of genesis, of which two were new. With the novel ideas involved in the terms 'superimposed drainage' and 'antecedent drainage' were associated the broader idea that the physical history of a region might be read in part from a study of its drainage system in relation to its rock structure. Another broad idea, that since the degradation of the land is limited downward by the level of the standing water which received its drainage, the types of land sculpture throughout a drainage area are conditioned by this limit, was formulated by means of the word 'base level'. These two ideas, gradually developed by a younger generation of

FIG. 115. *A dinner party given by Major Powell in the Spring of 1890 or 1891.*
G. K. Gilbert is seated third from the left
(U.S. Geological Survey)

students, are the fundamental principles of a new subscience of geology sometimes called geomorphology, or physical geography. (*Gilbert, 1902, p. 638*)

There is no doubt that of all Powell's ideas, the conception of base level has proved the most stimulating to subsequent workers.

Powell's conception of the base level of erosion did not pertain so much to the level base with respect to which subaerial erosion is performed, but rather to a delicately warped imaginary surface passing through all the stream lines of a worn-down land area, and therefore permissibly standing well above ocean level, especially in the upper reaches of a large river system like that of the Mississippi, even though somewhat underlying the visible surface of degradation. It is, I believe, not so much to Powell as to Gilbert that we owe the generally accepted idea of base level as a level base; in general, the extension of the ocean or geoid surface under the continents. (*Davis, 1925, p. 183*)

He grasped the concept of recent uplift; he foresaw the slowing down as the height should be worn away; he sensed the inevitable result – that the plateau must be reduced to a low surface, a uniform plain – ; and he recognized that erosion must then wait upon renewed activity of lifting forces. Thus he came to recognize what he afterwards always taught that the processes of erosion follow a cycle, comprising uplift, degradation, and sedimentation; and that such a cycle is but a brief interval between long periods when erosion is about at a standstill. (*Willis, 1942, p. 168*)

Previous to Powell no one had ventured to carry the carving of the land by rain and streams beyond what today we would call late maturity in the geomorphic cycle, but Powell recognized that the processes of erosion operating undisturbed upon the land would eventually reduce it to a lowland little above sea level. It remained for Davis to suggest later the name peneplain for such an area, but the idea of the peneplain was to a large degree anticipated by Powell. (*Thornbury, 1954, p. 10*)

POWELL'S CLOSING YEARS

As Director of the newly formed Survey Powell found his charge burdensome. His project of mapping the whole of the United States in detail had set himself and his staff a tremendous task. The quantity of mapping involved was enormous and, in addition, he had to strive constantly for congressional grants sufficient to meet his plans. In his early days in office he showed a flair for manipulating both the legislature and the executive. By employing senator's sons or favourite nephews on his expeditions he encouraged generous appropriations and in the expenditure of these he had a completely free hand. In 1885 an investigation into the efficiency of the Survey resulted in increased confidence in Powell. But in 1890 there was a disastrous drought in the West and appropriations for the Survey were cut drastically. This cut expressed Congress's dissatisfaction with Powell's progress in mapping the area and their irritation at his insistence that only a small part of the West was irrigable. In the same year the *New York Herald* printed an attack on Powell by Cope. The charges were nepotism, employment of part-time staff and wasteful or improper use of government funds. Despite these criticisms the appropriation in the following year was the largest ever.

Powell's arm now began to give pain again and a third operation was contemplated. Then in 1892 a wave of economy cut the normal grant by half, and, more important still, Powell's discretionary powers of expenditure were taken away from him. He fought against this decision but had lost much of his old vitality and two years later resigned a saddened man. In 1902 he died in Haven, Maine, far from the scenes of grandeur which he had loved and served so well.

CHAPTER TWENTY-EIGHT

The Landscape Studies of Grove Karl Gilbert

GROVE KARL GILBERT (1843–1918)

Grove Karl Gilbert is without doubt the most outstanding of Powell's assistants – more talented, indeed, than Powell himself. Upon his analyses of his own surveys, and his geomorphic researches of a less arduous nature, a large part of modern geomorphology rests.

He was born at Rochester in 1843 and graduated at the local University in 1862 (Mendenhall, 1920; Davis, 1918 & 1922; Birot, 1958). After graduation he went to Jackson, Michigan,

> Where he attempted to teach in order to pay off his college debt; but he was neither happy nor successful as a teacher . . . (*Mendenhall, 1920, p. 29*)

From 1863 to 1868 he acted as assistant geologist in his old university. He then took up surveying and for two years helped Newberry with the Ohio Survey, and between 1871 and 1874 worked under Lieut. Wheeler on a reconnaissance of south-east California, Nevada, Utah, Arizona and New Mexico. Gilbert's thoroughness of approach did not fit in with the rapidity of movement of the military party whose principal aim was geographical and topographical and he continually complained of the unnecessary haste which made accurate geological observation difficult. He was also dissatisfied with the restrictions that prohibited him from publishing reports in the *American Journal of Science* and by Wheeler's rigid application of military regulations.

> At a very lively rapid an attempt was made to drag up an otherwise empty boat with S . . . and me aboard, and we were swamped and upset. Shore was near at hand and we swam to it and hung our clothes on the rocks to dry. Here I found the inconvenience of having no change of raiment . . . In the first accident three carbines were lost and one of them mine. I do not feel very sorry unless I am called on to pay for it; which would be highly unjust. (*Diary of 18 October 1871; quoted by Davis, 1922, p. 23*)

Gilbert's main theoretical contribution during this period was his recognition that the form of the topography of the region being studied was due to faulting in a series of horsts and grabens and not, as geologists had formerly imagined, to an original folded structure. This period also provided material for his later conclusions on Lake Bonneville.

He first met Powell and Dutton in Washington in 1872;

Powell, with his originality and remarkable fertility of ideas, must have recognized a splendid foil in Gilbert's steadiness of thought and accuracy of observation, and Gilbert in turn must have been greatly stimulated by the Major's brilliancy in conception and in generalization. (*Mendenhall, 1920, pp. 35–36*)

Gilbert, however, had a strikingly different personality from that of the tough, one-armed Major:

Probably few who, in those years of long ago, knew Gilbert only on the scientific side realized his inner nature to be so emotional that if, while reading aloud, he came upon a pathetic passage, even his strong self-control could not wholly master tear ducts and vocal chords; his eyes would overflow and his voice would choke, so that he must hand the book to another to continue the story. For the same reason, while enjoying the theatre, he avoided

FIG. 116. *G. K. Gilbert in Colorado in 1894*
(*U.S. Geological Survey*)

distressing melodramas, as he did not like to 'make a spectacle of himself' in public. (*Davis, 1922, p. 68*)

He did not finally join the Powell Survey until 1874, the year when he married Fanny Loretta Porter, the sister-in-law of Marvine, whom he probably met at a dance given by Powell (Davis, 1922).

During the period 1875–6 Gilbert worked on the Aquarius and Kaiparo-wits Plateaus (16 June–12 September 1875) and in the Henry Mountains (19 August–2 September 1875 and August–November 1876). These surveys inspired his greatest conclusions on structure and erosional processes, including his explanations of laccolithic mountains; the influence of climate on land sculpture; the concept of 'grade'; the law of diminishing slopes; lateral planation; the formation of pediments, and of the interdependence and stability of drainage lines. It is interesting that, although he employed the terms 'consequent', 'antecedent' and 'superimposed', Gilbert never made much direct use of Powell's great concept of base level.

From 1877 to 1879 he worked on basic triangulation and irrigation studies in Utah and contributed two chapters to Powell's report on the arid regions. In 1881, on Powell's elevation to the Directorship, Gilbert was called to Washington as one of the six senior geologists and for about a decade his field talents were largely lost in a jungle of administrative duties. His first great task while working on the Lake Bonneville monograph (published in 1890) was to reorganize the Great Basin part of the survey which had suffered duplication and confusion due to the competition between the earlier four separate surveys. However, from about 1890 he relinquished many of his administrative duties in order to complete some of his early studies, and almost until his death in 1918 he continued to produce books and essays of geomorphological importance, including his monumental *The Transportation of Debris by Running Water*.

GILBERT'S EARLY GEOMORPHIC WRITINGS

The merit of Gilbert's work came not from his description of physical erosive processes but from his ability to deduce therefrom a system of laws governing their progress from initial to adjusted forms. The formulation of these laws gave immediate cohesion to processes which before had seemed only loosely related to form.

Even in his early work with Wheeler, Gilbert provides indications of his later ideas. Chapter I of the *Report on the Geology of Portions of Nevada, Utah, California and Arizona* contains his explanation of the basin-and-range topography. As already mentioned, for the traditional explanation that the topography was the residuary result of erosion acting upon a series of anticlines and synclines, he substituted the idea that it was due to a special type of block-faulting, now known as the horst and graben complex. He emphasized the difference between the two explanations:

The ridges of the system occupy loci of upheaval, and are not mere residua of denudation; the valleys of the system are not valleys of erosion, but mere intervals between lines of maximum uplift. (*Gilbert, 1875, p. 41*)

Chapter 2, which was entitled 'Valleys. Cañons. Erosion', gives several hints of the many ideas that were to be presented with his *Report* on the Henry Mountains. He contrasts the effects of erosion between the basin-and-range province and the plateau region. He showed how in the former subaerial agents had filled up the original fault-valleys while in the latter the rivers and rain were actually enlarging the valleys (p. 63). Later he notes how in the plateau the cliffs of horizontal strata have been worn back tens of miles by the action of subaerial agents (p. 68). He also describes how erosion of the stream bed progresses and in doing so speaks of 'the current cycle of events' within the Grand Canyon (p. 71). The meaning is made clearer a little farther on:

> There have been distinguished in rivers the torrent portion, in which the descent is comparatively rapid, and the bed is sinking by erosion into the subjacent rock; the river proper, in which the bed holds a constant mean level, and erosion is diverted to the increase of the width of the valley; and the delta, in which the bed is rising by deposition. (*Gilbert, 1875, p. 74*)
>
> As the work of denudation in the Grand and Marble Cañons progresses, and the river sinks deeper below the plateau, there will accompany a gradual diminution of the inclination of its bed, of the velocity of its current, and, in consequence, of its erosive power, until finally it can no longer clear its bottom of introduced detritus, and, its downward progress being arrested, the widening of its channel will begin. (*Gilbert, 1875, p. 75*)

So far there is little to distinguish Gilbert's ideas from Powell's and in the emphasis on the amount of strata denuded, there is a marked similarity between the two colleagues.

> Of the immensity of the denudation that has reduced the plateaus to their present condition, we have unmistakable, and at the same time unexpected evidence, in the existence of insular masses of strata, remote from the mesas of which they once formed part . . . If a line be drawn to connect these several islands, and be compared with the line of the present Triassic outcrop, the included area will be found to approximate 10,000 square miles; and from all this the Lower Trias has been denuded since the eruption of the older lavas of the Uinkaret and San Francisco groups. (*Gilbert, 1875, pp. 81–82*)

In addition, Gilbert's work for the Wheeler survey caused him to recognize the great unconformity beneath the Palaeozoic rocks in the Colorado Plateaus which he attributed to marine overlap; the characteristic topographic forms associated with faulting and folding (i.e. the Kaibab monocline and Zuni uplift); and the fact that, if steep slopes are worn down more rapidly than gentle ones, a uniformity of landscape will be produced. He also discussed the formation of alluvial fans, and made somewhat vague references to fluvial planation, and to the concept of 'grade' (p. 75).

In his report on *The Colorado Plateau Province* (1876) Gilbert described in general terms the structural geology, stressing how the arid conditions make this one of the best regions in the world for the observation of structural features:

> From a commanding eminence one may see spread before him, like a chart, to be read almost without effort, the structure of many miles of country, and in a brief space of time may reach conclusions, which, in a humid region, would reward only protracted and laborious observation and patient generalization. (*Gilbert, 1876, pp. 18–19*)
>
> In the regions of cañons a single bed can be followed, upon one continuous outcrop, for hundreds of miles, and every modification that it undergoes can be traced step by step. Moreover, by reason of the ramifications of cañons, it is frequently possible to trace a bed toward all points of the compass, so as to learn its changes, not merely along a simple line, but throughout an extended area. With such exposures, unconformity cannot escape detection, and the history of a system of sediments can be made out with a completeness that surely cannot be excelled elsewhere. (*Gilbert, 1876, p. 88*)

There follows almost an exact preview of part of Chapter 5 of the *Henry Mountains Report* which we intend to examine later (pp. 551–67). Gilbert then proceeds to describe his concept of grade, which embodies some confusion with 'gradient':

> In general, we may say that a stream tends to equalize its work in all parts of its course. Its power inheres in its fall and each foot of fall has the same power. When its work is to corrade and the resistance is unequal, it concentrates its energy where the resistance is great, by crowding many feet of descent into a small space; and diffuses it, where the resistance is small, by using but a small fall in a long distance. When its work is to transport, the resistance is constant, the fall is evenly distributed by a uniform grade. When the work includes both transportation and corrasion, as is the usual case, its grades are somewhat unequal; and the inequality is greatest when the load is least. (*Gilbert, 1876, p. 100*)

Gilbert's early ideas of the graded condition were linked to the belief that the magnitude of eroding forces on the earth's surface depended mainly on slope or gradient. In later years, as a result of his experimental work (1914), he modified this view with reference to stream action. Here we must, however, notice two points. First, the concept of grade which Davis built into his cyclic synthesis was based simply on Gilbert's *earlier* ideas. Second, these same ideas explain why Gilbert never felt the need to employ the concept of base level which, for him, merely represented a completely graded surface.

GILBERT'S MASTERPIECE ON THE MECHANICS OF FLUVIAL PROCESSES

The common denominator of all successful theories is the author's ability to compress an assortment of separate parts and consolidate them so that they become components of a major unified scheme. Hutton did it with his theory of a succession of worlds; Agassiz achieved it with his theory of glaciation;

Powell succeeded with his theory of base level; and now Gilbert balances his whole explanation of physical processes upon the gossamer-like substantiality of his theory of 'grade' or dynamic adjustment. Gilbert's *Report on the Geology of the Henry Mountains* (1877) is the first major treatment by any geologist of the mechanics of fluvial processes. Something approaching it had been done in Europe by engineers like Du Buat and Taylor but in America there had been no geomorphic work of such importance since Dana.

Gilbert deals first with erosion of the surface. In this, although he makes a few novel observations, he seldom does more than repeat what Powell has already written and gives the same threefold division of erosion processes.

> Stated in their natural order, the three general divisions of the process of erosion are (1) *weathering*, (2) *transportation*, and (3) *corrasion*. The rocks of the general surface of the land are disintegrated by *weathering*. The material thus loosened is transported by streams to the ocean or other receptacle. In transit it helps to corrade from the channels of the streams other material, which joins with it to be transported to the same goal. (*Gilbert, 1877, pp. 99–100*)

Gilbert's analysis of the various ways in which rocks break down is probably more comprehensive than anything written before outside Germany.

> In weathering the chief agents of disintegration are solution, change of temperature, the beating of rain, gravity, and vegetation.
> The great solvent of rocks is water, but it receives aid from some other substances of which it becomes the vehicle. These substances are chiefly products of the formation and decomposition of vegetable tissues. Some rocks are disintegrated by their complete solution, but the great majority are divided into grains by the solution of a portion; and fragmental rocks usually lose by solution the cement merely, and are thus reduced to their original incoherent condition.
> The most rigid rocks are cracked by sudden changes of temperature; and the crevices thus begun are opened by the freezing of the water within them. The coherence of the more porous rocks is impaired and often destroyed by the same expansive force of freezing water.
> The beating of the rain overcomes the feeble coherence of earths, and assists solution and frost by detaching the particles which they have partially loosened.
> When the base of a cliff is eroded so as to remove or diminish the support of the upper part, the rock thus deprived of support is broken off in blocks by gravity. The process of which this is a part is called cliff-erosion or *sapping*.
> Plants often pry apart rocks by the growth of their roots, but their chief aid to erosion is by increasing the solvent power of percolating water. (*Gilbert, 1877, pp. 100–1*)

Transportation, he explains, serves a treble purpose. It carries away the eroded material and in so doing erodes the river bed and comminutes the material, thus enabling the river to carry in suspension a greater load.

> Streams transport the combined products of corrasion and weathering. A part of the debris is carried in solution, and a part mechanically. The finest

of the undissolved detritus is held in suspension; the coarsest is rolled along the bottom; and there is a gradation between the two modes. There is a constant comminution of all the material as it moves, and the work of transportation is thereby accelerated. Bowlders and pebbles, while they wear the stream-bed by pounding and rubbing, are worn still more rapidly themselves. Sand grains are worn and broken by the continued jostling, and their fragments join the suspended mud. Finally the detritus is all more or less dissolved by the water, the finest most rapidly. (*Gilbert, 1877, pp. 101–2*)

He recognized that the three main physical processes of erosion were assisted by a number of factors, an increase in which could produce a maximum rate of erosion. For instance an increase in slope will generally produce an increase in the rate of weathering, transportation and corrasion. However, he stressed that although the rule is usually applicable, there are qualifications because of the presence of additional factors in particular circumstances.

In general erosion is most rapid where the slope is steepest; but weathering, transportation, and corrasion are affected in different ways and in different degrees.

With increase of slope goes increase in the velocity of running water, and with that goes increase in its power to transport undissolved detritus.

The ability of a stream to corrade by solution is not notably enhanced by great velocity; but its ability to corrade by mechanical wear keeps pace with its ability to transport, or may even increase more rapidly. For not only does the bottom receive more blows in proportion as the quantity of transient detritus increases, but the blows acquire greater force from the accelerated current, and from the greater size of the moving fragments. It is necessary however to distinguish the ability to corrade from the rate of corrasion, which will be seen further on to depend largely on other conditions.

Weathering is not directly influenced by slope, but it is reached indirectly through transportation. Solution and frost, the chief agents of rock decay, are both retarded by the excessive accumulation of disintegrated rock. Frost action ceases altogether at a few feet below the surface, and solution gradually decreases as the zone of its activity descends and the circulation on which it depends becomes more sluggish. Hence the rapid removal of the products of weathering stimulates its action, and especially that portion of its action which depends upon frost. If however the power of transportation is so great as to remove completely the products of weathering, the work of disintegration is thereby checked; for the soil which weathering tends to accumulate is a reservoir to catch rain as it reaches the earth and store it up for the work of solution and frost, instead of letting it run off at once unused. (*Gilbert, 1877, pp. 102–3*)

Favourable conditions include softness of or lack of cohesion in the rock.

Transportation is most favored by those rocks which yield by disintegration the most finely comminuted debris. (*Gilbert, 1877, p. 103*)

An increased discharge is favourable to transportation because it increases both the river's velocity and its carrying capacity.

Transportation is favored by increasing water supply as greatly as by increasing declivity. When the volume of a stream increases, it becomes at the same time more rapid, and its transporting capacity gains by the increment to velocity as well as by the increment to volume. Hence the increase in power of transportation is more than proportional to the increase in volume.

It is due to this fact chiefly that the transportation of a stream which is subject to floods is greater than it would be if its total water supply were evenly distributed in time. (*Gilbert, 1877, p. 104*)

However, total rates of erosion are subject to rather more complex controls.

In arid regions of which the declivities are sufficient to give thorough drainage, the absence of vegetation is accompanied by absence of soil. When a shower falls, nearly all the water runs off from the bare rock, and the little that is absorbed is rapidly reduced by evaporation. Solution becomes a slow process for lack of a continuous supply of water, and frost accomplishes its work only when it closely follows the infrequent rain. Thus weathering is retarded. Transportation has its work so concentrated by the quick gathering of showers into floods, as to compensate, in part at least, for the smallness of the total rainfall from which they derive their power.

Hence in regions of small rainfall, surface degradation is usually limited by the slow rate of disintegration; while in regions of great rainfall it is limited by the rate of transportation. There is probably an intermediate condition with moderate rainfall, in which a rate of disintegration greater than that of an arid climate is balanced by a more rapid transportation than consists with a very moist climate, and in which the rate of degradation attains its maximum. (*Gilbert, 1877, p. 105*)

Gilbert then examines the flow of a river. He realizes that it is more complex than a simple relationship of velocity against material to be carried. Friction, for instance, absorbs much of the river's energy and also acts as a lever to raise particles from the river bed by creating turbulence.

A stream of water flowing down its bed expends an amount of energy that is measured by the quantity of water and the vertical distance through which it descends. If there were no friction of the water upon its channel the velocity of the current would continually increase; but if, as is the usual case, there is no increase of velocity, then the whole of the energy is consumed in friction. The friction produces inequalities in the motion of the water, and especially induces subsidiary currents more or less oblique to the general onward movement. Some of these subsidiary currents have an upward tendency, and by them is performed the chief work of transportation. They lift small particles from the bottom and hold them in suspension while they move forward with the general current. The finest particles sink most slowly and are carried farthest before they fall. Larger ones are barely lifted, and are dropped at once. Still larger are only half lifted; that is, they are lifted on the side of the current and rolled over without hitting the bottom. And finally there is a limit to the power of every current, and the largest fragments of its bed are not moved at all. (*Gilbert, 1877, p. 106*)

The ability of rivers to carry particles of a given size, he explains, has been mathematically expressed by Hopkins' Law but adds that this Law shows only the maximum size of particles and not the *amount* of material that a river can carry. In other words, Gilbert recognized the difference between 'competence' and 'capacity', as will be seen more clearly later.

> There is a definite relation between the velocity of a current and the size of the largest bowlder it will roll. It has been shown by Hopkins that the weight of the bowlder is proportioned to the sixth power of the velocity. It is easily shown also that the weight of a suspended particle is proportioned to the sixth power of the velocity of the upward current that will prevent its sinking. But it must not be inferred that the total load of detritus that a stream will transport bears any such relation to the rapidity of its current. The true inference is that the velocity determines the size-limit of the detritus that a stream can move by rolling, or can hold in suspension. (*Gilbert, 1877, p. 106*)

A clear stream, he continues, demonstrates friction in its maximum role, for as soon as material falls into it its flow is retarded and the rate of retardation is proportionate to the amount of load. Yet even this is not a simple relationship for the introduction of fine material will not cause the flow to lessen so readily.

> Again, the energy of a clear stream is entirely consumed in the friction of flow; and the friction bears a direct relation to its velocity. But if detritus be added to the water, then a portion of its energy is diverted to the transportation of the load; and this is done at the expense of the friction of flow, and hence at the expense of velocity. As the energy expended in transportation increases, the velocity diminishes. If the detritus be composed of uniform particles, then we may also say that as the load increases the velocity diminishes. But the diminishing velocity will finally reach a point at which it can barely transport particles of the given size, and when this point is attained, the stream has its maximum load of detritus of the given size. But fine detritus requires less velocity for its transportation than coarse, and will not so soon reduce the current to the limit of its efficiency. A greater percentage of the total energy of the stream can hence be employed by fine detritus than by coarse. (*Gilbert, 1877, p. 107*)
>
> Thus the capacity of a stream for transportation is enhanced by comminution in two ways. Fine detritus, on the one hand, consumes less energy for the transportation of the same weight, and on the other, it can utilize a greater portion of the stream's energy.
>
> It follows, as a corollary, that the velocity of a fully loaded stream depends (*ceteris paribus*) on the comminution of the material of the load. When a stream has its maximum load of fine detritus, its velocity will be less than when carrying its maximum load of coarse detritus; and the greater load corresponds to the less velocity. (*Gilbert, 1877, pp. 107–8*)

From this examination of flow Gilbert concludes that slope is the main factor governing the rate of transportation, provided always that the other factors of friction, load and discharge are equal.

> . . . since the energy which each stream expends in transportation is the residual after deducting what it spends in friction from its total energy, it is evident that the stream with the greater declivity will not merely have the greater energy, but will expend a less percentage of it in friction and a greater percentage in transportation.

FIG. 117. *The north flank of the Henry Mountains, Utah, from the air (U.S. Geological Survey)*

Hence declivity favors transportation in a degree that is greater than its simple ratio. (*Gilbert, 1877, p. 108*)

Another factor which controls flow he discovered from the physical laws concerning friction. The energy consumed in bed friction is directly related to the character and surface area of the river bed and therefore, if the area of the bed does not increase greatly, and this is not likely, an additional discharge will produce an increase in velocity which in return will be able to raise the rate of transportation.

A stream's friction of flow depends mainly on the character of the bed, on the area of the surface of contact, and on the velocity of the current. When the other elements are constant, the friction varies approximately with the area of contact. The area of contact depends on the length and form of the channel, and on the quantity of water. For streams of the same length and same form of cross-section, but differing in size of cross-section, the area of contact varies directly as the square root of the quantity of water. Hence, *ceteris paribus*, the friction of a stream on its bed is proportioned to the square root of the quantity of water. But as stated above, the total energy of a stream is proportioned directly to the quantity of water; and the total energy is equal to the energy spent in friction, plus the energy spent in transportation. Whence it follows that if a stream change its quantity of water without changing its velocity or other accidents, the total energy will change at the same rate as the quantity of water; the energy spent in friction will change at a less rate, and the energy remaining for transportation will change at a greater rate.

Hence increase in quantity of water favors transportation in a degree that is greater than its simple ratio.

It follows as a corollary that the running water which carries the *débris* of a district loses power by subdivision towards its sources; and that, unless there is a compensating increment of declivity, the tributaries of a river will fail to supply it with the full load it is able to carry. (*Gilbert, 1877, pp. 109-10*)

The effect of these rules upon the transport of debris Gilbert translated into technical terms, differentiating between a river's 'competence' and a river's 'capacity'.

A stream which can transport *débris* of a given size, may be said to be *competent* to such *débris*. Since the maximum particles which streams are able to move are proportioned to the sixth powers of their velocities, competence depends on velocity. Velocity, in turn, depends on declivity and volume, and (inversely) on load.

In brief, the capacity of a stream for transportation is greater for fine *débris* than for coarse.

Its capacity for the transportation of a given kind of *débris* is enlarged in more than simple ratio by increase of declivity; and it is enlarged in more than simple ratio by increase of volume.

The competence of a stream for the transport of *débris* of a given fineness, is limited by a corresponding velocity. (*Gilbert, 1877, p. 110*)

He then summarized all the factors affecting transportation.

Thus rate of transportation, as well as capacity for transportation, is favored by fineness of *débris*, by declivity, and by quantity of water. It is opposed chiefly by vegetation, which holds together that which is loosened by weathering, and shields it from the agent of transportation in the very place where that agent is weakest.

When the current of a stream gradually diminishes in its course – as for example in approaching the ocean – the capacity for transportation also diminishes; and so soon as the capacity becomes less than the load, precipitation begins – the coarser particles being deposited first. (*Gilbert, 1877, p. 111*)

Gilbert possessed an intellect which allowed him to go further than this. He was able to combine these separate ideas into a theory which visualized all the processes working in unison towards the disciplined or regulated reduction of the whole land surface. His first statement is his concept of 'grade'. Gilbert realized that rivers could exist in three states. There was a state when the river flowed swiftly and had more energy than was necessary to move the material in suspension; when this was so the river would corrade its bed. There was the converse state when the river was flowing sluggishly and depositing material. The intermediate state occurs when the carrying energy of the river exactly balanced the amount of material it had to transport. This state of equilibrium when the river neither corraded nor deposited he termed 'grade' and he believed it was the ideal adjustment or final stage towards which all rivers were working. He pointed out that a river might be at 'grade' throughout its length but have within its course small unadjusted reaches where it was still corroding or depositing.

Where a stream has all the load of a given degree of comminution which it is capable of carrying, the entire energy of the descending water and load is consumed in the translation of the water and load and there is none applied to corrasion. If it has an excess of load its velocity is thereby diminished so as to lessen its competence and a portion is dropped. If it has less than a full load it is in condition to receive more and it corrades its bottom.

A fully loaded stream is on the verge between corrasion and deposition. As will be explained in another place, it may wear the walls of its channel, but its wear of one wall will be accompanied by an addition to the opposite wall. (*Gilbert, 1877, p. 111*)

Let us suppose that a stream endowed with a constant volume of water, is at some point continuously supplied with as great a load as it is capable of carrying. For so great a distance as its velocity remains the same, it will neither corrade (downward) nor deposit, but will leave the grade of its bed unchanged. But if in its progress it reaches a place where a less declivity of bed gives a diminished velocity, its capacity for transportation will become less than the load and part of the load will be deposited. Or if in its progress it reaches a place where a greater declivity of bed gives an increased velocity, the capacity for transportation will become greater than the load and there will be corrasion of the bed. In this way a stream which has a supply of *débris* equal to its capacity, tends to build up the gentler slopes of its bed and cut away the steeper. It tends to establish a single, uniform grade. (*Gilbert, 1877, p. 112*)

This tendency of the river to work towards a general slope depends upon differences in declivity. A steep slope, by inducing a quicker flow of water, accelerates the rate of erosion which proceeds to lessen the steepness of the slope, until it has reached a state of uniformity with the slope downstream.

> Let us now suppose that the stream after having obliterated all the inequalities of the grade of its bed loses nearly the whole of its load. Its velocity is at once accelerated and vertical corrasion begins through its whole length. Since the stream has the same declivity and consequently the same velocity at all points, its capacity for corrasion is everywhere the same. Its rate of corrasion however will depend on the character of its bed. Where the rock is hard corrasion will be less rapid than where it is soft, and there will result inequalities of grade. But so soon as there is inequality of grade there is inequality of velocity, and inequality of capacity for corrasion; and where hard rocks have produced declivities, there the capacity for corrasion will be increased. The differentiation will proceed until the capacity for corrasion is everywhere proportioned to the resistance, and no further, – that is, until there is an equilibrium of action. (*Gilbert, 1877, p. 113*)

By the same principle, the declivity of a main stream is less steep than that of its tributaries as its larger discharge wears its bed lower at a faster rate.

Gilbert saw that the principle need not be confined to river beds but could be extended to all transportational slopes. From this he evolved his *law of uniform slopes*. As he explains by this law, topography would tend towards uniformity in appearance were it not for the fact that the erosive agents do not act upon the surface in equal proportions.

> It is evident that if steep slopes are worn more rapidly than gentle, the tendency is to abolish all differences of slope and produce uniformity. The law of uniform slope thus opposes diversity of topography, and if not complemented by other laws, would reduce all drainage basins to plains. But in reality it is never free to work out its full results; for it demands a uniformity of conditions which nowhere exists. Only a water sheet of uniform depth, flowing over a surface of homogeneous material, would suffice; and every inequality of water depth or of rock texture produces a corresponding inequality of slope and diversity of form. The reliefs of the landscape exemplify other laws, and the law of uniform slopes is merely the conservative element which limits their results. (*Gilbert, 1877, p. 115*)

Durability of the rock he recognizes is another factor affecting topography but this must be balanced against the rule that steep declivities will hasten the rate of erosion. He relates his *laws of declivities and structure*.

> Erosion is most rapid where the resistance is least, and hence as the soft rocks are worn away the hard are left prominent. The differentiation continues until an equilibrium is reached through the law of declivities. When the ratio of erosive action as dependent on declivities becomes equal to the ratio of resistances as dependent on rock character, there is equality of action. (*Gilbert, 1877, pp. 115–16*)

Another general rule that Gilbert developed was the *law of divides*. As the greatest concentration of water was to be found at its mouth and the least at

FIG. 118. *Gilbert's (1877) illustration of a relief model of the Henry Mountains, Utah*

its head a river profile was flat at its mouth and gradually steepened towards its source. As this shape was the same for all river profiles it followed that the mountain divides have the same form. And as the essence of this shape lay in its creation by erosion, it was independent of, and often contrary to, the original structural form.

We have seen that the declivity over which water flows bears an inverse relation to the quantity of water. If we follow a stream from its mouth upward and pass successively the mouths of its tributaries, we find its volume gradually less and less and its grade steeper and steeper, until finally at its head we reach the steepest grade of all. If we draw the profile of the river on paper, we produce a curve concave upward and with the greatest curvature at the upper end. The same law applies to every tributary and even to the slopes over which the freshly fallen rain flows in a sheet before it is gathered into rills. The nearer the water-shed or divide the steeper the slope; the farther away the less the slope.

It is in accordance with this law that mountains are steepest at their crests. The profile of a mountain if taken along drainage lines is concave outward as represented in the diagram; and this is purely a matter of sculpture, the uplifts from which mountains are carved rarely if ever assuming this form.

Under the *law of Structure* and the *law of Divides* combined, the features of the earth are carved. Declivities are steep in proportion as their material is hard; and they are steep in proportion as they are near divides. (*Gilbert, 1877, p. 116*)

In areas of dense vegetation the *law of structure* will not operate freely, and topography will largely depend upon the *law of divides*;

We have seen that vegetation favors the disintegration of rocks and retards the transportation of the disintegrated material. Where vegetation is profuse there is always an excess of material awaiting transportation, and the limit to the rate of erosion comes to be merely the limit to the rate of transportation. And since the diversities of rock texture, such as hardness and softness, affect only the rate of disintegration (weathering and corrasion) and not the rate of transportation, these diversities do not affect the rate of erosion in regions of profuse vegetation, and do not produce corresponding diversities of form.

On the other hand, where vegetation is scant or absent, transportation and corrasion are favored, while weathering is retarded. There is no accumulation of disintegrated material. The rate of erosion is limited by the rate of weathering, and that varies with the diversity of rock texture. The soft are eaten away faster than the hard; and the structure is embodied in the topographic forms.

Thus a moist climate by stimulating vegetation produces a sculpture independent of diversities of rock texture, and a dry climate by repressing vegetation produces a sculpture dependent on these diversities. With great moisture the law of divides is supreme; with aridity, the law of structure. (*Gilbert, 1877, p. 119*)

This enunciation of general principles was a major advance on any previous work. Some of Gilbert's statements had been preceded or paralleled in Britain and on the Continent and he undoubtedly received much guidance from Powell's studies but, as a whole, the analysis had no equal in completeness and lucidity. The remainder of the report on the Henry Mountains is equally important but as it deals with more specific features these will be mentioned separately. Gilbert postulated that the more rapid retreat of sleeper slopes explained the central location of inter-stream watersheds. He also examined the valley-heads.

The heads of the secondary drainage lines are in nature tolerably definite points. The water which during rain converges at one of these points is there abruptly concentrated in volume. Above the point it is a sheet, or at least is divided into many rills. Below it, it is a single stream with greatly increased power of transportation and corrasion. The principle of equal action gives to the concentrated stream a less declivity than to the diffused sheet, and – what is especially important – it tends to produce an equal grade in all directions upward from the point of convergence. The converging surface becomes hopper-shaped or funnel-shaped; and as the point of convergence is lowered by corrasion, the walls of the funnel are eaten back equally in all directions – except of course the direction of the stream. The influence of the stream in stimulating erosion above its head is thus extended radially and equally through an arc of 180° of which the centre is at the point of convergence. (*Gilbert, 1877, p. 121*)

In the badlands Gilbert noticed an apparent anomaly to his *law of divides*. Instead of the concave slopes ending in a sharp arête or a cliff wall there was a tendency for the slope to flatten near the divide. Gilbert suggested that this might be due to an unknown factor, and thirty years later, in 1909, he himself demonstrated that soil creep was the missing factor.

There is one other peculiarity of bad-land forms which is of great significance, but which I shall nevertheless not undertake to explain. According to the law of divides, as stated in a previous paragraph, the profile of any slope in bad-lands should be concave upward, and the slope should be steepest at the divide. The union or intersection of two slopes on a divide should produce an angle. But in point of fact the slopes do not unite in an angle. They unite in a curve, and the profile of a drainage slope instead of being concave all the way to its summit, changes its curvature and becomes convex. (*Gilbert, 1877, p. 122*)

Thus in the sculpture of the bad-lands there is revealed an exception to the law of divides, – an exception which cannot be referred to accidents of structure, and which is as persistent in its recurrence as are the features which conform to the law, – an exception which in some unexplained way is part of the law. Our analysis of the agencies and conditions of erosion, on the one hand, has led to the conclusion that (where structure does not prevent) the declivities of a continuous drainage slope increase as the quantities of water flowing over them decrease; and that they are great in proportion as they are near divides. Our observation, on the other hand, shows that the declivities increase as the quantities of water diminish, up to a certain point where the quantity is very small, and then decrease; and that declivities are great in proportion as they are near divides, unless they are *very* near divides. Evidently some factor has been overlooked in the analysis, – a factor which in the main is less important than the flow of water, but which asserts its existence at those points where the flow of water is exceedingly small, and is there supreme. (*Gilbert, 1877, pp. 122-3*)

With his thoughts concentrated on processes concerned in the lowering of the general surface, he returns to the elements governing the rate of erosion. He emphasizes the reciprocity between declivity and the erosion rate (*see also* Gilbert, 1909) and points out that a variation of erosive power in any part of the river system is bound to be translated with varying effects throughout all its parts as form becomes universally adjusted to process;

> The tendency to equality of action, or to the establishment of a dynamic equilibrium, has already been pointed out in the discussion of the principles of erosion and of sculpture, but one of its most important results has not been noticed.
>
> Of the main conditions which determine the rate of erosion, namely, quantity of running water, vegetation, texture of rock, and declivity, only the last is reciprocally determined by rate of erosion. Declivity originates in upheaval, or in the displacements of the earth's crust by which mountains and continents are formed; but it receives its distribution in detail in accordance with the laws of erosion. Wherever by reason of change in any of the conditions the erosive agents come to have locally exceptional power, that power is steadily diminished by the reaction of rate of erosion upon declivity. Every slope is a member of a series, receiving the water and the waste of the slope above it, and discharging its own water and waste upon the slope below. If one member of the series is eroded with exceptional rapidity, two things immediately result: first, the member above has its level of discharge lowered, and its rate of erosion is thereby increased; and second, the member below, being clogged by an exceptional load of detritus, has its rate of erosion diminished. The acceleration above and the retardation below, diminish the declivity of the member in which the disturbance originated; and as the declivity is reduced the rate of erosion is likewise reduced.
>
> But the effect does not stop here. The disturbance which has been transferred from one member of the series to the two which adjoin it, is by them transmitted to others, and does not cease until it has reached the confines of the drainage basin. For in each basin all lines of drainage unite in a main line, and a disturbance upon any line is communicated through it to the main line and thence to every tributary. And as any member of the system may influence all the others, so each member is influenced by every other. There is an interdependence throughout the system. (*Gilbert, 1877, pp. 123–4*)

He describes how the attainment of grade can produce erosional peculiarities. When downward erosion ceases the river will cut its way sideways across its valley by eroding laterally one side and depositing a flood plain on the other. In this way also it will produce a planed surface, or pediment;

> It has been shown in the discussion of the relations of transportation and corrasion that downward wear ceases when the load equals the capacity for transportation. Whenever the load reduces the downward corrasion to little or nothing, lateral corrasion becomes relatively and actually of importance. The first result of the wearing of the walls of a stream's channel is the formation of a flood-plain. As an effect of momentum the current is always swiftest along the outside of a curve of the channel, and it is there that the wearing is performed; while at the inner side of the curve the current is so slow that part of the load is deposited. In this way the width of the channel remains the same while its position is shifted, and every part of the valley which it has crossed in its shiftings comes to be covered by a deposit which does not rise above the highest level of water. The surface of this deposit is hence appropriately called the *flood-plain* of the stream. The deposit is of nearly uniform depth, descending no lower than the bottom of the water-channel,

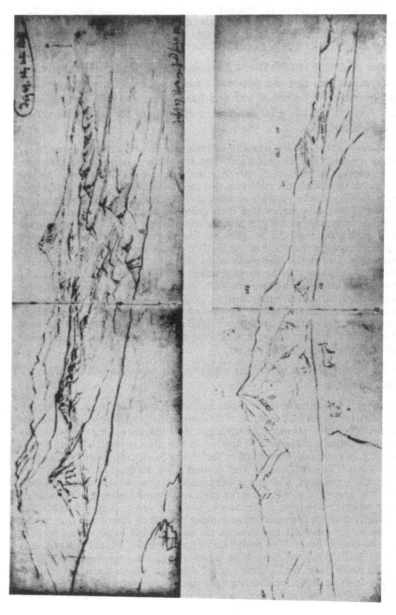

FIG. 119. *Sketches of the Henry Mountains reproduced from the field notebooks of G. K. Gilbert*
(U.S. Geological Survey)

and it rests upon a tolerably even surface of the rock or other material which is corraded by the stream. The process of carving away the rock so as to produce an even surface, and at the same time covering it with an alluvial deposit, is the process of *planation.* (*Gilbert, 1877, pp. 126–7*)

With an example from the Henry Mountains, he shows how the attainment of grade will ensure that the river has a smooth gradient even though it is passing over an alternation of soft and hard rocks. In other words, when grade is reached this has more control on the form of the slope than differences in rock resistance.

The slopes of the Henry Mountains illustrate the process in a peculiarly striking manner. The streams which flow down them are limited in their rate of degradation at both ends. At their sources, erosion is opposed by the hardness of the rocks; the trachytes and metamorphics of the mountain tops are carved very slowly. At their mouths, they discharge into the Colorado and the Dirty Devil, and cannot sink their channels more rapidly than do those rivers. Between the mountains and the rivers, they cross rocks which are soft in comparison with the trachyte, but they can deepen their channels with no greater rapidity than at their ends. The grades have adjusted themselves accordingly. Among the hard rocks of the mountains the declivities are great, so as to give efficiency to the eroding water. Among the sedimentary rocks of the base they are small in comparison, the chief work of the streams being the transportation of the trachyte *débris*. So greatly are the streams concerned in transportation, and so little in downward corrasion (outside the trachyte region), that their grades are almost unaffected by the differences of rock texture, and they pass through sandstone and shale with nearly the same declivity. (*Gilbert, 1877, pp. 127–8*)

A further factor operates where the supply of water and debris is subject to great fluctuations;

The streams which made these plains and which maintain them, accomplish their work by a continual shifting of their channels; and where the plains are best developed they employ another method of shifting – a method which in its proper logical order must be treated in the discussion of alluvial cones, but which is practically combined in the Henry Mountains, with the method of planation. The supply of detritus derived from the erosion of the trachyte is not entirely constant. Not only is more carried out in one season than another and in one year than another, but the work is accomplished in part by sudden storms which create great floods and as suddenly cease. It results from this irregularity that the channels are sometimes choked by *débris*, and that by the choking of the channels the streams are turned aside to seek new courses upon the general plain. The abandoned courses remain plainly marked, and one who looks down on them from some commanding eminence can often trace out many stages in the history of the drainage. Where a series of streams emerges from adjacent mountain gorges upon a common plain, their shiftings bring about frequent unions and separations, and produce a variety of combinations. (*Gilbert, 1877, p. 129*)

Gilbert went even further than Powell when he explained what would happen where the river's powers of erosion failed to keep pace with the rate of uplift;

When a mountain uplift crosses the course of a stream, it often happens that the rate of uplift is too rapid to be equalled by the corrasion of the stream, and the uprising rock becomes a dam over which the water still runs, but above which there is accumulated a pond or lake. Whenever this takes place, the pond catches all the *débris* of the upper course of the stream, and the water which overflows at the outlet having been relieved of its load is almost powerless for corrasion, and cannot continue its contest with the uplift unless the pond is silted up with detritus. As the uplift progresses the level of the pond is raised higher and higher, until finally it finds a new outlet at some other point. The original outlet is at once abandoned, and the new one becomes a permanent part of the course of the stream. As a rule it is only large streams which hold their courses while mountains rise; the smaller are turned back by ponding, and are usually diverted so as to join the larger. (*Gilbert, 1877, pp. 125–6*)

River terraces, he states, are unequivocally the result of river erosion acting upon an original flood-plain deposit. Alluvial cones he has already described as forming parts of a flood-plain. He now explains how they are created at the mouths of mountain valleys where these suddenly open out to a flat plain.

Wherever, as in Nevada and Western Utah, the valleys are the receptacles of the detritus washed out from the mountains, the foot-slopes of the mountains consist of a series of alluvial cones. From each mountain gorge the products of its erosion are discharged into the valley. The stream which bears the *débris* builds up the bed of its channel until it is higher than the adjacent land and then abandons it, and by the repetition of this process accumulated a conical hill of detritus which slopes equally in all directions from the mouth of the mountain gorge. At one time or another the water runs over every part of the cone and leaves it by every part of its base; and it sometimes happens that the opposite slopes of the cone lead to different drainage systems. (*Gilbert, 1877, p. 134*)

From the principles governing the formation of river valleys Gilbert goes on to discuss erosional slopes. He explains how the juxtaposition of a hard and soft bed may produce an escarpment which will gradually retreat under the action of basal sapping. This is really a process of undermining where the river running along the base of the scarp goes on removing the soft rocks underlying the hard cap until the rocks underneath can no longer provide support and part of the hard cap collapses. As the scarp slope in dipping strata retreats in this way, the divide which it forms will migrate across the surface by a process which is termed 'monoclinal shifting'.

The tendency of hard strata to rid themselves of waterways and of soft strata to accumulate them, is a prime element of the process which carves hills from the hard and valleys from the soft. Where hard rocks are crossed by waterways they cannot stand higher than the adjacent parts of the waterways; but where they are not so crossed they become divides, and the *law of divides* conspires with the *law of structure* to carve eminences from them.

The tendency of waterways to escape from hard strata and to abide in soft, and their tendency to follow the strike of soft strata and to cross hard at right angles, are tendencies only and do not always prevail. They are opposed by the tendency of drainage lines to stability. If the dip of the strata is small, or if the differences of hardness are slight, or if the changes of texture are gradual instead of abrupt, monoclinal shifting is greatly reduced. (*Gilbert, 1877, pp. 136–7*)

This lateral shifting of the divide will only cease when the declivities of the scarp and dip slopes become equal. Gilbert added to this the idea of lateral stream abstraction.

> It results also that if one of the waterways is corraded more rapidly than the other the divide moves steadily toward the latter, and eventually, if the process continues, reaches it. When this occurs, the stream with the higher valley abandons the lower part of its course and joins its water to that of the lower stream. Thus from the shifting of divides there arises yet another method of the shifting of waterways, a method which it will be convenient to characterize as that of *abstraction*. A stream which for any reason is able to corrade its bottom more rapidly than do its neighbours, expands its valley at their expense, and eventually 'abstracts' them. (*Gilbert, 1877, p. 141*)

Gilbert further expanded Powell's classification of rivers according to genetic principles;

> If a series of sediments accumulated in an ocean or lake be subjected to a system of displacements while still under water, and then be converted to dry land by elevation *en masse* or by the retirement of the water, the rains which fall on them will inaugurate a drainage system perfectly conformable with the system of displacements. Streams will rise along the crest of each anticlinal, will flow from it in the direction of the steepest dip, will unite in the synclinals, and will follow them lengthwise. The axis of each synclinal will be marked by a watercourse; the axis of each anticlinal by a watershed. Such a system is said to be *consequent* on the structure.
>
> If however a rock series is affected by a system of displacements after the series has become continental, it will have already acquired a system of waterways, and *provided the displacements are produced slowly* the waters will not be diverted from their accustomed ways. The effect of local elevation will be to stimulate local corrasion, and each river that crosses a line of uplift will inch by inch as the land rises deepen its channel and valorously maintain its original course. It will result that the directions of the drainage lines will be independent of the displacements. Such a drainage system is said to be *antecedent* to the structure.
>
> But if in the latter case the displacements are produced rapidly the drainage system will be rearranged and will become consequent to the structure. It has frequently happened that displacements formed with moderate rapidity have given rise to a drainage system of mixed character in which the courses of the larger streams are antecedent and those of the smaller are consequent.
>
> There is a fourth case. Suppose a rock series that has been folded and eroded to be again submerged, and to receive a new accumulation of un-conforming sediments. Suppose further that it once more emerges and that the new sediments are eroded from its surface. Then the drainage system will have been given by the form of the upper surface of the superior strata, but will be independent of the structure of the inferior series, into which it

will descend vertically as the degradation progresses. Such a drainage system is said to be *superimposed by sedimentation* upon the structure of the older series of strata.

Fifth. The drainage of an alluvial cone or of a delta is independent of the structure of the bed-rock beneath; and if in the course of time erosion takes the place of deposition and the alluvial formation is cut through, the drainage system which is acquired by the rocks beneath is not consequent upon their structure but is *superimposed by alluviation*.

Sixth. The drainage of a district of planation is independent of the structure of the rock from which it is carved; and when in the progress of degradation the beds favorable to lateral corrasion are destroyed and the waterways become permanent, their system may be said to be *superimposed by planation*. (*Gilbert, 1877, pp. 143–4*)

There is perhaps no need to point out here that a pleasing feature of Gilbert's work, as also of Powell's, is the clearness of approach. It is free of doctrinaire explanations and represents the author's untrammelled attempt to explain what he had actually seen. As a result it becomes a straightforward, scientific treatment of facts and does not degenerate into a philosophical soliloquy as did many of the geomorphic studies in Europe.

GILBERT'S LATER WORK

During his time in Washington Gilbert's major contribution to landscape research was his *History of Lake Bonneville*. In this monograph he demonstrated the main characteristics of marine processes and emphasized the distinction between marine and subaerial erosion (pp. 183–4).

Whoever makes a careful examination of a tract of sea coast, first before a great storm and again after the storm has passed, cannot fail to discover various modifications wrought by the storm in the form or character of the coast. Wherever the water margin is over-looked by a cliff it will be found that some portion of the rock or earth of the cliff has been washed away, and probably that fragments great or small have fallen from above. Wherever the shore is constituted by a beach or spit it will be found that additions have somewhere been made to the sand or shingle, and possibly that at other points losses have been sustained. It is universally recognized that these changes are the result of the conjoined action of waves and currents set in motion by the storm, and that they are limited to the immediate vicinity of the shore and to shallow water. There is a horizontal zone of activity practically corresponding with the zone which receives the force of the breakers, but extending somewhat farther downward, and along this zone nearly every portion of the coast either suffers abrasions or else grows by additions to its banks and bars. Where the coast is abraded, the zone of wave beating is carried progressively landward, and two features are wrought: the paring away of the land leaves a fringing terrace or shelf just beneath the water; and the same excavation produces a cliff at the landward margin of the terrace. Where the coast grows, its banks and bars are not merely enlarged, but if they did not before exist they are created. The sea-cliff and the terrace at its foot are the creatures of the wave and mark the spot where the nature of its action is destructive. (*Gilbert, 1880–1, pp. 171–2*)

His description of the results of fluvial erosion in Utah again lays emphasis on the formation of pediments. Gilbert was probably the first geologist to deal fully with this feature which happens to be extremely conspicuous in arid regions.

> The sculpture of a mountain by rain is a twofold process; on the one hand destructive, on the other constructive. The upper parts are eaten away in gorges and amphitheatres until the intervening remnants are reduced to sharp-edged spurs and crests, and all the detritus thus produced is swept outward and downward by the flowing waters and deposited beyond the mouths of the mountain gorges. A large share of it remains at the foot of the mountain mass, being built into a smooth sloping pediment. If the outward flow of the water were equal in all directions this pediment would be uniform upon all sides, but there is a principle of concentration involved whereby rill joins with rill, creek with creek and gorge with gorge, so that when the water leaves the margin of the rocky mass it is always united into a comparatively small number of streams, and it is by these that the entire volume of detritus is discharged. About the mouth of each gorge a symmetric heap of alluvium is produced – a conical mass of low slope, descending equally in all directions from the point of issue; and the base of each mountain exhibits a series of such alluvial cones, each with its apex at the mouth of a gorge and with its broad base resting upon the adjacent plain or valley. Rarely these cones stand so far apart as to be completely individual and distinct, but usually the parent gorges are so thickly set along the mountain front that the cones are more or less united and give to the contours of the mountain base a scalloped outline.
>
> The Bonneville Basin is surrounded by and interspersed with mountains, and from the summits of these down to the Bonneville shore the entire topography is of a rain-wrought type. From the shoreline downward to the valleys and plains its nature is composite, uniting the elements of wave sculpture with those of rain sculpture, but the manner of union is not indiscriminate and in it is written what we know of the pre-Bonneville history of the basin. All of the larger elements belong to the domain of rain, and upon these the elements derived from wave-work are lightly engraved and embossed. (*Gilbert, 1880–1, pp. 183–4*)

He follows Powell in his insistence that high mountains are necessarily young features.

> Each mountain therefore has a history, involving its inception, its growth, its decay, and finally its extinction. With slight exception all mountain growth is by subterranean action, and if we exclude volcanic mountains from consideration we may refer mountain building almost wholly to upheaval. With slight exception the destruction of mountains is by erosion, and in most mountains the formative and destructive processes at some time coexist. As soon as the formation of a mountain is initiated, slopes are created which enable rain and streams to wear away its surface, and since this wear never ceases until its final and complete destruction, it is contemporary with every stage of growth except the initiative. The magnitude of each mountain at any time represents the excess of formation over destruction, of upheaval over erosion. (*Gilbert, 1880–1, pp. 192–3*)

This relationship of altitude to the mountain's age, Gilbert realized, posed the question whether it was possible to tell if the mountain had ceased its movement upwards. Measuring this stability is a complex task because, if the land surface itself was unstable, there was little to act as a standard gauge or base line.

> . . . it is true in a general way that as mountain masses grow the rate of waste increases much more rapidly than the altitude. It was hence argued most cogently by Powell that all large mountains are young mountains, and from the point of view of the uniformitarian, it is equally evident that all large mountains must be growing mountains; for if the process of growth is continuous and if a high mountain melts with exceptional rapidity before the play of the elements, it is illogical to suppose that the uprising of any mountain which today is lofty has today ceased. If, therefore, it were possible to ask of all great mountains the question whether they are now growing, and obtain an answer, a solution might be reached of the problem which has divided investigators; and for this reason great interest attaches to any answer which can be obtained in the case of any mountain. (*Gilbert, 1880–1, pp. 193–4*)

In this instance he pointed out that successive lake-shorelines could act as a clue with regard to some of the mountains of Utah.

> The general results of the investigation are, first, that neither of the two shorelines is now horizontal, and, second, that the two are not parallel; whence it is evident that the region has been the scene of orographic movements both during the existence of the last high stage of the water and since the final subsidence. (*Gilbert, 1880–1, p. 195*)
>
> In the case of the Wasatch, however, our information is concise. The slopes interrupted by the escarpment are not simple alluvial slopes or rock slopes carved by subaerial agencies, but are slopes characterized by the peculiar sculpture of the waves; and the phenomena show not only that the last uplift of the Wasatch took place after the formation of the Bonneville and Provo shores, but that the water of Great Salt Lake has not since been even fifty feet higher than it is at present. It is therefore demonstrated that an actual uplift of the mountain occurred at so recent a date as to leave no reasonable suspicion that its growth has now ceased. (*Gilbert, 1880–1, pp. 199–200*)

About this time (1880) Gilbert completed and revised a manuscript left by Newton, a member of the government team of geologists who were sent to the Black Hills of Dakota to make a mineral survey as a result of the 1875 gold rush. Newton died of typhoid in 1877 at the early age of 32. He, and Gilbert his editor, explained the present topography of the Black Hills by assuming that the original surface was a simple dome from the central eminence of which fluvial erosion had removed a great thickness of strata. Thus, following Powell's ideas, the radial drainage was considered consequent upon an original dome.

> The courses of the creeks were fixed and their work of erosion was commenced when the uplift first exposed its summit to the rain. They are precisely as old as the Hills, and they have been the agents of the degradation of the uplift. (*Newton, 1880, p. 217*)

In a later work of his own on lake margins, Gilbert paid more attention to describing the finer details of the erosive processes. For instance he underlines the importance of the erosive action of the backwash in helping to continue the shoreward progress of wave erosion.

> The pulsating current of the undertow has an erosive as well as a transporting function. It carries to and fro the detritus of the shore, and, dragging it over the bottom, continues downward the erosion initiated by the breakers. This downward erosion is the necessary concomitant of the shoreward progress of wave erosion; for if the land were merely planed away to the level of the wave troughs, the incoming waves would break where shoal water was first reached and become ineffective at the water margin. In fact, this spending of the force of the waves where the water is so shallow as to induce them to break, increases at that point the erosive power by pulsation, and thus brings about an interdependence of parts. (*Gilbert, 1883-4, pp. 82-83*).

He even recognizes a state of beach equilibrium analogous to his theory of grade;

> In order that a particular portion of shore shall be the scene of littoral transportation, it is essential, first, that there be a supply of shore drift; second, that there be shore action by waves and currents; and in order that the local process be transportation only, and involve neither erosion nor deposition, a certain equilibrium must exist between the quantity of the shore drift on the one hand and the power of the waves and currents on the other. On the whole this equilibrium is a delicate one, but within certain narrow limits it is stable. That is to say, there are certain slight variations of the individual conditions of equilibrium which disturb the equilibrium only in a manner tending to its immediate readjustment. (*Gilbert, 1883-4, p. 101*)

In an article written in 1884 he explains how Ferrel's Law of Deflection will cause moving bodies to be deflected to the right in the northern hemisphere and thus may explain meandering. In a river this might produce lateral cutting with an inclination of the water surface towards the left bank, giving deeper water on the right. A river flowing at 45° north latitude at 3 metres a second will, he calculated, exert an excess pressure on its right bank of 1/63,539 of its weight. The form of a river's cross-section will otherwise control its velocity through the section of the valley affected. Once the channel has become curved, centrifugal force (proportional to the square of the river's velocity) occurs and transfers the locus of maximum velocity towards the outer bank. Thus the symmetry of the river profile is destroyed because the effect of lateral cutting becomes cumulative.

> Moreover there is no compensating tendency to restore an equilibrium, for the erosion of the outer bank increases the sinuosity of the channel instead of rectifying it. (*Gilbert, 1884, p. 429*)

There is therefore a certain amount of competition between the right and left banks.

Gilbert developed a formula showing the ratio between the selective influences that determine which bank shall be chosen.

$$\frac{R}{L} = \frac{V + \rho n \sin\lambda}{V - \rho n \sin\lambda}$$

where; λ = latitude; n = angular velocity of the earth's rotation; V = mean velocity of the stream; ρ = radius of stream curvature. When he applied the formula to the Mississippi he got the result that $\frac{R}{L}$ = 1·087 which meant that the tendency to erode the right bank was nearly 9 per cent greater than the tendency to erode the left.

The last article of Gilbert's to be quoted here represents one of his few methodological contributions, and is interesting because it reveals the technique he must have employed consciously or subconsciously in evolving his many geomorphic theories. It also stands in striking contrast to the methodological approach of W. M. Davis.

> Phenomena are arranged in chains of necessary sequence. In such a chain each link is the necessary consequent of that which precedes, and the necessary antecedent of that which follows. . . . If we examine any link of the chain, we find that it has more than one antecedent and more than one consequent. . . . Antecedent and consequent relations are therefore not merely linear, but constitute a plexus; and this plexus pervades nature. (*Gilbert, 1886, pp. 286–7*)
>
> It is the province of research to discover the antecedents of phenomena. A phenomenon having been observed, or a group of phenomena having been established by empiric classification, the investigator invents an hypothesis in explanation. He then devises and applies a test of the validity of the hypothesis. If it survives the test, he proceeds at once to devise a second test. . . . The process of testing is then the process of elimination, at least until all but one of the hypotheses have been disproved.
>
> In the testing of hypotheses lies the prime difference between the investigator and the theorist. The one seeks diligently for the facts which may overthrow his tentative theory, the other closes his eyes to these and searches only for those which will sustain it. (*Gilbert, 1886, p. 286*)
>
> The great investigator is primarily and pre-eminently a man who is rich in hypotheses. In the plenitude of his wealth he can spare the weaklings without regret; and having many from which to select, his mind maintains a judicial attitude. The man who can produce but one, cherishes and champions that one as his own, and is blind to its faults. With such men, the testing of alternative hypotheses is accomplished only through controversy. Critical observations are warped by prejudice, and the triumph of the truth is delayed. (*Gilbert, 1886, p. 287*)

No one could doubt that Grove Karl Gilbert was a great investigator, nor that he was a fruitful source of hypotheses. Of his hypotheses we have said much but as he published his remarkable investigations into *The Transportation of Debris by Running Water* in 1914 we must leave them for discussion in a further volume. Of the famous trio of geomorphologists, Powell, Gilbert and Dutton, he was the most brilliant and his deductions regarding stream and landscape mechanics have given new life to quantitative geomorphology in the twentieth century.

Indeed, some of the striking advances in geomorphology during the last fifty years had been foreseen in Gilbert's vivid phrases. Thus, his emphasis on the importance of lateral planation and of the stripping of weathered bedrock by sheet-floods in sub-arid and arid climates was not fully appreciated until the twentieth century. But it was Gilbert's concept of *grade*, employed in its widest context of form-adjustment, that made the most immediate impact upon contemporary geomorphologists.

Gilbert always thought in terms of equilibrium and ratios, particularly between force and resistance. He viewed topographic form as expressive of such relationships, whereby the mutual adjustment of form and process is achieved by a self-regulatory mechanism. Change of form through time concerned him, but was always subordinated to his primary consideration of 'equilibrium'.

Although an imaginative man, he was a deducer rather than an inventor of hypotheses. Physical investigation must precede formulation; the land-surface is excessively complex and the study of its evolution demands the spirit of a tireless investigator.

CHAPTER TWENTY-NINE

C. E. Dutton and other Western American Geologists

C. E. DUTTON

The outstanding member of Powell's survey group not yet discussed was Clarence Edward Dutton, who was born at Wallingford, Connecticut, in 1841. Dutton graduated from Yale University in 1860, where he won the Literary Prize and showed promise of the literary bent which was to be revealed in his descriptions of the Grand Canyon and in his naming of its topographical features, such as Vishnu's Temple and Shiva's Temple. On leaving Yale, Dutton was caught up in the Civil War and became an officer. After the war, he was posted to the arsenal at West Troy, New York, and, finding time heavy on his hands, developed an interest in geology. In 1871 he was transferred to Washington where he subsequently met Powell, who in 1875 engineered Dutton's appointment as leader of a field party in the Colorado Plateaus. During this period Dutton submitted his early papers on elevations and subsidences of the crust (1871-2), on the crustal contraction theory (1874) and on general theories of the earth's physical evolution (1876). When Clarence King became Director of the Geological Survey in 1879 Dutton was made head of the Division of the Colorado. Subsequent to the preparation of his monumental reports on the Grand Canyon region, he worked in the field of seismology (Diller, 1911), with particular reference to the Charleston earthquake, and on the volcanoes of Hawaii. His famous paper on isostasy appeared in 1889. He resumed active military duty in Texas in 1890 and studied the Mexican volcanoes during leave periods. He retired in 1901 and died in Englewood, New Jersey, on 4 January 1912.

Though Dutton's writings have a very personal style he was, perhaps, as a military man, closer in character to Powell than Gilbert was. Powell recognized this;

> Captain Dutton's assumption of the work was a source of special gratification because he entertained views substantially identical with those of the present Director concerning the physical laws and processes which have combined to produce the wonderful features of the region.

Like other geologists before him, Dutton could see the special advantages that the Colorado region offered to the study of geology. In his *Report on the Geology of the High Plateaus of Utah* (1880) he writes:

> The great value of the Plateau County is the certainty and fullness of the evidence. Nature here is more easily read than elsewhere. She seems at times

FIG. 120. *C. E. Dutton*
(*U.S. Geological Survey*)

amid those solitudes to have lifted from her countenance the veil of mystery
which she habitually wears among the haunts of men. Elsewhere an enormous
complexity renders the process difficult to study; here it is analyzed for us.
The different factors are presented to us in such a way that we may pick out
one in one place, another in another place, and study the effect of a single
variable, while the other factors remain constant. The land is stripped of its
normal clothing; its cliffs and cañons have dissected it and laid open its
tissues and framework, and 'he who runs may read' if his eyes have been
duly opened. As Dr Newberry most forcibly remarks: 'Though valueless to
the agriculturist, dreaded and shunned by the emigrant, the miner, and even
the adventurous trapper, the Colorado Plateau is to the geologist a paradise.
Nowhere on the earth's surface, so far as we know, are the secrets of its
structure so fully revealed as here.' (*Dutton, 1880, pp. 14–15*)

Dutton shared with Powell his artistic reaction to this environment and several of his descriptions are reminiscent of Powell's;

> The ascent leads us among rugged hills, almost mountainous in size, strewn with black bowlders, along precipitous ledges, and by the sides of cañons. Long detours must be made to escape the chasms and to avoid the taluses of fallen blocks; deep ravines must be crossed, projecting crags doubled, and lofty battlements scaled before the summit is reached. When the broad platform is gained the story of 'Jack and the beanstalk', the finding of a strange and beautiful country somewhere up in the region of the clouds, no longer seems incongruous. Yesterday we were toiling over a burning soil, where nothing grows save the ashy-colored sage, the prickly pear, and a few cedars that writhe and contort their stunted limbs under a scorching sun. Today we are among forests of rare beauty and luxuriance; the air is moist and cool, the grasses are green and rank, and hosts of flowers deck the turf like the hues of a Persian carpet. The forest opens in wide parks and winding avenues, which the fancy can easily people with fays and woodland nymphs. On either side the sylvan walls look impenetrable, and for the most part so thickly is the ground strewn with fallen trees, that any attempt to enter is as serious a matter as forcing an abattis. (*Dutton, 1880, p. 285*)

Dutton followed Powell also in his emphasis of the Colorado River's antecedent origin. He is very definite in his statement that the course of the river and of its tributaries was determined by its alignment upon the original form of the surface and had been maintained despite later structural movements.

Of the erosion which had produced these special features he makes two interesting observations. Erosion, he states, unlike in humid regions, is here concentrated on the scarps and the region is progressively reduced by recession of these scarps.

> It is a most significant fact that the brunt of erosion throughout the Plateau Country is directed against the edges of the strata and not against the surfaces. This is directly traceable to the fact that the strata are nearly horizontal, the dips rarely exceeding four or five degrees, and even then only where a great monoclinal flexure occurs. The rains wash and disintegrate most rapidly where the slopes are steepest, and where the strata are flat the steepest slopes are the valley sides and chasm walls. Thus the battering of time is here directed against the scarps and falls but lightly on the terrepleins. (*Dutton, 1880, p. 18*)

He is also impressed by the amount of material eroded, which in aggregate seems great enough to have a structural effect upon the crust beneath. As a large building will depress the soil beneath, so he saw that a great mass of alluvium can depress the earth's crust. In short, this was an early application of the principle of isostasy.

> We may note here another question which presents itself in connection with the differential movements among the various parts of the province. Those areas which have been uplifted most have suffered the greatest amount of denudation. Is it not possible in some cases and under certain restrictions to invert this statement and say that those regions which have been most

denuded have been most uplifted, thereby assuming the removal of the strata as a cause and the uplifting as the effect? May not the removal of such a mighty load as 6000 to 10,000 feet of strata from an area of 10,000 square miles have disturbed the earth's equilibrium of figure, and the earth, behaving as a quasi-plastic body, have reasserted its equilibrium of figure by making good a great part of the loss by drawing upon its whole mass beneath? Few geologists question that great masses of sedimentary deposits displace the earth beneath them and subside. (*Dutton, 1880, pp. 20–21*)

Dutton ends his general description by explaining the part volcanic activity has played in retaining the special characteristics of the region.

The average denudation of the Plateau Province since the closing periods of the local Eocene can be approximately estimated, and cannot fall much below 6000 feet, and may, nay, probably does, slightly exceed that amount. Of course this amount varies enormously, being in some localities practically nothing and in others nearly or quite 12,000 feet. It is a minimum in the High Plateaus. Within that district the average denudation will fall much below 1000 feet in the sedimentary beds. The enormous floods of volcanic emanations have protected them, and these have borne the brunt of erosion, and their degradation has given rise to local accumulations of sub-aerial conglomerates in all the valleys and plains surrounding the volcanic areas, thus increasing the protection.

The general cause which has enabled these strata to survive within the limits of the High Plateaus while they have been so terribly wasted elsewhere may be stated briefly. Until near the close of the Pliocene the High Plateaus were not only the theatre of an extended vulcanism, but those portions which never were sheeted over by lavas were low-lying areas, where alluvial strata tended to accumulate. They remained, in fact, base levels of erosion during the greater part of Tertiary time. (*Dutton, 1880, p. 23*)

Differential structural displacements, he repeats, had had comparatively little topographic influence.

In truth, there is but little 'structure'. The plateau is simply a remnant left by the erosion of the country around its southern and eastern flanks. A few of its minor features are due to displacements, and its western wall originated in a great fault or rather in several faults. The rest of the mass owes its preeminence to circumdenudation. We may gain some notion of the stupendous work which has accomplished this result by taking our position upon the south eastern salient at the verge of the upper platform.

It is a sublime panorama. The heart of the inner Plateau Country is spread out before us in a bird's eye view. It is a maze of cliffs and terraces lined off with stratification, of crumbling buttes, red and white domes, rock platforms gashed with profound cañons, burning plains barren even of sage – all glowing with bright color and flooded with blazing sunlight. Everything visible tells of ruin and decay. It is the extreme of desolation, the blankest solitude, a superlative desert. (*Dutton, 1880, pp. 286–7*)

Dutton's comments in this *Report* on river processes are not nearly so penetrating as those of Gilbert. He explains how the movement forward of the larger debris is not entirely dependent on flood periods but that there are a number of minor accelerations in the river's flow which are responsible for this progression downstream.

> An acceleration of the current removes the finer stuff and retardation replaces it with fresh. The washing out of the matrix of sand and grit which holds a pebble in its place leaves the pebble to the unobstructed energy of the current. If that energy is sufficient it will be carried along until the current slackens or until it finds a lodgment. If the energy is too small, the pebble will remain until the ceaseless wear of attrition reduces it and brings it within the power of the stream to move it. Nor are these movements dependent solely upon periodical floods. Any cause which alternately accelerates the movement of water may produce them, and these causes are many. Every stream and every shore current is affected by numerous rhythmical movements which produce these alternations in many ways and many degrees. The waves and surf, the undertow, the tides, the shifting of shore currents, the storms and monsoons, the ripples of the brook, the numberless surgings and waverings of rivers, the shifting of channels, the building and destruction of sand bars, the freshets – all are causes by virtue of which any spot at the bottom of the water is subject to alternate maxima and minima in the velocity of the water which passes over it. Sooner or later, then, the pebble must move on, provided any maximum of velocity in the water is sufficient to move it when subject to no other resistance than its own weight. (*Dutton, 1880, pp. 216–17*)

However, he is more thorough when describing how alluvial cones are formed.

> Reaching the valleys or plains, their velocity is at once checked by the diminished slope and the coarser debris comes to rest. These streams lie (within the mountains) in ravines usually profound, with steep flaring sides, and opening upon the valley bottoms or plains through magnificent gateways, and every long range or ridge has usually many such gateways opening at intervals of a very few miles along its flank. At the gateway the stream begins to surrender a part of its freight and to build up its channel. The check given to the velocity of the stream here is marked, indeed, but less incisive than might at first be supposed. The profile of the bed of the stream does not have an angle at this point, but is curved very gently, and is concave upward. Indeed, it is so throughout the entire course of the stream outside the gate and generally for a considerable distance inside the gate. Thus the velocity of the stream slows down gradually and not suddenly. As the velocity gradually diminishes so the stream gives up more and more of its load. But the stuff which it drops along any small part of its course is by no means of the same size; that is to say, there is no rigorous sifting of the material in such a manner that the stones or particles at any given place are of uniform size, while finer ones are carried on to be scrupulously selected where the slope and velocity are less. On the contrary, all sorts are deposited everywhere. Nevertheless there is a tendency to sorting. Higher up the slope there is a greater proportion of coarser deposit; lower down there is a larger proportion of finer deposit; but everywhere the coarse and the fine are commingled.

> Where the stream is progressively building up its bed outside of the gate, it is obvious that it cannot long occupy one position; for if it persisted in running for a very long time in one place it would begin to build an embankment. Its position soon becomes unstable, and the slightest cause will divert

it to a new bed which it builds up in turn, and which in turn becomes unstable and is also abandoned. The frequent repetition of these shiftings causes the course of the stream to vibrate radially around the gate as a center, and in the lapse of ages it builds up a half-cone, the apex of which is at the gate. The vibration is not regular, but vacillating, like a needle in a magnetic storm; but in the long run, and after very many shiftings, the stream will have swept over a whole semicircle with approximately equal and uniform results.

The formation thus built up is an 'alluvial Cone'. (*Dutton, 1880, pp. 219–21*)

Dutton's next reports, the *Physical Geology of the Grand Canyon District* (1880–1) and the *Tertiary History of the Grand Canyon Region* (1882), are more detailed than his first *Report*. At the outset he again agrees with other geologists in emphasizing the special advantages which the Colorado has for the geologist. He then proceeds to point out the main features of the region; the canyons; their enlargement by the continual erosion of the canyon walls; and the uniformity of structure. Like Powell, Dutton had an artistic sense which sometimes raised him to peaks of lyrical exaltation. His word-pictures are a constant source of delight.

Those who have long and carefully studied the Grand Cañon of the Colorado do not hesitate for a moment to pronounce it by far the most sublime of all earthly spectacles. If its sublimity consisted only in its dimensions, it could be sufficiently set forth in a single sentence. It is more than 200 miles long, from 5 to 12 miles wide, and from 5000 to 6000 feet deep. There are in the world valleys which are longer and a few which are deeper. There are valleys flanked by summits loftier than the palisades of the Kaibab. Still the Grand Cañon is the sublimest thing on earth. It is not so alone by virtue of its magnitudes, but by virtue of the whole – its ensemble. (*Dutton, 1880–1, p. 144*)

As the sun moves far into the west the scene again changes, slowly and imperceptibly at first, but afterwards more rapidly. In the hot summer afternoons the sky is full of cloud-play and the deep flushes with ready answers. The banks of snowy clouds pour a flood of light sidewise into the shadows and light up the gloom of the amphitheatres and alcoves, weakening the glow of the haze and rendering visible the details of the wall faces. At length as the sun draws near the horizon the great drama of the day begins.

Throughout the afternoon the prospect has been gradually growing clearer. The haze has relaxed its steely glare and has changed to a veil of transparent blue. Slowly the myriads of details have come out and the walls are flecked with lines of minute tracery, forming a diaper of light and shade. Stronger and sharper becomes the relief of each projection. The promontories come forth from the opposite wall. The sinuous lines of stratification which once seemed meaningless, distorted, and even chaotic, now range themselves into a true perspective of graceful curves, threading the scallop edges of the strata. The colossal buttes expand in every dimension. Their long narrow wings, which once were folded together and flattened against each other, open out, disclosing between them vast alcoves illumined with Rembrandt lights tinged with the pale refined blue of the ever-present haze. A thousand forms, hitherto unseen or obscure, start up within the abyss, and stand forth in strength and animation. All things seem to grow in beauty, power, and

FIG. 121. *Holmes' sketches which accompanied the reports of Clarence Dutton on the Grand Canyon:*
a. *The Grand Canyon at the foot of the Toroweap, looking east.*

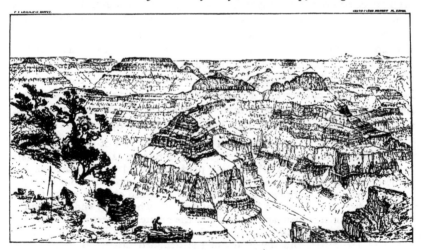

b. *Panorama from Point Sublime, looking east.*

dimensions. What was grand before has become majestic, the majestic becomes sublime, and, ever expanding and developing, the sublime passes beyond the reach of our faculties and becomes transcendent. The colors have come back. Inherently rich and strong, though not superlative under ordinary lights, they now begin to display an adventitious brilliancy. The western sky is all aflame. The scattered banks of cloud and wavy cirrhus have caught the waning splendor, and shine with orange and crimson. Broad slant beams of yellow light, shot through the glory-rifts, fall on turret and tower, on pinnacled crest, and winding ledge, suffusing them with a radiance less fulsome, but akin to that which flames in the western clouds. The summit band is brilliant yellow; the next below is pale rose. But the grand expanse within is a deep, luminous, resplendent red. The climax has now come. The blaze of sunlight poured over an illimitable surface of glowing red is flung back into the gulf, and, commingling with the blue haze, turns it into a sea of purple of most imperial hue – so rich, so strong, so pure that it makes the heart ache and the throat tighten. However vast the magnitudes, however majestic the forms, or sumptuous the decoration, it is in these kingly colors that the highest glory of the Grand Cañon is revealed. (*Dutton, 1880–1, pp. 154–5*)

The structure laid bare taught the observer many lessons. The first was Powell's phenomenon of antecedence, which Dutton termed the 'persistence of rivers'. He again explains how a river may be able to maintain its course across a rising landmass and claims that wherever examples are found of rivers flowing through mountain chains the principles of antecedence must be assumed to be the cause.

From the peculiarities of the Grand Canyon itself Dutton goes on to give a more general description of the geological history of the entire region. The deep penetrations of the Canyon help geologists in two ways. They bare the strata down to the earliest levels and by exhibiting the succession of rocks give a clearer idea of what had been eroded. By looking at the thickness and character of the beds the geologist was able to tell how long the area had experienced a particular set of conditions and from the junction of a lower bed with an upper bed of a different formation he could tell whether the succession had been merely uninterrupted or had suffered an intervening period of erosion. Where certain geological formations were entirely or partly absent he was able to judge what quantity of rock had been removed by erosion.

> The geologist, seeing that around a considerable part of the periphery of the Grand Cañon district the Eocene and Mesozoic strata suddenly terminate in great cliffs facing the Carboniferous platform, would at once conclude that these strata formerly reached beyond their present boundaries. But how far? The answer may be proposed at once. They extended over the entire Grand Cañon district and reached into central Arizona, where they ended along the shore of an ancient mainland, from which their materials were in part derived. The distance of that shore-line from the summit of the Pink Cliffs is from 130 to 180 miles, and the width of the denuded district is from 120 to 140 miles. From the base of the Vermilion Cliffs the distance is from 25 to 30 miles less. The area of maximum denudation is from 13,000 to 15,000 square miles, and the average thickness of the strata removed from it was about 10,000 feet. (*Dutton, 1882, p. 61*)

Dutton explains the removal of 10,000 feet of strata by supposing a rising of the landmass from beneath the sea.

> If it be true that the Grand Cañon district received between the close of the Carboniferous and the close of the local Eocene 10,000 feet of deposits

FIG. 122. *An air view of Marble Canyon of the Colorado River. The Echo Cliffs are in the right foreground, the Vermilion Cliffs in the middle distance, and the Paria Plateau in the background (U.S. Department of the Interior, National Park Service)*

averaged over its entire surface, it follows that at the latter epoch the summit of the Carboniferous lay at least 10,000 feet below sea-level and was much more nearly horizontal than it is at present. And if such was its position and configuration, the great faults and displacements which traverse it must be of Tertiary age, and there must have been an enormous amount of uplifting, ranging from 12,000 to 18,000 feet, in various portions of the district. These are some of the consequences of the great denudation.

The emergence of this new landmass exposed it to erosion by the elements and here Dutton enters into a very full statement of Powell's theory of base level. However, in Dutton's treatment he does not regard base level merely as a standard towards which erosion is aiming but as a positive phenomenon which will eventually be achieved. Both the terms of the statement itself and its context are much nearer the later theory of peneplanation than anything Powell wrote.

Whenever a smooth country lies at an altitude but little above the level of the sea, erosion proceeds at a rate so slow as to be merely nominal. The rivers cannot corrade their channels. Their declivities are very small, the velocities of their waters very feeble, and their transporting power is so much reduced that they can do no more than urge along the detritus brought into their troughs from highlands around their margins. Their transporting power is just equal to the load they have to carry, and there is no surplus energy left to wear away their bottoms. All that erosion can now do is to slowly carry off the soil formed on the slopes of mounds, banks, and hillocks, which faintly diversify the broad surrounding expanse. The erosion is at its base level or very nearly so. An extreme case is the State of Florida. All regions are tending to base levels of erosion, and if the time be long enough each region will, in its turn, approach nearer and nearer, and at last sensibly reach it. The approach, however, consists in an infinite series of approximations like the approach of a hyperbola to tangency with its asymptote. Thus far, however, there is the implied assumption that the region undergoes no change of altitude with reference to sea-level; that it is neither elevated nor depressed by subterranean forces. Many regions do remain without such vertical movements through a long succession of geological periods. But the greater portion of the existing land of the globe, so far as is known, has been subject to repeated throes of elevation or depression. Such a change, if of notable amount, at length destroys the pre-existing relation of a region to its base level of erosion. If it is depressed it becomes immediately an area of deposition. If it is elevated new energy is imparted to the agents and machinery of erosion. The declivities of the streams are increased, giving an excess of transporting power which sweeps the channels clear of debris; corrasion begins; new topographical features are literally carved out of the land in high relief; long rapid slopes or cliffs are generated and vigorously attacked by the destroying agents; and the degradation of the country proceeds with energy.

It is not necessary that a base level of erosion should lie at extremely low altitudes. Thus a large interior basin drained by a trunk river, across the lower portions of which a barrier is slowly rising, is a case in point. For a time the river is tasked to cut down its barrier as rapidly as it rises. This occasions slackwater in the courses above the barrier and stops corrasion, producing temporarily a local base level. (*Dutton, 1882, pp. 76–77*)

Dutton applies this principle to the region he is studying and shows how, together with changes of climate, there has been a succession of emergences and reductions to base level.

> The general tenor of the facts is to the effect that the Miocene was a humid period and the Pliocene a dry one throughout the greater part of the West. This is one of the reasons which lead us to the very probable conclusion that the age of the Grand Cañon is not older than the beginning of Pliocene time. We might also draw a similar inference from a consideration of the enormous erosion which took place here before the excavation of the chasm was begun. The denudation of the Mesozoic system was an incomparably greater work, and yet that denudation could not have begun until the last strata (the Lower Eocene) were deposited. If these inferences are well founded, we may assign the greater part of Eocene and the whole of Miocene time for the principal denudation of the Mesozoic, and the Pliocene and Quaternary for the excavation of the entire cañon. The proportion thus suggested between the portions of the work done and the divisions of time required to accomplish them seems very fair and reasonable. (*Dutton, 1880–1, p. 120*)

Further verification of these great periods of alternation he sought in the presence of unconformities which he saw as fossil erosion surfaces or old base levels.

> The meaning of this great unconformity obviously is that after a vast body of early Paleozoic strata had been laid down they were distorted by differential vertical movements, were flexed and faulted, and were elevated above the sea. They were then enormously eroded. Across the belt of country bounded on the east by the longitude of Cape Final and extending as far west as the lower end of the Grand Cañon, a recti-linear distance of nearly 110 miles, and, for aught we know, indefinitely further westward, nearly the whole mass of these strata was denuded. A few, and perhaps many, small remnants at the base of the series were preserved, but over most of the area the Archaean schists were laid bare and suffered erosion. (*Dutton, 1882, pp. 180–1*)

Dutton also realized that erosion does not simply depend on river action but is a combination of the corrasion of the river beds and weathering of the side slopes. Thus the steeper the cliffs or slopes, the more effective will be the weathering and the transportation of the weathered material.

> The geologist . . . has had impressed upon his mind the general fact that the most rapid waste takes place on the edges of the strata exposed in vertical wall-faces. Every year the rains wash away something from the mural fronts. In a single year it may be a mere film, but in the lapse of thousands of centuries the amount whittled off becomes a vast aggregate. Like the motion of the fixed stars the change is not perceptible to a generation; but a million years would change the aspect of a denuding country as profoundly as they would change the aspect of the heavens. How long in terms of years this 'Recession of the Cliffs' has been going on, the geologist does not know, though he presumes the period to have been certainly hundreds of thousands of years and very probably some millions. (*Dutton, 1880–1, pp. 95–96*)
>
> We must not conceive of erosion as merely sapping the face of a straight serried wall a hundred miles long; the locus of the wall receding parallel to its

former position at the rate of a foot or a few feet in a thousand years; the terrace back of its crest line remaining solid and uncut; the beds thus dissolving edgewise until after the lapse of millions of centuries their terminal cliffs stand a hundred miles or more back of their initial positions. The true story is told by the Triassic terrace ending in the Vermilion Cliffs. This terrace is literally sawed to pieces with cañons. There are dozens of these chasms opening at intervals of two or three miles along the front of the escarpment and setting far back into its mass. Every one of them ramifies again and again until they become an intricate net-work, like the fibres of a leaf. Every cañon wall, throughout its trunk, branches, and twigs and every alcove and niche, becomes a dissolving face. Thus the lines and area of attack are enormously multiplied. The front wall of the terrace is cut into promontories and bays. The interlacing of branch cañons back of the wall cuts off the promontories into detached buttes, and the buttes, attacked on all sides, molder away. The rate of recession, therefore, is correspondingly accelerated in its total effect . . .

For if the rate of recession of a wall fifty feet high is one foot in a given number of years, what will be (*ceteris paribus*) the rate of recession in a wall a thousand feet high? Very plainly the rate will be the same. If we suppose two walls of equal length, composed of the same kind of rocks, and situated under the same climate, but one of them twice as high as the other, it is obvious that the areas of wall-face will be proportional to their altitudes. In order that the rates of recession may be equal, the amount of material removed from the higher one must be double that removed from the other, and since the forces operating on the higher one have twice the area of attack, they ought to remove from it a double quantity, thus making the rates of recession equal. (*Dutton, 1882, pp. 62–63*)

In order that disintegration may go on rapidly the debris must be carried away as rapidly as it forms. But the efficiency of transportation depends upon the declivity. The greater the slope the greater the power of water to transport. When the slope is greater than 30° to 33° (the angle of repose), loose matter cannot lie upon the rocks, and shoots down until it finds a resting place. Hence the greater the slope the more fully are the rocks exposed to the disintegrating forces, and the more rapidly do they decay. (*Dutton, 1882, p. 64*)

This description of the enlargement of canyon walls, though containing little that Powell or Gilbert had not already said, was nevertheless a very thorough analysis showing an intimate knowledge of the region's special conditions. The recession of the cliffs led Dutton naturally to a description of how the weathered material, while it awaited transport by water, would form into talus cones below the harder cap-rocks which proved more resistant to erosion.

It appears, then, that the recession of the hard beds is accelerated by undermining, while the recession of the soft beds is retarded by the protection of the talus. The result is the final establishment of a definite profile, which thereafter remains very nearly constant as the cliff continues to recede. Thus the talus is the regulator of the cliff profile. (*Dutton, 1880–1, pp. 163–4*)

As already noticed, Dutton paid great attention to the formation of these cones, far more than Powell or Gilbert did, and he was able to demonstrate

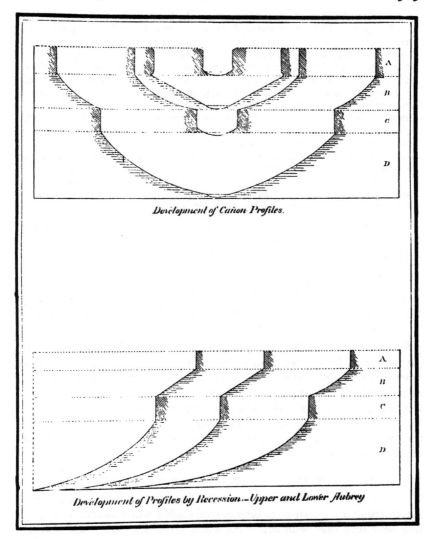

Development of Cañon Profiles.

Development of Profiles by Recession.—Upper and Lower Aubrey

FIG. 123. *Dutton's (1882) illustrations showing the recession of canyon and cliff slopes*

how their slopes were concave upwards like those of a river profile and he gave the reasons for this shape.

. . . the talus is the regulator of the cliff-profile; for it checks the rate of recession in the softer beds, keeping their recession *down* to the mean rate, while, by undermining, the recession of the hard beds is brought *up* to the mean rate. As soon as the talus is established on the lowest slope (D), the cliff may be said to have attained its normal profile, and in all subsequent recession that profile undergoes little change. The only modification it receives is a decrease in the slope D, which becomes longer and also takes the form of a curve, concave upwards. The cause of this curvature is as follows. If the rate of recession in the soft beds, as due to the protection of talus, were proportional in a simple ratio to the height above the bottom, the slope would be straight, but would gradually decrease its inclination as the cliff recedes. But, in fact, the law governing the rate of recession is more complex. The protection given to the lower beds increases downwards in a higher ratio than a simple one, being as the square of the distance below the base of C, or, perhaps, in a still higher ratio. This arises from the fact that not only is the quantity of debris and soil greater in the lowest beds, but it is finer and more compact. Hence the rate of recession becomes inversely proportional to the square (?) of the distance below the base of C, and the curve becomes a segment of a hyperbola. (*Dutton, 1882, pp. 252–3*)

A later work by Dutton shows his preoccupation with the immensity of fluvial erosion as his attention was turning more and more to reflections on isostasy;

> Imagine the Cretaceous strata which have been denuded from this valley of the Puerco to be restored. We must replace in imagination fully 1200 feet, and possibly much more, of the sandstones and shales in order to restore the sedimentary beds to the condition they were in when these necks were molten lava. (*Dutton, 1884–5, p. 167*)

Of the three great Western geologists Dutton was the least important, partly because he antedates and repeats many of Powell's ideas. We should however notice the beauty of his exposition, the occasional addition of an extra detail and the remarkable understanding of structural movements. In fact, Dutton's ideas on isostatic compensation fostered those common in Europe in the 1920's. Moreover, by elaborating and illustrating the base level concept, he prepared the way for the introduction of the idea of peneplanation into geomorphic literature.

OTHER WESTERN GEOLOGISTS: HAYDEN AND MARVINE

The necessity of not separating the individual contributions of the Powell survey group has forced us to omit several studies which should in point of time have been dealt with before the works of Powell, Gilbert and Dutton. There were, of course, other survey teams and, as already mentioned, before the consolidation of the surveys no less than four groups worked independently of and often in competition with each other. Of these the Hayden Survey was of especial importance.

The geomorphic work of this group is associated mainly with Hayden himself and Marvine. The brilliance of the great Western trio tends to cause the contributions of Hayden to be under-rated. We have seen how towards the end he was inclined to sacrifice accuracy and detail in order to gain priority over the rival groups (pp. 507–12). Yet as early as 1862 Hayden recognized the tremendous amount of erosion of which subaerial forces were capable. He recognized, as did Powell and Dutton, that as soon as a land surface rose above the sea the rain would collect and run off along what were the natural channels of descent. If the land continued to rise and the initial channels were firmly fixed then he also realized that the surface could become dissected by deep canyons.

> The erosion would go on as the mountains continued slowly rising at an almost imperceptible rate, and in the process of time the stupendous channels . . . would be formed. (*Hayden, 1862*)

In the same work he emphasized how with an increase in 'the barometric profile' (by which he meant slope) there was a corresponding increase in the power of erosive forces.

In 1867 Hayden stopped working on his own account and became a government geologist. In the years in which we are particularly interested the Hayden group surveyed much of the Colorado Rockies. In one of his reports of this work he repeats his belief that erosion acts immediately upon a newly-raised land surface and he places great emphasis on the total amount of erosion that can occur.

> (The hogback ridges at the base of the Front Range). . . . rise to a height of 1500 feet above the plains at their base . . . That these ridges, all along the base of the mountains, were originally much higher and may have even extended far up the mountain-sides, but have been removed by erosion, there is ample evidence. (*Hayden, 1874–5, p. 215*)
>
> . . . we cannot resist the conclusion that, prior to the elevation, the sedimentary beds extended in unbroken continuity across the area now occupied by the metamorphic central mass; the missing portions having been removed by erosion during the slow, long-continued process of erosion. (*Hayden, 1874–5, p. 217*)
>
> Indeed, we may presume that the work of degradation commenced as soon as the area now occupied by the mountain ranges rose above the sea; and, therefore, the work of destruction of the original forms has been going on for an almost unlimited period of time, dating back nearly or quite to the Carboniferous period. (*Hayden, 1874–5, p. 217*)

Another feature which attracted Hayden's attention was the presence of terraces in association with the rivers and the fact that the relative elevation of these terraces increased toward the source of the rivers. This made him believe that the region had experienced a decrease in the amount of rainfall.

> They, the deposits, assume importance from the fact that they date back to a period when there was much more water in the streams than at the present time, and in consequence, the aqueous forces were much more marked than they are now. (*Hayden, 1876, p. 200*)

In his *Report* of the following year he draws attention to river antecedence and quotes the Gallatin River as a striking example of a feature which he realized was common in the West.

The drainage was undoubtedly marked out at an early period. . . . The drainage (of the Gallatin river) seems to have originated in a sort of depression or sag, in the sedimentary crust; for the gorge has there worn through these rocks for nearly the entire distance, and the inclination of the strata on either side is towards the cañon. Nothing seems even to have deflected the river from its course, but it has worn its narrow way directly through highest mountains. . . . The erosion of the channel, or cañon, must have commenced with the elevation of the crust, and continued on, keeping pace with the elevating forces. Obstructions from time to time have occurred, which produced in part the numerous lake-basins which we find at the present time in the valley of all the mountain streams. (*Hayden, 1876, p. 200*)

The geomorphic work of Archibald Robertson Marvine was rather fuller of ideas and of a slightly higher quality than Hayden's. Born at Auburn, New York, in 1848, he eventually became an instructor at the Harvard Mining School (Powell, 1876B). In 1869 he first visited the Colorado Rockies on a university field trip and in 1870 went as geologist to the Santo Domingo Expedition. In 1871 his academic connexions ended and he served as astronomer to the Wheeler Expedition and wrote reports on the geology of southern Nevada, southern California and north-west Arizona. It was presumably during this period that he became friendly with Gilbert to whom he later became related by marriage. This friendship is recorded by Gilbert;

Afternoon Marvine and I threw ball near the market (in Washington) until stopped by the police. (*Note in diary for 28 March 1872; quoted by Davis, 1922, p. 67*)

Marvine became a member of the Hayden Survey in 1873 and worked in the Middle Park area of the Colorado Rockies. He died on the 2 March 1875.

In his report on the stratigraphy of the east slope of the Front Range he describes the region as having the following characteristics;

. . . as a whole the mountain-zone lying between the main divide and the plains certainly impresses one as being, with a few exceptions, a region of very uniform or gently undulating general elevation, carved by the powers of erosion, perhaps partly glacial but mostly by streams, into a mountain area of which portions are exceedingly rugged. (*Marvine, 1874, p. 89*)

Marvine explains the historical sequence of the land being raised, then eroded down to a low level by subaerial elements, but, following the European writers, he supposed that the final irregularities were planed off by the sea. The land being raised once more, streams will develop on the overlapping marine formations until, these being eroded away, the drainage will be superimposed on the truncated earlier structures;

. . . the erosion producing the present surface features of the mountain region had the direction of its action determined by movements of the surface which were not closely connected with the extended plications of its rocks; and, moreover, since this erosion has not long been active among these rocks, there appears no well-defined connection between the topography and the

FIG. 124. *Aerial view of the Colorado Front Range taken from an elevation of 9300 feet looking south-west from Castle Rock, Colorado. In the distance Pike's Peak rises above the accordant lower summit levels*
(*Photograph by T. S. Lovering, U.S. Geological Survey*)

structural geology. The ancient erosion gradually wore down the mass to the surface of the sea, and while previously to this it was no doubt directed by the structure, yet the mass was finally leveled off irrespective of structure or relative hardnesses of its beds by the encroaching ocean, which worked over its ruins and laid them down upon the smooth surface in the form of the Triassic and other beds. The recent great uplift, which probably added new plications to the accumulated plications of the past in the ancient rocks, was quite simple with respect to their total plication, and left the upper Triassic and other sedimentary beds comparatively simply structured, they having been affected alone by the later movements.

As the mass appeared above the sea and surface erosion once more commenced, but which now acts upon the recent rocks covering probably in greater part the complex underlying rocks, it was directed off from the line of greater uplift down the long slopes of the rising continent to the retiring sea. The channels of drainage started were directed solely by the structure and characters of the upper rocks, and when they gradually cut down through these and commenced sinking their cañons into the underlying complicated rocks, these cañons have no relation whatever to their complications. It is but recently that the upper rocks have been completely removed from the summits of the mountain spurs, the ancient level of subaqueous erosion being still indicated by the often uniform level of the spurs and hill-tops over considerable areas, and large plateau-like regions which become very marked from certain points of view. (*Marvine, 1874, pp. 144–5*)

The main aspects of importance in his description are first, the account of the formation of a surface of low relief and the emphasis on how prolonged erosion will often result in landforms unrelated to the structure on which they occur; and second, the explanation of the presence of superimposed drainage.

Two great facts appear: first, that the range is composed of metamorphic schists and granites having a detailed structure independent of the grand topographic forms now existing, but related to a topography antecedent to the present and which was buried by encroaching waters prior to the up-heaval of what we now know as the great Colorado Range: and, second, that the great orographic movements producing the present grand features of the country brought up once more that ancient and buried land; and the present drainage system, determined by these later upheavals, while conforming to the later structure, was superimposed on the earlier; . . . (*Marvine, 1874*)

A third aspect, which is equally important, was his perception of accordant summits, the last remnants of earlier denudational surfaces.

These rugged ridges, in their easternmost portions, present a pretty uniform general elevation, and as the northern ridge expands at its end into an even-surfaced table-like mass of rock, the impression is given that all of these sharp ridges are but remnants left from the cutting away of a plateau-like step which once followed along the mountain-face. (*Marvine, 1874, p. 188*)

The subaqueous erosion, in smoothing all to a common level, destroys all former surface expression of geological character. (*Marvine, 1874, p. 145*)

Davis (1909) comments on this theory of subaerial base-levelling to account for the reduction of great areas of raised land, but correctly draws attention to Marvine's use of the sea in the later stages of the levelling process.

CHAPTER THIRTY

The Effect of the Western Explorations on Geomorphic Thought

INFLUENCE OF THE WESTERN SURVEYS ON GEOMORPHIC THOUGHT IN THE UNITED STATES

Not since the time of Lyell had a new body of thought had such an immediate effect on geomorphic thinking in general as was produced by the work of the Western Surveys. This was, of course, particularly true in the United States, and its influence affected geologists in the humid East as well as Western workers. Most landscape studies began to admit readily the power of sub-aerial erosion.

Lewis (1877), examining the rivers of Long Island, felt himself obliged to allow some deepening of the valleys by post-glacial stream action;

> The surface valleys were evidently water-courses for sub-glacial streams supplied by the ever wasting ice while it still lay upon the coast. (*Lewis, 1877, p. 143*)
> There is some reason for the conclusion, however, that some deepening of the principal surface valleys has taken place subsequently to the disappearance of the ice, which may be attributed to streams. (*Lewis, 1877, p. 144*)

One of the first textbooks to incorporate the Western Explorations was Joseph Le Conte's *Elements of Geology*, New York, 1879. Le Conte, of the new University of California, obviously placed a good deal of faith in the efficacy of subaerial erosion, referring to the importance of the work of Newberry, Powell, Dutton and Gilbert and giving several of his own examples of erosion now in progress. He repeats what is known of the physical properties of erosive agents and explains how running water is able to overcome the cohesion of particles as the square of its velocity. He also follows Hopkins in stating that its transporting power is such that it is able to overcome the weight of particles as the sixth power of its velocity. All this led up to Le Conte's obvious acceptance of the idea of base level.

> The general effect of erosion is leveling. If unopposed, the final effect would be to cut down all lands to the level of the sea, at an average rate of about one foot in 5000 years. But the immediate local effect is to increase the inequalities of land-surface, deepening the furrows, gullies, and gorges, and increasing the intervening ridges and peaks. The effect, therefore, is like that of a graver's tool, constantly cutting at every elevation, but making trenches at every stroke. (*Le Conte, 1879, p. 11*)

He believed that a land surface could be largely levelled by the uninterrupted action of subaerial forces. He followed both Ramsay and Powell in the supposition that erosion could remove thousands of feet of rock from areas within its influence, but was careful to make the point that such quantities of erosion were mainly restricted to land surfaces and did not occur in submarine areas. He only gave limited importance to the ability of waves to produce shore platforms by continued lateral planation and omitted altogether the possibility that they were able to erode wide platforms. Waves he decided had far less erosive power than rivers and far less opportunity to use that power on land surfaces. For the same reasons he declared that ocean currents had very small erosive force.

Le Conte also drew attention to the variety of topographic forms produced by fluvial erosion, and discussed how these differences in form could be related to differences in mountain structures. At one time, he states, structure and fluvial erosion as controls of topography were regarded by geologists as two opposing ideas but now were understood to be complementary.

The importance of Le Conte's publication lies not in its content, which though it provides a good analysis of contemporary knowledge has nothing new to offer in ideas, but in the fact that, except for a brief mention in Dana's *Manual*, this is the first reference made in an American textbook to the work of the Western Explorers; and certainly the first popular appreciation of the work of Powell, Gilbert and Dutton. It fixes with some accuracy the date for the beginning of their general influence on academic thought.

THE EFFECT OF THE WESTERN EXPLORATIONS ON GEOMORPHIC THOUGHT IN BRITAIN

The impact of the geomorphological work of the Western Surveys was not as immediate in the British Isles. Indeed there was a general lack of interest and effort directed towards physiographic ends in Britain during the 1870's and 1880's. This was partly a result of the direction overseas of geographical interest fostered by the Royal Geographical Society and partly due to reorganization of the Geological Survey by Archibald Geikie. Particularly following the public interest which was aroused in March 1872 by the report of the discovery of David Livingstone at Ujiji on Lake Tanganyika by the *New York Herald* journalist H. M. Stanley, the Royal Geographical Society was able to obtain financial backing for an intensified series of foreign geographical explorations, all of which tended to stress the reconnaissance rather than the systematic branches of the subject (Mill, 1930).

The geologists, too, had conflicting interests. On the first of February 1882 Archibald Geikie, previously the head of the Scottish section of the Survey, took over the Directorship of the Geological Survey of Great Britain from the ailing Ramsay. Geikie immediately stepped up the primary survey of the British Isles, which had lagged under Ramsay's administration, and during the next ten years the production of one-inch geological sheets reached 68, comparing with 34 produced under a decade of Ramsay's directorship, and with 56 during 1862–71 under Murchison (Bailey, 1952). However, this publication went ahead largely without explanatory memoirs and thus Geikie, perhaps unwittingly, threw emphasis on the structural aspects of British geology. He also directed attention away from the classic fluvial regions of southern Britain by halving the number of geological surveyors in England (1881–5) and by sending the outstanding geologists to work on structural problems in Scotland. This soon produced significant results, as witnessed by the description of the Glen Coul thrust (Calloway, 1883), the recognition of the general structural characteristics of the North-West Highlands (Lapworth, 1883) and the appreciation of the ten-mile lateral movement of the Moine thrust (Peach and Horne, 1884 and 1888; Bailey, 1935 and 1952). Thus Geikie, who retired as Director in 1901 but continued geological work until his death in 1924 at the age of 89, is partly responsible for the interest in structural problems, tectonics and glacial phenomena which revived in Britain just at the time when the geomorphological results of the American Western Surveys were becoming known in Europe. The British reaction to the Western Surveys must be seen against this native geological background. It can be safely judged by contemporary British publications.

Green (1876) favoured subaerial forces at the expense of marine and expresses surprise at the continued ignoring of fluvial ideas by some geologists.

> It seems almost past belief that the importance of subaerial denudation should have been so long over-looked, and that truths so simple and apparently so self-evident as those, of which we have just given an abstract, should not have forced themselves on the notice of geologists from the very first birth of the science. (*Green, 1876, pp. 115–16*)

Yet there is still no mention of any American discoveries.

Woodward (1876) similarly emphasizes how subaerial forces carve out the landscape. He also follows Topley and Jukes when he utilizes Ramsay's plains of marine denudation, supposing this to be the form of the surface when the land is first exposed to erosion. Again there is no reference to American work.

Cooley's *Physical Geography* of 1876 shows much the same views. The author has little to say on marine erosion and places more emphasis on its depositional function.

> When the sea ravages a shore, it never carries the detritus to a distance, but spreads it out in front of its former position, and forms a bank or shoal, which ultimately breaks its force and sets a limit to its invasion. (*Cooley, 1876, p. 423*)

He definitely accepted that rivers could erode their beds, though his following statement about 'wind and weather' seems a little uninformed considering what was already known about the erosive action of these two elements:

> We cannot believe that wind and weather wear away the hills and plains covered with vegetation; . . . Surely the snow-clad summits of Gaurisankar and of Chimborazo are preserved intact from denudation, and the grassy

slopes of the Abassinian mountains, 14,000 feet high, are not liable to be washed or blown away. (*Cooley, 1876, p. 423*)

His work on the factors affecting the movement of a river is more advanced and he relates the velocity of a river to its slope and the depth of its channel.

> The velocity of a river depends on its declivity or rate of fall in its channel, the depth and volume of water being supposed constant. The declivity and volume remaining the same, the velocity will increase with the depth and freedom of the channel. The greatest velocity of a stream is at its surface, immediately over the line of greatest depth. The mean velocity is about one tenth of the depth below the surface: . . . At the bottom the river's velocity is least, but most important to be determined, as upon it depends the transport of materials and consequent changes in its channel. (*Cooley, 1876, p. 255*)

He expands this point by giving a table of particle sizes moved by different velocities. One other point he considered was the liability of rivers in the northern hemisphere to be deflected to the right.

Medlicott and Blanford, both strong fluvialists, combined to write a *Manual* on the Geology of India, in which they make absolutely clear their strong belief in the fluvial theory:

> Rivers are one of the most palpable, most wide-spread and ceaseless agents of geological changes; . . . Yet we have hardly outlived the time when much of this river action was ignored. (*Medlicott and Blanford, 1879, II, p. 675*)

The two authors rejected the idea that the valleys were associated with faults and used the phenomenon of antecedence to account for the deepness of the gorges and the origin of the main rivers in a region of relief lower than that crossed in their middle courses.

> . . . all the great Himalayan rivers have their sources well to the north of the line of greatest elevation in the main snowy range; . . . the three longest of them . . . run for great distances in longitudinal courses within the mountains . . . We have no direct observations upon the gorges of these rivers in the high mountains, as to whether they can be regarded as lines of fissure; but all the collateral evidence is decidedly against this assumption . . . it is quite certain that prodigious erosion has taken place in those (transverse gorge) positions through the agency of the rivers. It can then be absolutely asserted (if we ignore the supposition of original fissures) that throughout that process of erosion, and at its origin, the whole ground to the north was higher than these transverse drainage lines . . . The facts further point to the probability that the present longitudinal character of the Himalayan drainage may have been more or less brought about from an original transverse type . . . (*Medlicott and Blanford, 1879, II, p. 677*)

Yet it is Jukes' ideas which they follow, and not the American discoveries with which the Himalayan rivers were more comparable.

The results of the Western Explorations, when they became known in England, did not settle the marine-fluvial conflict. This had already happened largely through the efforts of Jukes, Geikie and Topley. What the results did do was to hasten and confirm the conversion to fluvial ideas which had already begun.

Jukes in his *School Manual* (1863) mentions the report of Newberry on the Ives Expedition and uses one of Newberry's illustrations of the Grand Canyon as frontispiece.

The first British review of Powell's Colorado Expeditions appeared anonymously in 1876. Much of the emphasis is concentrated on the scenic details and the reviewer gives a vivid description of the dangers and difficulties which the expeditions had to face. He comments on the antecedent origin of the Green River but fails to make any mention at all of the concept of base level.

Cadell's review (1887) showed a much clearer understanding of what the new concepts involved and gave a complete synopsis of Powell's *Report* (1875) and Dutton's *Tertiary History* (1882), beginning with a long analysis of the major structural and topographical divisions of the Colorado Plateaus. Quoting Newberry, he stressed how the region was probably the finest example of the erosive power of running water; how the slowness of the great upheavals of land had permitted the rivers to maintain their courses in opposition to structure; and illustrated the special topographic forms that were to be found in a dry climate. Unlike the other reviewer, Cadell both noticed and accepted the concept of base level:

> Rivers are born with the land, and their first courses are shaped by the prevailing slopes. While the land remains above sea-level its drainage areas are being constantly worn away, and this process will only be arrested when the rivers reach their 'base level of erosion' – the slope at which their transportation power ceases, and they begin to deposit their load. (*Cadell, 1887*)

T. H. Huxley's *Physiography* of 1878 refers to Powell's report, but it is old-fashioned in outlook, belonging to the compromise school that grew out of the marine *versus* fluvial controversy. Huxley is impressed by the ability of rivers to erode and transport great quantities of material and he makes a great point of citing Powell's report as evidence of what could be done. The size of a river he emphasized puts no limitation on its power to erode provided that erosion can continue without interruption.

> At first sight, it may seem incredible that a great river-system, like that of the Thames, should have been shaped by the action of instruments which seem so insignificant (e.g. running water) . . . If it be admitted that the little stream has worn out the gutter in which it runs, it is hard to deny that the larger stream has not done similar work on a larger scale. The whole affair is indeed merely a question of time. The smallest cause can produce a vast effect if it is only allowed to work long enough. (*Huxley, 1878, pp. 138–9*)

A more retrograde outlook appears when he put forward the view that waves could cut out an erosion platform of 'perhaps a hundred fathoms' in depth. By this means he believed that uninterrupted wave erosion could reduce the surface of Britain to a plain of marine denudation. Like Topley, Ramsay and Geikie he regarded this plain of denudation as the surface upon which the subaerial elements began to act after the land had been elevated above the sea. But he also felt that the topographic forms eroded by the sea and those eroded by the subaerial elements could be distinguished, and he refers to the 'different geometrical results' produced. Thus it is obvious that only a few English geologists had reached the stage of believing that almost all topographic forms were the result of subaerial erosion. Huxley belonged to the main body who retained the belief that at least a small part of the erosion was due to marine action.

The two Geikie brothers were staunch upholders of the new ideas. Archibald Geikie, the elder, in his influential *Text-Book of Geology* (1882) mentions the work of Newberry and Powell, and has one of the American illustrations of the Grand Canyon as the frontispiece. He deals in detail with the weathering processes that make up erosion and gives particular emphasis to the action of rain and rivers. With the overwhelming support provided by the results of the Western Explorations he is not slow to decry the effects of marine denudation.

James Geikie (1886) also refers to the Western discoveries and notes how the region provides clear evidence of a vast amount of erosion in the post-Tertiary period. Yet though he knew and approved of the new methods of American geomorphologists, in his triple classification of mountain forms (accumulation; elevation; circumdenudation) he persisted in adopting an over-simplified genetic basis which was no advance on Powell's, and a good deal more elementary.

Another important textbook, published by Jukes-Browne in 1884, shows a complete absorption of and an almost complete dependence upon American ideas. He argues strongly against marine erosion and in favour of its alternative, subaerial erosion. He refers constantly to the work of Newberry, King, Gilbert (*Wheeler Report*), Powell (1875) and Dutton (1882). His description of the erosive sequence is founded on the base level concept and almost seems to go farther than anything claimed by the three Western workers.

> Let us assume a tract of country with a central water-shed of a given elevation above the sea, and a river flowing down one of its slopes; this river will continue to deepen its channel until the inclination of its bed is so lowered throughout its course that vertical erosion ceases to be possible, and the power of the flood waters is diverted to the process of lateral erosion.
> When a river ceases to have any further capacity of vertical erosion, it may be said to have reached its base level of erosion. (*Jukes-Browne, 1884, p. 110*)
> The valley will gradually be widened at the expense of the intervening ridges, which will come to assume the character of hill ranges; and, finally, instead of an upper plain with deeply excavated valleys, we shall have a lower plain with chains of hills more or less isolated from one another. (*Jukes-Browne, 1884, p. 112*)

Prestwich, who had devoted most of his energies to studying the formation of river terraces, appears to have absorbed some of the Western ideas, subject to a few of his own personal qualifications. In his 1886 textbook he devotes much space to the power of rivers to erode and to transport material and, as might have been expected, gives a detailed discussion on river terraces. When he comes to mention the findings of Hayden and Powell, however, he is surprisingly retrograde and expresses doubt about ascribing all the features of the work of erosion (partly because no remnants of fluvial deposits had been found on the canyon walls!). He feels that the depth and pattern of the canyons must have been assisted by the existence of surface faults and explains the present lack of evidence for faults by assuming that there had been no marked displacement along them.

Though this qualification is unfortunate, Prestwich does not mar his reputation by insisting too strongly upon it:

> Yet, notwithstanding that the evidence is, I think, insufficient to assign the whole work to the one cause, it is impossible to resist the conviction, looking at the present force of the rivers, the rapid degradation of the rocks, and the vast depth of the chasms, that prolonged water-action has been the one main agent, although its operation may have been facilitated, the time shortened, and the magnitude of the work enhanced by the rents and fissures which were so likely to be formed in consequence of the strain to the surface caused by the great dome-shaped elevation of a 'massif' of rocks rising to the height of 12,000 to 14,000 feet. (*Prestwich, 1886, p. 97*)

Thus by 1890 most geomorphic thought in Britain was strongly affected by the Western Surveys. There remained, however, a considerable belief in marine planation and, as always, a small group who fought stoutly to maintain conservative or ultra-traditional ideas. Among the latter was G. H. Darwin, who in 1879 used the following reason to suggest why marine denudation could formerly have been more effective.

> The moon-earth system is, from a dynamical point of view, continually losing energy from the internal tidal friction.
> There are other consequences of interest to geologists which flow from the present hypothesis . . . the ellipticity of figure of the earth must have been continually diminishing, and thus the polar regions must have been ever rising and the equatorial ones falling; but, as the ocean always followed these changes, they might quite well have left no geological traces.
> The tides must have been very much more frequent and larger, and accordingly the rate of oceanic denudation much accelerated.
> The more rapid alternations of day and night would probably lead to more sudden and violent storms, and the increased rotation of the earth would augment the violence of the trade winds, which in their turn would affect oceanic currents.
> Thus there would result an acceleration of geological action. (*Darwin, 1879, p. 532*)

THE EFFECT OF WESTERN EXPLORATIONS ON GEOMORPHIC THOUGHT ON THE CONTINENT

On the mainland of Europe the climate of geological thought militated against the immediate acceptance of the concepts of fluvial action so convincingly propounded by the American geologists. Despite the contributions to landscape knowledge made by Heim, whose studies on the denudation of mountains are said to have influenced the ideas of Walther Penck; by Löwl (1884), to whose book on valleys Davis refers; and by Albrecht Penck (1887), continental geologists had other more pressing interests. In addition to the long-held bias in favour of marine erosion, work on Alpine tectonics tended to shift the geomorphic emphasis towards considerations of structural controls over landforms and to glacial processes. The detailed study of Alpine structures, which reached a wide audience at the First International Congress of Geology at Bologna in 1878, had in fact begun about the middle of the century (Bailey, 1935; Heritsch, 1929). In 1859 Von Richthofen, influenced by earlier ideas on Alpine thrusting, presented a paper on the subject; in 1867 Eduard Suess of Vienna demonstrated the thrusting of crystalline rocks over Tertiary deposits, and in his later massive contributions to geology devoted much space to the problems of Alpine tectonics. In France, too, Bertrand (1887) drew geological thought towards structural problems.

Von Richthofen, in a monumental work on China (1882), came to the conclusion that the planes of unconformity found between different formations were produced by marine erosion and decided that the most favourable condition for such erosion was a subsiding landmass. He noted how sedimentary formations were invariably deposited not on a mountainous surface but upon one which was almost flat, and that, even where the rocks were tilted, the surface had been planed off to a uniform level which more often than not bore no relation to the underlying structure. He reasoned that whatever force had denuded this platform must of necessity have been very powerful as it had subdued the structural differences:

> Among the formative factors which play an important role in the geological history of North China, the periodic appearance of transgressive deposition is marked in a special manner. In the majority of cases, the beds of the overlapping formation are not, as would be expected in the extension of sea over land, laid down on a mountainous floor formed by high ranges and erosion valleys; but they rest, widespread and uniform, on a surface especially prepared as it were for the deposit, being flattened out for the most part, occasionally wavy, and sometimes in terraces. There also project from it, more or less, a number of high, resistant, mountain ridges which have been partially or completely covered by the horizontal beds. The surface of deposition cuts across the formations antedating the transgressing system, regardless of their position. As their beds are highly tilted and folded, these folds are planed down along the surface, while all parts which once projected above the latter have disappeared. At many places, the amount of

FIG. 125. *Baron F. von Richthofen*

denudation is extraordinarily great. A few troughs still preserved are often the only remnants of sedimentary formations which had many thousand feet of thickness and were folded together in a series of arches. Not only have the arches formed by these been cut away, but such parts of still older rocks as projected among them were removed in the same way. (*Von Richthofen, 1882; quoted by Mather and Mason, 1939, pp. 512–14*)

Von Richthofen discusses the various erosive agents and finally selects the sea as the only possible cause. His whole argument is against the effects of river erosion:

Therefore, an enormous cutting away of the then-existing land preceded every period of transgression, forming a surface which has the tendency to

approach, as much as possible, a rolling plain – although this is only accomplished rather imperfectly. If we seek the agents which might bring about
such an extraordinary great amount of erosion and produce denudation surfaces of the described form, those offered by the atmospheric activities of
weathering and by the mechanical force of flowing water are entirely inadequate. Where the first alone are active, they produce, as we have seen,
a decomposition of the soil, the depth of which depends first on climatic
factors, then on the shape of the surface, and third on the nature of the rock.
The unevenness of the contact between the solid and decomposed rock would
be in no way lessened by this, but considerably increased. Then, if agents
which are in a position to take away this cover become active, the form of
this generally very uneven surface will appear. Just as little would flowing
water have the power to denude great mountainous areas into flat plains.
The base form of its erosion is the groove. In the beginning it strives to
deepen the channels and to increase the unevenness in a lateral direction.
Only when this has been accomplished to a certain extent, will it tend to
widen the groove, to smooth the gradient, to destroy the side walls and, in
further sequence, to connect the bottoms of neighbouring grooves by removing the separating ridges. But this stage is only reached locally and to a
negligibly small extent. The change of level between land and sea works
constantly against it.

The cutting down of a widespread mountainous land to a surface approaching a plain would thus never occur through the separate or united activities
of land agents, and would remain an unattainable goal of their activity.
Neither atmospheric force, flowing water, nor ice would ever be able to
achieve that goal on a great scale.

Among all the mechanically destructive agents, there is only one which
might accomplish regional abrasion on a wide scale. We have indicated it
earlier in the description of individual cases. It is the work of waves of the
surf working towards the interior of a continent. (*Von Richthofen, 1882;
quoted by Mather and Mason, 1939, pp. 514–15*)

Having established the cause he concluded that the most favourable conditions necessary to produce this type of erosional surface would be a slowly
– and steadily – subsiding coastline.

Regional abrasion can therefore be accomplished only by the advancing
surf line. Where the abraded material is not constantly conveyed to great
distances by other agents, transgressive beds must be formed as a consequence. Therefore, as a rule, where transgression is very regular over wide
stretches, the surface of deposition is formed by regional abrasion. (*Von
Richthofen, 1882; quoted by Mather and Mason, 1939, pp. 516–17*)

In much the same way, Cornet and Briart in 1877 when noticing the truncation of the Palaeozoic rocks in Belgium, though they considered subaerial
forces competent to wear down a rough surface to a plain, preferred to conclude that the planing action had been produced by encroaching waves.
Similarly, Oscar Peschel's posthumous *Physical Geography* (*Physische Erdkunde*, 1879–80) stressed the destructive action of waves.

This popularity of wave-action and marine planation denotes clearly
enough either a dislike or, what is more probable, an ignorance of the Western

Explorations. Hence it is not surprising that European writers continued to follow their own ideas on fluvial erosion. Thus the Swiss geologists Heim and Bruckner followed Rutimeyer's thesis that gorges were formed by backward cutting of valley heads through an existing chain rather than Powell's theory of antecedence.

The typical European tendencies of emphasis on tectonics and on marine action are seen in Von Richthofen's other major contribution to geomorphology, a generic/genetic classification of mountains:

I.   TECTONIC MOUNTAINS
    A Block mountains.
        1. Scarp or tilted block mountains.
            (i)   Table blocks.
            (ii) Abraded blocks.
            (iii) Blocks of unconformable strata.
        2. Flexure mountains.
        3. Symmetrical block mountains.
    B Fold mountains.
        1. Homomorphic fold mountains.
        2. Heteromorphic fold mountains.

II. TRUNK OR ABRADED MOUNTAINS

III. ERUPTIVE MOUNTAINS

IV. MOUNTAINS OF ACCUMULATION

V.   PLATEAUS
    A Abraded plateaus.
    B Plains of marine erosion.
    C Horizontally stratified tableland.
    D Lava plain.
    E River plain.
    F Plains of aeolian formation.

VI. MOUNTAINS OF EROSION

(Von Richthofen, 1886, pp. 652–85)

A comparison of this classification with those of Powell (pp. 540–2) shows vividly the importance attached on the Continent to tilting, fracturing and folding and the addition, under Plateaus, of *Plains of Marine Erosion*, introduces phenomena quite outside the ken of the three great Western explorers.

If geologists in Western Europe were slow to appreciate the new fluvial emphasis emanating from America, we would expect Eastern European writers to be virtually unaffected. Actually the contact between thinkers at the opposite ends of the Continent seems remarkably small, partly no doubt because of language difficulties. This may largely account for the total neglect in France and Britain of the work of Lomonosov who, it is claimed, developed the principle of uniformitarianism in 1763, or prior to the work of Hutton. Strong claims are also put forward on behalf of Jan Czerski who was deported to Siberia in 1863 and developed an interest in natural science as a result of having chanced to find a copy of Lyell's *Principles*. Czerski is said to have

postulated 'the progressive development of erosional relief 11 years before
the celebrated work of Davis' (Sobolev, 1946; Markov, 1948; Dylik, 1953).
Whatever the truth of these claims, one is on surer ground in drawing attention
to the studies of Dokučaiev on ravines and river valleys in European Russia
(1877–8). Dokučaiev, best known for his important later work on pedology
and particularly for his famous *Russian Chernozem* (1883), apparently possessed a delightful, if somewhat eccentric, personality. When asked by his
Professor: 'Tell me, young man, what are you occupied with primarily?',
he answered: 'Playing cards and drinking', and received the reply: 'Great!
Continue and do not spoil life with dry science.' Fortunately, Dokučaiev,
like many other students, did not follow his Professor's advice completely,
although it is recorded that his 'erratic tendencies' contributed to his death
in 1903 at the age of 57 (Joffe, 1936, p. 24).

CHAPTER THIRTY-ONE

Quantitative, Dynamic and Chronological Advances

During the latter part of the nineteenth century it becomes increasingly difficult to assess, on a purely contemporary basis, the relevance to landscape studies of much of the quantitative work then being undertaken, particularly by hydraulic and civil engineers. In the period 1846 to 1875, already discussed in Part 3, the quantitative investigations, and especially those relating to the transportation mechanics of streams, were of immediate physiographic significance because of the support they gave to the fluvialistic cause; but the geomorphic importance of much of the work now to be described was obscured until recently by the qualitative veil which W. M. Davis threw over geomorphology for half a century. It is because of this obscurity that we feel it imperative to mention some of the remarkable advances in quantitative and dynamic topics made between 1875 and 1890.

WEATHERING

Hitherto the study of weathering had been neglected, compared with that of running water. However, now that subaerial forces were accepted as the main cause of erosion, it was gradually realized that the study of weathering would repay closer attention. The elder Geikie (1880) discussed the rate of weathering caused by the Edinburgh atmosphere on a number of buildings and tombstones. Italian marble, composed of calcite grains of about $\frac{1}{100}$ inch diameter, showed solution on exposed surfaces at the rate of $\frac{1}{8}$ inch every century. Internal disintegration was also important and occurred from solution along crystal faces and cleavage planes. A third type of disintegration occurred where marble slabs were set in frames of other types of stone; after a time the slabs tended to bulge out and eventually to fracture. This swelling, Geikie believed, was mainly due to the freezing of interstitial water. The memorial slab to Joseph Black (Hutton's friend) had been affected in this way. Sandstone which was more resistant, particularly when it had a hard matrix, was found to suffer only roughening after as much as 200 years of exposure. Where however the matrix was not so resistant, rapid disintegration resulted and, in one example where the matrix was ferruginous, Geikie calculated that $\frac{1}{8}$ inch was removed in only 16 years. He was unable to give an accurate estimate for granite because, when examined, the rock had not been exposed long enough to give a proper knowledge of its weathering. He did however quote the estimate of a German chemist that disintegration of the granite surface would occur at the rate of 0·0085 to 0·0076 mm. annually.

Charles Darwin, at one time a convinced marine erosionist, now admits his change of view:

> Until the last twenty or thirty years, most geologists thought that the waves of the sea were the chief agents in the work of denudation; but we may now feel sure that air and rain, aided by streams and rivers, are much more powerful agents, – that is if we consider the whole area of the land. (*Darwin, 1882, p. 234*)

His work on worm casts demonstrated how even a small organism can activate a high rate of erosion if considered over a large area or period of time. Darwin (1882) from worm casts collected over one square yard at the bottom of a chalk valley, estimated that worms ejected here 18·12 tons/acre/year. They thus aided the work of erosion by bringing to the surface erodible material which would be either washed down the slope or blown away. For a grass terrace, he estimated the ejection rate to be 7·56 tons/acre/year and for an area near Leith Hill Tower 16·1 tons/acre/year. He also estimated the rate at which this cast-material would be washed away, and calculated that ordinary surface wash would annually carry away material weighing 11·56 pounds across a line 100 yards long measured along a slope with an angle of 9° 26'. This flow, he decided, would be at a maximum when the material at the surface was charged with water (Darwin, 1882, p. 271).

THE MASS MOVEMENT OF SURFACE DEBRIS

During this period, C. W. Thomson and the younger Geikie put forward a novel view of mass movement. It had long been accepted without question that soil or rock would be broken up by subaerial action and eventually carried away by rainwash. The new theory that these two authors propounded was that in certain circumstances fragmented material would move downslope without the direct intervention of surface runoff.

Thomson (1877) described the rock glaciers found in the Falkland Islands which extend down from the 6° to 8° hillside slopes to the valley bottoms having 2° to 3° slopes. Although these slopes were far too gentle for the rocks to roll down under the force of gravity, tongue-like masses of fragmented debris appeared to be creeping steadily down them. Thomson explained this by assuming that the weathered rock fragments came to rest on the hillside, gradually became embedded in the soil and then the whole mass crept down the slope. There were several possible mechanics for this, the principal being, he thought, the expansion and contraction of the soil, particularly of the included vegetable matter, and the removal of some of the soil by rainwash. He concludes by supposing that by these means the soil was always moving downslope:

It seems to me almost self-evident that wherever there is a slope, be it ever so gentle, the soil-cap must be in motion, be the motion ever so slow. (*Thomson, 1877, p. 360*)

J. Geikie (1877) disagreed with Thomson's general conclusions that the soil cap moves *en masse*. Geikie had often observed that partly weathered strata was found to have become bent downslope. He believed that this was due mainly to frost action which forced the particles apart and caused a movement downslope which was the line of least resistance. He also carefully noted that rainwash also moved the particles downslope and sometimes, when the soil became saturated, it 'flowed' downhill of its own accord. Geikie thought that this mass-flowage occurred only under extraordinary conditions of saturation, and that rainwash was a far more important cause of the movement of individual weathered particles:

. . . the mere surface-action of rain would suffice to carry away the whole soil, particle by particle, long before the power of frost could have moved it bodily more than an inch or two. (*Geikie, 1877, p. 397*)

His explanation of the characteristics of *rock glaciers* was that movement had occurred under damper climatic conditions than those prevailing at present. When the climate became drier the finer particles had already been washed out leaving only the immobile larger blocks.

These geologists of about 1880 who were studying the movements of fragmented material were apparently unaware of earlier work on soil mechanics. (Cooling, 1945; Krynine, 1947; Singer, *et al.*, 1958). As far back as 1729 Belidor had developed the concept of angle of repose of fragmented material, as well as early ideas on earth pressure; Coulomb (1773), the great French physicist and military engineer, had contributed the wedge theory of earth pressure and the concept of a critical height for clay banks; Français had published on the same subjects in 1820; Collin (1846) had discussed the shearing resistance of clays and given the first description of the 'Swedish Break' as applied to slips in clay slopes (i.e. the predictable sliding surface developed); and Rankine (1857), the Professor of Civil Engineering at Glasgow, had applied the theories of bearing capacity and earth pressure to sands.

AMOUNT OF MATERIAL CARRIED BY RIVERS, AND RATES OF SUBAERIAL DENUDATION

Calculations and measurements of the matter carried by rivers and the rates at which drainage basins were lowered continued to be popular in the late nineteenth century. More stress, however, was put upon the less obvious material carried in solution.

Ewing (1885) working in Pennsylvania calculated the rate of erosion of a limestone valley. He found that the stream discharge from the basin, much of which was floored by Silurian limestone, was 18,172 cubic metres per hour.

This discharge carried 2,905,974 grams of dissolved material per hour – or 25,456,560 kilos per year – which represented 282 tons of material removed per square mile annually, or 275 tons of limestone after allowing for a small amount of other substances also carried in solution. Calculating it in another dimension, Ewing found:

> Hence to lower the surface to the extent of 1 m. by this process would require 29,173 years, or about 9000 years to remove one foot from the surface. (*Ewing, 1885, p. 31*)
>
> A simple further deduction shows that accordingly Nittany valley has been 1,000,000 years in the process of formation. (*Ewing, 1885, p. 31*)

Reade (1885) made similar estimates for some of the great American rivers, including the Mississippi, the Plate, St Lawrence and the Amazon. However, the most significant point he made was that Geikie in his famous comparison between land and sea erosion, had omitted the vast amount of eroded material which is removed in solution, and so had actually belittled his striking conclusions.

THE RELATION BETWEEN VELOCITY AND THE TRANSPORT OF DEBRIS BY STREAMS

The relation between velocity and size of particle carried also continued to excite hydraulic engineers and geologists. Partiot (1871) examined this relationship as it affected sand movement on the bed of the Loire. At low velocities the entire bottom load was moved by a gentle and progressive shifting of the dune forms whereas at high velocities the grains were swept from crest to crest and the sandbanks were gradually reduced in height (*see* Gilbert, 1914, p. 32).

Cooley, as already mentioned (p. 594), included in his textbook (1876) a table showing the velocities required to shift particles of various sizes. Similarly, Du Boys in 1879 wrote an article which concentrated on the factors involved in the movement of bed material. He termed the minimum force necessary to move given particles resting on a stream-bed, *la force d'entrâinement*, and discussed in detail the mathematical nature of the factors involved. He even took his discussion to the stage of evolving the well-known *Du Boys' Tractive Force* formula, which is still used today. He was particularly interested in the precise relationships between stream velocity, particle size and slope at the stage when a state of bed equilibrium had been reached. He defined the equilibrium stage as being the condition when the entraining force was equal to the resistance to movement of the particles on the bed of the river. This was exactly what Gilbert had envisaged in his concept of grade. When he examined this concept in more detail Du Boys (1879, p. 165) was able to draw the following conclusions:

1. The entraining force is proportional to the slope of the water surface and the depth of flow.

2. The total mass of entrained (moving) material is greater for small sizes than for large, with a given slope and depth.
3. The thickness of the moving bed layer is proportional to the magnitude of the entraining force.
4. Small particles move faster than large, under given fluvial conditions.
5. If the entraining force decreases, the largest particles are dropped first, as a coarse lower bed layer.
6. The bottom of the bed arranges itself naturally in such a way that the entraining force is constant in passing from one cross-section to another.

It is fairly obvious that Du Boys arrived at his deductions independently, although conclusions 2 to 5 are contained in Gilbert's more qualitative ideas of competence and capacity; while conclusion 1 was given prominence by both Powell and Gilbert and conclusion 6 is an almost exact statement of Gilbert's concept of grade. As Gilbert's great work on erosive processes was only published two years before, it is understandable that Du Boys should not be aware that he was merely elaborating mathematically some of Gilbert's ideas. Nor need we point out that the previous investigations of European hydrologists contained all the ingredients Du Boys required.

Another notable publication in 1879 was Daubrée's monumental *Experimental Geology*, in which he described a series of experiments designed to gauge the effect of the transport of material by running water. From his experiments he discovered that sand of mean diameter $\frac{1}{10}$ mm. will float in a current when the water is only feebly agitated and that the grains lose little of their angular characteristics. Larger grains of $\frac{1}{2}$ mm. mean diameter, if moved along the river bed by a current flowing at 1 metre a second, would lose by attrition about $\frac{1}{10,000}$ of their weight for each kilometre travelled. Continued attrition would give the particles a rounded shape. This latter calculation is particularly important because it gave reality and mathematical expression to the principle of comminution already developed by Sternberg.

A British writer, Shelford (1885A), supplied data about the material in suspension in the River Tiber during the years 1872 to 1877 and postulated that the stream's total ability to scour varied as the cube of the slope of its water surface. Therefore he concluded that the growth of a delta, by lengthening the river-profile and so lessening its average inclination, automatically caused deposition in the lower part of the river's course.

THE NATURE OF RIVER-FLOW

The progress in the scientific expression and evaluation of river-flow is nicely seen in J. Thomson's work (1876–7) on the windings of rivers in alluvial plains. He disagreed with the simple popular theory that a river bend caused a mass deflection of water towards the outer or undercut bank, so allowing deposition against the inner bank. Thomson found by experiment that the flow was slower along the outside, and faster on the inside of the meander. He demonstrated that the meandering process was natural and that in curves the surface water moves towards the outer bank and the bottom water towards the inside bank. This is the first recorded explanation of the *helicoidal* flow of water round a bend. Thomson went on to attempt to explain why erosion occurred along the outer bank where the flow was slower. The water pressure, he suggests, increases from the inner to the outer side because of centrifugal force associated with a greater depth as the water piles up at the bend. The lower velocity of the bottom layers, due to friction with the bed, gives them a lower centrifugal force. Consequently the bottom layers of water tend to move towards the inner bank of the bend where they deposit mud, while the swift surface water rushing and subsiding against the outer bank causes erosion. It is interesting to notice that in the twentieth century Einstein tackled this problem from this starting point.

About the time of Thomson's researches, a great deal of engineering work was proceeding on irrigation canals in India, the ultimate results of which were to have a profound influence on modern hydraulic geometry. From canals constructed from 1821 onwards, Jackson gave details of the number and length of meander curves in 'rivers of fixed regimen' (1875, II, p. 6). However the construction of the modern irrigation system in the Indian sub-continent dates from the 1880's when the Upper Bari Doab Canal was engineered on sizes and channel slopes derived from the Kutter flow formula (Blench, 1957). This work led to Kennedy's idea (1895) that in an adjusted self-regulating channel the flow velocity (V) could be expressed as a function of depth (d):

$$V = Cd^n$$

where $C = 0.84$ and $n = 0.64$.

This, of course, was a purely empirical relationship which tended to ignore as variables types of bed material, slope, and the automatic adjustment of channel width; but, although n has since been determined as nearer 0.5, Kennedy's formulation began a programme of empirical studies which culminated in the 1930's with the important work of Lacey on channel hydraulics and the meandering of alluvial channels.

An equally significant and long-lived contribution was made by Reynolds who in 1883 attempted to define the general nature of fluid flow so as to determine what inter-relationship of factors caused the transition from laminar to turbulent flow. He decided that flow-state depends on four factors, which he related in his famous formula:

> ... the general character of the motion of fluids in contact with solid surfaces depends on the relation between a physical constant of the fluid and the product of the linear dimensions of the space occupied by the fluid and the velocity. (*Reynolds, 1883, p. 935*)

It seemed, however, to be certain if the eddies were owing to one particular cause, that integration would show the birth of eddies to depend on some definite value of:

$$\frac{c\rho V}{\mu}$$

(*Reynolds, 1883, p. 938*)

(where c = radius of the tube; ρ = the density of the fluid; V = mean velocity; μ = the viscosity of the fluid). This value, the famous REYNOLDS' NUMBER, expresses the state of fluid flow as the ratio of the forces of inertia to those of viscosity; the formula is important because it marks the birth of the first dimensionless number.

THE GEOMETRY OF THE RIVER PROFILE

The longitudinal profile of a river had, as we have already discussed, long attracted much attention. In this topic, it was strange how often notable works by early hydraulic engineers remained un-read by later writers. Thus G. K. Gilbert, in developing his concept of grade, was little influenced by the previous ideas of European hydrologists such as Surell and Dausse. Gilbert, as we have already discussed (pp. 554–7), also developed the notion that a given stream has a maximum capacity for transportation. Above all, he linked the stability of the longitudinal profile with an equilibrium between transportation, erosion and deposition on the stream bed, brought about by an equality between stream discharge and the expenditure of energy in friction and transportation. In Germany, Philippson (1886), drawing on the work of the French engineers, of Gilbert and, no doubt, of Rutimeyer, postulated that, under given conditions, a stable longitudinal profile of equilibrium (the *terminant*) would be developed, by an action which worked headward along the stream. This stable curve he believed to have a flat profile in humid regions and a steep one in arid areas (Woodford, 1951).

In England, Oldham also considered the factors controlling river development and was able to make an addition to Gilbert's concept of grade. Like Gilbert he understood that the development of a river depended upon a number of variants and the manner in which they mutually reacted. He also understood that these variants would go on adjusting themselves to each other until a state of equilibrium was reached between them all.

> As it seems to me, the law is not that the river tends to preserve or obtain a constant velocity, but that at any point of its course the velocity of current will tend to become such that the stream can just carry its solid burden. (*Oldham, 1888, p. 734*)

But this must not be taken to mean that the current is directly governed by the nature of the burden cast upon the stream. Gradient and shape of channel are of course the two principal factors which govern the velocity of current, and in a subsidiary degree the nature of the sides of the channel; of these the two former frequently, the last generally, are to a greater or less degree the product of the stream itself, and by deposition or erosion they tend to become such that, when equilibrium has been established, the stream is just able to transport its solid burden. (*Oldham, 1888, p. 734*)

. . . suppose the velocity of current to be greater than that due to the law propounded, the immediate consequence would be erosion of the channel and transport of debris when the stream reached the sea, if not sooner, the velocity would be checked and the transported debris would be deposited. In this way the lower end of the reach would be raised, the upper lowered, and the gradient diminished; this, leaving out of consideration the effect of the shape of the channel, would result in a diminished velocity and equilibrium when the stream could just transport its solid burden. Similarly, if the velocity were less than that demanded, deposition would commence, and the gradient below the deposit would increase until the velocity of the stream reached the required limit.

But it is not only by an alteration of gradients that the stream can adjust its velocity, for, with the same slope and sectional area of channel, the velocity varies with the square root of the hydraulic mean depth, or of the sectional area divided by the wetted perimeter. From this it follows that the gradient required to produce a given velocity will vary with the shape of the channel, being greater where this is broad and shallow, and less where it is narrow and deep. (*Oldham, 1888, p. 734*)

However, whereas Gilbert had failed in his early work to distinguish between these factors and had merely stressed that they were working towards a common end, Oldham emphasized the factor of friction;

. . . the velocity of the stream is ultimately regulated by the work it has to do, and not its work by the velocity it possesses. (*Oldham, 1888, p. 738*)

This statement was a very important clarification of Gilbert's concept.

DENUDATION CHRONOLOGY

Between 1875 and 1890 great progress was also made in the main branches of denudation chronology. In recent chronology, such as Pleistocene or post-Pliocene, the fundamental advances were probably least, largely because there was less left to do. The chronological value of river terraces was no longer in doubt and the terrace-tier continued to form a main determinant in tracing the recent stages in a river's evolution. Miller (1881–3) gave an excellent survey of earlier theories on the formation of river terraces but in his own explanations tended to ignore the possibilities of varying discharge and of a relative rise of the landmass.

On the other hand, in older, or pre-Pleistocene, chronologies the advance was enormous. We have already noticed how Ramsay and Jukes in a sense inaugurated denudation chronology by destroying the marine dissection theory of Lyell who conceived that landscapes were virtually created *ab nova* and *in toto* at their emergence. To the marine planationists the emerged landscape began as a plain or undulating lowland on which subaerial forces gradually etched the present topography. Their chronology was based mainly on accordant summit-levels and on the development of the drainage patterns.

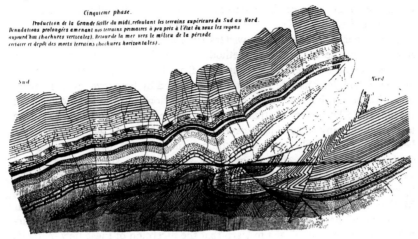

Cinquième phase.

*Production de la Grande faille du midi, refoulant les terrains supérieurs du Sud au Nord.
Dénudations prolongées amenant nos terrains primaires à peu près à l'état où nous les voyons
aujourd'hui (hachures verticales). Retour de la mer vers le milieu de la période
crétacée et dépôt des morts terrains (hachures horizontales).*

Sud

Nord

AA. Faille de Boussu. BB. Cran de retour d'Anzin. CC. Grande faille du midi.

FIG. 126. *The idea advanced by Cornet and Briart (1877) to explain the present
topography and geology (vertical shading) of the Belgian coalfield. A complex
sequence of folding and faulting was followed by almost complete peneplanation*

They dealt largely with skyline profiles and river patterns in relation to
uplands and vales. In Europe these ideas still predominated in the 1870's as
can be illustrated from the work of Cornet and Briart on the topography of
Belgium. They refer to Lyell and De Beaumont but do not mention any
American geologist. In spite of this omission, and their strong belief in
marine denudation, they also accept the potency of stream erosion. In fact,
they go so far as to state that subaerial forces can bevel off the land surface:

> Not only have streams of water flowing with a given velocity a tendency to
> deepen or enlarge their channels, but the action of pluvial water, combined
> in certain localities with frost action, results in the continual degradation of
> the summits and the carrying of the disaggregated particles into the lower
> parts of the valleys and into seas and lakes. This destruction, which never
> stops, certainly produces, as an ultimate result, the complete removal of the
> surface of the earth, providing that up-doming does not occur to oppose
> this action which seems, up to the present time, to have been predominant.
> (*Cornet and Briart, 1877, p. 72*)

The meaning is not in doubt when Cornet and Briart refer to past evidence
of fluvial erosion which has been able to bevel off mountain ranges to a 'plain
nearly perfectly flat'. Time, they realize, is the important factor and it is the
dual action of time and erosion that largely determines the changes in erosional
forms. The other important aspect of their work is their attempt to date this
denudation. In working out a time-sequence of erosion, they included a
method to be employed later by Dutton – the use of geological unconformities
as boundary marks at the end of each denudational stage. However, it should

be noticed that most of these findings need not be directly related to the present landscape.

The principles employed by writers such as Cornet and Briart form the link between the old and new methods in denudation chronology. But the new methods came from America and not from Europe. In America the work of Powell had demonstrated the control exercised by base level over the form of landscapes in the later stages of fluvial reduction, and Gilbert's concept of stream grade had enabled stream profiles to be used as indicators of base level at almost any stage of landscape reduction by running water. Thus there became available in the late 1870's the two fundamental tools of the modern student of denudation chronology – late-stage erosion surfaces and graded stream-profiles. The two, we must note, are inseparable and can be discerned by erosional and depositional landscape features. They were used by Le Conte, McGee and Willis either before, or contemporaneously with, the first clear exposition of the *geographical cycle* by Davis in 1889. It often comes as a surprise to the geomorphologist to learn that the inception of these new methods in the chronological analysis of landscape evolution just pre-dates the cyclic concept, but the fact is that Davis's work represents a steady logical development of contemporary thought. His synthesis relied heavily on the ideas of base level and grade and his idealized development of landscapes or landforms during one cycle contained other ingredients such as the growth of river patterns and the evolution of slopes and interfluves, some of which would have been heartily approved by Ramsay, Jukes, and Dutton. Before praising the master synthesist we ought to do credit to his immediate predecessors.

Le Conte (1880) invoked the principle of grade to explain the main features of river beds in the mining area of Middle California. He noticed the presence of abandoned river channels choked with fluvial debris which had been capped and preserved in this condition by lava flows. Yet in the same area the present streams were deeply trenching and eroding the landscape. He asks why, then, should these ancient rivers have aggraded their channels. It cannot be that the streams formerly had a lower velocity for much of the abandoned material is extremely coarse and high velocities would have been necessary to transport it. He decided that the proper explanation was contained in Gilbert's principle of grade, in accordance with which the ancient aggradation would have occurred because the total stream load exceeded its capacity for transportation. Le Conte explained the trenching of the recent streams as due to progressive elevation, *en masse*, of the Sierra Nevada Mountains, into which the gorges are cut.

> There is a certain definite relation between the slope and the amount of detritus which determines the depth of the cañons. If this relation be disturbed by increase of slope, the stream will strive to re-establish it. All deep cañons have been cut in rising ground and for the purpose of re-establishing this relation. (*Le Conte, 1880, p. 187*)

He pursued this idea in a later paper (1886), and estimated from the depth of the canyons that the recent uplift of the Sierra Nevada was about two to three thousand feet. He went on to stress how in this way river beds and their deposits can serve as accurate indicators of crustal movements:

> River-beds are most important and accurate indicators of crust-movements. . . . Suppose then, that a river in its lower course, has reached or nearly reached its base-level. If now the land rises the river immediately begins to cut its bed to lower and lower level until it again reaches its base-level. If on the contrary the land sinks the river will build up by sedimentation until it again finds the level of equilibrium. (*Le Conte, 1886, p. 168*)

When he discusses the principle of base level, Le Conte seems to confuse it with the concept of grade:

> In a steady crust a stream never completely reaches its base-level in any part because the base-level is constantly changing. The position of the base-level of a stream is determined by the relation of the velocity of the current to the freight of sediment and the freight of sediment depends on the slope of the upper tributaries. These upper tributaries are always above their base-levels and therefore always cutting and lowering their slopes. But the lowering of the slopes in the upper parts diminishes the freight of sediments and therefore lowers the base-level of the lower parts of the stream, which therefore also renews its cutting . . . The only complete limit to erosion – the only final and absolute base-level is, of course, the sea level. But although a river never completely reaches its base-level until it reaches the sea level; yet in a steady crust, it so nearly reaches it in its lower parts that for our purposes we may regard this level as being often practically reached. (*Le Conte, 1886, p. 169*)

He is much surer of his reasoning when he describes the various stages in the erosion of a landmass. Finally, he makes the important suggestion that erosional forms can be correlated with previous or continuing movements of the land, an idea which is very close to the theoretical framework so fully developed by Davis.

> Suppose then a country to be raised and then to remain steady . . . the streams will cut until in their lower parts they reach or nearly reach their base-level. Now they cut no longer or only at extremely slow rate. As soon as the down-cutting ceases, the stream begins to sweep from side to side under-cutting its banks and widening its channel. But meanwhile the valley slopes on either side and the separating divides are being worn down lower and lower by rain – erosion or weathering. Thus in a country which has remained steady for a long time the topography consists of wide trough-like river-beds separated by low rounded divides. On the contrary, in a country rising or recently risen we have deep, narrow cañons and high, sharp divides, or else high table-lands, between. Therefore deep, narrow cañons are evidence of recent and rapid rising; the deeper the cañon the more rapid has been the rising. But since the widening is meanwhile also progressing, the narrower the cañons, the more recent has been the rising. (*Le Conte, 1886, p. 169*)

About this time W. J. McGee was also beginning to consider landscape evolution from a chronological viewpoint. Born in 1853, McGee joined the

FIG. 127. *W. J. McGee*
(*U.S. Geological Survey*)

United States Geological Survey in 1882 and retired in 1893 (Darton, 1913). He became closely associated with Powell and a great friendship arose between them. Powell on his death bequeathed his brain to McGee, as a result of an argument between the two as to whose brain was the larger. When he died in 1912 McGee willed that their two brains should be sent to a specialist, Dr Spitzka. The doctor found that Powell's brain weighed 1,488 grains and McGee's 1,410. Scientifically it is a little doubtful what the experiment proved! It certainly has no significance for us, as McGee's work is of undoubted importance. His main contribution was to demonstrate clearly that geologic history could be read from topographic forms, as well as from sedimentary deposits. In 1888, making an early use of the term *geomorphology*, he emphasized how much a consideration of the geomorphic forms could aid the study of geologic history.

> Of the two great geologic processes involved in aqueous action only that of deposition has hitherto been utilized in interpreting local and general geologic history. Yet it is evident that the period represented by each deposit must be represented elsewhere by the complementary degradation; and in some cases the history of a region can be interpreted from the topographic forms resulting from degradation, as well as from the formations, with their contained fossils, resulting from the correlative process. Such genetic study of topographic forms (which has been denominated geomorphology) is specifically applicable in the investigation of the Cenozoic phenomena of the eastern United States . . . (*McGee, 1888A, p. 547*)

By this method McGee identified additional stages in the geological evolution of the Appalachian region, and explained how the intervening unconformities showed that there had been a succession of phases of marine deposition followed by periods of subaerial denudation. He used river patterns as indications of superimposition and sequential history:

> . . . the symmetrically dendritic forms of the drainage and the consequent relation between the value of local relief and the size of streams proves that the hydrography was autogenetically developed on an approximately plane surface, while the absence of alluvium and of confining highlands alike show that it was not super-imposed through alluviation. It follows that the drainage here is super-imposed through planation or through base-level degradation so perfect that the old water-ways were completely obliterated. (*McGee, 1888A, p. 563*)

From the trenching of the V-shaped valleys in the Piedmont Region, he deduced a post-Quaternary uplift.

The strength of McGee's thesis was re-emphasized in his article on the formations of the Middle Atlantic Slope (1888B).

> The Piedmont region comprises an area of highly inclined crystalline rocks, abundantly diked, veined and faulted; its surface is a rather strongly undulating plain without conspicuous eminences, inclined seaward, and everywhere graven deeply by the larger and to proportionately less depths by the smaller water-ways, which thus give origin to endless mazes of minor hills; its hydrography comprises the great rivers which meander irregularly through it, and a widely-branching dendritic system of secondary and tertiary drainage in which the individual members have no uniformity in direction, in which the basins are irregularly rounded or pyriform, and in which the divides are low and inconspicuous and constantly curving and recurving in labyrinthine convolutions. (*McGee, 1888B, p. 121*)

He repeats his statement that the study of topographic forms can greatly assist in deciphering geologic history.

> Conditions of deposition were inferred from deposits, and continent-movement was in turn inferred from evident conditions of deposition superinduced thereby; conditions of degradation in unsubmerged areas were

inferred from the topographic forms thereby developed, and, since degradation is pre-eminently dependent on base-level, another means of inferring
continent-movement was thus evolved; the record of events interpreted from
earth-forms fashioned in accordance with determinate principles on the one
hand was compared with the record interpreted from correlative deposits on
the other hand, and history was thus deduced from independent but consistent and cumulative testimony; and final correlations were made through
deposits regarded not only as rocks but also as indices of continent-movement,
and at the same time through the correlative topographic forms. In short, the
methods, standards, and criteria have been of necessity physiographic rather
than paleontologic or petrographic. (*McGee, 1888A, p. 124*)

In the case of the Middle Atlantic Slope he is able to relate the unconformity
of the ancient rocks, the deeply ravined plain, and the accompanying deposits
to a complicated succession of marine submergence and uplift:

From the relations of the formation to the foundation upon which it rests,
from structure and composition and indirectly from the conditions of deposition indicated thereby, the physiographic conditions attending the deposition
of the Potomac formation may be inferred. The surface upon which the
deposits rest is formed of dislocated strata of Archean, Cambrian, Silurian,
Triassic and Rhaetic age, all degraded to a plain as uniform as the Piedmont
zone of today – a plain destitute of noteworthy eminences despite the great
heterogeneity of the rocks, and one which accordingly must have been reduced to base-level; yet the unequal altitude of the deposits about the waterways indicate that this plain was ravined as deeply as is the present Piedmont
plain; and the slight sinuosity of the shore line, despite the depth of the
ravines, is proof of pronounced seaward inclination of the surface.

Thus the structure, composition, and stratigraphic relations of the Potomac
formation, when freely interpreted, give the outlines of an intelligible and
harmonious picture of the Atlantic slope during and for some time antecedent to the Potomac period. Before the initiation of Potomac deposition,
but subsequent to the accumulation of the Triassic and Rhaetic deposits and
to the displacement and diking by which they are affected, there was an eon
of degradation during which a grand mountain system was obliterated and
its base reduced to a plain which, as its topography tells us, was slightly
inclined seaward and little elevated above tide – the Piedmont zone alike of
the later Mesozoic and the present; and over this plain meandered the proto-
types of the Delaware, the Susquehanna, the Potomac, the James and the
Roanoke, within a few miles at most of the present courses and but a few
hundred feet above their present channels, flowing slack and in shallow
valleys because at base-level. There followed a slight elevation of the land,
when the rivers attacked their beds and excavated valleys as deep as those
today intersecting the Piedmont plain; but whether or not there was concomitant tilting of the land, the phenomena thus far fail to indicate. Then
came the movement by which the deposition of the Potomac formation was
initiated – the deeply ravined base-level plain was at the same time submerged and tilted oceanward; its waterways became deep but short estuaries;
deep oceanic waters extended quite to the intermediate shores; the declivity
and transporting power of the rivers was increased; and the accumulation
of coarse delta and littoral deposits progressed rapidly. With continued
deposition the sea gradually shoaled, the declivity of the land decreased,

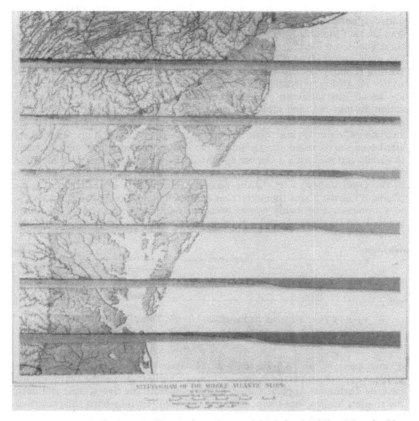

FIG. 128. *McGee's (1888) illustration of a model of the Middle Atlantic Slope*

the materials became finer and finer; there was probably temporary emergence of the land about the middle of the Potomac period, followed by renewed submergence without seaward tilting during which the clays of the upper members were laid down; and the period was finally closed by an emergence represented by the unconformity between the upper Potomac and the glauconitic deposits of the Maryland Cretaceous. (*McGee, 1888B, pp. 141–2*)

In the same way McGee was able to provide a likely explanation for the uniformity of surface that exists between all the rivers that intersect the region:

A remarkable topographic characteristic is displayed by the Piedmont and Appalachian regions in the middle Atlantic slope, which has only been interpreted – or indeed recognized – within the decade. The entire area is but a gently undulating plain diversified throughout by deeply incised waterways and, in the Appalachian zone, by bosses and ridges of obdurate strata which are narrowed and truncated by erosion but not planed off. The cross-section of the Susquehanna [*Fig.* 129], with its gently undulating plain bounded by mountains and dissected by a steep bluffed gorge, is representative of the entire Appalachian zone; it is constantly repeated along each principal water-way of that zone, and – save that the bounding mountains are absent – throughout the Piedmont region; and lines drawn in any direction through the area give ever-varying but harmonious combinations of this profile. During recent years this peculiar configuration has attracted the attention of nearly all geologists who have worked in the area. Stevenson has attributed the broad intermontane plains of the Pennsylvania Appalachians to wave-action during, and their minor irregularities to spasmodic elevation following, a general submergence, and ascribed the incised valleys to the action of the streams now occupying them during a recent epoch of high land; Kerr attributed the corresponding plains of the Piedmont region in North Carolina to glacial action during a remote epoch; G. F. Wright ascribes certain of the plains along the western slope of the Appalachians to a temporary ice-dam in the Ohio Valley; I. C. White recently referred the deposits upon these plains, if not the plains themselves, as exhibited along the Appalachian rivers, to submergence probably coeval with northern glaciation; but Gilbert has pointed out (orally) that in Virginia and North Carolina, at least, the system

FIG. 129. '*Base-leveled surfaces*' in the Susquehanna Valley
(*From McGee, 1888*)

of intermontane plains represents an old base-level of erosion. The com-posite Appalachian profile indeed indicates clearly that at some period of the past the Piedmont–Appalachian area stood low until the rivers, their affluents, the rivulets leading into these, and even the minutest rain-born rills, cut their channels to base-level and planed all the rocks except the obdurate quartzites and sandstones to the same level; and that afterward the land was lifted until the waters attacked their channels, cut out the labyrinth of recent gorges, and reduced the valleys, but not the hills, to a new base-level. This degradation-record is as definite and reliable as any found within deposits; and while so little is known of the physical relations of the clastic deposits of the Coastal plain (though they have been systematically classified repeatedly upon other bases) that they tell us less than the Piedmont hills of the evolution of the continent, and while it is yet impossible for that reason to correlate the records of land and sea, it will eventually be shown that the broad base-level plain corresponds to an important marine formation somewhere in the Coastal plain series. The unconformity in deposits corresponding to the rise of land closing the base-level period has not been certainly identified, but it seems probable that the deep and broad estuaries of the Coastal plain were then excavated, or at least deepened; and their depth suggests that for a time the land stood higher than now.

The Columbia formation reposes upon the less elevated portion of the Piedmont–Appalachian base-level plain and within the newer gorges dis-secting it, as well as upon the Coastal plain and within its estuaries to considerable depths (generally undetermined but known to exceed 140 feet to Chesapeake Bay). It is evident from the relations of deposit to sub-terrane in the Piedmont region that the deposition of the formation occurred long posterior to the rise of the land by which the old base-level was disturbed; for despite the high declivity of the stream post-Columbia erosion has not sufficed to lay bare the bottom of the pre-Columbia gorge or to remove more than half or two-thirds of the Columbia deposits in the Susquehanna and Delaware Valleys; and the post-Columbia erosion of the Potomac is measured by a gorge but 15 miles long, half a mile wide, and 75 feet in average depth, while the post-base-level erosion is represented by an outer gorge more than 200 miles long, over a mile in width, and fully 200 feet in average depth, and by corresponding gorges extending to the very sources of all its tributaries. Indeed, when post-glacial erosion is measured in yards and post-Columbia erosion in rods, post-base-level erosion must be measured in furlongs, if not in miles. (*McGee, 1888B, pp. 463–5*)

McGee also wrote clear reviews on contemporary attempts to classify topographic features, such as had been attempted by Powell and Richthofen and he refers with approval to the genetic approach used by Davis in 1884.

> . . . the implied classification in all these (above) cases is morphologic rather than genetic and is based upon superficial and ever-varying if not accidental characters; and if it were extended to the endless variety of forms exhibited in the topography of different regions it would only lead to the discrimination of a multitude of meaningless and unrelated topographic elements. The classification of topographic forms proposed by Davis . . ., and Richthofen's arrangement of the categories of surface forms . . . are more acceptable, since they are based in part on conditions of genesis. (*McGee, 1888A, p. 559*)
>
> Scientific progress may be measured by advance in the classification of phenomena. The primitive classification is based on external appearances, and is a classification by analogies; a higher classification is based on internal as well as external characters, and is a classification by homologies; but the ultimate classification expresses the relations of the phenomena classified to all other known phenomena, and is commonly a classification by genesis. (*McGee, 1888C, p. 27*)
>
> . . . the stages of geologic development are best interpreted in terms of geography. So a genetic classification of geologic phenomena . . . will apply equally to geography, whether observational or of the more philosophic nature which Davis proposes to call Systematic Geography, and which Powell has called Geomorphology. (*McGee, 1888C, p. 27*)

As an example of what he meant, McGee compiled a genetic classification of landforms based upon the processes which control their evolution. It includes the sub-divisions: (1) deformation, (2) gradation (i.e. in deposition and erosion), (3) vulcanism, (4) alteration (i.e. lithification, etc.), (5) glaciation and (6) wind action (air and waves). Though this division is not followed exactly at the present day, most textbooks on geomorphology tend to use somewhat similar classifications.

McGee's writings are especially important because they stand closest to the later work of Davis. In fact it is probable that his own conclusions were to some extent influenced by Davis's early writings. Davis, however, in his *Geographical Essays* (1909), mentions how McGee made use of the base level concept in describing the Chesapeake Bay area and how he was the first to recognize the amount of subaerial denudation that had taken place on the Atlantic slope during Mesozoic time. Gregory (1918) also emphasizes McGee's recognition of denudation surfaces on the Atlantic slopes, but goes further than Davis when he stresses that the main importance of McGee's work lay in his use of physiographic criteria to solve stratigraphic problems.

We have yet to mention Willis, who in 1889 followed McGee's method in an attempt to solve the topographic problems of the Asheville region in the southern Appalachians. He too sees the topographic surfaces as erosion surfaces and the incised valleys as evidence of the disturbance of stability by subsequent uplift:

> We have recognized that dissected plain, the level of the Asheville amphitheatre, now 2400 feet above the sea; it was a surface produced by subaerial erosion, and as such it is evidence of the fact that the French Broad River, and such of its tributaries as drain this area, at one time completed their work upon it, reached a base level. That they should have accomplished this the level of discharge of the sculpturing streams must have been constant during a long period, a condition which implies either that the fall from the Asheville plain to the ocean was much less than it is now, or that through local causes the French Broad was held by a natural dam, where it cuts the Unaka chain. (*Willis, 1889, p. 297*)
>
> And thus in thought I have looked from Big Bald out on a gently sloping plain which covered the many domes of nearly equal height and stretched away to merge on the horizon in the level of the sea. That, I conceive, was the first base level plain of which we have any evidence in the Appalachians and from that plain our present valleys have been eroded. The continental elevation must then have been 3000 or 4000 feet less than it is now, and the highest hills were probably not more than 2500 feet above the sea. This was perhaps a period of constant relation between sea and land, but it was succeeded by one during which the land slowly rose. The rivers, which had probably assumed nearly their present courses, were revived; the important channels soon sank in cañons, the tributaries leaped in rapids and cut back into the old base level. The region continued to rise during a period long enough to produce the essential features of the mountain ranges of today; then it stood still in relation to the sea or perhaps subsided somewhat, and the French Broad and probably other rivers made record of the pause in plains like that round about Asheville. Again the land rose slowly; again it paused, and rivers, working always from their mouths backward, carved a base level into the limestones of the great valley; but before that level could extend up through gorges in the Unakas, the continent was raised to its present elevation, the streams responded to the increased fall given them and the rivers in the valley began to cut their still incomplete cañons. (*Willis, 1889, p. 299*)

CHAPTER THIRTY-TWO

The Geographical Cycle

The face of places, and their forms decay;
And that is solid earth, that once was sea:
Seas, in their turn, retreating from the shore,
Make solid land, what ocean was before.
(Ovid: *Metamorphoses*, Book 15th, *c.* A.D. 8)

It is not part of the present volume to analyse the importance of the work of William Morris Davis, which will be dealt with in detail in the second of the three volumes that we intend to devote to the history of physical landscape studies. However, having led the more persistent reader through the maze of eighteenth- and nineteenth-century thought on landscape evolution, it would be a distinct anticlimax if we omitted to outline the early work of the man who was to become for many the embodiment of all past excellence in geomorphology and whose work formed the mainspring of half a century of research. The term *genius* has often been applied to Davis and, although his cyclic theory has been severely criticized during the past thirty years, we see no strong reason for disagreeing with this accolade. His genius was of a particular sort. He created nothing – nothing, that is, except coherence and vitality; and his synthetic means of so doing was as simple and obvious as the manner in which Lazarus was raised from the dead. Bowman (1934, p. 180) gives an interesting sidelight on Davis's evaluation of his own work:

> A manuscript on the analysis of land forms submitted to him late in 1925 referred to the work of Powell, Gilbert, Dutton, and Davis. It ascribed to Davis more credit than he wished to take for the development of physiography. He edited a paragraph to read as follows, . . . 'Powell's work was noteworthy for the forceful ideas it conveyed of base-levelled surfaces; Gilbert and Dutton excelled in their analyses of individual features, Davis systematized the sequence of forms through an ideal cycle and provided a terminology.'

DAVIS'S EARLY LIFE

William Morris Davis was born in Philadelphia of Quaker stock in 1850. As a young man he showed none of his later ebullience, and was studious and retiring. He graduated from Harvard in 1869, receiving the degree of Master of Engineering in 1870. During these two years Davis began his lifetime of qualitative field observation by attending field courses to Colorado and to the Lake Superior region, under Raphael Pumpelly the Harvard Professor of Mining (1866–75). From 1870 to 1873 he was employed in the survey and meteorological service of the National Observatory at Cordoba in the Argentine, and, after more travels, became an Assistant in Geology at Harvard under Professor N. S. Shaler in 1876. Especially notable was his subsequent association with Pumpelly on the 1883 survey for the Northern Pacific Railroad route in Montana, where he developed some of his early geomorphic ideas. He recognized a broad bench, or temporary base level, 250–300 feet above the present Missouri river, and, referring to the plains between Fort Benton and the Sun River, wrote:

> All these we consider the work of a river system now extinct, a system that, when in activity, cut and carried away an unknown thickness of overlying strata and brought the original surface down near its base level of drainage, thus producing the comparative smoothness of the existing plains. (*Davis, 1886C, p. 710*)

In the nearby mountains he noticed:

> Old base levels, cut by narrow stream-channels at many points, mark recent massive changes in elevation. (*Davis, 1886C, p. 712*)

Davis, always critical of the work of others, was evidently critical in later years of his own early work as the following anecdote from one of his students illustrates.

> On one happy day he interrupted my discourse while I was reading a descriptive quotation from an author whose name I had not yet announced. 'What does the man mean?' said Davis. 'I cannot conceive,' he continued, 'what the author had in mind when he wrote those words.' He dissected the quotation at some length. Poor Martin had been hanging on to his chair and holding his tongue all this time. Then I said politely: 'What did you have in mind when you wrote those words?' My quotation was from W. M. D. himself in a report on the Northern Transcontinental Survey under Pumpelly in 1877,* some 28 years before, and my teacher had forgotten his own child. All he said, however, after a long awkward silence, uninterrupted by me of course, was 'Go on.' (*Martin, 1950, p. 175*)

In 1885 Davis was appointed Assistant Professor of Physical Geography at Harvard, and in the following year was awarded the H. H. Warner Medal for scientific discovery. Already his publications were reaching massive proportions, and by 1890 they totalled almost 100 works, including about 50 on meteorology, 12 on the Connecticut Valley, half a dozen on glacial topics and one on wasp stings. In later years he used to advise his students, 'If it's worth doing it's worth printing' (*Bowman, 1934, p. 180*).

However, Davis's early academic career was not entirely smooth progress, as the following quotation shows:

> In 1876, in consequence of attending Shaler's summer school in Kentucky, Davis was invited to take charge of the field work in Shaler's classes, and two years later was made Instructor in Physical Geography. For this new work

* Presumably 1886 is intended here.

Cam. June 1 – 1882 –

Dear Mr. Davis –

The Corporation offer you a reappointment as instructor in geology at a salary of $1200 a year, and with work as follows :– To give the course called Nat. His. 1 :– and to assist Prof. Shaler in giving Nat. His. 4 ; and an advanced course in geological field-work (as in 79 & 8) or to render an equivalent amount of assistance in the department of geology. The Corporation are quite aware that this position is not suitable for you as a permanency; but it is all that they are able to offer you now, with their present resources, and all that they expect to be able to afford for some time to come. In considering whether it is your interest to accept this offer temporarily, I hope that you will look in the face the fact, that the chances of advancement for you are by no means good, although the Corporation have every reason to be satisfied with your work as teacher.

very truly yours,
Charles W. Eliot.

Mr. W. M. Davis ––

copy

FIG. 130. *Copy of the letter sent by President Eliot of Harvard to W. M. Davis on 1st June, 1882. This copy was made by Davis himself and was kept by him until he died, being found among his personal effects*
(*By permission of R. Mott Davis*)

he was ill prepared, and an unfacile lecturer. His teaching, though carefully, indeed laboriously prepared, was so little successful that in 1882 President Eliot wrote him a discouraging letter, advising him to look for a place elsewhere, as his chance for promotion at Harvard was small. By good fortune, his former teacher, Raphael Pumpelly, who was then conducting a geological survey for the Northern Pacific Railway, invited him to undertake a study of the stratigraphic series below the coal-bearing Cretaceous in Montana. There Davis came upon the first fruitful idea that the physical geography of the lands should be presented as exemplifying the natural history of rivers, because rivers are the chief agency in land sculpture; and in the following year, building on the work of Powell, Gilbert, and Chamberlin, he developed the scheme of the cycle of erosion. This greatly fortified his physiographic teaching by introducing a rational or evolutionary motive in the study of land forms which, following the empirical methods of tradition, he had previously taught. (*Davis and Daly*, *1930*, *pp. 314–15*)

If Davis was criticized for his poor lecturing ability, this was certainly a gross miscalculation of his ultimate potential; however, if he had been threatened with dismissal because of his lack of fundamental research, some unkind geomorphologists might credit the Harvard President with greater insight than was to be vouchsafed to students of geomorphology for at least another half century!

DAVIS'S EARLY WORK

One of Davis's early articles was on the structure and petrology of the Triassic trap rocks of the Connecticut region (1882). In the following year he showed an interest in current geomorphological problems by reviewing an article by F. Löwl in which the German postulated that water-gaps are never formed by a pre-existing river keeping pace with a slowly-rising mountain fold or fault. Löwl was certain that river erosion could not keep pace with mountain building, and he explained transverse valleys as due to either the down-cutting of a lake overflow or the headward back-cutting of a powerful river. Davis disagreed with this view and believed that water-gaps, particularly in the Appalachians, were often antecedent in origin. In this review he also stressed the need for a uniform genetic terminology, pointing out that unnecessary problems and misunderstandings arose when geologists gave the same name to forms of differing origin:

... no science has so loose, inaccurate, and insufficient a terminology as geography. (*Davis*, *1883*, *p. 326*)

An early use of the idea of the cycle is made in an 1884 article on gorges and waterfalls, which he considers seldom occur in the maturity and old age of a valley.

Narrow gorges and abrupt waterfalls are seldom found in old countries of flat rocks and moderate elevation. For a gorge gapes wider and wider in the course of its natural growth, and becomes a broad valley in its maturity and old age; a waterfall decreases in height as it is worn farther and farther back, until at last it remains only as a faint ripple in the almost uniform down-grade of its stream. (*Davis, 1884A, p. 123*)

These stages of river erosion and their accompanying forms Davis here relates to the ideas of grade and base level. He explains that in youth no part of a river has reached equilibrium between its velocity and the amount of material to be carried. Consequently, the characteristic forms are narrow valleys, gorges, waterfalls and similar features of maladjustment. As the bed is lowered and river processes begin to balance, most of these youthful features are eliminated or confined to the headward stretches only. The approach of equilibrium between a river's transporting power and load he terms *maturity* and the features associated with it are the broadening of valleys and the dis-appearance of falls and gorges. When a river has reached a state of equili-brium throughout all its tributaries the land will have been reduced to a general plain, with a surface-level unrelated to the underlying structure.

The important concepts of grade and base level are of course not new but their intimate relation to the evolution of the landscape and associated land-forms in a series of stages which Davis called youth, maturity and old age was a development of the first importance.

Another article written in 1884 develops the theme of the cycle of evolution in some detail. Davis begins by explaining that topographic forms can be classified according to either their structural origin, like volcanoes or fold mountains, or according to their erosional history;

> The relations of a number of geographic forms may be exhibited by means of a classification based first on the peculiarities of structure and second on the amount of erosion that they have suffered. Conditions of accumulation, elevation and deformation give subheadings under the first class, while the stages of advance in the destructive processes of erosion that cut all land forms alike down to a base-level plain of denudation provides further opportunity for subdivision. (*Davis, 1884B, p. 428*)

However, he is quick to stress that erosion has far more effect on the final landscape forms than has structure and, therefore, an erosional classification is more valuable to the geomorphologist in deciphering the landscape. This leads Davis to the first real statement of his cycle of erosion; for the first time he associates the terms 'cycle' and 'stage'. He also adds detail to his descrip-tions of the three stages in the cycle and instead of concentrating on river valley forms he begins to include within the cycle the characteristics of the general relief. For instance he explains that maturity is characterized as the era of greatest relief when the river valleys have reached the stage of land dissection at which nothing of the old plain remains except hilltops and ridge crests. Similarly he characterizes old age as the stage when these divides are worn down and the level of the relief is progressively diminished:

> But of greater geographical importance than these early and constructional characteristics, are the later, destructional ones, determined by erosion,

inasmuch as they comprehend the topographical form that we actually observe ... the surface of the deposit rises above its base-level of erosion ... a smooth, unbroken plain is revealed ... The smoothness of the surface and the shallow lakes are indeed truly infantile features ... Rivers establish their courses, the smooth plain is trenched by their meandering channels, and all the lakes disappear. This is adolescence. The channels will be narrow and steep walled in regions of relatively rapid elevation, but broadly open in regions that have risen slowly, and I believe that rate of elevation is thus of greater importance than climatic conditions in giving cañon form to a valley ... Adolescence as thus defined includes that part of a plateau's early life in which the stream channels are narrow and well marked; but as the valleys increase in number and open widely so as to consume a good share (one-sixth to one-fifth) of the plateau mass measured above the existing base-level of discharge, then adolescence merges into maturity. Maturity may be said to last through the period of greatest diversity of form, or maximum topographic differentiation, until about three-quarters of the original mass are carried away; ... During maturity no vestige of plain surface remains ... Long lines of cliffs are the most pronounced features; their upper edge is of constant altitude ... Every cliff marks the outcropping of a more resistant stratum ... The cliffs are separated by sloping surfaces ... ; but with the recession of the cliffs, the talus proper is restricted to a narrow space around their base, and the intervening bench land has a more gentle inclination ... During the maturity of a plateau all the streams cut down nearly to their base-level and give a maximum relief between the upper levels and the valley bottoms ... the stream courses having been selected as nearly horizontal surfaces, and finding no lateral constraint in the attitude of the strata through which they cut, are necessarily meandering ... One of the latest phases of maturity reveals the higher levels of the plateau broken through by the ever increasing number of ravines and valleys, producing in some cases the peculiar pinnacled topography ... On leaving this stage, the plateau passes from maturity to old age; the relief of the surface diminishes, for while the pinnacles dwindle away and the country loses in altitude, the valleys do not correspondingly gain in depth. The cliff faces are forced farther and farther back and the valleys increase greatly in width, the slopes become gentler ... At last ... these remnants are carried away, reducing the once rugged country to almost as low, flat and featureless a surface as it was at birth; but its perfected system of meandering rivers separates it from new-born plains, and although now smooth when compared to the ruggedness of its maturity, still it rises gently between the streams in faint swells which distinguish it from the more nearly perfect level of unworn surfaces ... This is simple old age, a second childhood in which infantile features are imitated and thus the decrepit surface must wait either until extinguished by submergence below the sea, or regenerated by elevation into a new cycle of life. (*Davis, 1884B, pp. 429–32*)

Though it is generally accepted that Davis's first full statement of the cycle was made in 1889 there is very little in the later articles which had not basically been covered in 1884.

Davis's next articles revert to a purely structural account of the Triassic formations of the Connecticut valley, and not until 1886 did he return to his geomorphic principles. Then in an orthodox description of the structural geology of the Connecticut valley he refers to a base level surface and we realize that the germ of the peneplain concept is still maturing in his mind:

> The Triassic strata were laid down on the edges of these steeply inclined slabs after they had been bevelled off to a tolerable even surface by pre-Triassic erosion. (*Davis, 1886B, p. 349*)

In 1887 in a classification of lakes he repeats the successive phases in his cycle, and in the following year again mentions the base level concept:

> . . . the only ultimate form of land-sculpture is the base-level plain, down to which every surface must be reduced, whatever its structure, if time be allowed. (*Davis, 1888, p. 320*)

From the assumption that every land surface will eventually be levelled to a plain he argues that structure does not primarily determine erosional forms, but that in a folded structure the forms will depend upon the relationship of the existing surface and structure to the base level.

THE CYCLE CONCEPT IN EUROPE

Before Davis's full statement of the erosional cycle is discussed we must mention here an important European work, *Les Formes du Terrain*, produced jointly in 1888 by Lt.-Col. G. De La Noë of the Geographic Service of the French army and Emm. de Margerie. In it they refer frequently to the early papers of Ramsay, Gilbert, Dutton, Philippson and Richthofen and occasionally to works by Powell, Dana and Jukes but W. M. Davis is mentioned only once, and then in connexion with glaciation! Davis's article of 1884 on the cycle is actually given in the bibliography but it obviously had made little impression on the authors.

The volume is divided into three sections: (1) Elements and principles, which mainly concern weathering, the fashioning of slopes and fluvial landscapes; (2) factors determining the pattern of river systems; and (3) the rôle of causes foreign to subaerial erosion, which are given as ice, sea, wind and vulcanism.

In their treatment of river- and slope-processes, the ideas of De La Noë and De Margerie are more like those of Gilbert than of Davis. They recognized runoff as one of the chief factors controlling the angle of land-slope, in as much as it would progressively decrease the slope angle with time:

> THE FASHIONING OF SLOPES. The effect of rain on slopes is to lessen increasingly their gradient: for a given rock this decrease will be hastened in proportion to the rapidity of disintegration of the surface, the fineness of the decomposed debris, the steepness of the slope, and the volume of water. Furthermore, as the least slope on which the materials can be moved depends on their size, if there is no limit to the breaking up of these particles *the only limit to the gradient of the slope will be that necessary for the flow of water supercharged with the finest particles*. Obviously whatever may have been the size of the materials when first detached from the bedrock, they can eventually, under the repeated action of the agents of decomposition and also of

collisions with each other during even a short journey, be reduced in their turn to such fine particles that we should similarly admit that all rocks, irrespective of their mineral composition, are liable to reach *a very feeble slope, on the sole condition that the action of rain is sufficiently prolonged.* In brief, the action of the rain finally flattens all slopes. (*De La Noë and De Margerie, 1888, pp. 20–21*)

They go on to discuss the effect of disintegration (weathering) and to remind the reader that the greatest runoff of water was concentrated towards the base of a slope and therefore the slope base was likely to have the gentlest profile. They recognized that steep slopes were a sign of youth, which would disappear when erosion had acted for any long period of time.

If we consider the slopes of valleys cut through rocks of different kinds and subjected during the same period to the same action of disintegration and removal, it is clearly evident that the gradient of the slopes of each of these valleys depends solely on the rapidity with which the rock composing it is disintegrated: the faster this rate, the gentler the slope. If we consider now the case of a slope on which rocks of different kinds outcrop, each part of the profile will present a slope, which will differ in inverse relation to the speed of disintegration of the rock to which that part corresponds. (*De La Noë and De Margerie, 1888, p. 24*)

We must remark here that the degradation of easily decomposed rock, instead of producing a gentle gradient directly uniting the top and bottom of the two adjacent resistant beds, appears . . . as a fairly deep excavation of the profile which has the effect of leaving the higher rock overhanging.

In reality, the law which governs the profile of slopes is not quite so simple as we have said, because of the greater volume of water that the lowest beds receive; as a result of this excess, if the same rock forms two distinct beds, one at the summit and the other toward the base of the slope, the first would present a slope steeper than the second. (*De La Noë and De Margerie, 1888, pp. 24–25*)

The authors even carried their discussion to the point where they speak of a slope of equilibrium for any given material, and they state that the regular or balanced slope profile will be initiated at the base, as with the profile of a river.

The number of different factors which act in the production of a slope is too great, however, to permit any reliable classification of rocks based on the angles of their debris-covered slopes.

It is impossible to make a general classification of rocks according to the inclination of their talus, since these latter vary with altitude, latitude, and even the method of destruction that the rocks undergo. We must say, however, that in general slopes are steeper in dry climates and gentler in wet climates, a fact which is easily understood by referring back to the arguments considered above concerning the rôle of rain water in the reduction of the profile of slopes; thus the Quaternary alluvium of the Sahara, outcropping in the *gour* or isolated buttes and on the sides of the dry wadis appears upon gradients rarely attained by the same rocks in our climates. (*De La Noë and De Margerie, 1888, pp. 45–46*)

There is also one further factor which has an important control over the angle of equilibrium.

> We know that following the deforestation of a slope, rain water immediately carries away the moveable materials on its surface and causes disasters of a type previously quite unknown. Therefore, the vegetation intervenes to maintain the surface of the soil at a certain gradient; without it the disintegration, and especially the transport of weathered materials by the turbulent waters, would continue until the levelling of continental relief would be almost complete. But, in order to permit vegetation to establish itself, the rapidity of its development must be greater than the ease with which the movable materials are carried away. This particular state of equilibrium is reached with varying rapidity according to the nature of the subsoil and to all the external circumstances that in some degree favour vegetation. Thus, in warm, damp climates, vegetation will struggle, as it were, against the abundance of rain, in such a way as to maintain a notable slope of the talus in spite of the (destructive) tendency of the rain. In a general way, the vegetation has thus given to the rocks a slope of equilibrium and has assured the stability or fixation of the landforms which without it would continue indefinitely to be worn down and flattened. *The concept* of equilibrium of slopes is thus inseparable from the idea of the vegetation which determines it. (*De La Noë and De Margerie, 1888, p. 47*)

Their study of slopes also produced other interesting observations. They were able to say that on a homogeneous valley-side the slopes do not retreat by the removal of parallel slices because then the material falling at the foot of the valley-side would not be removed on so feeble a basal slope; the valley-sides flatten as they retreat. They also state that when opposing valley-sides are asymmetrical this may be due to the greater exposure of one of the slopes to erosional agents. Where slopes intersect different outcrops, they consider that weathering of the slope ultimately proceeds independently of the stratification.

> In establishing itself in the form of a variably inclined talus, the surface of the valley sides, at least where the beds are nearly horizontal, intersect the stratification planes at angles which increase in acuteness as the gradient of the slopes flattens, hence producing a bevel . . . Rainwater, acting on the material weathered from the rock surface, fashions the slopes at the expense of the fresh rock, without being notably influenced by changes in the arrangement of the stratification (thickness and dip), just as a plane shaves a piece of wood, without changes in the dip of the concentric beds bringing about any modification in the result. Generally speaking, we can therefore say that *the declivity of slopes is independent of stratification*. (*De La Noë and De Margerie, 1888, p. 32*)

They also combined Gilbert's ideas on mechanical equilibrium of slopes with those relating to slope evolution currently being developed by Davis:

> Geometric Definition of Slope Surfaces. If we wish to give slope-forms geometric definition, which it is true can never be rigorously applied in nature, but which has the advantage of lending precision to the above

considerations and systematizing them, we can say that *the surface of slopes is produced by a force which should have at each point the maximum slope corresponding to the rock outcropping at that point, this force being always coincident with the line of greatest slope of the surface, and constantly depending on the line forming the foot of the slope as a control.* Let us recall that by *maximum value* of the slope we mean the greatest slope which the rock has as yet attained in its given state of weathering and mobility in which it occurs.

It is important to recall in effect that landforms, as it were, are always undergoing modification; we see them in a state of *transition,* which is variably advanced according to the resistance of the rocks, the activity of the modifying forces, and the duration of their work. A similar rock in a dry climate will have undergone after an equal period of time considerably less weathering than in a humid one; in the first case it will appear with a steep talus and sharp arêtes, which will be replaced in the second by the rounding of angles and the flattening of relief. It is by considerations of this sort that, given the same climate, it would be possible to determine the relative age of different landforms. (*De La Noë and De Margerie, 1888, p. 39*)

De La Noë and De Margerie then proceed to discuss the development of river profiles. To them, the ultimate principle governing the development of a river was the concept of time.

First Law. The form of the (river) profile is independent of the weight of the load and its size, provided that the length of the action is sufficiently prolonged. (*De La Noë and De Margerie, 1888, p. 52*)

To this they added the control of base level and of bed-load comminution:

Application of this first law to the case of a stream.

We have seen that the inclination from the horizontal of this section of the bed is precisely equal to the natural equilibrium of the talus (slope) of the material forming the current: in other words, the experiment proves that *the current has lowered the slope of its bed almost to the limit beyond which it can no longer flow.* From experiments, this limit depends on the speed of the current, always lessened by friction; the speed itself depends on the height of the fall, being least as one approaches the source; it is therefore wholly natural that the ultimate profile will be a concave curve of such a type that the tangents to each point are the more steeply inclined to the horizontal in proportion to its proximity to the highest point; whereas in the part of the stream considered where we have supposed the volume and speed to be constant, the slope of the bottom of the bed must be everywhere the same: the long profile is a straight line, of which the inclination to the horizontal, according to the findings of our experiment, is the least slope necessary for the flow of load material; but if one supposes a sufficiently long period of time, such that the continuously comminuted material is reduced to excessively fine particles, this limit is none other than – or very nearly – that which is necessary for the flow of water. One sees therefore that in major streams, which generally have a considerable length, the slope of the bed near the mouth must be excessively small, since at those points the material brought from upstream has undergone sufficiently long comminution to be transported in particularly fine particles.

It is for the same reason that most tributary streams at a certain stage of development have near a confluence gradients which are generally very small. (*De La Noë and De Margerie, 1888, pp. 53–54*)

We notice in effect that as long as the mouth of a stream is susceptible of deepening, all the points on the profile upstream must equally be the object of modifications in their height; in consequence, these points only attain their final position after the mouth itself has been fixed. It is in this sense that we are justified in saying that the equilibrium profile of a stream establishes itself from the mouth upwards.

The profile of this bed is hence a curve concave to the sky, which is not due any longer to an increase in speed, . . . but to a progressive decrease upstream in the volume of water and in the load, from which it inevitably results that erosion lessens in duration and in intensity on the bottom of the bed, and consequently the gradient steepens, in proportion to the distance upstream of the point in question. This diminution in the volume of water and of debris is obvious; it is enough to examine a map to prove that the number of tributaries diminishes in proportion as we ascend toward the source and that there is an equal diminution in the surface area of the basin drained, from which the materials transported by rill-wash must pass by the point under consideration on the stream. All these materials, on the contrary, must eventually pass through the mouth: thus there evidently erosion will be most energetic and sustained. Consequently, it is there that we are sure to find the gentlest slope. (*De La Noë and De Margerie, 1888, pp. 55–56*)

They return to the all-important time factor.

IMPORTANCE OF THE CONSIDERATION OF THE DEGREE OF DEVELOPMENT OF STREAMS. The preceding considerations are not immediately brought out by an examination of streams as they are for the most part today, because they have attained a state of equilibrium very different from the successive steps through which they must have passed during the period of incision; this equilibrium state (as clearly stated by Dausse in 1872) has special characteristics and deserves its name in the sense that in that state the bottom of the bed varies relatively little.

During the period of incision the gradients of the stream were noticeably steeper and the current had an excess of speed which allowed it to deepen its bed most energetically: the slopes at each point would be inversely proportional to the velocities. Today, on the other hand, wherever the bed has ceased to be deepened, we see the gentlest slopes corresponding usually to the lowest speeds, and this must at first sight seem contradictory to the experimental conclusions described above, knowing that the gentlest slopes correspond to the greatest speeds. But we must not lose sight of the fact that it is not speed alone which determines the deepening of the bed and the lessening of the slope; in the final analysis we must consider the intensity of erosion; and this depends not only on the speed of the current, but on the quantity of load acting on the same part of the bed, and on the time during which the action is exercised. But the nearer the point in question lies towards the mouth, the more the materials that will pass it, and the longer it will be subjected to their attacks, since, as we have seen, the equilibrium curve establishes itself from below upstream, and consequently the total erosion at this point will be greater than for all the points lying upstream, in spite of the diminution of speed resulting from greater friction due to the lessening of the gradient, providing simply that the stream being considered has arrived at *a sufficiently advanced stage in its evolution*. We see then the importance of

the consideration of the *time* factor, or, more exactly, of the *degree of development* and *the age* of the stream, in the analysis of the phenomena which we
are studying. (*De La Noë and De Margerie, 1888, pp. 56–57*)

Second law: the form of the ultimate profile of a stream is independent of
the nature of its bed. In all previous discussions we have assumed that the
rock crossed by the streams was everywhere the same. We will see now what
differences are produced when the terrain is composed of beds of unequal
resistance. For this, we refer to the experiment . . . which showed that *the
form of the profile of a stream is independent of the nature of the bed*, provided
that the length of the experiment was long enough for the profile to develop
its ultimate form. (*De La Noë and De Margerie, 1888, p. 57*)

When De La Noë and De Margerie come to examine the erosion of the
landscape as a whole, a few of their views resemble those of Davis but their
main inspiration has come from the works of Gilbert and Powell and from
their own experiments. In addition, they express close agreement with
Philippson's work on watersheds and slopes although they derived their
findings quite independently. Their description of a newly-emerged landmass closely resembles that of Du Buat and strongly suggests the cyclic
concept.

Let us consider a continent which has just emerged from the sea: the rainwaters falling on its surface collect together along a series of channels determined by the general form; consequently they may present the most diverse
dispositions, but they will all converge strongly towards the lowest point or
line on the emerged continent, that is to say, in a general way, towards the
shoreline. Meanwhile, as the general slope is not necessarily continuous, the
run-off, before flowing to the sea, will collect, in this case, in enclosed hollows
where they rise until finding an outlet, across the lowest part, towards lower
regions. The general stream-pattern thus formed will consist of a series of
channels inter-connecting various lakes, but finally draining into the sea.
Under the continuous action of water erosion, the parts of the surface not
covered with lake waters will provide detritus for the filling up of the depressions, and when the detritus has accumulated to the level of the channels
draining the lakes into each other, it will, in its turn, be used in the deepening
of those channels which as they are gradually lowered will allow the successive removal of the loose materials deposited upstream. The bed of the channels will usually be finally lowered to the level of the original depression;
then all the deposited detritus will disappear, and the stream will establish
itself on the surface which originally formed the floor of the depression. In
the contrary case, where because base-level is not low enough, the stream
bed cannot deepen itself sufficiently, the detritus deposited in the depression
will remain in part as evidence of the action of in-filling . . ., and the stream
will establish itself on the surface of the corresponding plain, over which it
will continue to wind. (*De La Noë and De Margerie, 1888, pp. 113–14*)

When the authors describe the consequent water-courses as 'original' they
are, of course, following Powell. They go on to emphasize the genetic distinction between these consequent streams and those of subsequent origin
which they term 'subordinate'.

We must distinguish between the different streams which furrow the land. Indeed, if it is true that the main streams of a region – such as the Seine, Marne, Oise and Yonne in the Paris Basin – present a pattern concordant with the general slope of the original surface, in such a way that we can reconstruct the latter by examining the underlying rocks, there also exist in this region other streams whose pattern, on the contrary, no longer depends on the original slope but solely on the declivity acquired by the general surface in the *neighbourhood* of the first (main) streams after they have created adjacent valley-slopes. Accordingly a distinction can be made between *original* streams, that is to say, those which bear a direct relationship with the general slope of the surface at the time of emergence, and *subordinate* streams, of which the pattern depends only on that of the original streams. Subordinate streams are considered to be much less important, and can only be born at the time when the first have already formed their beds and appreciably lowered their original level. (*De La Noë and De Margerie, 1888, p. 116*)

Their description of drainage density in permeable rocks, and the control exercised by rainfall intensity, is more reminiscent of the Western geologists than of Davis's work. Finally, they mention the general occurrence of what they call 'abrasion platforms', the origin of which, in typical European fashion, they seem prepared to ascribe equally to action by the waves and erosion by subaerial agents:

Abrasion platforms. One of the most remarkable and at the same time one of the most frequently observed facts which has been discovered by examining the geological structure of continents consists of the levelling-off, along an almost plane surface of beds outcropping there with variable dips: this characteristic often persists over vast areas, and imparts a special character to whole regions, such as the Ardennes and Rhenish Prussia. Several geologists (notably Ramsay and Richthofen) have attributed the formation of these *abrasion platforms* to the action of waves, slowly invading a subsiding continent; this opinion can, in fact be reconciled very well with that which we have outlined on the present formation of littoral platforms by marine action; and we can often evoke in support the fact of the superposition upon the planed beds of more recent marine deposits which evidently prove that since the planing the ocean has again covered the corresponding areas.

But it remains to know to what extent the abrasion surfaces result not merely from the modification by the waves of surfaces very little different, resulting from the prolonged action of streams that have long reached their equilibrium level, and have, with the aid of atmospheric agents, almost completely removed the previously existing inequalities: this is actually, as we have seen, the tendency of agents of subaerial erosion; and the case which must often occur, especially in regions at a height near sea level. It would thus be wrong to attribute the creation of abrasion platforms exclusively to the sea. (*De La Noë and De Margerie, 1888, pp. 187–8*)

It is difficult to evaluate the mutual influences of the ideas of Davis and of De La Noë and De Margerie, because their writings are so contemporaneous. This perhaps explains why the work of the French authors is so frequently overlooked, despite the stress they laid on the importance of the time element in the sequential development of landforms. It is possible that some impetus for their studies was provided by the early work of Davis, for a few years later De Margerie referred to Davis as 'our dear friend and master' (*De Margerie, 1915, p. 107*), but even if this is so, it should not detract from the great value of their contribution, particularly to the study of slope development.

We prefer to think that they were not immediately impressed by Davis's early ideas on the cycle, which probably did not appeal to their practical minds. On stages and base level they write as follows:

> It is necessary to distinguish in the phenomenon of the deepening of river-channels two very distinct periods: in the first, the rivers have an excess of energy and of speed; they vigorously deepen their bed, which shows rapids and waterfalls in, other things being equal, increasing frequency toward their source. During this period, the tributaries nearly have accordant junctions; when this accordance occurs accidentally it retards, up to a certain point, the rate of deepening on the minor stream. Soon the excess of velocity lessens; the main river increasingly regularizes the curve of its bed; the slope of which is increasingly smoothed, working upstream from the mouth.
>
> Then the second period begins, in which the longitudinal profile undergoes almost no further change; the tributaries with a level not yet in accordance with that of the main river continue to be deepened and eventually achieve accordance, and thus, provided that the work of deepening is sufficiently prolonged, there is established the accordance of level that we have noticed and that can be seen over the greater part of continental surfaces. (*De La Noë and De Margerie, 1888, p. 75*)
>
> It is during the second period, when the final profile is definitely attained, that the rivers which can no longer deepen their bed, widen it. Then finally the speed at each point diminishes, under the influence of friction, with the flattening of the gradients, and thus the *état de régime* (normal state) is established. (*De La Noë and De Margerie, 1888, p. 76*)

Here the authors add an interesting footnote.

> The ideas which have recently been expounded on the rôle of base-level, the upstream migration of incisions (nick-points), and the successive stages in the deepening of the bed, are not new. In fact, we find them indicated, at least in substance, in several works. However, it seems that no one up to the present had the idea of verifying them experimentally, as we have done. Among the authors who appear to have had the best understanding of the mechanics of fluvial erosion, we will mention Surell, Dana, Dausse, Gilbert, Heim, Dutton and Philippson. (*De La Noë and De Margerie, 1888, p. 76*)

DAVIS'S AMPLIFICATION AND ADVERTISEMENT OF THE CYCLE

By 1889, as we have seen above, the fluvial stages of youth, maturity and old age had not been widely accepted in Europe. Nor were they any more popular in the United States. But that year marks the launching by Davis of an advertisement campaign for the 'geographical cycle'. It was a fruitful year for him in many ways. He began by publishing an example of river-piracy (river-capture) in east Pennsylvania. Here he considered that an old eroded surface

FIG. 131. *William Morris Davis at the age of 12*
(*By permission of W. M. Davis II*)

had been disturbed and raised and its existing streams had had their gradients increased unequally so that they eroded at different rates, in places in competition with each other.

The country hereabouts was in ancient times a surface of faint relief, at a lower stand than now, traversed by idle streams; but, in consequence of elevation to a greater altitude, the streams have revived their lost activities, and set to work to sink their channels and open out their valleys in the process of reducing the land to its proper level again, even with the sea; . . . In the processes of adjustment thus called forth, every stream struggles for its own existence . . . the steeper streams have gnawed more quickly into the landmass than the flatter ones, and the divide between the pair of contesting streams has consequently been pushed in the direction of the fainter descent. (*Davis, 1889A, p. 108*)

Already a literary affectation is beginning to form part of his style. Valleys and rivers and hills are described as if they had human attributes and were capable of acting in the same way as men and women. In another section describing a mature drainage basin's reaction to uplift, there is a good deal of literary exaggeration and a tendency to use the concept of base level as a guiding fairy god-mother to the wilder excursions of his literary creation. There seems no doubt that he is now applying the concept very adeptly to a large number of varying situations but the close juxtaposition of scientific deduction and literary whimsy strikes us, at least, as sometimes unhappy.

When the district was lifted from its former lowly estate, the streams found a new task set before them. They at once set to work at it with the best disposition in the world. But in their immaturity, they accepted without question such guidance as the faint relief of the surface afforded, only to discover later on that the primitive division of territory was inadvisable as a permanency, because it was not adapted to the best accomplishment of the work assigned to them . . . (*Davis, 1889A, p. 109*)

Davis, who had concentrated many of his studies on the geology of the Connecticut valley, now utilized certain anomalies in the southern part of that river's course to develop his cyclic theory one stage further. For the first time the term *peneplain* appears in literature.

Given time enough, and the faulted ridges of Connecticut must be reduced to a low base-level plain. I believe that time enough has already been allowed, and that the strong Jurassic topography was really worn out somewhere in Cretaceous time, when all this part of the country was reduced to a nearly featureless plain, a 'peneplain', as I would call it, at a low level; a plain that was broadly lifted in early Tertiary time – or thereabouts – and thus thrown into another cycle of destructive development, and whose elevated remnants are now to be recognized in the crystalline upland on either side of the present Triassic valley of Connecticut and Massachusetts . . ., and in the crest-line summits of the main trap ridges. (*Davis, 1889B, pp. 429–30*)

Where the land surface was divided into two levels or platforms he explained these forms by supposing a succession of cycles; the higher platform represented an old peneplain while the lower platform formed the near base level surface of the new cycle:

Like mountains of repeated growth, this topography may be called 'polygenetic'. The present form of the region is modelled with reference to at least two base-levels. (*Davis, 1889B, p. 431*)

The basic essentials of Davis's famous cycle of erosion have now all been stated and it only remains for Davis to popularize his idea among other geomorphologists. This he tries to do in an article on geographic methods. He begins by suggesting that geography should include, as one of its branches, a study of geomorphological forms.

> . . . it is clearly advisable in this case to take such steps as shall hasten a critical and minute examination of the form of the earth's surface by geographers, and to this end it may serve a useful purpose to enlarge the limited definition of geography . . . and insist that it shall include not only a descriptive and statistical account of the present surface of the earth, but also a systematic classification of the features of the earth's surface, viewed as the results of certain processes, acting for various periods, at different ages, on diverse structures. (*Davis, 1889C, p. 11*)

Taking an example from the Appalachians he explains how the concept of base level can give a third dimension to the interpretation of topography. Before the evolution of the base level concept a study of the topography could show the various transgressions of the sea, what had been eroded and how this erosion had been influenced by the underlying structure. Since the adoption of the cyclic concept it was possible to discover a great deal more. There is much in Davis's argument that recalls the arguments of McGee.

> Topography revealed structure, but it did not reveal the long history that structure has passed through. The anticlinal valleys, hemmed in by the even-topped sandstone mountains of middle Pennsylvania, were found to tell plainly enough that vast erosion had taken place, and that the resulting forms depended on the structure of the eroded masses, but it was tacitly understood that the land stood at its present altitude during the erosion. The even crest lines of the mountains and the general highland level of the dissected further west did not then reveal that the land had stood lower than at present during a great part of the erosion, and thus the full lesson of topography was not learned. The systematic relation of form to structure, base-level and time; the change of drainage areas by contest of headwaters at divides; a revival of exhausted rivers by massive elevations of their drainage areas; all these consequences of slow adjustments were then unperceived. (*Davis, 1889C, p. 14*)
> . . . the age of mountains may be deduced from their form as well as from their rocks; the altitudes at which a district has stood may be determined by traces of its old base-levels. . . . (*Davis, 1889C, p. 17*)

Passing on from the idea of base level to the relative age of surface forms, he emphasizes how each is a guide to the stage which erosion has reached in that area.

> Young mountains possess structural lakes and are drained largely by longitudinal valleys; old mountains have no such lakes and have transverse drainage, formed as the growing headwaters of the external streams lead out much water that formerly filled the longitudinal valleys. Young rivers may have falls on tilted beds, but such are short lived. . . . All falls disappear in old rivers, provided they are not resuscitated by some accident in the normal,

simple cycle of river life. The phases of growth are as distinct as in organic forms. (*Davis, 1889C, p. 16*)

(One may consider) . . . the features of the land as the present stage of a long cycle of systematically changing forms, sculptured by processes still in operation. Now recognizing the sequence of changing forms, we may deduce the place that any given feature occupies in the entire sequence through which it must pass in its whole cycle of development. (*Davis, 1889C, p. 18*)

In a further article in 1889 Davis published his most polished and complete account yet of the 'cycle' concept. In essence it differed little from his paper in 1884 except that it introduced (in the third paragraph below) the idea of a long period of stand-still, the validity of which was increasingly challenged in later years.

Rivers are so long-lived and survive with more or less modification so many changes in the attitude and even in the structure of the land, that the best way of entering on their discussion seems to be to examine the development of an ideal river of simple history, and from the general features thus discovered, it may then be possible to unravel the complex sequence of events that lead to the present condition of actual rivers of complicated history.

A river that is established on a new land may be called an original river. It must at first be of the kind known as a consequent river, for it has no ancestor from which to be derived. Examples of simple original rivers may be seen in young plains, of which southern New Jersey furnishes a fair illustration. Examples of essentially original rivers may be seen also in regions of recent and rapid displacement, such as the Jura or the broken country of southern Idaho, where the directly consequent character of the drainage leads us to conclude that, if any rivers occupied these regions before their recent deformation, they were so completely extinguished by the newly made slopes that we see nothing of them now.

Once established, an original river advances through its long life, manifesting certain peculiarities of youth, maturity, and old age, by which its successive stages of growth may be recognized without much difficulty. For the sake of simplicity, let us suppose the landmass, on which an original river has begun its work, stands perfectly still after its first elevation or deformation, and so remains until the river has completed its task of carrying away all the mass of rocks that rise above its base-level. The lapse of time will be called a cycle in the life of a river. A complete cycle is a long measure of time in regions of great elevation or of hard rocks; but whether or not any river ever passed through a single cycle of life without interruption we need not now inquire. Our purpose is only to learn what changes it would experience if it did thus develop steadily from infancy to old age without disturbance.

In its infancy the river drains its basin imperfectly, for it is then embarrassed by the original inequalities of the surface, and lakes collect in all the depressions. At such time the ratio of evaporation to rainfall is relatively large, and the ratio of transported land waste to rainfall is small. The channels followed by the streams that compose the river as a whole are narrow and shallow, and their number is small compared to that which will be developed at a later stage. The divides by which the side streams are separated are poorly marked, and in level countries are surfaces of considerable area and not lines at all. It is only in the later maturity of a system that the divides are reduced to lines by the consumption of the softer rocks on either side. The difference between constructional forms and these forms that are due to the action of denuding forces is in a general way so easily recognized that immaturity and maturity of a drainage area can be readily discriminated. In the truly infantile drainage system of the Red River of the North, the inter-stream areas are so absolutely flat that water collects on them in wet weather, not having either original structural slope or subsequently developed denuded slope to lead it to the streams. On the almost equally young lava blocks of southern Oregon, the well-marked slopes are as yet hardly channelled by the flow of rain down them, and the depressions among the tilted blocks are still undrained, unfilled basins.

As the river becomes adolescent, its channels are deepened and all the larger ones descend close to base-level. If local contrasts of hardness allow a quick deepening of the downstream part of the channel, while the part next upstream resists erosion, a cascade or waterfall results; but, like the lakes of earlier youth, it is evanescent and endures but a small part of the whole cycle of growth; but the falls on the small headwater streams of a large river may last into its maturity, just as there are young twigs on the branches of a large tree. With the deepening of the channels, there comes an increase in the number of gulleys on the slopes of the channel; the gulleys grow into ravines and these into side valleys, joining their master streams at right angles (La Noë and Margerie). With their continued development the maturity of the system is reached; it is marked by an almost complete acquisition of every part of the original constructional surface by erosion under the guidance of the streams, so that every drop of rain that falls finds a way prepared to lead it to a stream and then to the ocean, its goal. The lakes of initial imperfection have long since disappeared; the waterfalls of adolescence have been worn back, unless on the still young headwaters. With the increase of the number of side streams, ramifying into all parts of the drainage basin, there is a proportionate increase in the surface of the valley slopes, and with this comes an increase in the rate of waste under atmospheric forces; hence it is at maturity that the river receives and carries the greatest load; indeed, the increase may be carried so far that the lower trunk stream, of gentle slope in its early maturity, is unable to carry the load brought to it by the upper branches, and therefore resorts to the temporary expedient of laying it aside in a flood plain. The level of the flood plain is sometimes built up faster than the small side streams of the lower course can fill their valleys, and hence they are converted for a little distance above their mouths into shallow lakes. The growth of the flood plain also results in carrying the point of junction of tributaries farther and farther downstream, and at last in turning lateral streams aside from the main stream, sometimes forcing them to follow independent courses to the sea (Lombardini). But although thus separated from the main trunk, it would be no more rational to regard such streams as independent rivers than it would be to regard the branch of an old tree, now fallen to the ground in the decay of advancing age, as an independent plant; both are detached portions of a single individual, from which they have been separated in the normal processes of growth and decay.

In the later and quieter old age of a river system, the waste of the land is yielded more slowly by reason of the diminishing slopes of the valley sides; then the headwater streams deliver less detritus to the main channel, which, thus relieved, turns to its postponed task of carrying its former excess of load to the sea, and cuts terraces in its flood plain, preparatory to sweeping it away. It does not always find the buried channel again, and perhaps settling down on a low spur a little to one side of its old line, produces a rapid or a low fall on the lower slope of such an obstruction (Penck). Such courses may be called locally superimposed.

It is only during maturity and for a time before and afterwards that the three divisions of a river, commonly recognized, appear most distinctly; the torrent portion being the still young headwater branches, growing by gnawing backwards at their sources; the valley portion proper, where longer time of work has enabled the valley to obtain a greater depth and width; and the lower flood-plain portion, where the temporary deposition of the excess of load is made until the activity of middle life is past.

'Maturity' seems to be a proper term to apply to this long-enduring stage; for as in organic forms, where the term first came into use, it here also signifies the highest development of all functions between a youth of endeavour towards better work and an old age of relinquishment of fullest powers. It is the mature river in which the rainfall is best led away to the sea, and which carries with it the greatest load of land waste; it is at maturity that the regular descent and steady flow of the river is best developed, being the least delayed in lakes and least over-hurried in impetuous falls.

Maturity past, and the power of the river is on the decay. The relief of the land diminishes, for the streams no longer deepen their valleys, although the hill-tops are degraded; and with the general loss of elevation, there is a failure of rainfall to a certain extent; for it is well known that up to certain considerable altitudes rainfall increases with height. A hyetographic and hypsometric map of a country for this reason show a marked correspondence. The slopes of the headwaters decrease and the valley sides widen so far that the land waste descends from them more slowly than before. Later, what with failure of rainfall and decrease of slope, there is perhaps a return to the early imperfection of drainage, and the number of side streams diminishes as branches fall from a dying tree. The flood plains of maturity are carried down to the sea, and at last the river settles down to an old age of well-earned rest with gentle flow and light load, little work remaining to be done. The great task that the river entered upon is completed. (*Davis, 1889D, pp. 203–206*)

This description of the cycle is very effective, and accentuates skilfully all the main criteria. The whole work rightly received a great deal of attention. It clearly illustrates that Davis's concept of the cycle of landscape evolution was based on his belief that the geometrical forms of topography change systematically in a progressive and irreversible manner. Once a landmass has been raised, the cycle begins and the landscape will progress through a pre-destined series of forms until either the ultimate form of the peneplain is achieved or the landmass is affected by other events. It seems probable that Davis absorbed his evolutionary tendency from the general climate of Darwinism. Darwin's principles of 'evolution', 'progressive divergence' and 'laws of change' were certainly very similar to Davis's mode of thinking and the link between the two is suggested in the following passages from T. H. Huxley: 'What I mean by "evolution" is consistent and thoroughgoing uniformitarianism' (quoted by Judd, 1911, p. 23); and 'Consistent uniformitarianism postulates evolution, as much in the organic as in the inorganic world . . .' (*Letter to Darwin*; *Darwin, 1887, II, p. 190*)

Apart from the merits of his ideas Davis was fortunate in the age in which he worked. The development of his cycle of erosion came at a time when there was a rapid increase in the number of new scientific journals. *Science* began in 1883, *The National Geographic Magazine* in 1888, *The Bulletin of the Geological Society of America* in 1890 and *The Journal of Geology* in 1893. Fenneman paints the popular picture of him.

'Had Davis never lived, the combined work of Powell, Gilbert, McGee, and others would have left the general picture of landmasses being reduced to base-level by streams of various origins, life histories, and characteristic courses, the forms of unconsumed masses being determined by structure, climate, and other factors, even including age. But this is only a rough frame and largely empty, and the lives of these men were spent largely in other phases of geology, and it was left to Davis to paint the panorama of changes from initial forms to their final disappearance. This is the theme of the geomorphic cycle, the most distinctive element of modern physiography. (*Fenneman, 1939, p. 355*)

We intend in a later volume to describe how Davis's cyclic concept grew in popularity until it eliminated almost every other school of thought, and how its attractive ideas tended to lull twentieth-century geomorphologists into a too easy acceptance of its partial view of landscape development. But if the present volume has taught only one lesson, it is that stable equilibrium cannot exist long in a developing science, and, even before the time of the flood tide of Davis's popularity, we can confidently look forward to the inevitable ebb, when 'the old order changeth, yielding place to new . . . lest one good custom should corrupt the world'.

References: Part Four

ANON. (1876) 'Review of Powell's exploration of the Colorado' *Geol. Mag.*, N.S., Dec. 2, Vol. 3, pp. 365–70.
— (1937) 'C. E. Dutton, explorer, geologist and nature writer' *Sci. Monthly*, Vol. 45, July.
AIRY, W. (1885) 'Discussion on non-tidal rivers' *Proc. Inst. Civil Engineers*, Vol. 82, pp. 25–26.
BAILEY, E. B. (1935) 'Tectonic Essays, mainly Alpine' (Oxford), 200 pp.
— (1952) 'Geological Survey of Great Britain' (Murby, London), 278 pp.
BALL, R. S. (1882) 'On the occurrence of great tides since the commencement of the geological epoch' *Nature*, Vol. 27, pp. 201–3.
BEARD, C. A. and BEARD, M. R. (1943) 'The Rise of American Civilization.'
BECKER, G. F. (1912) 'Major C. E. Dutton' *Amer. Jour. Sci.*, Ser. 4, Vol. 33, pp. 387–8.
BELIDOR, B. F. DE (1729) 'La Science des Ingénieurs.'
BERTRAND, M. (1887) 'La Chaîne des Alpes, et le formation du continent européen' *Bull. Soc. Géol. France*, 3rd Ser., Vol. 15, 1886–7, p. 423.
BIROT, P. (1958) 'La leçon de Grove Karl Gilbert' *Ann. de Géog.*, No. 362, pp. 289–307.
BLENCH, T. (1957) 'Regime Behaviour of Canals and Rivers' (Butterworths, London), 138 pp.
BOWMAN, I. (1934) 'William Morris Davis' *Geog. Rev.*, Vol. 24, pp. 177–81.
BREWER, W. H. (1902) 'Obituary of J. W. Powell' *Amer. Jour. Sci.*, 4th Ser., Vol. 14, pp. 377–82.
CADELL, H. M. (1887) 'The Colorado River of the West' *Scot. Geog. Mag.*, Vol. 3, pp. 441–60.
CALLOWAY, C. (1883) 'The age of the new gneissic rocks of the northern Highlands' *Quart. Jour. Geol. Soc.*, Vol. 39, p. 348.
CATE, A. (1959) 'J. Peter Lesley, a biographical sketch' *Geotimes*, Vol. 4, No. 4, pp. 18–19 and 45.
COLLIN, A. (1846) 'Recherches Expérimentales sur les Glissements spontanés des Terrains Argileux' (Carilian-Gœury, Paris). Translation, with a memoir by A. W. Skempton, Toronto, 1956.
COOLEY, W. D. (1876) 'Physical Geography' (London), 429 pp.
COOLING, L. F. (1945) 'Development and scope of soil mechanics' in *The Principles and Application of Soil Mechanics*, Publication of the Inst. of Civil Engineers, pp. 1–30.
CORNET, F. L. and BRIART, A. (1877) 'Sur le relief du sol en Belgique' *Ann. de la Soc. Géol. de Belgique*, Tome 4, pp. 71–115.
COULOMB, C. A. (1773) 'Essai sur une application des règles de maximis et minimis a quelques problèmes de statique, relatifs a l'architecture' *Mém. Math. Phys.*, *Acad. Royale des Sciences*, Paris, Vol. 7, pp. 343–82 (published in 1776).
DANA, J. D. (1886) 'A dissected volcanic mountain' *Amer. Jour. Sci.*, 3rd Ser., Vol. 32, pp. 247–55.

DARRAH, W. C. (1951) 'Powell of the Colorado' (Princeton), 426 pp.

DARTON, N. H. (1913) 'Memoir of W. J. McGee' *Ann. Ass. Amer. Geog.*, Vol. 3, pp. 103-10.

DARWIN, C. (1882) 'The Formation of Vegetable Mould, through the action of worms' (London), 328 pp.

DARWIN, F. (1887) 'The Life and Letters of Charles Darwin', 3 vols. (London).

DARWIN, G. H. (1879) 'On the precession of a viscous spheroid, and on the remote history of the earth' *Phil. Trans. Roy. Soc. London*, Vol. 170, pp. 447-538.

DAUBRÉE, A. (1879) 'Études Synthétiques de Géologie Expérimentale' (Paris), 828 pp.

DAVIS, W. M. (1882) 'Brief notice of observations on the Triassic trap rocks of Massachusetts, Connecticut and New Jersey' *Amer. Jour. Sci.*, 3rd Ser., Vol. 24, pp. 345-9.

— (1883) 'The origin of cross-valleys' *Science*, Vol. 1, pp. 325-7 and 356-7.

— (1884A) 'Gorges and waterfalls' *Amer. Jour. Sci.*, 3rd Ser., Vol. 28, pp. 123-32.

— (1884B) 'Geographic classification, illustrated by a study of plains, plateaus and their derivatives' *Proc. Amer. Assn. Adv. Sci.*, Vol. 33, pp. 428-32.

— (1885-6) 'Triassic formation of the Connecticut Valley' *U.S. Geol. Surv.*, 7th Annual Rept., pp. 461-90.

— (1886A) 'Mechanical origin of the Triassic monoclinal in the Connecticut Valley' *Proc. Amer. Assn. Adv. Sci.*, Vol. 35, pp. 224-7.

— (1886B) 'The structure of the Triassic formation of the Connecticut Valley' *Amer. Jour. Sci.*, 3rd Ser., Vol. 32, pp. 342-52.

— (1886C) 'Relation of the coal of Montana to the older rocks' *Report on the Mining Industries of the United States*, Dept. of the Interior, Tenth Census of the United States, XV, Washington.

— (1887) 'The classification of lakes' *Science*, Vol. 10, pp. 142-3.

— (1888) 'Synclinal mountains and anticlinal valleys' *Science*, Vol. 12, p. 320.

— (1889A) 'A river-pirate' *Science*, Vol. 13, pp. 108-9.

— (1889B) 'Topographic development of the Triassic formation of the Connecticut Valley' *Amer. Jour. Sci.*, 3rd Ser., Vol. 37, pp. 423-34.

— (1889C) 'Geographic methods in geologic investigation' *Nat. Geog. Mag.*, Vol. 1, pp. 11-26.

— (1889D) 'The rivers and valleys of Pennsylvania' *Nat. Geog. Mag.*, Vol. 1, pp. 183-253.

— (1909) 'Geographical Essays' (Ginn, Boston), 777 pp.

— (1915) 'Biographical memoir of John Wesley Powell 1834-1902' *Nat. Acad. of Sci., Washington, Biog. Memoirs*, Vol. 8, pp. 9-83.

— (1918) 'Grove Karl Gilbert' *Amer. Jour. Sci.*, 4th Ser., Vol. 46, pp. 669-81.

— (1922) 'Biographical memoir of Grove Karl Gilbert (1843-1918)' *Nat. Acad. of Sci. Washington*, Vol. 21, 5th Memoir, 303 pp.

— (1925) 'The progress of geography in the United States' *Ann. Ass. Amer. Geog.*, Vol. 14, pp. 159-215.

DAVIS, W. M. and DALY, R. A. (1930) 'Geology and Geography 1858-1928, in *The Development of Harvard University, 1869-1929*, ed. by S. E. Morison (Harvard University Press), Chapter 19, pp. 307-28.

DE LA NOË, G. D. and DE MARGERIE, E. (1888) 'Les Formes du Terrain' (Paris), 205 pp.

DELLENBAUGH, F. S. (1902) 'The Romance of the Colorado River' (Putnam's, N.Y.), 399 pp.

DE MARGERIE, E. (1915) 'The debt of geographical science to American explorers' Mem. Vol. of The Trans-continental Excursion of 1912, *Amer. Geog. Soc.*, pp. 105–13.

DILLER, J. S. (1911) 'Major Clarence Edward Dutton' *Bull. Seism. Soc. Amer.*, Vol. I, No. 4, Dec., pp. 137–42.

DOKUČAIEV, V. V. (1877) 'Ovragi i ikh značene' *Trudy Volnogo Ekonom.* Obšč., t. 3.

— (1878) 'Sposoby obrazovania rečnykh dolin Evropeiskoi Rosii', SPb.

DU BOYS, M. P. (1879) 'Le Rhône et les rivières à lit affouillable' *Ann. des Ponts et Chaussées*, 5th Ser., Tome 18, 2nd Part, pp. 141–95.

DUTTON, C. E. (1871–2) 'The causes of regional elevations and subsidences' *Proc. Amer. Phil. Soc.*, Vol. 12, pp. 70–72.

— (1876) 'Critical observations on theories of the earth's physical evolution' *The Penn Monthly*, Philadelphia, May–June.

— (1880) 'Report on the geology of the high plateaus of Utah' U.S. Geog. and Geol. Surv. of the Rocky Mt. Region (Washington), 307 pp.

— (1880–1) 'The physical geology of the Grand Canyon district' U.S. Geol. Surv., 2nd Ann. Rept. (1880–1), pp. 47–166.

— (1882) 'Tertiary history of the Grand Canyon region' Monographs of the U.S. Geol. Surv., Vol. 2 (Washington), 264 pp.

— (1884–5) 'Mount Taylor and the Zuni Plateau' U.S. Geol. Sur., 6th Ann. Rept., pp. 111–98.

— (1889) 'On some of the greater problems of physical geology' *Bull. Phil. Soc., Washington*, No. 11, pp. 51–64.

DYLIK, J. (1953) 'Caractères du développement de la géomorphologie moderne' *Bull. de la Soc. des Sci. et des Lettres de Lödz*, Classe III, Vol. 4, 3, 40 pp.

ENDLICH, F. M. (1878) 'On some striking products of erosion in Colorado' *Bull. U.S. Geol. and Geog. Survey of the Terr.*, Vol. 4, pp. 831–64.

EWING, A. L. (1885) 'An attempt to determine the amount and rate of chemical erosion taking place in the limestone valley of Center Co., Pennsylvania' *Amer. Jour. Sci.*, 3rd Ser., Vol. 29, pp. 29–31.

FENNEMAN, N. M. (1931) 'Physiography of Western United States' (McGraw-Hill, N.Y.), 534 pp.

— (1939) 'The rise of physiography' *Bull. Geol. Soc. Amer.*, Vol. 50, pp. 349–60.

FREEMAN, L. R. (1923) 'The Colorado River' (Heinemann, London), 451 pp.

GEIKIE, A. (1880) 'Rock-weathering, as illustrated in Edinburgh churchyards' *Proc. Roy. Soc. Edinburgh*, Vol. 10, pp. 518–32.

— (1882) 'Text Book on Geology' (London), 971 pp.

GEIKIE, J. (1877) 'The movement of the soil cap' *Nature*, Vol. XV, pp. 397–8.

— (1886) 'Mountains: their origin, growth and decay' *Scot. Geog. Mag.*, Vol. 2, pp. 145–62.

GILBERT, G. K. (1875) 'Report on the Geology of portions of Nevada, Utah, California and Arizona (1871–72)' in *Report upon Geographical and Geological Explorations and Surveys West of the One Hundredth Meridian*; in charge of First Lieut. G. M. Wheeler (Washington), Vol. 3, Part 1, pp. 21–187.

— (1876) 'The Colorado Plateau Province as a field for geological study' *Amer. Jour. Sci.*, 3rd Ser., Vol. 12, pp. 16–24 and 85–103.

— (1877) 'Report on the Geology of the Henry Mountains' (Washington), 160 pp.

— (1880–1) 'Contributions to the history of Lake Bonneville' *U.S. Geol. Surv. 2nd Ann. Rept. (1880–81)*, pp. 167–200.

GILBERT, G. K. (1883) 'Drainage system and loess deposition of Eastern Iowa' *Science*, Vol. 2, pp. 762–3.

— (1883) Review of Whitney's 'Climatic Changes' III, *Science*, Vol. 1, pp. 192–5.

— (1884) 'The sufficiency of terrestrial rotation for the deflection of streams' *Amer. Jour. Sci.*, 3rd Ser., Vol. 27, pp. 427–32.

— (1885) 'The topographic features of lake shores' *U.S. Geol. Surv., 5th Ann. Rept. (1883–84)*, pp. 65–123.

— (1886) 'The inculcation of the scientific method by example' *Amer. Jour. Sci.*, 3rd Ser., Vol. 31, pp. 284–99.

— (1902) 'John Wesley Powell' *Ann. Rept. Smithsonian Institution for 1902*, pp. 633–40.

— (1909) 'The convexity of hill-tops' *Jour. Geol.*, Vol. 17, pp. 344–50.

— (1914) 'The transportation of debris by running water' *U.S. Geol. Surv., Prof. Pap. 86*, 263 pp.

GOETZMANN, W. H. (1959) 'Army Exploration in the American West 1803–1863' (Yale), 509 pp.

GREEN, A. H. (1876) 'Geology' (London), 552 pp.

GREGORY, H. E. (1918) 'Steps of progress in the interpretation of land forms' *Amer. Jour. Sci.*, Vol. 46, pp. 104–32.

HAYDEN, F. V. (1861) 'Sketch of the geology of the country about the head-waters of the Missouri and Yellow Stone rivers' *Amer. Jour. Sci.*, 2nd Ser., Vol. 31, pp. 229–45.

— (1862) 'Some remarks in regard to the period of elevation of those ranges of the Rocky Mountains, near the sources of the Missouri River and its tributaries' *Amer. Jour. Sci.*, 2nd Ser., Vol. 33, pp. 305–13.

— (1874–5) 'Notes on the surface features of the Colorado or Front Range of the Rocky Mountains' *Bull. U.S. Geol. and Geog. Surv. of the Terr.*, Vol. 1, pp. 215–20.

— (1876) 'Notes descriptive of some geological sections of the country about the headwaters of the Missouri and Yellowstone rivers' *Bull. U.S. Geol. and Geog. Surv. of the Terr.*, Vol. 2, pp. 197–209.

HEIM, A. (1871–8) 'Untersuchungen über den Mechanismus der Gebirgs-bildung' 2 Vols. (Basle).

— (1879) 'Ueber die Verwetterung im Gebirge' (Basle).

HERITSCH, F. (1929) 'The Nappe Theory in the Alps' (Methuen, London), 228 pp. (Translated by P. G. H. Boswell).

HETTNER, A. (1898) 'Die Entwickelung der Geographie im 19 Jahrhundert' *Geog. Zeit.*, Vol. 4, pp. 305–20.

HUXLEY, T. H. (1878) 'Physiography', 2nd edn (London), 384 pp.

IVES, J. C. (1861) 'Report upon the Colorado River of the West' (Washington), Part I, 131 pp.

JACKSON, L. D'A. (1875) 'Hydraulic Manual' (Allen, London), 3rd edn.

JOFFE, J. S. (1936) 'Pedology' (Rutgers University Press), 575 pp.

JUDD, J. W. (1911) 'The Coming of Evolution' (Cambridge), 171 pp.

JUKES, J. B. (1863) 'The School Manual of Geology' (Edinburgh), 362 pp.

JUKES-BROWNE, A. J. (1884) 'The Student's Handbook of Physical Geology' (London), 541 pp.

KENNEDY, R. G. (1895) 'The prevention of silting in irrigation canals' *Min. Proc. Inst. Civil Engineers*, Vol. 119.

KESSELI, J. E. (1941) 'The concept of the graded river' *Jour. Geol.*, Vol. 49, pp. 561–88.

KRYNINE, D. P. (1947) 'Soil Mechanics' (McGraw-Hill, N.Y.), 2nd Edn, 511 pp.
LAPWORTH, C. (1883) 'On the secret of the Highlands' *Geol. Mag.*, N.S., Vol. 10, p. 121.
LE CONTE, J. (1879) 'Elements of Geology' (Appleton, N.Y.), 588 pp.
— (1880) 'The old river-beds of California' *Amer. Jour. Sci.*, 3rd Ser., Vol. 19, pp. 176–90.
— (1886) 'A post-Tertiary elevation of the Sierra Nevada shown by the river beds' *Amer. Jour. Sci.*, 3rd Ser., Vol. 32, pp. 167–81.
LEWIS, E. (1877) 'On water courses upon Long Island' *Amer. Jour. Sci.*, 3rd Ser., Vol. 13, pp. 142–6.
LOMONOSOV, M. V. (1763) 'On the strata of the earth' (2nd supplement to the 'Principles of Metallurgy and Mining', St. Petersburg. Vol. 5 of the collected works of M. V. Lomonosov publ. by Acad. Sci. U.S.S.R., 1954).
LÖWL, F. (1884) 'Ueber Thalbildung' (Prague).
LURIE, E. (1960) 'Louis Agassiz' (Univ. of Chicago), 449 pp.
MCGEE, E. R. (1915) 'Life of W. J. McGee.' (Farley, Iowa).
MCGEE, W. J. (1888A) 'The geology of the head of Chesapeake Bay' *U.S. Geol. Surv., 7th Ann. Rept.* (1885–6), pp. 537–646.
— (1888B) 'Three formations of the middle Atlantic slope' *Amer. Jour. Sci.*, 3rd Ser., Vol. 35, pp. 120–43, 328–30, 367–88 and 448–66.
— (1888C) 'The classification of geographic forms by genesis' *Nat. Geog. Mag.*, Vol. 1, pp. 27–36.
MARKOV, K. K. (1948) 'Osnovnye problemy geomorfologii' (Moscou), 343 pp.
MARTIN, L. (1950) 'William Morris Davis: Investigator, teacher, and leader in geomorphology' *Ann. Ass. Amer. Geog.*, Vol. 40, pp. 172–7.
MARVINE, A. R. (1874) 'The stratigraphy of the east slope of the Front Range' *U.S. Geol. and Geog. Surv.* (Hayden Survey), Annual Rept. for 1783 (Washington).
MATHER, K. F. and MASON, S. L. (1939) 'A Source Book in Geology' (McGraw-Hill, N.Y.), 702 pp.
MEDLICOTT, H. B. and BLANFORD, W. T. (1879) 'A Manual of the Geology of India', 2 parts (Calcutta).
MENDENHALL, W. C. (1920) 'Memorial to Grove Karl Gilbert' *Bull. Geol. Soc. Amer.*, Vol. 31, pp. 26–64.
MERRILL, G. P. (1924) 'The First One Hundred Years of American Geology' (Yale), 773 pp.
MILL, H. R. (1930) 'The Record of the Royal Geographical Society 1830–1930' (Royal Geographical Society, London), 288 pp.
MILLER, H. (1881–3) 'River terracing: its methods and their results' *Proc. Roy. Physical Soc. of Edinburgh*, Vol. 7, pp. 263–306.
MOORE, R. (1957) 'The Earth We Live On '(Jonathan Cape, London), 348 pp.
NEWBERRY, J. S. (1861) 'Report upon the Colorado River of the West' (Washington), Part III, Geological Report.
— (1862) 'Colorado River of the West' *Amer. Jour. Sci.*, 2nd Ser., Vol. 33, pp. 387–403.
— (1876) 'Geological Report of the Exploring Expedition from Santa Fé, New Mexico, to the junction of the Grand and Green Rivers of the Great Colorado of the West in 1859' (Washington), 152 pp.
NEWTON, H. (1880) 'Geology of the Black Hills of Dakota', Chapter 1 in 'Report on the Geology and Resources of the Black Hills of Dakota' (Washington), pp. 1–222.

NICKERSON, L. (1865) 'Periodic action of water' *Amer. Jour. Sci.*, 2nd Ser., Vol. 39, pp. 151–6.

OLDHAM, R. D. (1888) 'On the law that governs the action of flowing streams' *Quart. Jour. Geol. Soc.*, Vol. 44, pp. 733–9.

PARRY, C. C. (1862) 'Physiographical sketch of that portion of the Rocky Mountain Range, at the headwaters of South Clear Creek, and east of Middle Park' *Amer. Jour. Sci.*, 2nd Ser., Vol. 33, pp. 231–7.

PARTIOT, H. L. (1871) 'Sur les sables de la Loire' *Ann. des Ponts et Chaussées*, 5th Ser., Tome 1, pp. 233–92.

PEACH, B. N. and HORNE, J. (1884) 'Report on the geology of the north-west of Sutherland' *Nature*, Vol. 31, p. 31.

PEACH, B. N., HORNE, J., et al. (1888) 'Report on the recent work of the Geological Survey in the North-West Highlands of Scotland' *Quart. Jour. Geol. Soc.*, Vol. 44, p. 378.

PENCK, A. (1887) 'Ueber Denudation der Erdoberflache' *Ein Vortrag gehalten im Vereine zur Verbreitung naturwissenschaftlichen Kentnisse in Wien, in-18*, Vol. 27, p. 431.

PESCHEL, O. (1879–80) 'Physische Erdkunde' 2 vols.

PHILIPPSON, A. (1886A) 'Studien über Wasserscheiden' *Ver. Erdkunde*, Leipzig, 163 pp.

— (1886B) 'Ein Beitrag zur Erosionstheorie' *Petermann's geogr. Mitteilungen*, Band 32, pp. 67–79.

POWELL, J. W. (1875) 'Exploration of the Colorado River of the West' (1869–72); (Washington), 291 pp. Reprinted in 1957 by University of Chicago and University of Cambridge Press.

— (1876A) 'Report on the geology of the eastern portion of the Uinta Mountains' (Washington), 218 pp.

— (1876B) 'Biographical notice of Archibald Robertson Marvine' *Bull. Phil. Soc. Washington*, Vol. 2, Appendix X, 8 pp.

— (1895) 'Canyons of the Colorado', 400 pp. Dover Edn 1961.

— (1896) 'Physiographic features' in 'The Physiography of the United States' by J. W. Powell et al. (American Book Co., N.Y.), 345 pp.

PRESTWICH, J. (1886) 'Geology: Chemical, Physical, and Stratigraphical' (Oxford), 2 vols.

RANKINE, W. J. M. (1857) 'On the stability of loose earth' *Phil. Trans. Roy. Soc.*, Vol. 147.

READE, T. M. (1885) 'Denudation of the two Americas' *Amer. Jour. Sci.*, 3rd Ser., Vol. 29, pp. 290–300.

REYNOLDS, O. (1883) 'An experimental investigation of the circumstances which determine whether the motion of water shall be direct or sinuous, and of the law of resistance in parallel channels' *Phil. Trans. Roy. Soc. London*, Vol. 174, pp. 935–82.

SHELFORD, W. (1885A) 'On rivers flowing into tideless seas, illustrated by the river Tiber' *Proc. Inst. Civil Engineers*, Vol. 82, pp. 2–19.

— (1885B) 'Discussion on non-tidal rivers' *Proc. Inst. Civil. Engineers*, Vol. 82, pp. 47–50.

SINGER, C., et al. (1958) 'A History of Technology', Vol. 4 (Oxford), 728 pp.

SMITH, G. O. (1924) 'A century of government geological surveys' in 'A Century of Science in America', a symposium by E. S. Dana et al. (Yale), 458 pp., pp. 193–216.

SOBOLEV, S. S. (1946) 'V. V. Dokučaiev i geomorfologia' in 'V. V. Dokučaiev i geografia', *Akad. Nauk SSSR*.

STEPHENSON, J. J. (1893) 'John Strong Newberry' *Amer. Geol.*, Vol. 12, pp. 1–15.

— (1903) 'Memoir of J. Peter Lesley' *Bull. Geol. Soc. Amer.*, Vol. 15, pp. 532–541.

SUESS, E. (1875) 'Die Entstehung der Alpen' (Vienna).

— (1883–5 and 1888) 'Das Antlitz der Erde' Vols. 1 and 2 (Vienna).

THOMSON, C. W. (1877) 'The movement of the soil-cap' *Nature*, Vol. XV, pp. 359–60.

THOMSON, J. (1876–7) 'On the origin of windings of rivers in alluvial plains, with remarks on the flow of water round bends in pipes' *Proc. Roy. Soc. London*, Vol. 25, pp. 5–8.

THORNBURY, W. D. (1954) 'Principles of Geomorphology' (Wiley, N.Y.), 618 pp.

VON RICHTHOFEN, F. F. (1859) 'Die Kalkalpen von Vorarlberg und Nord-Tirol' *Jahr. geol. Reichsanst.*, Vol. X (Vienna), (Published in 1862).

— (1882) 'China' (Berlin), Vol. 2.

— (1886) 'Führer für Forschungreisende' (Berlin).

VON ZITTEL, K. (1901) 'History of Geology and Palaeontology' (London), 562 pp.

WARREN, G. K. (1859) 'Preliminary report of explorations in Nebraska and Dakota' *Amer. Jour. Sci.*, 2nd Ser., Vol. 27, pp. 378–80.

WILLIS, B. (1889) 'Round about Asheville' *Nat. Geog. Mag.*, Vol. 1, pp. 291–300

— (1942) 'American geology 1850–1900' *Science*, Vol 96, pp. 167–72.

WOODFORD, A. O. (1951) 'Stream gradients and Monterey sea valley' *Bull. Geol. Soc. Amer.*, Vol. 62, pp. 799–852.

WOODWARD, H. B. (1876) 'The Geology of England and Wales' (London), 476 pp.

Informative Index

For biographical information on the numerous workers who have contributed to the knowledge of landforms, the researcher with limited time should look first in the national collections of biographies such as the British *Dictionary of National Biography*, the *Dictionary of American Biography* and the *Dictionnaire de Biographie Française*; and secondly for obituaries in the journals of the various learned societies to which the individuals in question contributed or belonged. A few countries have notable bibliographies of natural history in addition to those contained in the vast bibliographies compiled by the national libraries; among these useful selective references are N. H. Darton's *Catalogue and Index of Contributions to North American Geology, 1732–1891* (Washington, 1896; pp. 1045), H. Barth's *Bibliographie der Schweizer Geschichte* (3 vols., Basle, 1914–15) and Max Meisel's incomparable *Bibliography of American Natural History* (3 vols., New York, 1924–9). Two very helpful general references are K. Lambrecht and W. and A. Quenstedt, *Fossilium Catalogus*. I: *Animalia Palaeontologi Catalogus bio-bibliographicus* (s'Gravenhage, 1938) and John W. Wells & George W. White, 'Biographies of Geologists', *Ohio Jour. Sci.*, Vol. 58, 1958, pp. 285–98. In addition works such as those by F. D. Adams, the Fentons, A. Geikie, Mather & Mason and Von Zittel are indispensable for geological matters generally. The obtaining of accurate biographical details of civil engineers and hydrologists proved especially difficult and laborious, particularly as our study seemed always outside or on the outer fringe of Professor Steponas Kolupaila's *Bibliography of Hydrometry* (University of Notre Dame Press, Indiana, 1961). If, however, our findings provide a small prologue to that notable volume, we would feel highly gratified.

References to illustrations are given in italic at the end of the line.

Where helpful the more important page references are given in bold type. References to individual lakes, mountains, rivers, vales and river valleys are grouped alphabetically under Lake, Mountain, River, Vale and River Valley.